Chirality in Supramolecular Assemblies

Chirality in Supramolecular Assemblies

Causes and Consequences

Edited by

F. RICHARD KEENE

Department of Chemistry,
School of Physical Sciences,
University of Adelaide, Australia

WILEY

This edition first published 2017

Registered Office
John Wiley & Sons, Ltd, The Atrium, Southern Gate, Chichester, West Sussex, PO19 8SQ, United Kingdom

For details of our global editorial offices, for customer services and for information about how to apply for permission to reuse the copyright material in this book please see our website at www.wiley.com.

Library of Congress Cataloging-in-Publication data applied for

A catalogue record for this book is available from the British Library.

ISBN: 9781118867341

Set in 10/12pt Times by SPi Global, Pondicherry, India

10 9 8 7 6 5 4 3 2 1

Contents

List of Contributors

David B. Amabilino, School of Chemistry, University of Nottingham, Nottingham, United Kingdom

Paul D. Beer, Inorganic Chemistry Laboratory, University of Oxford, Oxford, United Kingdom

Roger Bishop, School of Chemistry, The University of New South Wales, Sydney, Australia

Jean-Claude Chambron, Institut de Chimie Moléculaire de l'Université de Bourgogne, Dijon, France

Pedro Cintas, Departamento de Química Orgánica e Inorgánica, Facultad de Ciencias, Universidad de Extremadura, Badajoz, Spain

Jack K. Clegg, School of Chemistry and Molecular Biosciences, The University of Queensland, Brisbane, Australia

Steven De Feyter, Division of Molecular Imaging and Photonics, Department of Chemistry, KU Leuven-University of Leuven, Leuven, Belgium

Iris Destoop, Division of Molecular Imaging and Photonics, Department of Chemistry, KU Leuven-University of Leuven, Leuven, Belgium

Christopher J. Easton, Research School of Chemistry, Australian National University, Canberra, Australia

Nicholas H. Evans, Department of Chemistry, Lancaster University, Lancaster, United Kingdom

Daniel Fankhauser, Research School of Chemistry, Australian National University, Canberra, Australia

Philip A. Gale, Chemistry, University of Southampton, Southampton, United Kingdom

Declan P. Gavin, Department of Chemistry, Synthesis and Solid State Pharmaceutical Centre, University College Cork, Cork, Ireland

F. Richard Keene, Department of Chemistry, School of Physical Sciences, University of Adelaide, Adelaide, Australia

Jonathan A. Kitchen, Chemistry, University of Southampton, Southampton, United Kingdom

Edward G. Latter, Surface Science Research Centre and the Department of Chemistry, University of Liverpool, Liverpool, United Kingdom

Anita R. Maguire, School of Pharmacy and Department of Chemistry, Analytical and Biological Chemistry Research Facility, Synthesis and Solid State Pharmaceutical Centre, University College Cork, Cork, Ireland

John C. McMurtrie, School of Chemistry, Physics and Mechanical Engineering, Queensland University of Technology, Brisbane, Australia

Rasmita Raval, Surface Science Research Centre and the Department of Chemistry, University of Liverpool, Liverpool, United Kingdom

David R. Turner, School of Chemistry, Monash University, Clayton, Australia

Preface

The origins of what is now called *supramolecular chemistry* have been somewhat disparate, arising in part from studies of the chemistry of macrocycles (a development of naturally occurring analogues), spherands and carcerands, and cryptates . . . but the award of the 1987 Nobel Prize to Donald Cram, Charles Pedersen and Jean-Marie Lehn in many ways gave it a consolidated focus and led to its emergence as a field that retains vigorous and distinctly multidisciplinary activities. Supramolecular chemistry – defined by Lehn as "the chemistry of molecular assemblies and of the intermolecular bond" – deals with the organization of molecules into defined assemblies using noncovalent interactions, including weaker and reversible associations such as hydrogen bonds, π-π interactions, dispersion interactions, hydrophobic and solvophobic effects, and metal-ligand interactions. The aspect of stereochemistry within such chemical architectures, and in particular *chirality*, is of very special interest as it impacts on considerations of molecular recognition, the development of functional materials, the vexed question of homochirality, nanoscale effects of interactions at interfaces, biocatalysis and enzymatic catalysis, and applications in organic synthesis.

This book is intended to address the nature of the phenomenon of chirality in its broadest sense, noting the change in its nuances and subtlety in the progression from simple individual molecules to molecular assemblies, and to show the manifestations of chirality in the synthesis, properties, and applications of supramolecular systems, emphasizing their multidisciplinary importance.

The book is essentially divided in to four broad parts. The first constitutes an introduction to chirality: Chapter 1 develops the concept of chirality from rigid isolated molecules through to assemblies of molecules (in supramolecular entities), to topological chirality. Chapter 2 discusses chirogenesis and the phenomenon of homochirality (loss of parity) in the development of naturally occurring polymers (including nucleic acids and polypeptides) – and its consequences for the formation of artificial supramolecular aggregates. Chapter 3 provides an overview of chiral aspects arising in the crystallization of small organic molecules – principles that are applicable to all classes of molecules, including supramolecular assemblies.

The second part is predominantly (but not exclusively) centered on metallosupramolecular chemistry. By the use of examples, Chapter 4 addresses the diversity of supramolecular assemblies – and in particular metallosupramolecular assemblies – and describes the complexity of chiral structures and their construction through self-assembly procedures. Chapter 5 describes the role of chirality in molecular recognition and host-guest systems. Chapter 6 develops the notion that unique characteristics can be built into supramolecular assemblies because of features of chirality – characteristics that can lead to functional properties of such materials. Chapter 7 addresses bulk homochiral solids formed using chiral reagents – either by direct incorporation, or by templating or induction, during synthesis. Chapter 8 considers the basic design principles that underpin the construction of metallosupramolecular polyhedra.

The third part is devoted to chirality at interfaces. Chapter 9 focuses on chirality expression and amplification at solution / solid-state interfaces, and applications such as heterogeneous catalysis and chiral separations. Chapter 10 addresses the initiation of chiral suprastructures on surfaces, and their modeling by high-resolution experimental methods and theoretical calculations.

The fourth part addresses chirality in organic hosts, and in biological / enzymatic systems: organic hosts are used in analytical chemistry to separate racemic guest mixtures or simply to distinguish enantiomers, and chiral hosts can function as catalysts in asymmetric reactions – Chapter 11 reviews particular features and applications of chiral organic host systems based primarily on cyclodextrins, calixarenes, and crown ethers in this regard. Chapter 12 stresses the enormous potential of microorganisms and enzymes as catalysts in asymmetric synthesis for controlling the stereochemical outcome of reactions, and discusses the use of whole cells and isolated enzymes as an attractive option for the chemical industry.

It is always understood that supramolecular chemistry is so diverse that one book cannot be totally equitable in its coverage of all aspects of the field. This book attempts to address some of the major aspects authoritatively and highlight important current thrusts. It will be useful to researchers working with chiral supramolecular assemblies, and will hopefully draw others with an existing interest in supramolecular systems to a further appreciation of the importance of chirality in the field, as seen through contributions of experts in their respective parts of that firmament.

F. Richard Keene
Adelaide, Australia

1

Principles of Molecular Chirality

Jean-Claude Chambron and F. Richard Keene

1.1 General Introduction

Chirality is probably one of the most significant topics in chemistry. The strong connection between chirality and symmetry has made it appealing from the mathematical and aesthetic viewpoints, and the recent interest in topologically chiral interlocked and knotted molecules has increased its intellectual attraction, raising the concept of a hierarchy in chirality [1]. The most fascinating aspect of chirality stems from the dynamic properties of molecules and supramolecular assemblies, rather than their static properties, because they are the cause of many intriguing and sometimes paradoxical issues. At the same time, dynamic chirality is also the most useful topic because of the numerous applications it underpins, from chiral recognition to molecular motors.

Historically, chirality is rooted into crystallography (the concept of hemiedry), and the first breakthrough into the field of molecular chirality was Louis Pasteur's hypothesis that the dissymmetry of a crystal was a consequence of dissymmetry at the molecular level [2]. The second milestone was the Le Bel and van't Hoff model of the tetrahedral carbon atom, which accounted for the chirality of the organic compounds known at that time, and several years later Werner was the first to study and provide evidence for the chirality of metal complexes. The discovery of organic molecules that did not owe their chirality to tetrahedral carbon atoms carrying four different substituents (e.g., allenes, biphenyls, cyclophanes), and of helical structures in nucleic acids and proteins, finally led Cahn, Ingold, and Prelog to establish a general system for the description of chiral structures. Since then, many novel chiral molecules have been reported, and most of them could be described in the frame of the CIP rules. The most notable developments in chirality in recent decades concern aspects of the generation and control of chirality: transfer by supramolecular

Chirality in Supramolecular Assemblies: Causes and Consequences, First Edition. Edited by F. Richard Keene.

interactions; chirality of molecular assemblies (chirality at the supramolecular level or "supramolecular chirality"); and finally, the concept of "topological chirality" brought forward by the development of interlocked and knotted molecules.

This chapter constitutes an introduction to molecular chirality from the rigid geometrical model to the topological model, but also from the isolated molecule to assemblies of molecules. As the first chapter in this book on the causes and consequences of chirality in supramolecular assemblies, it will, nevertheless, not cover all the aspects of chirality transfer – in particular those resulting from a covalent bond formation.

1.2 Geometrical Chirality

A chiral object is the one that does not coincide with its mirror image. The source object and its mirror image are called enantiomorphs. From the point of view of symmetry, enantiomorphic objects can have only rotation axes C_n, $n \geq 1$, as symmetry elements: they are either asymmetric (C_1) or dissymmetric (C_n, $n \neq 1$). There are many natural examples of enantiomorphic objects, the prototypical one being the human hand, the Greek word for which (χειρ) has been used to create the English word "chiral." Molecules are objects at the nanometer scale that are made of atoms connected by chemical bonds. If molecules are considered as rigid nanoscale objects, the definition given above can be very easily transposed to the molecular level, with the term "enantiomorph" being replaced by "enantiomer." However, molecules differ from macroscopic objects according to two criteria: (i) they are not rigid and can encompass a great variety of shapes called conformations, the distribution of which depends on time, temperature, and solvent; (ii) they are not usually handled as a single object, but as populations of very large number of individuals (~ Avogadro number). These two unique characteristics make the definition of molecular chirality not as simple as that of a rigid object (such as a quartz crystal), and therefore it needs further developments in order to be refined [2].

The object molecule can be described at different levels of complexity, which are represented by models [3, 4]. The chemical formula, which uses atomic symbols for the atoms and lines for the bonds (traditionally, dashed lines for the weakest bonds), is no more than what has been termed a molecular graph, a concept derived from mathematics that has been introduced and used fruitfully in various areas of chemistry, in particular in molecular topology (see section 1.3). The structural formula is more informative because it shows the spatial relationships between the atoms and the bonds, which can be, for example, probed by nOe effects in NMR spectroscopy. The most accomplished description of the molecule as a rigid object is the 3D representation resulting from an X-ray crystal structure analysis, as it gives the distances between the atoms (bond lengths), and the angles between bonds. This points to the fact that the image of the molecule we have depends on the observation technique – in particular its timescale, observation conditions such as temperature, but also the state of the observed molecule (solid, solution, gas) [5]. In fact, a large number of molecules, including chiral ones, can be described using the approximation of rigidity (i.e., a rigid model) because fluctuations of atom positions are averaged around a thermodynamic equilibrium value at the observation timescale. In that approximation, as pointed out by Mislow [4], the chirality of the molecule is the chirality of the model, which depends only on the atomic positions, so that in principle the bonds can be ignored. However, the

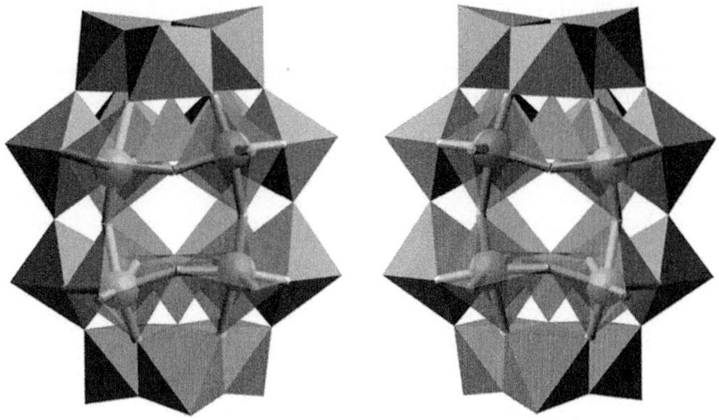

Figure 1.1 The enantiomers of the chiral Keggin polyoxometallate α-$[P_2Mo_{18}O_{62}]^{6-}$ **1**. The chirality of this molecule has a dynamic character, which allows the dynamic thermodynamic resolution of a given enantiomer of this hexaanion by interaction with enantiomerically pure cations

presence of a bond between two atoms indicates that these atoms are closer to each other than if they were not bonded, so that, in practice the bond formalism is very useful for assessing, in a straightforward manner, the chirality of a rigid molecular model. This is the case where, for example, within two identical sets of atoms symmetry-related bonds have different lengths, leading to a distortion of the entire structure. Such an example of chirality due to alternating bond lengths is illustrated by the Keggin polyoxometallate α-$[P_2Mo_{18}O_{62}]^{6-}$ **1** of Figure 1.1 [6].

1.2.1 Origins and Description of Chirality within the Rigid Model Approximation

1.2.1.1 General Considerations

This section will deal with general considerations relating to the description and origins of chirality. Examples selected for their unique chirality properties will be then discussed in more detail in the following sections. Rigidly chiral molecules can only undergo rotations about bonds. They belong to one of the following point groups: (C_1, asymmetric), C_n, D_n, *T*, *O*, and *I* – the latter three being quite rare (see section 1.2.1.3) – which contain only proper symmetry axes as symmetry operations (Table 1.1). Molecular chirality concerns molecules or molecular assemblies featuring a 3D structure. The latter is determined by the interplay between molecular constitution, atom bonding geometry, and intramolecular and intermolecular interactions – including repulsions resulting from strain and steric hindrance. These factors then translate into arrangements of atoms that are either asymmetric (no symmetry element is present) or dissymmetric (with $C_{n>1}$ symmetry elements only) in the 3D space – the necessary but not sufficient (see below) criteria for chirality [2].

The conversion of a planar object into a 3D object can be achieved by either of two possible pathways. It is illustrated in Figure 1.2, starting from a rectangle as an example of a 2D object. Of course the rectangle, lying horizontally, is achiral (D_{2h} symmetry). In the first pathway let us take one of the points of the rectangle, for example its center, and pull

Table 1.1 *Symmetry elements of chiral point groups, the corresponding geometries they are generated from, and maximal symmetries*

Point group (achiral geometrical figure)	Symmetry elements	Symmetry properties
C_1 (general polyhedron)	None	Asymmetric
C_n, n≠1 (cone)	C_n	Dissymmetric
D_n, n≠1 (cylinder)	C_n, n × C_2	Dissymmetric
T (tetrahedron)	4 × C_3, 3 × C_2	Dissymmetric
O (octahedron and cube)	3 × C_4, 4 × C_3, 8 × C_2	Dissymmetric
I (icosahedron and dodecahedron)	6 × C_5, 10 × C_3, 15 × C_2	Dissymmetric

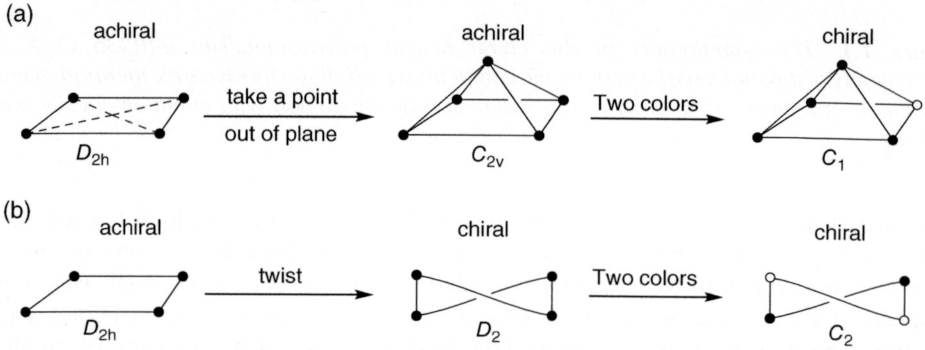

Figure 1.2 *Two pathways for the conversion of a planar object into a 3D object, exemplified by a rectangle. (a) Taking a point out of the plane of the rectangle generates an achiral C_{2v}-symmetric pyramid, of which the desymmetrization to a C_1-chiral object requires the use of two colored vertices (black and white). (b) Twisting converts the rectangle into a D_2-symmetric chiral object, the symmetry of which can be decreased to C_2 by coloring (black and white disks) of selected vertices*

it out of the plane along the vertical direction. This will generate a C_{2v}-symmetric pyramid. This achiral pyramid can be made chiral by changing its constitution – e.g. by coloring selected vertices: a minimum of two colors is required, as shown in Figure 1.2a, which produces an asymmetric (C_1) pyramid. The second pathway arises from a twist to the rectangle along its principal C_2 axis, which makes it a propeller with D_2 symmetry (Figure 1.2b). Hence, unlike the former case, the generation of chirality is simultaneous with the generation of a 3D object. Next, the symmetry is decreased to C_2 by color-differentiation of any two vertices out of the four. Of course, making three vertices of the same color would further decrease the symmetry of the propeller to C_1.

Stacks of an achiral planar object (such as an isosceles triangle, as shown in Figure 1.3) can produce an achiral D_{3h} symmetrical column (b), which upon a regular twist of the individual components is converted into a chiral wreathed column, either left- (a) or right- (c) handed.

In chemical vocabulary, the deformation applied to the rectangle of Figure 1.2a corresponds to a constitutional change as the rectangle (four vertices) has been changed to a

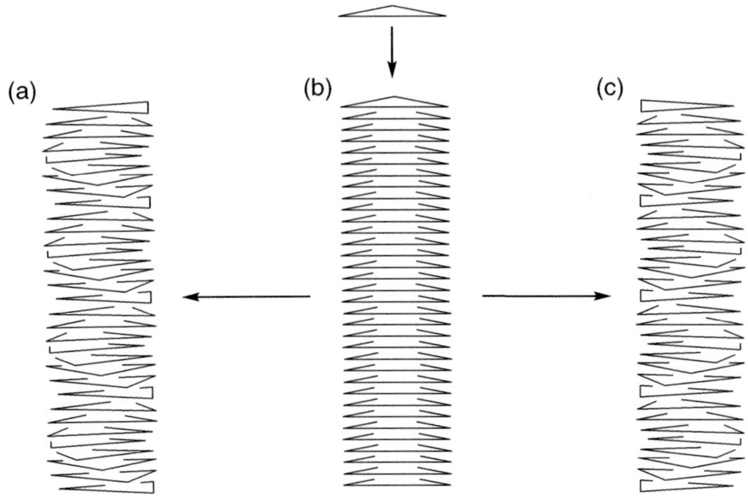

Figure 1.3 *Generating chirality by making (b) stacks of a planar triangular figure, followed by twisting of the resulting column either anticlockwise (a) or clockwise (c)*

pyramid (five vertices), whereas in the case of Figure 1.2b it corresponds to a conformational change as the twisted object has the same number of vertices and faces. Twisting may result from various mechanisms, such as rotations about bonds or variations in bond lengths – and in the case of molecular assemblies, from the generation of a curvature because of intermolecular attractions or repulsions.

The two basic processes of Figure 1.2 can be illustrated in the construction of the mirror-image molecular parallelepipeds $[Zn_2\text{-}\mathbf{2}]_4$ shown in Figure 1.4 by self-assembly of Zn(porphyrin) covalent dimers (*R*)-Zn_2-**2** and (*S*)-Zn_2-**2** driven by the Zn-pyridyl interaction [7]. The vertices of the cubes are occupied by Zn porphyrin (ZnPor) subunits, whereas four parallel edges are formed either by *meso* C–C single bonds or the *meso* C–(4-pyridyl)–ZnPor bond sequence. The bis(porphyrin) subunits are twisted by 90° with respect to each other, while each Zn^{2+} cation has a pyramidal N5 environment in the assembly.

The specification of chirality was formalized by Cahn, Ingold, and Prelog using (in the first instance) the "chirality model," which involves three stereogenic elements of chirality: the center, the axis, and the plane [8]. The chirality model of molecules is based on the tetrahedron, which is also the minimal 3D polyhedron [9]. In the first case (asymmetry, Figure 1.5) the perfect tetrahedron of T_d symmetry needs four different achiral vertices (A, B, C, and D) to be C_1 chiral (asymmetric constitution). Another possibility is to consider a tetrahedron of C_1 symmetry, in which all six edges have different lengths (asymmetric arrangement of the atoms). In practice, the asymmetric tetrahedron results both from asymmetric constitution and atom arrangement (Figure 1.5d).

In the second case (dissymmetry, Figure 1.6), elongation along one of the C_2 symmetry axes of the tetrahedron of Figure 1.5a decreases its symmetry to D_{2d}, and therefore only two different achiral substituents (A and B) are now needed to make it C_2-symmetric chiral. In addition, the D_{2d} elongated tetrahedron can also be made chiral without the need of substituents, by differentiating another pair of edges that are related by the main C_2 axis (z direction). This is done by compressing the tetrahedron of Figure 1.6a in the y direction,

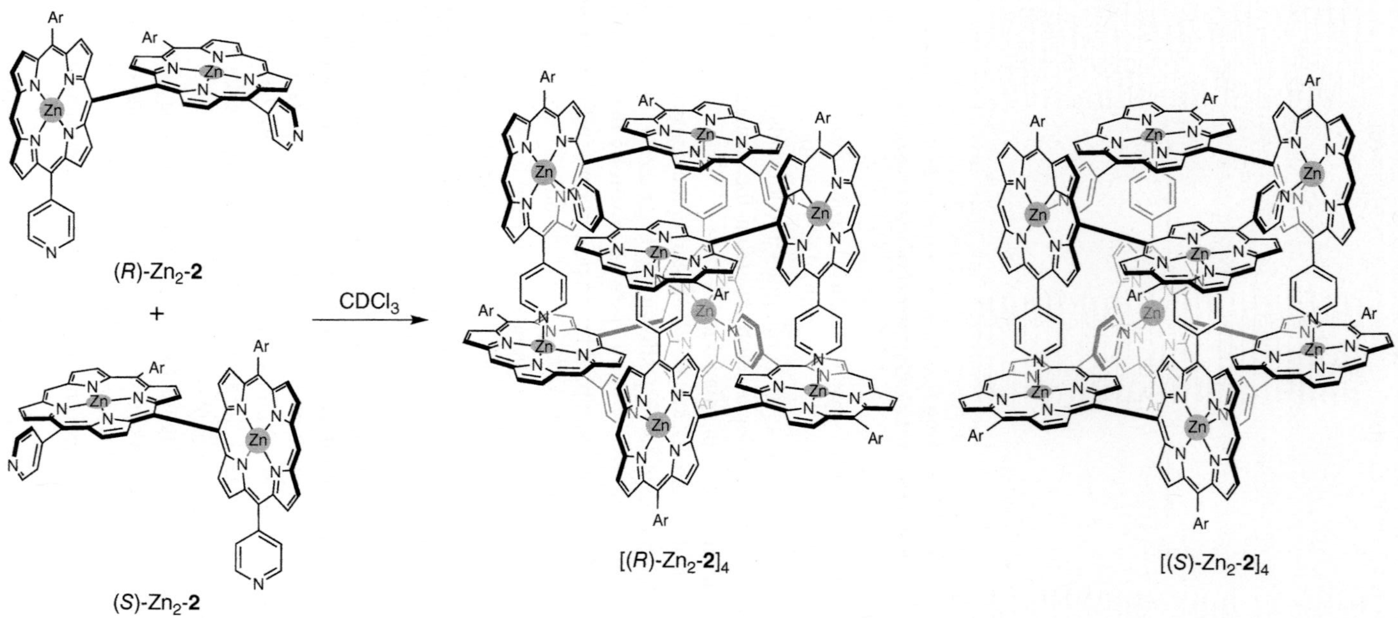

Figure 1.4 *Formation of homochiral assemblies [(R)-Zn$_2$-**2**]$_4$ and [(S)-Zn$_2$-**2**]$_4$ from twisted Zn(porphyrin) covalent dimers (R)-Zn$_2$-**2** and (S)-Zn$_2$-**2**, based on the Zn^{2+}-pyridyl interaction*

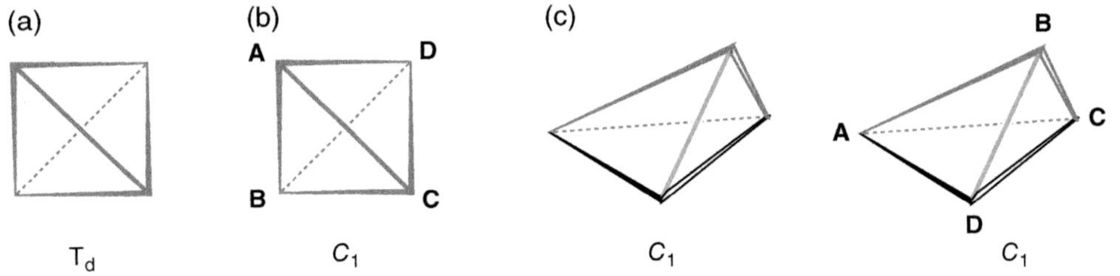

Figure 1.5 *Making the regular, T_d symmetric, tetrahedron (a) asymmetric: (b) by assigning the vertices four different labels; (c) by differentiating the lengths of all six edges using six different "colors"*

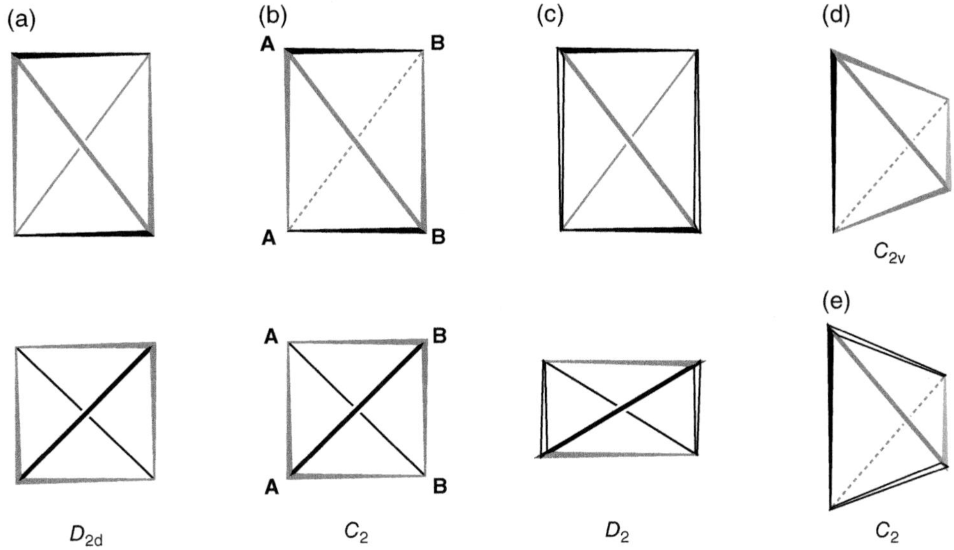

Figure 1.6 *Desymmetrization of the regular tetrahedron. (a) Elongation along one of the C_2 symmetry axes makes the two edges that are perpendicular to it (colored in black) different from the others. A view from the top is shown below the side view. (b) This D_{2d}-symmetric tetrahedron is made C_2-symmetric by labeling the four edges with two different labels, A and B. (c) It can be made D_2-symmetric by further coloring (in white) two edges that are symmetry related by the main C_2 axis. As shown in the top view below, this corresponds to a second elongation, along the C_2' axis. (d) The symmetry of tetrahedron (a) is further decreased to C_{2v} by differentiating a third edge (colored in light gray). (e) The latter is made C_2-chiral by coloring in white two edges that are related by the C_2 symmetry axis, leaving the two others in dark gray*

which removes its symmetry planes. The resulting tetrahedron (Figure 1.6c) is D_2-symmetric. Decreasing the symmetry of the D_{2d} tetrahedron further by moving symmetrically two vertices closer to each other as shown in Figure 1.6d, produces a C_{2v}-symmetric tetrahedron, which is made C_2-symmetric chiral by differentiating a pair of C_2 symmetry-related edges (Figure 1.6e).

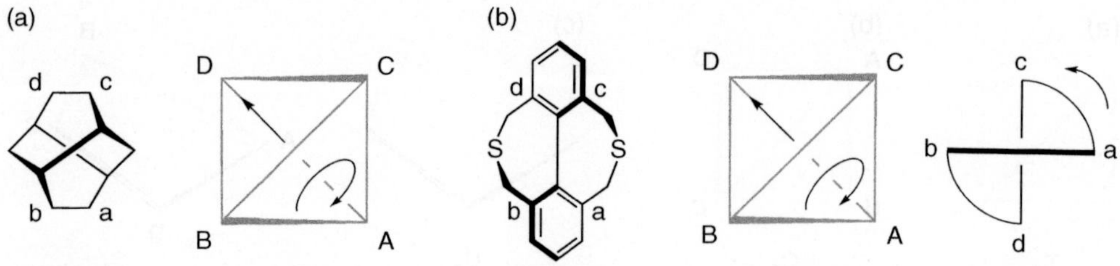

(*R*)-**3** R_a (*R*)-**4** R_a *M*

Figure 1.7 *Description of chirality using the chirality axis as stereogenic unit (a, b), and comparison with the description of chirality by identification of a twist (b). a) (+)-Twistane **3**. The chirality axis bisects [a, b] and [c, d]. b) A* D_2*-symmetric doubly bridged biphenyl **4**. The chirality axis is the biphenyl Ar-Ar bond. In both cases the positions of a and b are arbitrary, however the CIP rules govern those of c and d. Biphenyl (b) is also a molecular propeller, the conformation of which is* M

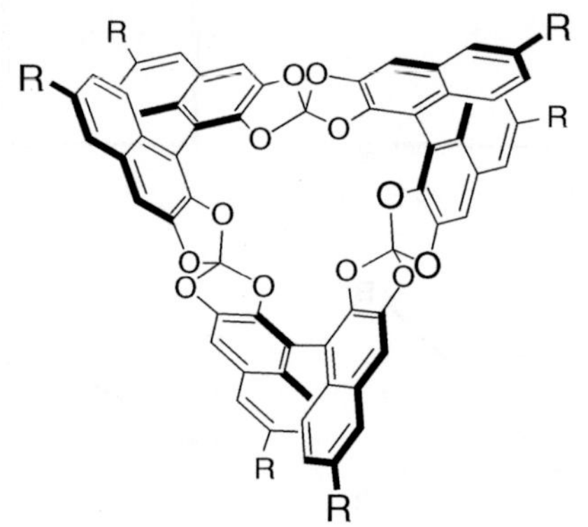

(*R*,*R*,*R*,*R*,*R*,*R*)-**5** (R = H or *n*Bu)

Figure 1.8 *The 3D triangular Janus cyclophane **5** is made by connecting three homochiral binaphthol-derived subunits by three carbon bridges. The configuration of all six chirality axes is* R

Figure 1.7 illustrates how two molecules, the 3D structures of which arise from different factors, are described using the same formalism (the chirality model) – in this particular case, the chiral axis. (+)-Twistane **3** (Figure 1.7a) owes its chirality to a highly symmetrical arrangement of sp^3 carbon atoms in space. The ansa-biphenyl **4** of Figure 1.7b is D_2-symmetric chiral due to strain-relieving twisting. Both molecules have the same configuration (R_a), which is obtained from the chirality model. In addition, the biphenyl can also be considered as a molecular propeller, and as its 3D structure is of conformational origin, it is best described using the *M*/*P* nomenclature. From the CIP rules, it is the *M* conformation that corresponds to the R_a configuration.

An additional illustration of the chirality axis is given in Figure 1.8, which shows a tris(spiroorthocarbonate) cyclophane (**5**) made in low yield by condensation of (*R*)-2,2′,3,3′-tetrahydroxy-1,1′-binaphthyl with dichlorodiphenoxymethane as the carbon source in refluxing toluene [10]. The resulting D_3-symmetric cyclophane has six chirality axes, three

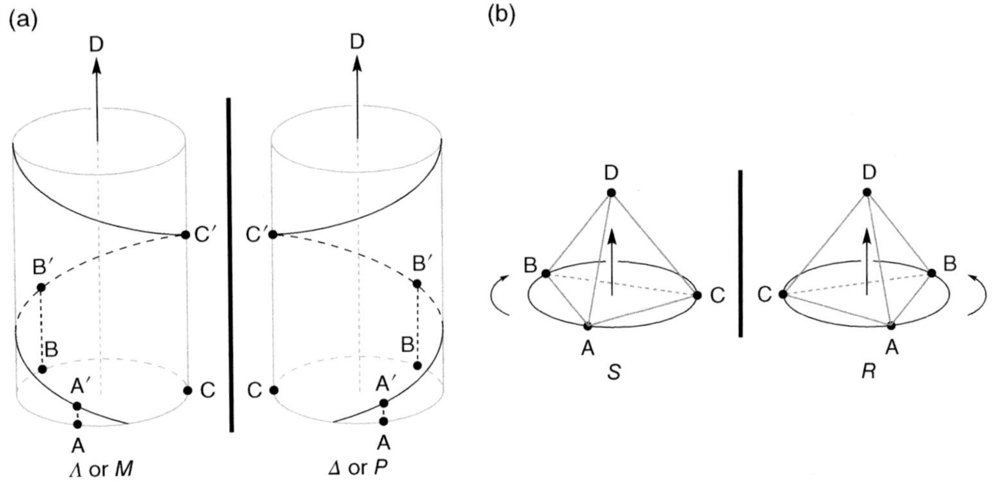

Figure 1.9 *Two different ways to define and orient the 3D space and the analogies between them. (a) Definition and orientation of the 3D space within the helicity model: generation of a helix and description of helical chirality using the Λ, Δ or* M, P *descriptors. (b) Chirality model: reduction of the stereogenic unit to a tetrahedron substituted with four different substituents (descriptors* S *and* R*). The vertical arrows are oriented towards the face from which the ABC plane must be seen. In (a) the D point has been sent to the infinite. Note that, when both models can be equally applied, there is no relationship between the helicity and chirality descriptors, except in the case of the biaryls, where* M *and* P *correspond respectively to* R *and* S

of conformational origin from the binaphthyl components, and three of configurational origin from the spiroorthocarbonate connections, which are interdependent. This molecule features two back-to-back aromatic concavities, which were shown by X-ray crystallography to be able to complex two C_{60} guests via multivalent π-π interactions.

The other model that was devised by Cahn, Ingold, and Prelog is the "helicity model," which proved subsequently to be extremely relevant in describing the chirality of a great variety of molecules and polymers, in spite of the fact that – at the time it was proposed – examples of helical nanoscale objects were rare [8]. From the mathematical viewpoint, a helix results from the combination of a rotation and a translation, and can be cylindrical (C_2 symmetry) or conical (C_1 symmetry). Once a helical structure is clearly identified, for example as a secondary structure, the sense of chirality is given by the helical path. If a clockwise rotation produces a translation away from the observer (following the sequence A′, B′, C′ in Figure 1.9a), the sense of chirality is *P* or *Δ*; if the same effect is produced by a counterclockwise rotation, the sense of chirality is *M* or *Λ*. Note that *P* and *M* descriptors generally apply to conformations and to the so-called secondary structures, and that the *Δ* and *Λ* descriptors are used for the configurations of transition metal complexes.

Natural macromolecular compounds such as DNA, polypeptides, and amylose, as well as synthetic examples such as polyacetylenes and polyisocyanates, can take up helical shapes [11]. This is also the case with molecular compounds like foldamers [12], helicenes [13], and helicates [14] (Figure 1.10). Larger structures encompass at least a full helix turn. By contrast, the smaller members of these families of molecules do not incorporate a 360° turn

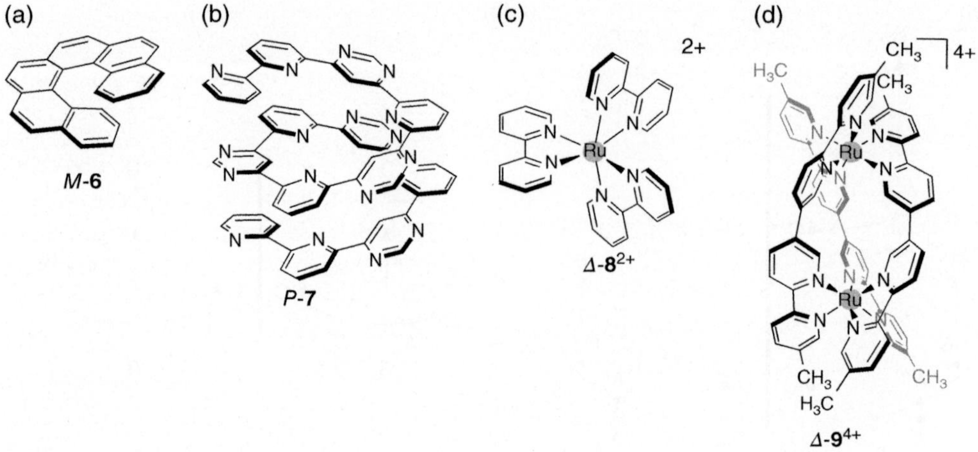

*Figure 1.10 Examples of helically chiral molecules and molecular propellers. (a) [6]Helicene **6**. (b) Foldamer **7** based on alternating pyridine and pyrimidine subunits. (c) The [Ru(bipy)$_3$]$^{2+}$ coordination complex (**8**$^{2+}$), where bipy is 2,2′-bipyridyl, is a C$_3$-symmetric propeller. (d) Connecting two homochiral [Ru(bipy)$_3$]$^{2+}$ subunits through the positions 4 and 4′, respectively, of the bipy ligands produces a fragment of the triple helical dinuclear complex **9**$^{4+}$ in which each quaterpyridine ligand has the same helical conformation*

and actually represent helical fragments: This is notably the case of the so-called molecular propellers [15] (Figure 1.10c), or of molecules that feature a simple twist (Figure 1.7b). Helicity can also manifest itself at the supramolecular level, for example in the case of helical stacks of achiral molecules. It is important to note at this stage that the formation of hierarchically organized chiral supramolecular structures can make the connection between nanoscopic and microscopic or macroscopic chirality (e.g., chiral molecular gels or chiral mesophases). The highest symmetry molecular propellers belong to the D_n symmetry point groups. Among D_n-symmetric propellers, those belonging to the D_2 symmetry point group are worth highlighting because they make the connection between the helicity model and the chirality model, as both models apply in that case (see Figure 1.7b).

As is the case for DNA, many helically chiral molecular compounds feature double or triple helices. This is particularly the situation for the helicates in which polychelate ligands take up helical conformations upon bridging at least two metal cations. This is illustrated in Figure 1.10d by the dinuclear Ru^{2+} complex of a quaterpyridine ligand (**9**$^{2+}$) [16].

After this short overview of the origins and description of chirality we shall detail several examples that illustrate the two basic principles of formation of chiral structures in the 3D space shown in Figure 1.2 – that is, desymmetrization by constitution and desymmetrization by twisting.

1.2.1.2 Desymmetrization by Constitution

Figure 1.11 shows the grid-type tetranuclear metal-ligand assembly $[Os_2Fe_2(\mathbf{10})_4]^{8+}$ made from a “fused” bis(terpyridine)-like ligand (**10**) (in which two 2,2′-bipyridine moieties are bridged by a central pyrimidinyl fragment), and two different metal ions (Os^{2+} and Fe^{2+}),

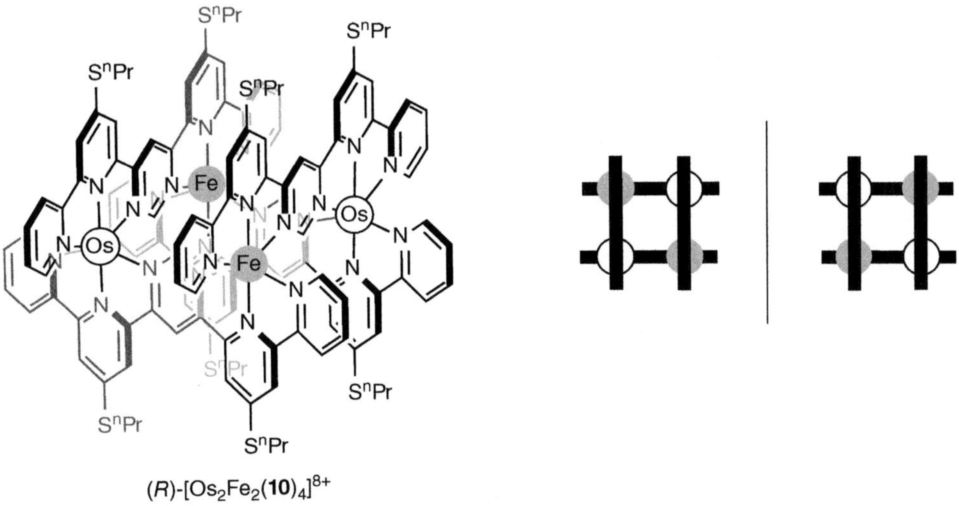

Figure 1.11 *The chirality of the grid-type tetranuclear complex $[Os_2Fe_2(\mathbf{10})_4]^{8+}$ of the "fused" bis(terpyridine)-like ligand **10***

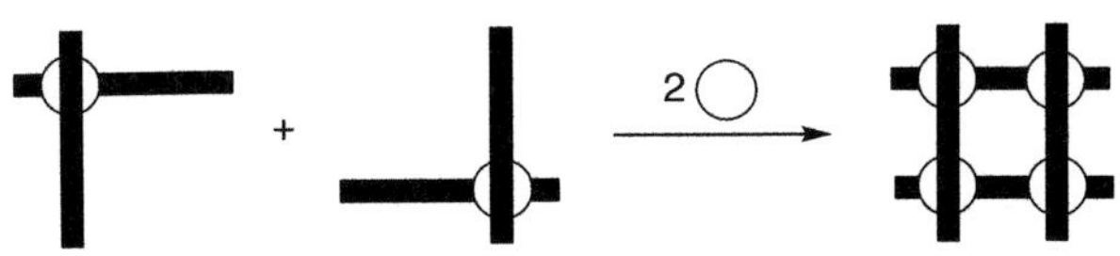

Figure 1.12 *The achiral D_{2h}-symmetric molecular grid is formed from two homochiral halves of mononuclear corner complexes with the bischelate ligands (black elongated rectangles) by addition of two metal cations that are identical to those involved in the starting homochiral complexes*

the pairs of identical metal centers being located on a diagonal [17]. This was done in a straightforward manner by introducing the metal centers in the order of increasing lability – reacting at first the di-chelate with NH_4OsCl_6 in 1 : 1 ratio, thus generating a corner-type chiral mononuclear complex, followed by the addition of $Fe(BF_4)_2$ (2 equivalents). Interestingly, the reaction proceeded stereoselectively to produce the chiral D_2-symmetric tetranuclear complex, as only corner-type precursors of the same handedness react with each other, excluding the formation of achiral *meso* C_{2v} assemblies. It is noteworthy that the tetra-homonuclear assembly represents a stereochemical curiosity, as it can be disconnected into two homochiral mononuclear di-chelate complex subunits. This illustrates the stereochemical paradox called "la coupe du roi" (Figure 1.12) [18].

Another remarkable case of desymmetrization by molecular constitution is offered by the higher order fullerenes. Fullerenes were unprecedented examples of molecules featuring a closed-shell structure. C_{60} itself has icosahedral I_h symmetry and is therefore achiral, but several higher order fullerenes such as $\mathbf{C_{76}}$, have been isolated and characterized. $\mathbf{C_{76}}$, which derives from C_{60} by incorporation of 16 additional C atoms, has D_2 symmetry, as shown by ^{13}C NMR (19 lines of equal intensity), and its chirality arises from its oblong,

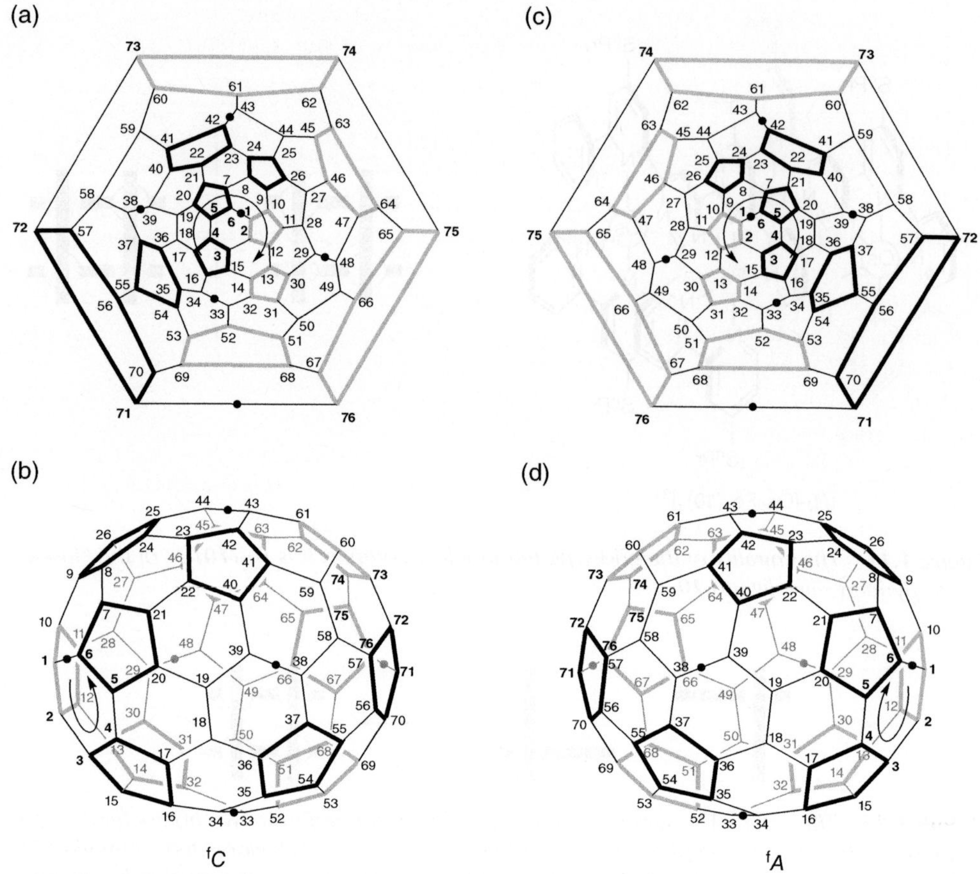

Figure 1.13 *Schlegel diagrams (a and c) and perspective representations (b and d) of the corresponding enantiomers of* $\mathbf{C}_{76}$*. The double bonds have been omitted for clarity. Five-membered rings have been highlighted in bold (black for the front ones, light gray for the rear ones in (b) and (d). The Schlegel diagram is obtained by opening the C71 to C76 six-membered ring and looking down the C1 to C6 analog (bold labels). The descriptor is* ^{f}C *if the C1 to C6 sequence is clockwise and* ^{f}A *if it is the opposite. The intersection of the three* C_2 *axes with the bonds have been materialized by the black dots: the vertical axis crosses C43–C44 and C33–C34, one horizontal axis crosses C1–C6 and C71–C76, and the other crosses C38–C39 and C29–C48*

helically twisted structure (Figures 1.13b to 1.13d) [19]. The enantiomers of $\mathbf{C}_{76}$ were resolved through the HPLC separation of the two diastereomers obtained by regioselective functionalization of C_{76} with an optically active malonate, followed by an electrochemical retro-Bingel reaction performed on each isolated diastereomer to release each optically pure $\mathbf{C}_{76}$. In principle, as all carbon atoms are pyramidalized, the configuration of the fullerene can be described by listing the absolute configuration (*R* or *S*) of each stereogenic center. The latter is obtained by developing the corresponding hierarchic directed graph which, however, is a cumbersome task.

Therefore a simplified procedure, which uses a single descriptor, has been developed which relies on the fact that the numbering schemes of fullerenes are helically chiral (Figures 1.13a to 1.13c), and can be used to differentiate between enantiomeric fullerenes. Whereas two isometric mirror-symmetric numbering schemes can be applied to an achiral parent fullerene such as C_{60}, a unique one is associable with a specific enantiomer of an inherently chiral carbon spheroid. Depending on whether the path traced from C(1) via C(2) to C(3) of this numbering is clockwise (C) or anticlockwise (A), the descriptors are defined as ${}^{f}C$ and ${}^{f}A$. Figures 1.13a and 1.13c show the Schlegel diagrams of the enantiomers of $\mathbf{C_{76}}$ viewed through the opening of the six-membered C71–C72–C73–C74–C75–C76 cycle in the direction of its C1–C2–C3–C4–C5–C6 analog. The sense of the latter sequence (clockwise or anticlockwise) gives the chirality descriptor ${}^{f}C$ or ${}^{f}A$. As $\mathbf{C_{76}}$ is D_2-symmetric, it has three C_2 symmetry axes that are orthogonal to each other.

Concave, bowl-shaped molecules represent a very important family of receptors and precursors of receptors that may display chirality [20]. Examples are resorcinarenes, calixarenes, cyclotribenzylenes and cyclotriveratrylenes, tribenzotriquinacenes [21], sumanenes [22], subphthalocyanines [23] and receptors built from these compounds – such as the cryptophanes made by dimerization of functionalized cyclotribenzylenes [24], or molecular capsules assembled by hydrogen bonding between urea-functionalized calix[4]arenes [25]. As concave molecules are nonplanar, they can be made chiral just by rim orientation. The simplest geometrical model of a concave molecule is a tetrahedron with an "empty" ABC face opposed to the D vertex [26]. Calix[4]arenes carrying at least two different substituents in the *para* positions of the phenol rings, or having even a single *meta* substituent, such as **11** (Figure 1.14a) [27], cryptophanes carrying two different substituents at the *meta* positions of the phenylene rings, such as **12** (Figure 1.14b) [24] – just to mention a few – are examples of concave molecules that owe their chirality to rim orientation. These compounds have been qualified as "inherently chiral," because their chirality (which does not depend on the presence of chiral substituents) is a property of the overall structure [26]. However, this expression may be misleading as bowl inversion, when it is possible, reverses the sense of chirality: therefore concave molecules are better described under the heading of conformational chirality [28]. The recommended descriptors to characterize these molecules are *P* and *M* [8]. Rim orientation of achiral concave molecules may also result from the concerted orientation of substituents, for example by a directed network of hydrogen bonding. The self-assembled molecular capsule $(\mathbf{13})_2$ of Figure 1.14c is obtained by Et_4N^+-templated head-to-head dimerization of two urea-substituted calix[4]arene (**13**) components [25].

Cyclodextrins are concave macrocyclic oligomers of *D*-glucose, and are therefore enantiomerically pure compounds. The recent development of efficient methods for the selective functionalization of their primary rim has led, in particular, to the synthesis of α-cyclodextrins carrying three different substituents [29]. Figure 1.15 shows an example in which the original primary alcohol functions have been replaced by $-PPh_2$, -OBn (Bn is CH_2Ph) and -Me groups that alternate twice, which imparts an orientation to the primary rim. Therefore, the modified cyclodextrin has two diastereomers **14a** and **14b**, because the chirality due to rim orientation is superposed on the chirality of the native cyclodextrin backbone. The resulting molecule can be considered a diphosphine ligand, and indeed it was used in the Tsuji–Trost allylation reaction. It was shown that opposite orientations of the primary rim led to opposite enantioselectivities, albeit rather low (30%), whereas the

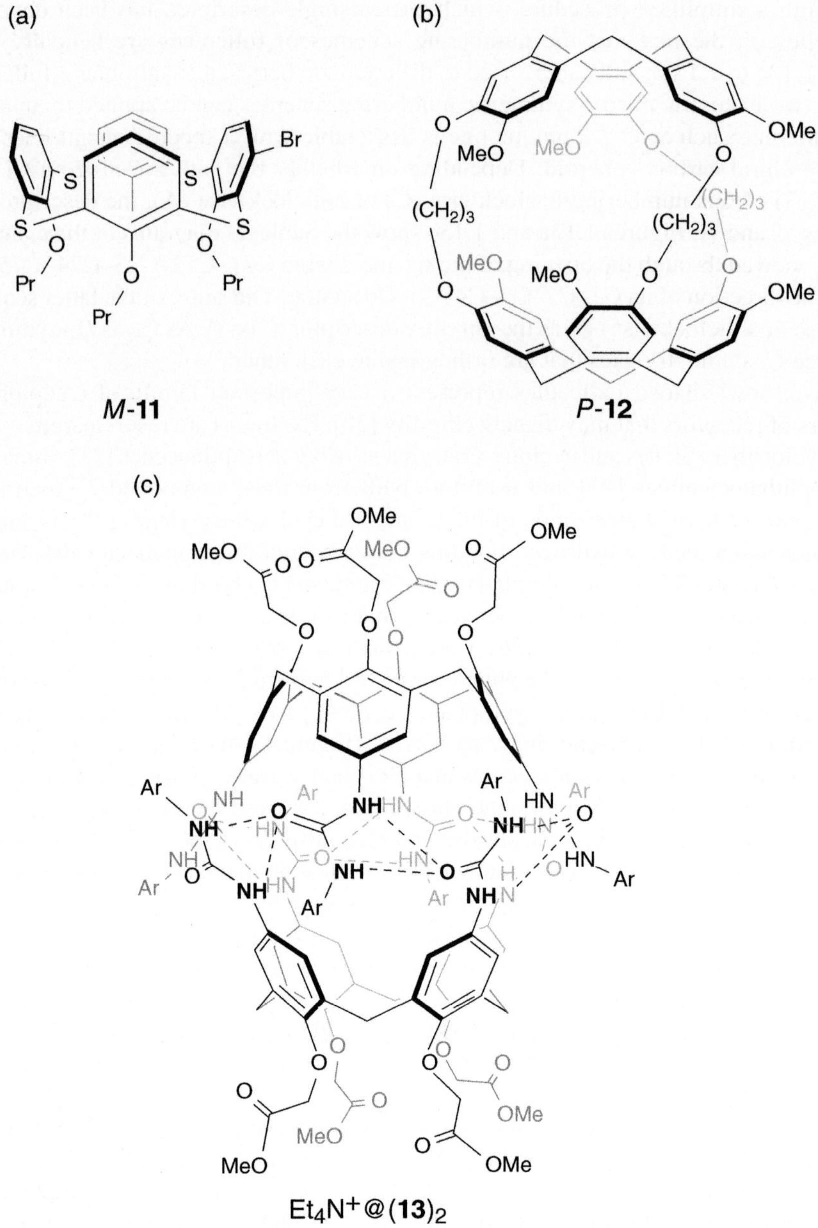

Figure 1.14 *Examples of concave chiral molecules. (a) One of the phenyl rings of thiacalix[4]arene **11** bears a bromine atom in the* meta *position, which destroys the* C_{4v} *symmetry of the parent compound, and makes the corresponding system asymmetric. The propyl groups prevent ring inversion at ambient temperature. (b) Cryptophane-A (**12**) in the chiral,* anti*-configuration (*P*). (c) A head-to-head calix[4]arene dimer* $(\mathbf{13})_2$ *via hydrogen bonding between arylurea substituents, that encapsulates* EtN^+ *(removed for clarity; Ar = p-tolyl). The methyl acetate substituents maintain the macrocycles in the cone conformation*

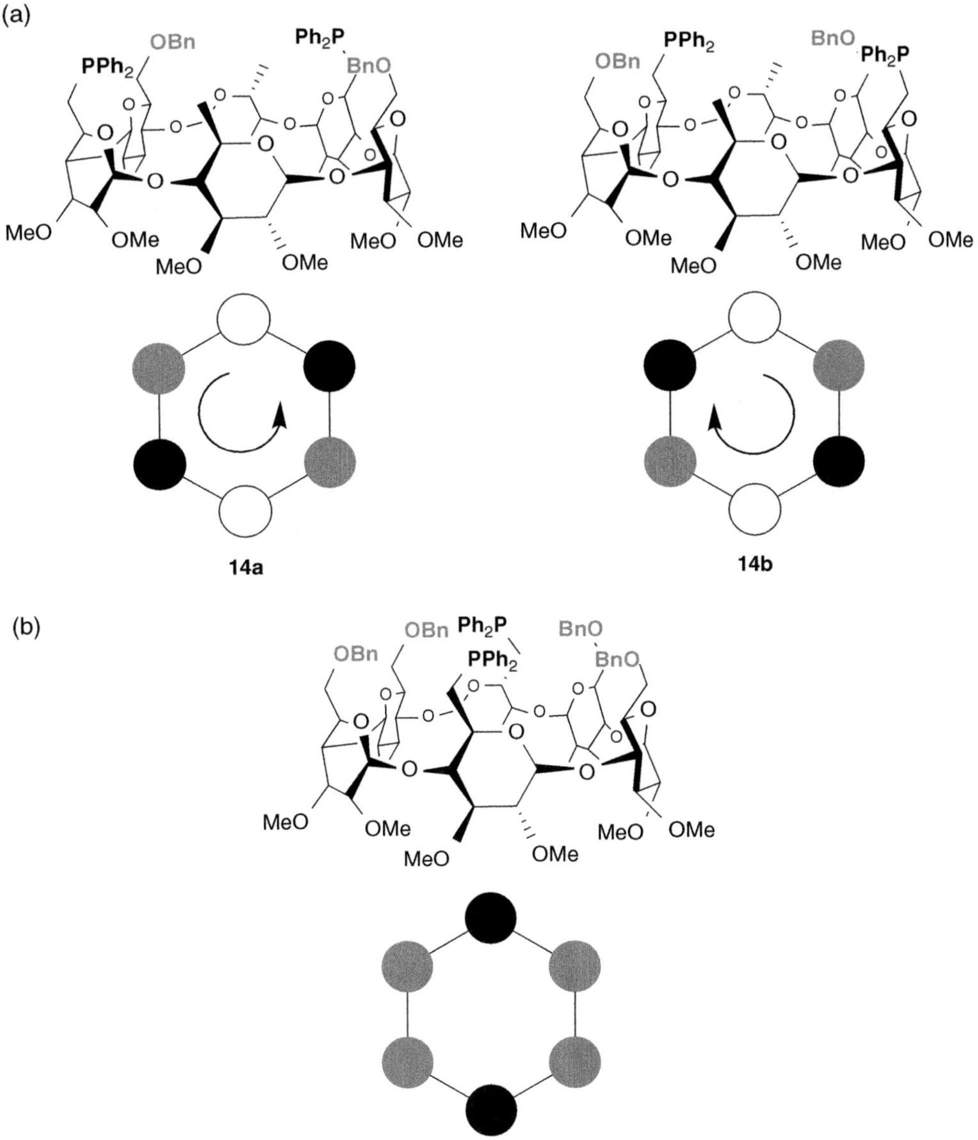

Figure 1.15 Examples of (a) a pair of diastereomeric α-cyclodextrins (**14a** and **14b**) that differ by the primary rim orientation; (b) Cyclodextrin **15** is chiral; however, the primary rim is not oriented

corresponding cyclodextrin (**15**) with a σ_h-symmetrical arrangement of two -PPh_2 and four -OBn groups led to no asymmetric induction at all. This pointed to the higher asymmetric character, with respect to the palladium-catalyzed allylation reaction, of the rim orientation by comparison with the cyclodextrin cavity.

1.2.1.3 Desymmetrization by Twisting

Examples of synthetic molecules exhibiting helical chirality are multidecker systems (cyclophanes, etc.) [30], helicates[14] or foldamers [12]. Whereas helicity of multidecker molecules results from constitutional features only, helicity of helicates and foldamers also involve intramolecular interactions as a twisting factor. Helicates are transition metal complexes containing at least two metal centers, and generally feature double- and triple-helical structures, single-helical cases being best described as foldamers. Helical folding of the multichelate strands is directed by the metal centers, the coordination geometry of which orients the ligands in well defined directions of space. Helicates containing at least two metals centers are linear but higher order helicates can show also circular [31] and other geometries, in particular tetrahedral, when they are made of four metal centers that occupy the vertices of a tetrahedron [32]. True helicates are chiral species that incorporate homochiral metal complex subunits. However, they can have diastereomers that differ by the sense of chirality at the metal centers. For example, in the case of dinuclear systems, there are two diastereomers, the enantiomeric pair of helicates (Δ,Δ and Λ,Λ), and the achiral, so-called mesocate (Δ,Λ), containing metal complex subunits that have opposite chirality senses. The formation of a helicate versus a mesocate has been shown to be highly dependent on the nature of the bridge and the connections between the chelates forming a polynucleating ligand strand – in particular *meta*-phenylene bridges strongly favor the helicate. This highlights the role of the ligand bridges in conveying the chiral information between stereolabile metal centers. Figure 1.16 shows two of the three possible diastereomers (excluding enantiomeric pairs) of the tetranuclear Fe^{2+} complex $[Fe_4(\mathbf{16})_6]^{8+}$ with a binucleating diimine ligand (**16**) built *in situ* from the corresponding amine and aldehyde, and bridged by a *p*-terphenylene spacer [33]. All four iron centers have the same chirality in the true helicate, which displays chiral *T* symmetry, whereas the *meso* form, incorporating two Δ and two Λ centers, has achiral S_4 symmetry. In the *T* isomer chelates of the same ligand display the *anti* orientation with respect to each other, and the three phenyl groups of the spacer are arranged with a helical twist that allows for the perfect stereochemical coupling between the metal centers, whereas in the case of the S_4 isomer four of the six bridging ligands have the *syn* orientation.

The diastereomeric ratio of the *T*, S_4, and C_3 diastereomers was ca. 1 : 1 : 1. However, it could be changed by modification of the terphenyl spacers connecting the diimine ligands. The *anti* orientation is favored by introduction of two methyl substituents *ortho* to the central ring, which induce a 60° dihedral angle between adjacent phenyl rings, as a result of a weak van der Waals interaction between the methyl group and the phenyl group of an adjacent ligand [33]. The *syn* orientation is favored by constraining the two phenyl rings attached to the chelate groups to be parallel to each other. This is done by permethylation of the central phenyl ring, which makes it perpendicular to the peripheral rings. Therefore, when dimethylated terphenyl spacers are used, the major diastereomer is the true helicate (*T*-symmetry), whereas when tetramethylated terphenyl spacers are used, the major diastereomer is the *meso* form of S_4 symmetry. Stereochemical coupling between the metal centers is increased in the first case, whereas it is decreased in the second case.

Foldamers are single stranded molecules that can fold in order to take up directed bent conformations. Helicity is the conformational response to avoid steric interactions between overlapping sections of strands. In the case of the helicenes (Figure 1.10a), folding is encoded in the molecular constitution but in true foldamers it is the result of (programmed)

(a)

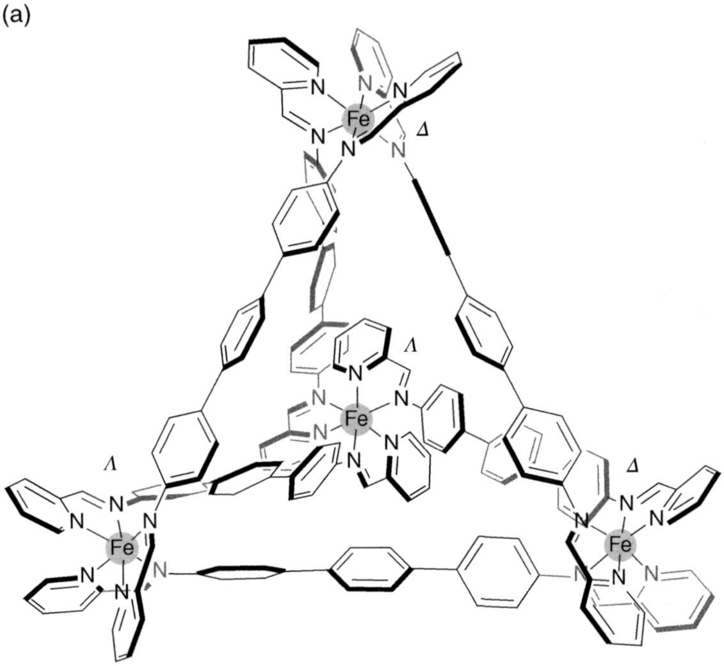

S_4-$[Fe_4(\mathbf{16})_6]^{8+}$

(b)

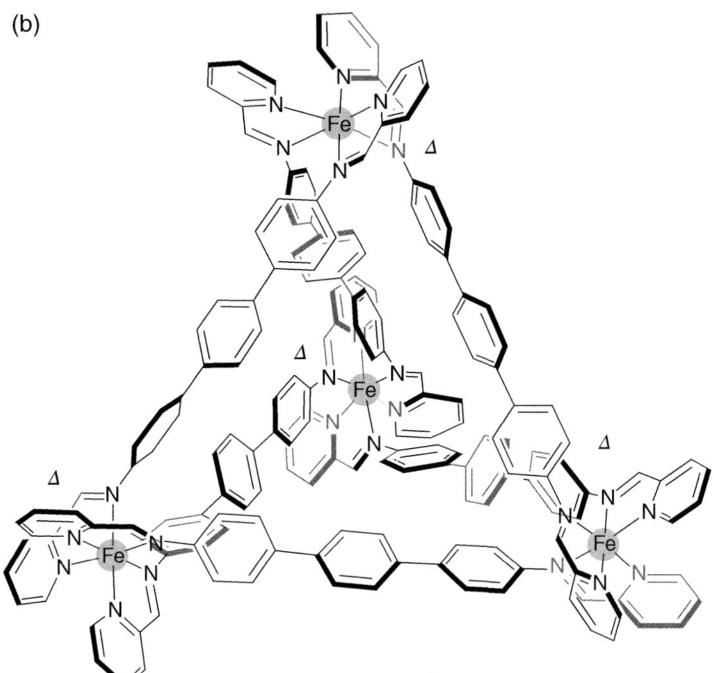

T-$[Fe_4(\mathbf{16})_6]^{8+}$

Figure 1.16 *Two diastereomers of the tetranuclear transition metal aggregate* $[Fe_4(\mathbf{16})_6]^{8+}$ *built on four* Fe^{2+} *cations and six binucleating diimine ligands* ***16****. (a) Achiral (Δ, Δ, Λ, Λ)* meso *form, with* S_4 *symmetry. (b) Chiral (Δ, Δ, Δ, Δ) form, with T symmetry*

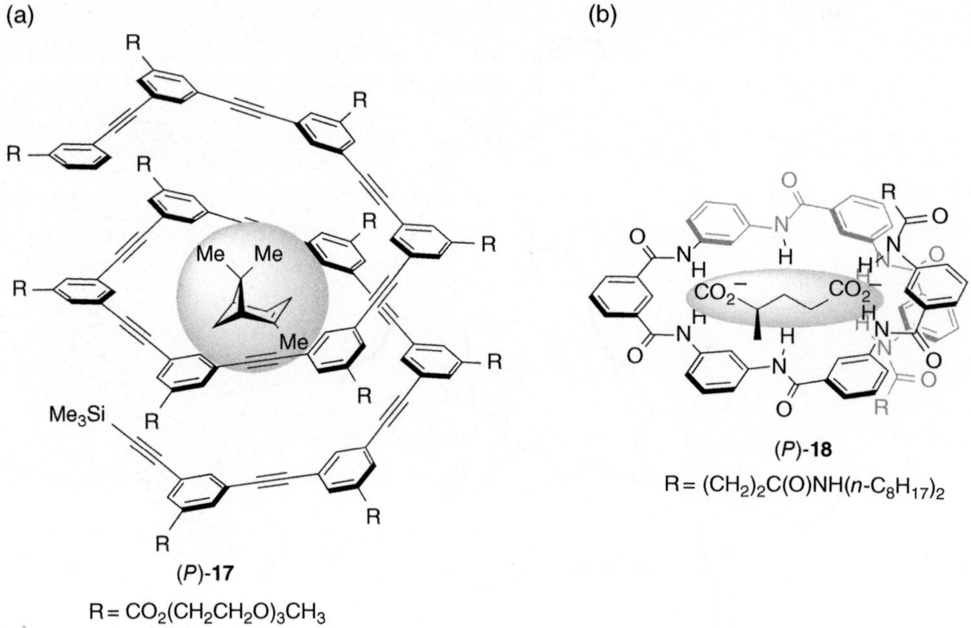

Figure 1.17 *Examples of foldamer hosts, the chirality sense of which is controlled by the chirality sense of the chiral guest. The preferred* P *conformation of the foldamers shown are controlled by (a) (–)-α-pinene in its complex with* ***17****, and (b) D-glutamate in its complex with* ***18***

intramolecular interactions – such as electronic repulsions (Figure 1.10b) or electronic attractions, and hydrogen bonding. When the strand wraps around a molecule, a cation or an anion, folding is the result of intercomponent interactions (such as van der Waals interactions), coordination bonds, and hydrogen bonds. Figure 1.17 shows two representative examples of foldamers that act also as hosts. The arylethynyl system **17** of Figure 1.17a folds around the neutral optically active (–)-α-pinene template [34] as a result of van der Waals and hydrophobic interactions, whereas the oligoarylamide system **18** of Figure 1.17b folds around the optically active D-glutamate dicarboxylate [35] as the result of hydrogen bonding interactions between the amide protons and the anionic carboxylate functions. Foldamers are in general dynamic systems. In the present cases the helicity sense of the foldamers is controlled by the absolute configuration of the optically active guest.

1.2.2 Dynamic and Supramolecular Chirality

1.2.2.1 Enantiomerization Pathways

There are molecules that can encompass enantiomeric conformations under the conditions of observation, and it is this very point that makes molecules unique nanoscale objects. Two cases can be distinguished, depending on the sequence of intermediates that are involved in the interconversion pathway, as they can involve achiral or only chiral species. The first case is classically illustrated by *cis*-1,2-difluorocyclohexane (**19**), the stable chair conformations of which have C_1 symmetry with one fluorine atom in equatorial position

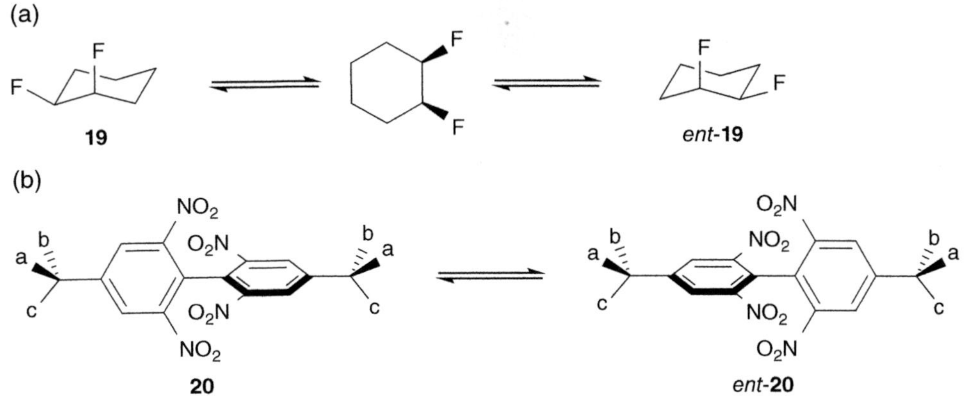

Figure 1.18 (a) Interconversion of the asymmetric chiral conformations of cis*-1,2-difluorocyclohexane **19** to their mirror-image involves an achiral intermediate of C_s symmetry. (b) The generic biphenyl **20** is chiral-asymmetric in all of its conformations. As a result, interconversion of mirror-image conformations takes place through a chiral pathway. Note that Cabc stands for the menthyl group*

and the other in axial position. Interconversion involves a higher energy achiral intermediate with C_s symmetry (Figure 1.18a). Other examples are provided by the mechanisms of racemization of octahedral transition metal complexes with bidentate ligands – i.e. the Bailar and the Ray–Dutt twists, which involve achiral trigonal prismatic species of D_{3d} and C_{2v} symmetry as intermediates, respectively. The second case is illustrated by the biphenyls of generic formula shown in Figure 1.18b (**20**) [36]. These molecules, composed of three rigid blocks (A, B, and A*; A* being mirror-image of A), are asymmetric in every conformation, because the symmetry plane exchanging A and A* is destroyed by the biphenyl block B. However, unrestricted rotation between the blocks A and B on the one hand, and B and A* on the other hand, allows for the smooth interconversion between mirror-image conformers without involving achiral intermediates. Such molecules mimic, at the nanoscale level, the reversal of a real rubber glove by peeling it off inside out from the hand. At no time during the process of interconversion does the glove achieve a shape with a symmetry plane. Molecules mimicking this process have later on been coined "molecular rubber gloves." The chirality of *cis*-1,2-difluorohexane can be detected by ^{19}F NMR spectrometry at low temperature, where the exchange of mirror-image conformations through the achiral intermediate is blocked.

Molecular rubber-glove molecules are important milestones in stereochemistry, because they can be used as the basis to express the sufficient condition for chirality: if a molecule is chiral, then there are no enantiomerization chiral pathways converting an enantiomer into its mirror image [2].

1.2.2.2 Controlling Chirality by Coordination and Supramolecular Interactions

In this section we consider molecules that are not rigidly chiral but can exist in the form of enantiomeric conformations, which are in the fast exchange regime under observation conditions (stereochemical lability). Such molecules are interesting platforms to elicit

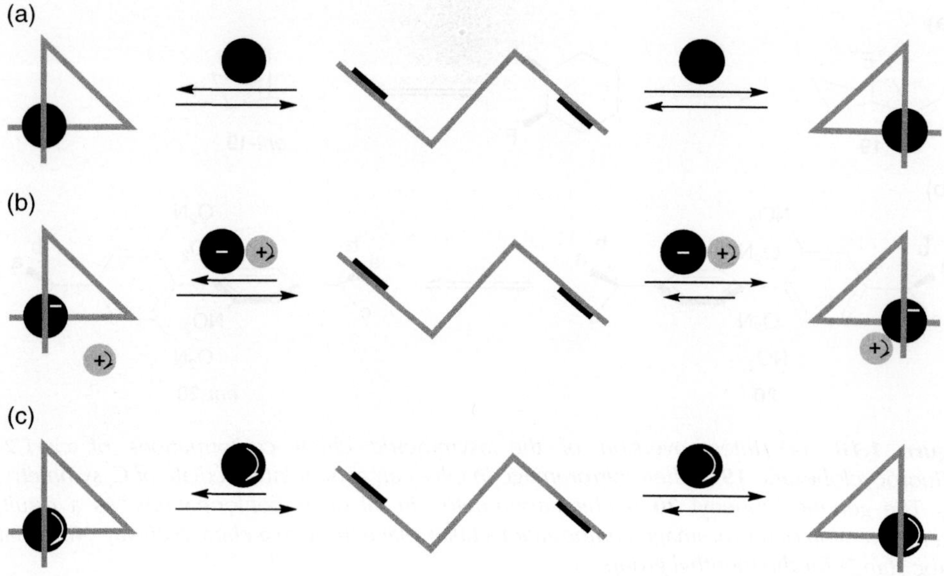

Figure 1.19 *Principle of freezing chirality by supramolecular or coordination interactions. A ditopic (black rectangles) linear receptor (central Z shape) can wrap around (a) an achiral guest (black disk) either clockwise (left) or anticlockwise (right). (b) If the guest is a charged species (e.g. an anion), the use of an optically active counter-ion (a cation) can lead to the control of host chirality by supramolecular ion-pair interactions. (c) The same result is obtained by using an optically active guest itself*

evidence of induced circular dichroism effects, to design chirality switches and molecular systems for chirality sensing (stereodynamic chemosensor), just to mention a few applications.

In order to control the chirality of a stereochemically labile molecule it is necessary to have an optically active chirality effector, which is able to interact with the former via supramolecular (e.g., hydrogen bonding, ion pair or van der Waals contact) or coordination interactions. Three cases of interest are reported schematically in Figure 1.19. Figure 1.19a represents the general case in which the achiral receptor, featuring two recognition elements, coils around an achiral guest. Assuming that the host-guest systems are in thermodynamic equilibrium, Figures 1.19b and 1.19c show two different approaches in order to control the sense of coiling of the host – either (Figure 1.19b) by using a ionic achiral guest in combination with an optically active counterion, or (Figure 1.19c) by using an optically active guest.

The phenomenon was independently described in the 1930s for the interaction of stereolabile transition metal complexes with optically active counterions by Pfeiffer [37] and Kuhn [38], who called it "asymmetric transformation." Most recent applications concern the design of receptors featuring chiral conformations, which can bind optically active guests. As a result, the dynamic equilibrium between diastereomeric host-guest pairs is biased towards the formation of the most stable one. Best evidence for the phenomenon was provided by induced circular dichroism. Figure 1.20 shows one of the numerous examples of the

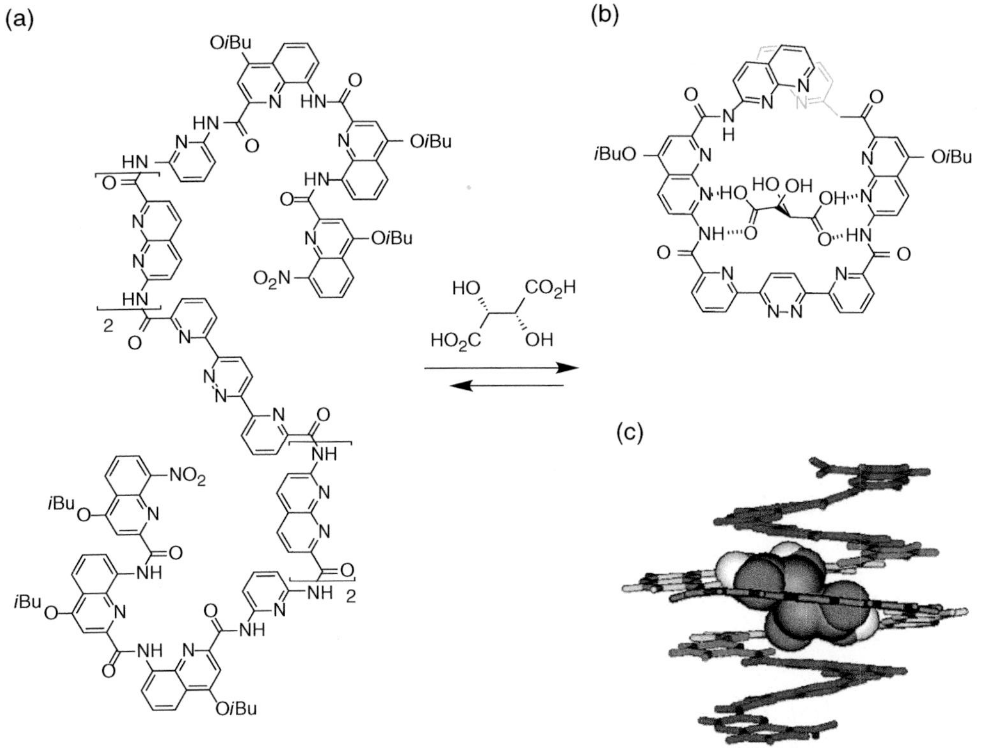

21

Figure 1.20 *Compound **21** (a), made of rigid aromatic hydrogen bond acceptors separated by carboxamide connectors, folds in a doubly conical, singly helical structure that features (b) a cavity able to host L-(+)-tartaric acid, as shown by (c) the view of the X-ray crystal structure. When the guest is optically pure, the diastereomeric purity of the host-guest ensemble is greater than 99%*

literature [39]. A foldamer (**21**) made of three different sequences symmetrically attached at the extremities of a central bis(pyridyl)pyridazine spacer forms a D_2-symmetrical bis-conical helical structure in $CHCl_3$/DMSO 98.8 : 0.2 v/v. This compact helical conformation features a central cavity, which is able to encapsulate various H-bond donor guests, such as L-(+)-threitol or L-(+)-tartaric acid (Figure 1.20). The latter, which shows the highest association constant (5300 M^{-1}), also produces the stronger chiral bias, as the diastereomeric excess is greater than 99%.

1.2.2.3 Memory-of-Chirality Effects

Another related phenomenon is the "memory of chirality" effect, for which there has been evidence in the case of many achiral polymers that feature chiral helical conformations, such as polyacetylenes [40]. In addition, as we shall see in the next paragraph, this phenomenon is also frequently observed in the case of self-assembled helically chiral assemblies of molecules [41], but seldom in the case of small molecules. In the example

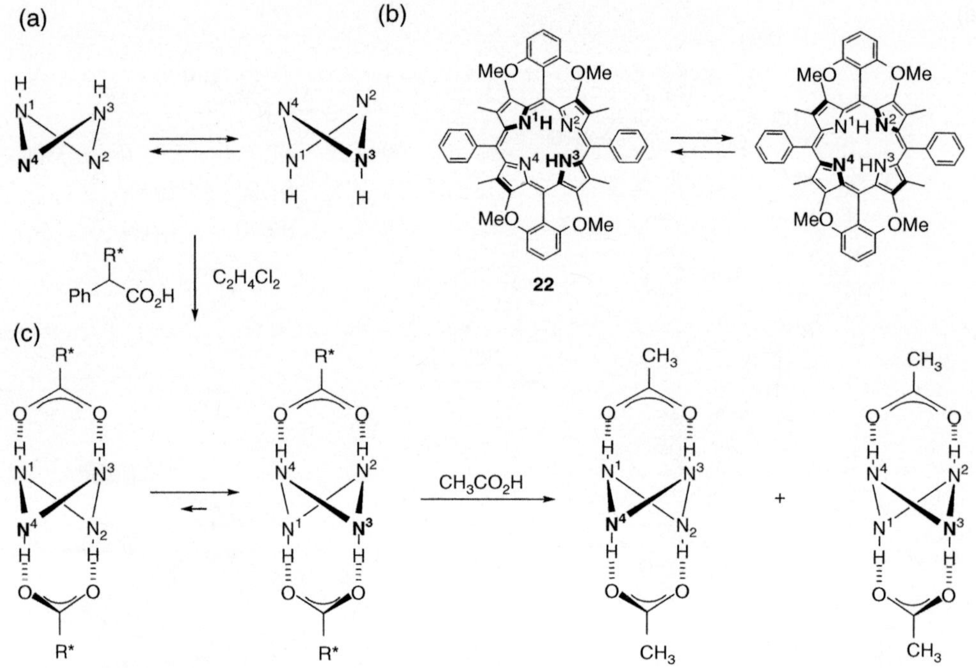

Figure 1.21 *Illustration of the chiral biasing of the equilibrium between mirror-image conformations of the ruffled porphyrin* ***22*** *depicted in (b). (a) Schematic view of the real system. (c) Addition of one equivalent of optically pure mandelic acid shifts the conformational equilibrium towards the formation of the most stable diastereomer. Replacement of mandelic acid by acetic acid, which is achiral, keeps the enantiomeric bias: this is the so-called memory-of-chirality effect*

shown in Figure 1.21 a sterically crowded porphyrin (**22**) has ruffled enantiomeric conformations in a 1 : 1 ratio. Binding of optically active carboxylic acid (host/guest = 1 : 2) by salt-bridge formation shifted the equilibrium to the most stable diastereomer (98% *de*), giving rise to an induced circular dichroism (ICD). Subsequently, the optically active carboxylic acid analyte was displaced by achiral acetic acid. The memory-of-chirality effect stems from the fact that there is no exchange between the acetic acid molecules bound to the hosts. As a consequence, two enantiomers (a major one and a minor one) are obtained. Remarkably, the ICD measured for the pair of enantiomers ($t_{1/2}$ = 200 h at 23 °C) is higher than the one measured for the pair of parent diastereomers [42].

Memory-of-chirality effects are actually more commonly observed in the case of chiral polymers. These are polymers taking up helical conformations, for example polyacetylenes, polyamides and polyisocyanates. Figure 1.22 shows schematically the experiments that could be carried out using a poly(aryl)acetylene carrying carboxylic acid functions [40]. The polymer takes up random-coil conformations that show domains with a well defined helical sense that are separated by reversal points. The polymer can be further functionalized by salt bridge formation with an optically active amine, which triggers the

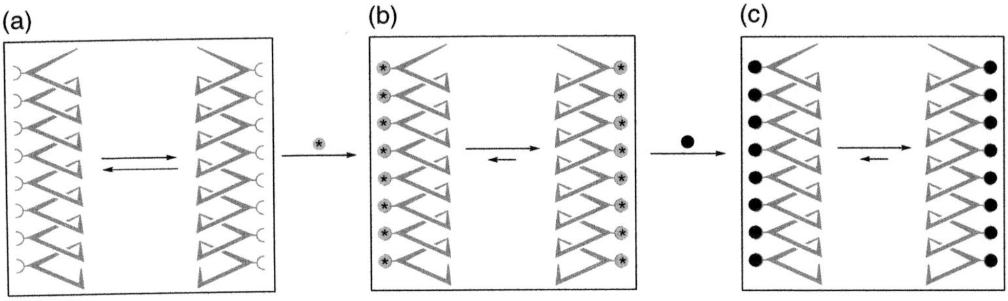

Figure 1.22 Control of the chirality of polymers taking up helical conformations (a) by supramolecular interactions. The polymer contains appended binding groups, which upon interaction with a chiral, optically active guest (gray disks marked with a star), allow the control of the chiral bias between the mirror-image polymeric helices (b). (c) The memory-of-chirality effect takes place upon exchange of the chiral guest by an achiral analogue (black disks)

homogenization of the favored chirality sense. The memory-of-chirality effect takes place upon exchange of the optically active amine for an achiral one.

1.2.2.4 Supramolecular Chirality

Supramolecular chirality must be distinguished from the previous cases, as it does not concern the chirality of an isolated molecule but the chirality of molecular assemblies constructed from achiral molecules held by noncovalent interactions (hydrogen bonding, coordination bonds, π-π stacking, dipole-dipole interactions, etc.) – such as aggregates of various shapes (in particular rods and ribbons), liquid crystals, solvent gels, and supramolecular polymers that are in dynamic equilibrium. The manifestation of supramolecular chirality is the formation of helical structures (see Figure 1.3), and these are characterized by circular dichroism. Remarkable phenomena have been observed (Figure 1.23). Chiral supramolecular polymers assembled from achiral molecular subunits (monomers) exist as racemates. Addition of an optically active chirality inducer (either a chiral version of the monomers, or a chiral molecule that is able to interact with the monomers) can bias the original equilibrium between the enantiomeric supramolecular polymers by formation of diastereomeric supramolecular polymers [41]. This phenomenon is a manifestation of the sergeant-and-soldiers principle, which was formulated for the first time in the case of covalent polymers made of achiral monomers plus a small amount of an optically active "dopant" [43]. The chiral "sergeant" molecule can be removed, without significantly altering the enantiomeric ratio between the mirror-image helices, because of kinetic effects. This is a manifestation, at the supramolecular level, of the memory-of-chirality effect [44].

Another chirality principle that was discovered in the case of covalent polymers, which also applies to supramolecular chirality, is the majority rule [45]. This rule applies when supramolecular polymerization uses only optically active chiral monomers. If an optically pure monomer is used, then an optically pure helical aggregate is obtained. However, if a racemate plus a slight excess of one of the enantiomer monomer is used, the majority rule tells us that the major supramolecular helical aggregate will be the one whose sense of chirality is dictated by the optically active monomer that is in excess [46].

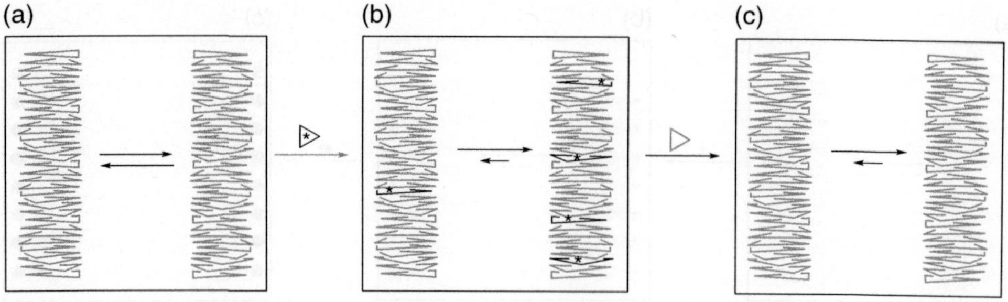

Figure 1.23 Sergeant-and-soldiers principle in supramolecular chirality. (a) Aggregation of certain "soldier" achiral molecules (symbolized here as triangles) can lead to helical supramolecular structures held by π-π stacking or hydrogen bonding interactions. (b) Introduction of a small percentage of chiral, optically active monomer ("sergeant," starred triangle) leads to biasing of the ratio between the left- and right-handed supramolecular helices. (c) The "sergeant" molecules are replaced by equilibration with an excess of "soldier" molecules (grey triangle) without significantly changing the ratio between mirror-image stacks

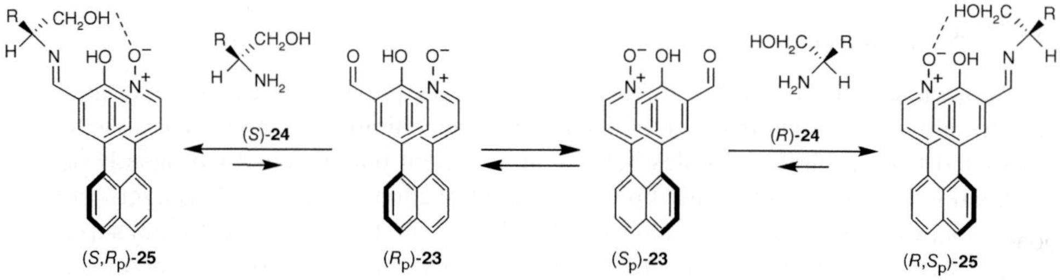

Figure 1.24 A chemosensor (**23**) for chiral aminoalcohols based on a planar chiral naphthalene platform carrying a salicylaldehyde recognition subunit and a pyridine oxide chromophore

1.2.2.5 Chirality Amplification and Switching of Chirality

The main consequence of dynamic chirality is the possibility of observing chirality amplification, because that chirality – be it of conformational or supramolecular origin – is never frozen once and for all. Chirality amplification translates into enhanced CD effects upon going from the optically active chirality inducer to the host-guest complex with the achiral receptor, because the latter may contain chromophores that are strongly CD active in a chiral environment [47]. As a consequence, the so-called induced circular dichroism (ICD) effect has been used as a principle of design for chirality sensors. Figure 1.24 shows one of numerous examples, in which a chiral aminoalcohol freezes out the enantiomeric conformations of a 1,8-diarylnaphthalene chromophore (**23**), by covalent and hydrogen bond formation between the analyte and the sensor [48]. In the case of supramolecular assemblies, positive nonlinear effects have been measured between the enantiomeric purity of the chiral dopant and the ICD of the assembly. Formation of supramolecular species (aggregates) could also explain nonlinear effects in asymmetric catalysis [49].

Chirality switching is not limited to molecular recognition and sensing. In some instances it has been induced by light irradiation [50] or solvent changes [51].

Finally, the phenomenon of supramolecular chirality resides at the interface between chirality at the nanometer level and chirality at the micrometer, if not millimeter, level. It is important to recall, in this respect, that it was the dissymmetric external shape of crystals of potassium tartrate examined with a microscope that suggested to Louis Pasteur that it could result from dissymmetric arrangements of the atoms of the salt. Hierarchical effects in chirality are observed in other materials than crystals. For example molecules forming liquid crystals can be combined with small amounts of chiral dopants that induce optically active liquid crystalline phases [52]. Similarly achiral molecular gelators can be doped with optically active analogs, leading to optically active solvent gels [53].

1.3 Topological Chirality

The chirality of a molecular object in the rigid geometry approximation is a property of the whole object, therefore it depends only on the positions of the atoms. This means that the bonds can be ignored, as first noted by Mislow [4]. This is exactly the opposite to the case of the topological character of chirality, which is invariant whatever the atomic positions, provided that all the connections (chemical bonds) between the atoms have been identified [3].

Molecular rigidity results from the stereochemical properties of atoms and bonds, from steric hindrance and strain, which limit the amplitude of bond rotations and elongations that could convert an enantiomer into its mirror image. It is at the very origin of geometrical chirality. There are cases, however, for which molecular rigidity is not a requirement for chirality, the molecule remaining chiral whatever the deformations – provided that no bonds can pass through each other. This is actually another form of molecular rigidity, which arises from topological constraints rather than geometrical ones. This section will focus on the concept of topological chirality, which, judging from the literature, has been somewhat overlooked.

1.3.1 The Molecular Graph

A graph G is an ordered pair (V, E) formed by a set of vertices V and a set of edges E connecting them. It is an abstract mathematical (topological) object; however, it can be embedded into and manipulated in 3D space. Whereas a geometrical figure is rigid and characterized by distances and angles, the embedded graph is endowed with complete flexibility: the vertices can be placed anywhere in space and the edges can be stretched or shrunk at will, the only requirement being that no two vertices can coincide and no two edges can cross each other – that is, edges and vertices represent boundaries that cannot be crossed. Embedded graphs in 3D space can be represented as a projection (the so-called presentation) in two dimensions, in which the relative positions of the projected edges (above or below) must be indicated [3]. Figure 1.25 shows different presentations of the $K_{3,3}$ graph, the so-called bipartite graph on three vertices, of which two are chiral. The topological property of $K_{3,3}$ is that it is a nonplanar graph – that is, it cannot be presented in the plane without any crossing. Examples of minimal presentations (a single crossing) are shown in Figure 1.25b and Figure 1.25c.

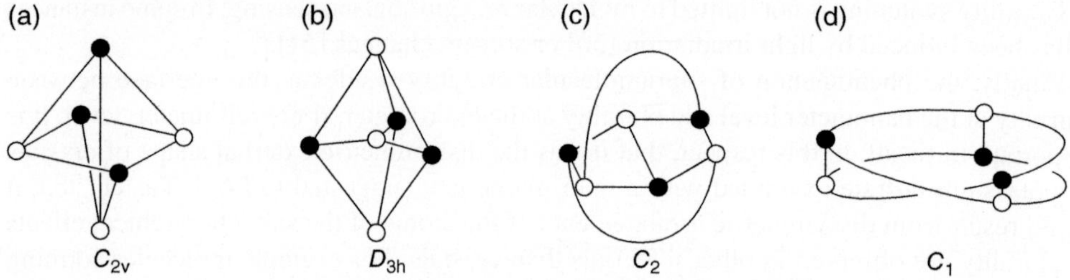

Figure 1.25 *Four presentations of the* $K_{3,3}$ *graph (the bipartite graph on three vertices) with their symmetry group. The presentations shown in (c) and (d) are geometrically chiral*

As the *graph of a molecule* is the set of atoms (vertices) and bonds (edges), it is equivalent to the molecular constitution. The *molecular graph* contains more information, because it has been defined as the embedded graph of a molecule, to take into account the fact that molecules are not abstract entities but real objects in 3D space [3, 4]. The molecular graphs of topological stereoisomers (including enantiomers) cannot be converted into each other by continuous deformations, whatever their amplitude. Walba, in his seminal paper about topological stereochemistry, defined an edge of the molecular graph to be a covalent bond and chose to leave open the question of a topologically significant bond. Topological stereoisomers are operationally distinct as long as, under the conditions of observation, the bonds involved in the graph are not labile – that is, cannot be broken and reformed [3].

Classical molecules exhibiting topological steroisomerism are those having the topology of knots and links [54]. Figure 1.26 shows the diagrams of the four first prime knots, including the unknot (a). The trefoil knot (b) has three crossings, the figure-of-eight knot (c) four, the pentafoil knot (d) five, etc. The simplest link is the Hopf link (f), which is made of a pair of singly interlocked rings. The Solomon link (g) is a pair of doubly interlocked rings. Of the links made from three subunits, the "Borromean rings" have a remarkable topology because they are formed from three interlocked rings that form disjointed pairs. Links and knots that have been realized chemically are the trefoil, the pentafoil, and the figure-of-eight knot on the one hand, and the Hopf and the Solomon links, and the "Borromean rings," on the other hand.

1.3.2 Topological Chirality

A chiral molecule is topologically chiral if and only if its enantiomers cannot be converted into each other by a continuous deformation [55, 56]. Such a property can be proved only by mathematical methods. On the other hand, topological achirality of a molecule is demonstrated by finding at least one achiral presentation of its molecular graph. For example, the trefoil knot has been shown to be topologically chiral, whereas the figure-of-eight knot is topologically achiral, because it has a S_4-symmetric presentation. However, the lack of the existence of at least one achiral presentation does not imply topological chirality, as exemplified by the so-called topological "rubber gloves" [57], which are topological analogs of the biphenyls mentioned in section 1.2.2.1.

The discovery that dynamic bonds (i.e., labile bonds that can be cleaved and reformed reversibly in certain conditions) can be used for the synthesis of topologically nonplanar molecules has opened a completely new field of investigations in molecular topology

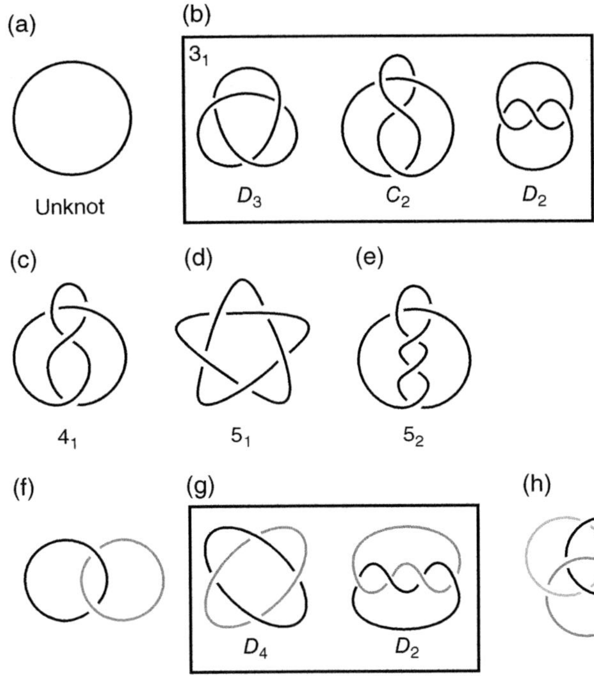

Figure 1.26 *Diagrams of the four first prime knots, (a) unknot, (b) 3_1 (trefoil knot, shown as its three common diagrams), (c) 4_1 (figure-of-eight knot; compare with the diagram of C_2 symmetry of 3_1), (d) 5_1 (pentafoil knot), (e) 5_2, (f) Hopf link, (g) Solomon link, (h) Borromean rings*

[58, 62]. Essentially, these bonds are either coordination bonds (e.g., palladium–nitrogen, gold–phosphorus), or covalent bonds (e.g., imine and disulfide being the most commonly used), and/or their combination (e.g., the formation of imine bonds under first-row transition metal templation). As the formation of these bonds operates under thermodynamic control, the reactions lead to mixtures of topological stereoisomers. *Sensu stricto*, only bonds that are not reversibly created and cleaved under the conditions of observation should be considered as topologically relevant, which limits them to classical covalent and some coordination bonds (those in the range above 30 kcal × mol^{-1}). In fact topological chirality can be considered from the formal or the operational viewpoint. Whereas the first case concerns the manipulation of graphs, the second concerns experiment. A molecule that is demonstrated to be topologically chiral based on a formal graph, may not be operationally (topologically) chiral because some of the bonds involved in the graph are reversible under the conditions of observation. Reciprocally, a molecule that is demonstrated to be chiral experimentally will obviously not be necessarily topologically chiral.

1.3.3 Topologically Relevant Molecules that are not Topologically Chiral

This section examines systems that are not topogically chiral, yet they have a nontrivial topology in relation to chirality. In fact they are topological analogues of the molecular rubber gloves discussed in section 1.2.2.1.

1.3.3.1 Topological Rubber-Glove Molecule

We have seen that a geometrical rubber-glove molecule was a molecule with a planar graph, which is rigidly chiral in all of its conformations, yet any conformation is converted into its mirror-image by a chemically realizable pathway, which means that no *meso* form is involved as intermediate in the conversion of a chiral conformer into its mirror-image enantiomer [3, 4]. To go a step further, a topological rubber-glove molecule adds to the definition above that it results from a graph embedding that has no planar presentation. It differs from the rubber glove itself, which could be physically stretched so as to bring it down to a simple flattened piece of rubber, thus making a real rubber glove an Euclidean rubber-glove. The concept of topological rubber-glove was illustrated by the [2]catenane shown in Figure 1.27 as its copper(I) complex (Cu**26**) [57]. This molecule is built from two different macrocycles based on the 2,9-diphenyl-1,10-phenanthroline chelate, one oriented by substitution of proton 4 by a *p*-tolyl group, the other incorporating a 1,5-binaphthol-derived fragment that is free to rotate about the aryl carbon-oxygen bonds. From the geometrical point of view the [2]catenane is chiral in all its conformations because the C_2-symmetry axis of the binaphthol-incorporating macrocycle, which is contained in the symmetry plane of the oriented macrocycle, cannot generate a second symmetry plane. However, the unhindered rotation of the binaphthol moiety allows the conversion of any conceivable conformation into its mirror image. As a consequence, the [2]catenane Cu**26** is not operationally chiral. However, as it cannot achieve any planar topological symmetry, it is a topological rubber glove.

1.3.3.2 The Figure-of-Eight Knot

The figure-of-eight knot is interesting for two reasons. Its classical presentation (diagram) is the one with rigid chiral C_2 symmetry, which must not be confused with the one of the trefoil knot (see Figures 1.26b and c). It is possible to convert the corresponding 3D structure into its mirror image by continuous deformation involving only chiral intermediates, which

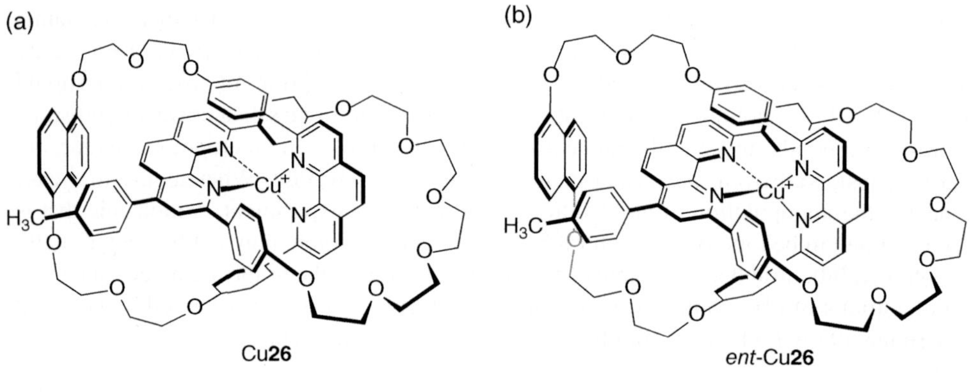

__Figure 1.27__ A topological rubber glove molecule. The [2]catenane __26__ represented as its copper(I) complex Cu__26__ is chiral in all its conformations, but none of its presentations can be drawn in the plane without crossings, that is, has a symmetry plane. To convince the reader that (a) and (b) are indeed mirror images conformations, just rotate them in the plane by 90° clockwise. The mirror plane will appear in the middle

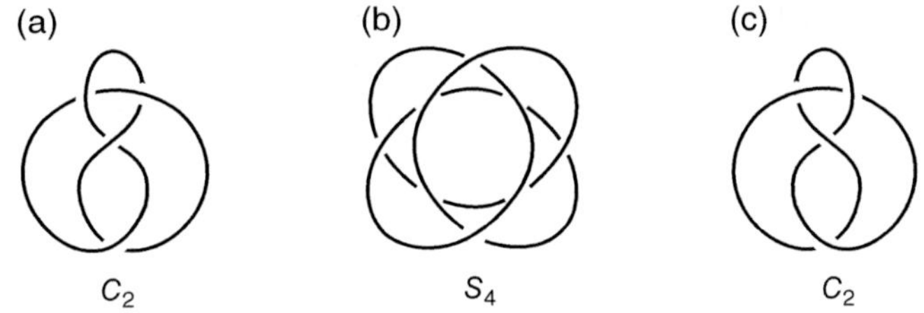

Figure 1.28 *Three diagrams of the figure-of-eight knot with their rigid symmetry point groups. Note that those shown in (a) and (c) are chiral and enantiomers of each other. The deformation pathway from one enantiomer to the other may involve only asymmetric presentations, which makes the corresponding transformation similar to that of an Euclidean rubber glove, however the existence of the S_4-symmetric presentation shown in (b), even if not attainable physically, precludes the figure-of-eight knot from being a topological rubber glove*

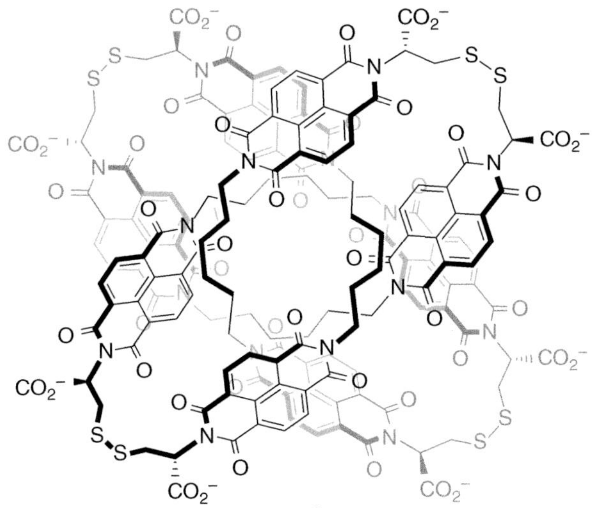

L-**27**$^{8-}$

Figure 1.29 *Molecular structure of the water-soluble figure-of-eight knot L-**27**$^{8-}$ viewed as a rigidly chiral, D_2-symmetrical conformation*

makes the figure-of-eight knot appear at first glance as a topological rubber glove. However, it is also possible to draw a diagram of the figure-of-eight knot with an achiral S_4 symmetry (Figure 1.28b), which confirms that this knot is not chiral in all of its presentations and is therefore not topologically chiral, and that in addition it is not a topological rubber glove.

So far the only reported figure-of-eight knot (**27**$^{8-}$) was obtained, among other topological isomers (a doubly interlocked [2]catenane, and a trefoil knot), by dynamic covalent synthesis involving disulfide bond formation [62]. The reaction used a disulfide library generated from an optically active building block, a $(CH_2)_6$-bridged naphthalene diimide (NDI) dimer, the outer nitrogen atoms of which were functionalized with *L*-cysteine residues. The rigid, geometrically chiral D_2-symmetrical conformation of this molecule is shown in Figure 1.29, and was ascertained by ^{1}H NMR, mass spectrometry and circular dichroism studies. The CD spectrum showed an intense negative couplet with an absolute maximum at 380 nm ($\Delta\varepsilon \sim -20 \times 10^4\, M^{-1}cm^{-1}$, water/acetonitrile, 75:25, v/v), which was independent of temperature and polarity, in agreement with the rigidly twisted character of the highly symmetric homochiral figure-of-eight knot. It must be stressed that if the carboxylate substituents were made planar (which is topologically allowed), the resulting molecule would have the topological achiral S_4-symmetry of diagram (b) of Figure 1.28.

Rather interestingly, a racemic library containing (*L,L*) and (*D,D*) naphthalene diimide (NDI) dimers led to the nearly exclusive formation (90%) of the *meso* form of the figure-of-eight knot. In principle, three diastereomers are possible, corresponding to the sequences *LL-DD-LL-DD*, *LL-LL-DD-DD*, and *LL-DD-DD-DD* of building blocks. However, only a single peak was observed by HPLC analysis, and ^{1}H NMR spectroscopy showed that the isolated compound exhibited twofold symmetry. This indicates that the *meso* figure-of-eight knot (*meso*-**27**$^{8-}$) is the *LL-DD-LL-DD* diastereomer, and that it exists as a mixture of mirror-image conformers having the geometrical C_2 symmetry, while the topological symmetry is S_4 (Figure 1.30). Quantitative 2D EXSY NMR showed that exchange between the latter had the following kinetic parameters: $\Delta H^{\ddagger} = -21\,kJ\,mol^{-1}$, $\Delta S^{\ddagger} = -302\,J\,K^{-1}\,mol^{-1}$, and $\Delta G^{\ddagger} = 69\,kJ\,mol^{-1}$ at 298 K, showing that it was controlled by entropy. This indicates that the transition state is highly destabilized, probably because the NDI hydrophobic surfaces, which are buried in the limiting conformers, are exposed to the aqueous solvent. Its diffusion coefficient is $2.0 \times 10^{-10}\, m^2/s$, which is slightly smaller than that of the homochiral figure-of-eight knot ($2.5 \times 10^{-10}\, m^2/s$). The ^{1}H NMR spectrum of the *meso* figure-of-eight knot shows sharper signals than its homochiral analogue, which was interpreted as the consequence of a perfect fit between its topological achirality and its *meso* character, whereas the latter is chiral from the geometrical viewpoint, yet it is topologically achiral.

1.3.4 Topologically Chiral Milestone Molecules (Based on Covalent Bonds)

1.3.4.1 Singly Interlocked [2]Catenanes

Catenanes are molecules made of two or more interlocking rings, and therefore they are the chemical realizations of the embedded graphs called links. The Hopf link, made of two interlocking rings, is the embedded graph of [2]catenanes. [2]Catenanes are topologically chiral if and only if they are made from oriented macrocycles – that is, macrocycles in which a bond sequence imparts an orientation [3].

1.3.4.1.1 From Copper(I) Templated Synthesis. The [2]catenanes shown in Figure 1.31a as their copper(I) complexes ([2]catenates) were the first of this type. Orientation is conferred to each macrocyclic polyether incorporating a 2,9-diphenyl-1,10-phenanthroline chelate, by aryl (phenyl in [Cu(**28**)]$^+$, *p*-tolyl in [Cu(**29**)]$^+$) substitution of the 4-position of the phenanthroline, making the resulting catenane C_2-symmetric [63, 64]. These compounds were synthesized as racemic copper complexes, and the corresponding catenanes were obtained by demetallation with cyanide. Partial analytical separation of the enantiomers of [Cu(**28**)](BF_4) was obtained by HPLC using amylose tris(3,5-dimethylphenylcarbamate)-coated silica gel and *n*-hexane/ethanol, 1:1 (v/v containing 0.1% of $LiPF_6$) as eluent. The first eluted enantiomer (70% ee) showed a positive Cotton effect centered around 340 nm, which corresponds to the MLCT band of the copper(I) complex. Its [α] was +1300° (ethanol, 0.015 g/L, 365 nm, 25 °C), quite comparable to the $[\alpha]_D$ of 1,3-diphenylallene,

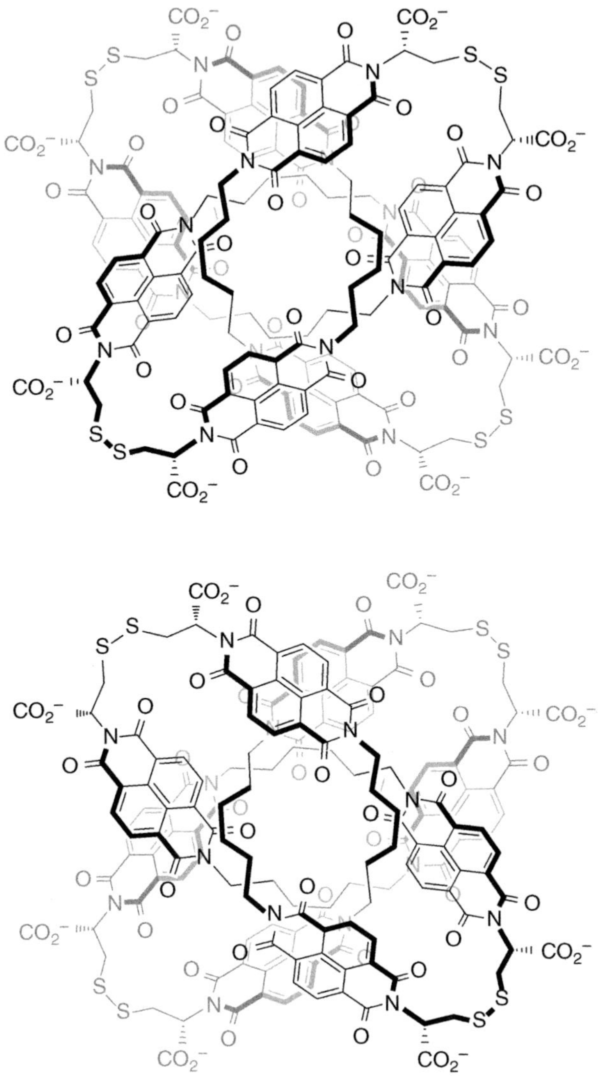

meso-**27**$^{8-}$

Figure 1.30 *Two mirror-image conformations of the achiral (*meso-**27**$^{8-}$*) figure-of-eight knot, which can be converted into each other by continuous deformation*

which has the same C_2-symmetry, and is a prototypical model of the chirality axis (see section 1.2.1 and Figure 1.7) [65].

1.3.4.1.2 From Hydrogen Bond Templated Synthesis. Other milestone examples of topologically chiral [2]catenanes are shown in Figure 1.31b. They were obtained by the amide bond formation method [66, 67]. These lactam-based catenanes differ from the earlier examples by the replacement of one of the four amide bonds by a sulfonamide bond

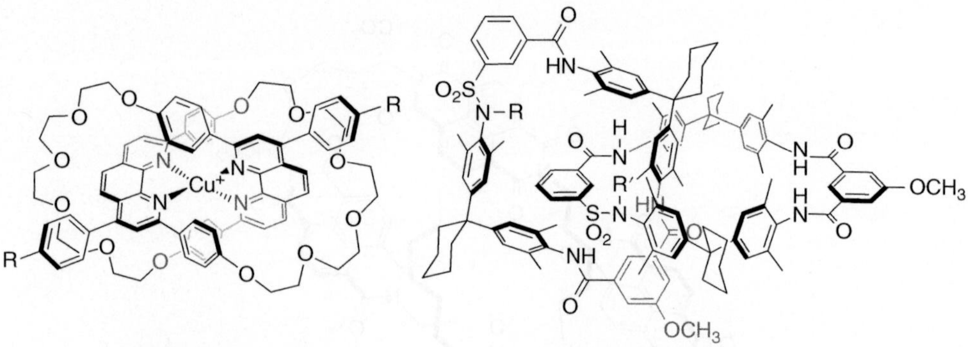

$[Cu(\mathbf{28})]^+$: R = H
$[Cu(\mathbf{29})]^+$: R = CH_3

30 : R = R′ = H
31 : R = R′ = CH_3
32 : R = H, R′ = CH_3

Figure 1.31 *Topologically chiral [2]catenanes. (a) Based on 2,9-diaryl-1,10-phenanthroline, as their copper(I) complexes, viewed along their* C_2 *symmetry axis. (b) From arylamide-based macrocycles incorporating one sulfonamide orientating group each*

in each macrocycle, which gives the latter an orientation that makes the corresponding catenanes topologically chiral. It must be noted that only a single conformer was obtained, the one in which the methoxy substituents display the *endo/exo* orientation, which renders these molecules asymmetric, contrary to the [2]catenanes of the previous section which are dissymmetric. The study of these asymmetrical catenanes (in terms of yield, conformation, X-ray crystal structure of racemic **32**) was very useful for establishing the mechanism of catenane formation via the "amide bond" formation. Catenanes **30** and **31** were resolved into their enantiomers by HPLC on Chiralcel OD column (eluent: hexane/2-propanol) with separation factors α of 6.95 and 2.39, respectively. (+)-**30** eluted at first, and showed an $[\alpha]_D$ value of 168° (compare with $[\alpha]_D$=282° measured in the same conditions for Tröger's base). The separated enantiomers of catenane **32** showed mirror-image CD spectra, with a strong couplet around 200 nm exhibiting large $\Delta\varepsilon$ values (± 130 $M^{-1}cm^{-1}$) [68].

1.3.4.2 A Doubly Interlocked Catenane

Unlike singly interlocked [2]catenanes, doubly interlocked [2]catenanes, which are the chemical realization of the Solomon rings (or links), are unconditionally topologically chiral, because orientation of the rings is not necessary to make them dissymmetric. There are several examples of doubly interlocked [2]catenanes, the earliest being the one developed by the copper(I) phenanthroline template strategy.

The doubly interlocked [2]catenane, represented in Figure 1.32 as its tri-copper(I) complex ($[Cu_3\mathbf{33}]^{3+}$), belongs to a family of compounds that differ by the nature of the bridges between the phenanthroline subunits [69, 70]. Its high yield synthesis relied on the successful formation of an unusual trinuclear double-helical complex of Li^+ with two phenanthroline-based tri-chelate ligands, in order to achieve the presentation of D_2 symmetry shown in Figure 1.26g, and on the use of ring closing alkene metathesis as "topologically fixing" step [71]. It was, however, purified as its tri-copper(I) complex obtained by transmetallation.

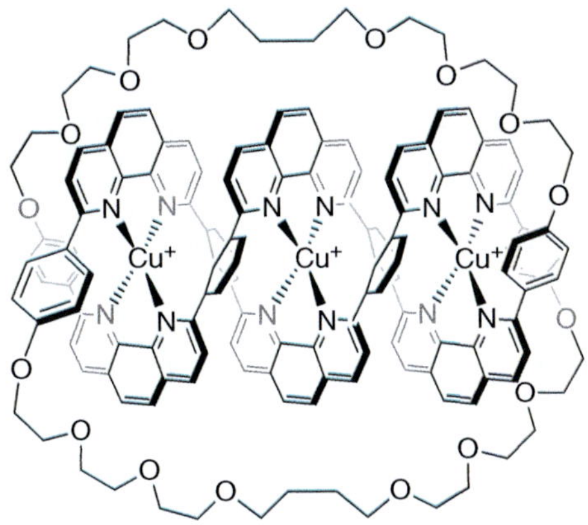

$[Cu_3(\mathbf{33})]^{3+}$

Figure 1.32 *A topologically chiral doubly interlocked [2]catenane ($[Cu_3\mathbf{33}]^{3+}$) based on copper(I) phenanthroline complex subunits. The view is shown along one of its three C_2-symmetry axes*

The metal-free doubly interlocked [2]catenane **33** was obtained quantitatively by treatment with an excess of potassium cyanide in acetonitrile, and was characterized by an ^1H NMR spectrum exhibiting broad features, thus pointing to restricted intramolecular motions occurring at the NMR timescale.

1.3.4.3 Molecular Knots

Since the pioneering work of Sauvage, Buchecker, and coworkers, several examples of molecular trefoil knots have been published. They have been synthesized according to various strategies [71]. It is noteworthy that the transition metal approach has used the three presentations shown in Figure 1.26b. This short review will be restricted to examples of molecular trefoil knots that were resolved into their enantiomers.

1.3.4.3.1 Trefoil Knots from Double-helical Precursors. This trefoil knot is a typical example of a trefoil knot, the construction of which follows a predicted sequence (kinetic template). The synthesis of this trefoil knot relied on the successful stabilization of a dinuclear double-helical complex of copper(I) with two 2,9-diphenyl-1,10-phenanthroline (dpp)-based di-chelate ligands, in order to achieve the presentation of D_2 symmetry shown in Figure 1.26b. In earlier syntheses, the chiral trefoil knot $[Cu_2(\mathbf{k\text{-}86})]^{2+}$ and the *meso* face-to-face complex $[Cu_2(\mathbf{m\text{-}43})_2]^{2+}$ were formed (Figure 1.33), however they could not be differentiated by standard ^1H NMR, as they both showed diastereotopic pairs of protons [72]. Internal correlation times for the protons of the central -$(CH_2)_4$- bridge were longer in the case of the knotted system by comparison with the other, pointing to a significant decrease of mobility of this methylenic chain in the case of the knot. The connection of the two phenanthroline subunits by a *meta*-phenylene spacer rather than a -$(CH_2)_n$- chain turned out to be crucial, as

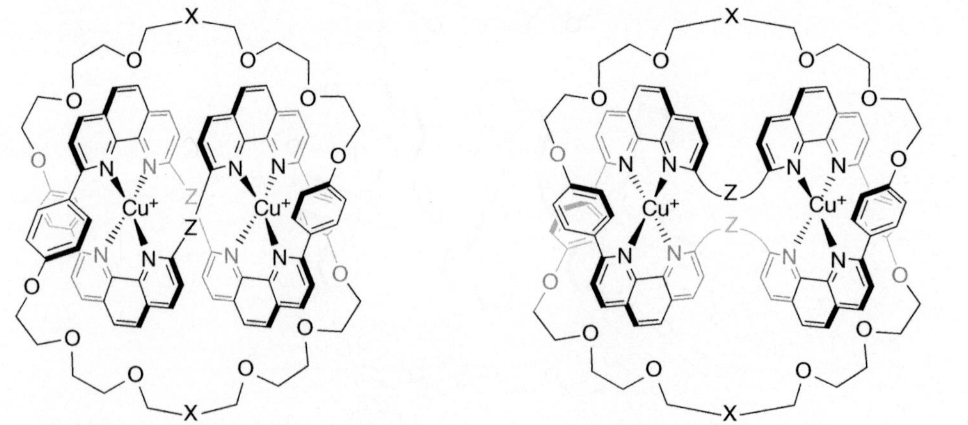

$[Cu_2(\mathbf{k\text{-}80})]^{2+}$: X = /, Z = $(CH_2)_4$	$[Cu_2(\mathbf{m\text{-}40})_2]^{2+}$: X = /, Z = $(CH_2)_4$
$[Cu_2(\mathbf{k\text{-}82})]^{2+}$: X = CH_2OCH_2, Z = $(CH_2)_2$	$[Cu_2(\mathbf{m\text{-}41})_2]^{2+}$: X = CH_2OCH_2, Z = $(CH_2)_2$
$[Cu_2(\mathbf{k\text{-}83})]^{2+}$: X = $(CH_2)_2$, Z = *m*-C_6H_4	
$[Cu_2(\mathbf{k\text{-}84})]^{2+}$: X = /, Z = $(CH_2)_6$	$[Cu_2(\mathbf{m\text{-}42})_2]^{2+}$: X = /, Z = $(CH_2)_6$
$[Cu_2(\mathbf{k\text{-}85})]^{2+}$: X = CH_2OCH_2, Z = *m*-C_6H_4	
$[Cu_2(\mathbf{k\text{-}86})]^{2+}$: X = CH_2OCH_2, Z = $(CH_2)_4$	$[Cu_2(\mathbf{m\text{-}43})_2]^{2+}$: X = CH_2OCH_2, Z = $(CH_2)_4$
$[Cu_2(\mathbf{k\text{-}90})]^{2+}$: X = CH_2OCH_2, Z = $(CH_2)_6$	$[Cu_2(\mathbf{m\text{-}45})_2]^{2+}$: X = CH_2OCH_2, Z = $(CH_2)_6$

Figure 1.33 *Structural formulae of trefoil molecular knots synthesized via the copper(I)-templated strategy and the corresponding face-to-face complexes. The numbers following k (for knot) and m (for macrocycle) correspond to the minimum number of atoms in each macrocycle*

it allowed for the transition metal-templated synthesis of the trefoil knot $[Cu_2(\mathbf{k\text{-}85})]^{2+}$ in significantly improved yield (29%), without the observation of face-to-face complex formation. In its most accomplished version, the cyclization reaction involved alkene metathesis and the yield of the corresponding knot $[Cu_2(\mathbf{k\text{-}83})]^{2+}$ was raised to 74%. The metal-free knots were obtained as metal-free catenanes by competitive reaction with cyanide.

Very gratifyingly, the di-copper complex of the trefoil knot $[Cu_2(\mathbf{k\text{-}86})]^{2+}$ crystallized as a conglomerate of enantiomers, therefore exhibiting spontaneous resolution [73]. However the first planned resolution of a molecular trefoil knot was performed by separation of ion-pair diastereomeric derivatives [74]. The knot $[Cu_2(\mathbf{k\text{-}85})]^{2+}$, one enantiomer of which is illustrated in Figure 1.34a, is a dication and its diastereomeric salt with the (+)-(*R*)-binaphthylphosphate (BNP) optically active anion was formed by anion exchange followed by fractional crystallization. This allowed for the resolution of the di-copper(I) complex into optically pure crystalline [(+)-(*M*)-Cu_2(**k-85**)]·2[(+)-(*R*)-BNP^-], while optically pure [(–)-(*P*)-Cu_2(**k-85**)]·2[(+)-(*R*)-BNP^-] could be isolated from a mother liquor. After exchange of [(+)-(*R*)-BNP^-] back for PF_6^-, the chiroptical properties of the resolved complexes were measured, and the circular dichroism spectrum was shown to exhibit pronounced Cotton effects. The $\Delta\varepsilon$ value reached a maximum of $\pm 20.7\,M^{-1}cm^{-1}$ in the visible, corresponding to the absorption maximum of the MLCT transition at 510 nm. The molar rotation $[M]_D$ was +73000 for [(+)-(*M*)-Cu_2(**k-85**)] $(PF_6)_2$.

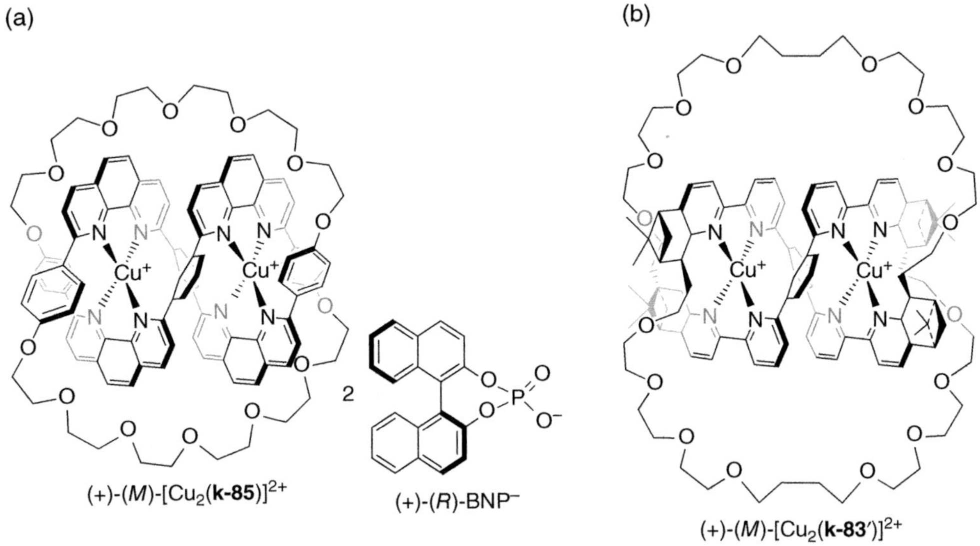

Figure 1.34 *Topologically chiral trefoil molecular knots based on large ring macrocycles incorporating four chelates derived from dpp ([Cu_2(**k-85**)]$^{2+}$, left) or bipy (Cu_2(**k-83'**)]$^{2+}$, right), shown as their D_2-symmetric bis-copper(I) complexes with M helicity. Attached to the [(+)-(M)-Cu_2(**k-85**)$^{2+}$] dication is the [(+)-(R)-BNP^-] optically active anion that was used for its resolution by fractional crystallization*

As an alternative, a related trefoil knot was obtained by stereoselective synthesis. Indeed, the use of enantiomerically pure di-chelate ligands (obtained by constructing (+)-(*R*)-α-pinene-fused 2,2′-bipyridyl ligands belonging to the so-called CHIRAGEN family), led to the stereoselective formation of the di-copper double helix having the *M* absolute configuration, as shown by a positive exciton couplet [75]. This helicity was further transferred to the knotted system ([Cu_2(**k-83'**)]$^{2+}$), which was obtained in 74% isolated yield by ring-closing metathesis reaction followed by a quantitative catalytic hydrogenation (Figure 1.34b).

The metal-free knots retain high molar rotations (e. g., +20000 for [(+)-(*M*)-(**k-85**)]), which are comparable to values found in the case of helicenes. The ^{1}H NMR spectrum of the molecular knot **k-84** in d^7-DMF shows broad features at room temperature, indicative of molecular motions that could be identified to a molecular "reptation" phenomenon. Variable temperature studies gave a ~2 s characteristic time for these motions. Upon heating the solution of the knot, a well resolved spectrum is obtained, which shows that the two phenanthroline subunits are again equivalent, as in the di-copper complex. A process of conformational change that has been called "reptation" allows all the atoms to occupy all the possible positions along the molecular knot **k-84**, which achieves in such a way the maximal topological D_3 symmetry. Interestingly, this latter symmetry is clearly apparent in the atomic force microscopy (AFM) image of a trefoil knotted polymer synthesized via ring expansion of a Sn(IV)-locked molecular knot precursor displaying the D_2 symmetry [76].

1.3.4.3.2 Trefoil Knots Based on Hydrogen Bonding Interactions. These trefoil knots are an illustration of systems obtained by serendipity (Figure 1.35). The first "knotane"

(a)

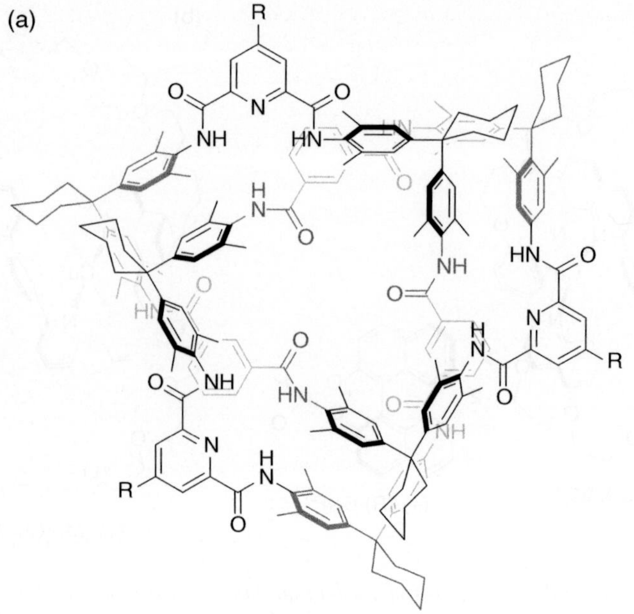

34: R = H
35: R = O-allyl

(b)

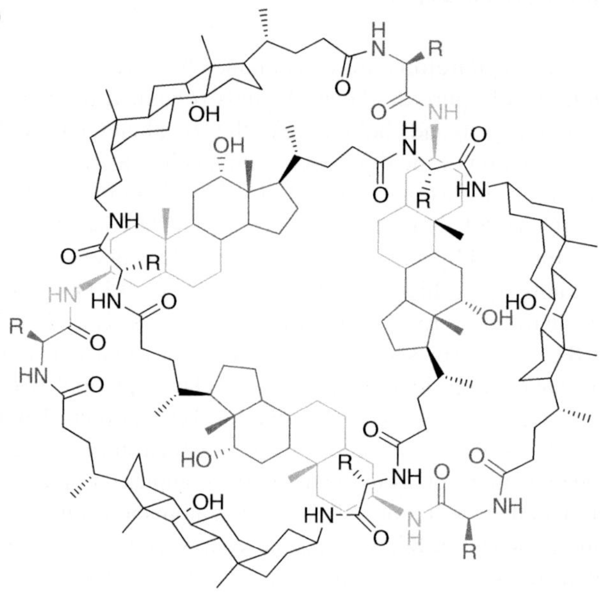

36

Figure 1.35 *Trefoil knots based on hydrogen bonding and other interactions. (a) General structure of amide-based knotanes:* ***34*** *(R=H) is the first member of a series of compounds;* ***35*** *(R=OCH_2CHCH_2=O-allyl) is the knotane that was used for further functionalization. (b) General structure of knots* ***36*** *based on optically active choline spacers between dipeptide fragments*

(**34** of Figure 1.35a) was obtained in the course of the synthesis of rigid macrocycles incorporating 2,6-pyridine and 1,3-benzenedicarboxamide subunits as precursors of catenanes. Corresponding to the knotted macrocyclic trimer, it was isolated in 20% yield alongside with the target large ring macrocyclic dimer (23%) and the macrocyclic monomer (20%) [77]. An X-ray crystal structure analysis of racemic crystals of knotane **34** showed that there were four intramolecular hydrogen bonds, and that the molecular structure had C_1 symmetry. This was confirmed in solution at the NMR timescale by the study of a more soluble analogue, knotane **35** (Figure 1.35a), in which R=OCH_2CHCH_2 (O-allyl). In $CDCl_3$, the ^1H NMR spectrum showed very broad features, pointing to the occurrence of conformational changes taking place at the NMR timescale. However, these were slowed either by cooling down the solution to 223 K, or by adding 10% d^6-DMSO at room temperature (and of course in pure d^6-DMSO at RT). Both experiments afforded a well resolved ^1H NMR spectrum corresponding to a C_1-symmetric species.

Knotane **34** was resolved on a Chiralpak-AD HPLC column using a mixture of *n*-hexane/$CHCl_3$/iPrOH, 30:20:1, v/v as eluent, with a large separation factor ($\alpha = 2.14$) [78]. The experimental CD spectra showed mirror image curves for the (+) and (−) enantiomers, with maxima at 240 nm ($\Delta\varepsilon = +100/-108\,M^{-1}cm^{-1}$), 272 nm ($\Delta\varepsilon = \pm 4\,M^{-1}cm^{-1}$), and 293 nm ($\Delta\varepsilon = \pm 8\,M^{-1}cm^{-1}$). The strong band at 240 nm, which results mainly from exciton-coupled $B_{a/b}$ states of the benzene/benzene-amide subunits, was stable with respect to variations in the calculations and could be used to assign the absolute configuration of the levorotatory enantiomer at 365 m, which showed a negative Cotton effect at 240 nm, as the (*P*,*P*,*P*) one.

The second case (**36**, Figure 1.35b) represents another example of the stereoselective synthesis of a molecular knot [79]. With the aim of obtaining macrolactams incorporating the choline scaffold, a linear trimeric aminoacid incorporating three choline fragments was subjected to peptide bond formation conditions. The desired macrocycle was obtained in 31% yield, and significant amounts (21% yield) of a compound were isolated that could be identified as a trefoil knot macrocycle incorporating six choline subunits (Figure 1.35b). The knotted nature of the system was identified without any ambiguity by an X-ray crystal structure analysis, which showed the threefold symmetry of the molecule and that the chirality of the starting aminoacid was fully transferred to the knot, which exhibited the *P* helicity. It is noteworthy that the molecular knot showed enhanced CD properties by comparison with the cyclic analogue, as its normalized Θ value (48 000° at 200 nm) was nearly five times higher.

1.3.4.3.3 A Pentafoil Knot from a Circular Helicate. Just as binuclear double helicates were found to be ideal precursors for the trefoil knot, pentanuclear circular helicates were used as precursors for the pentafoil knot. Lehn and coworkers had shown that a linear tris(2,2′-bipyridyl) molecular fragment (**37**) with short connections (ethylene linkers) between the central and the peripheral chelates favored the formation of the circular pentafoil pentanuclear helicate $[Fe_5(\mathbf{37})_5]^{10+}$ with respect to the corresponding linear triple-stranded trinuclear helicate in the presence of Fe^{2+} cations, and a chloride anion – which played the role of template (Figure 1.36a) [80]. Subsequently, Leigh and coworkers were able to show that the same kind of assembly could be obtained if the peripheral bipyridyl subunits were replaced by α-imino pyridyl chelates generated during the assembly process from a molecular thread incorporating a central bipyridyl subunit and 2-pyridylcarboxaldehyde end groups, and various aliphatic amines [81]. The use of enantiomerically pure

(a)

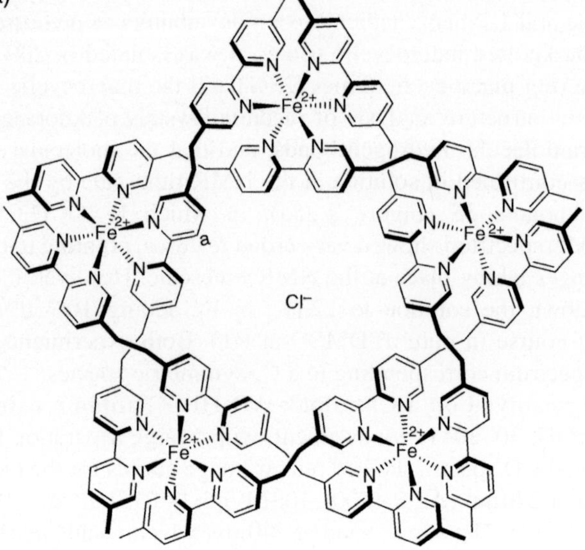

$[Fe_5(\mathbf{37})_5]^{10+} \supset Cl^-$

(b)

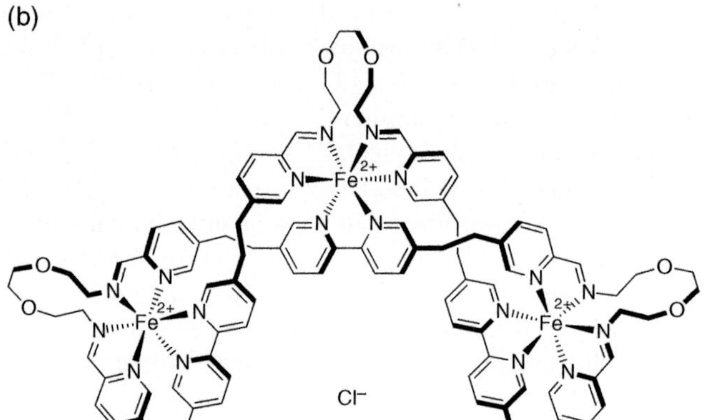

$[Fe_5(\mathbf{38})]^{10+} \supset Cl^-$

Figure 1.36 *(a) The original circular pentanuclear* Fe^{2+} *helicate* $[Fe_5(\mathbf{37})_5]^{10+}$ *as its chloride complex. (b) A pentafoil molecular knot* $[Fe_5\mathbf{38}]^{10+}$ *based ion a circular pentanuclear* Fe^{2+} *helicate precursor*

amines led to the complete control of the helical sense of the circular helicate, as demonstrated by circular dichroism studies. Thanks to the fact that the two terminal alkyl imines are close to each other (template effect), intramolecular bridging by using a diamine was envisaged in order to connect the five tri-chelate molecular threads so as to generate *in situ* a pentafoil knotted giant macrocycle. Simple alkyl chain diamines gave complex mixtures of oligomers but the use of the diamine deriving from triethylene glycol afforded the desired pentafoil knot $[Fe_5\mathbf{38}]^{10+}$ after two days reaction in 44% isolated yield (Figure 1.36b). The chirality and symmetry of the pentafoil knot were analyzed by ^{1}H NMR spectroscopy, which showed the presence of diastereotopic pairs of methylene protons, and by X-ray crystallography, which brought a definitive proof of the topology of the molecule. Removal of the chloride anion by treatment with a large excess of $AgPF_6$ produced the chloride-free trefoil knot, however no attempts were made to remove the templating metal cations.

1.4 Conclusion

Chirality, the nonsuperimposability of a molecular object or assembly with its mirror image, is one of the most ancient concepts of chemistry. However, in spite of its apparent simplicity it underlies many difficulties. These are principally due to the fact that molecules are complex nanoscale objects that cannot be described by a single model. We have shown that, starting from an achiral object, chirality arises either from a constitutional change or a twist, or both. The most significant advances in recent decades are the synthesis and resolution of topologically chiral molecules (in particular trefoil knots), and developments in the supramolecular transfer of chirality and the supramolecular chirality itself, which are strongly interconnected. Whereas the former explains effects as different as enantioselective catalysis by transition metals and organocatalysis, enantiomeric resolution and chirality detection by ICD effects, the latter has led to the development of chiral materials that have made molecular chirality tangible at the macroscopic level, as foreseen by Louis Pasteur.

References

[1] Liang, C., Mislow, K. (1994) Classification of topologically chiral molecules. *J. Math. Chem.*, **15** (1), 245–260.
[2] For an account on the historical development of chirality, see: O'Loane, J. K. (1980) Optical activity in small molecules, nonenantiomorphous crystals, and nematic liquid crystals. *Chem. Rev.*, **80** (1), 41–61.
[3] Walba, D. M. (1985) Topological stereochemistry. *Tetrahedron*, **41** (16), 3161–3212.
[4] Mislow, K. (1999) Molecular chirality. *Top. Stereochem.*, **22**, 1–82.
[5] Mislow, K., Bickart, P. (1976/1977) An epistemological note on chirality. *Isr. J. Chem.*, **15** (1–2), 1–6.
[6] Hasenknopf, B., Micoine, K., Lacôte, E., *et al.* (2008) Chirality in polyoxometallate chemistry. *Eur. J. Inorg. Chem.*, **32**, 5001–5013.
[7] Tsuda, A., Kakamura, T., Sakamoto, S., *et al.* (2002) A self-assembled porphyrin box from *meso-meso* linked bis{5-*p*-pyridyl-15-(3,5-di-octyloxyphenyl)porpyrinato zinc(II)}. *Angew. Chem. Int. Ed.*, **41** (15), 2817–2821.
[8] Cahn, R. S., Ingold, C., Prelog, V. (1966) Specification of molecular chirality. *Angew. Chem. Int. Ed.*, **5** (4), 385–415.
[9] Prelog, V. (1976) Chirality in chemistry. *Science*, **193** (4247), 17–24.
[10] Danjo, H., Iwaso, K., Kawahata, M., *et al.* (2013) Preparation of tris(spiroorthocarbonate) cyclophanes as back to back ditopic hosts. *Org. Lett.*, **15** (9), 2164–2167.

[11] Yashima, E. (2010) Synthesis and structure determination of helical polymers. *Polymer J.*, **42** (1), 3–16.
[12] Guichard, G., Huc, I. (2011) Synthetic foldamers. *Chem. Commun.*, **47**, 5933–5941.
[13] Gingras, M. (2013) One hundred years of helicene chemistry. Part 1: non-stereoselective syntheses of carbohelicenes. *Chem. Soc. Rev.*, **42**, 968–1006.
[14] Provent, C. and Williams, A. (1999) The chirality of polynuclear transition metal complexes, in *Transition Metals in Supramolecular Chemistry* (ed. J.-P. Sauvage), John Wiley & Sons, Ltd, Chichester, pp. 135–192.
[15] Laleu, B., Bernardinelli, G., Chauvin, R., Lacour, J. (2006) Trimesitylmethylphosphonium cation. Supramolecular stereocontrol and simple enantiomerization mechanism determination. *J. Org. Chem.*, **71** (19), 7412–7416.
[16] Glasson, C. R. K., Meehan, G. V., Clegg, J. K., *et al.* (2008) Microwave synthesis of a rare $[Ru_2L_3]^{4+}$ triple helicate and its interaction with DNA. *Chem. Eur. J.*, **14** (34), 10535–10538.
[17] Bassani, D., Lehn, J.-M., Fromm, K., Fenske, D. (1998) Toposelective and chiroselective self-assembly of [2×2] grid-type inorganic arrays containing different octahedral metallic centers. *Angew. Chem. Int. Ed.*, **37** (17), 2364–2367.
[18] Anet, F. A. L., Miura, S. S., Siegel, J., Mislow, K. (1983) La coupe du roi and its relevance to stereochemistry. Combination of two homochiral molecules to give an achiral product. *J. Am. Chem. Soc.*, **105** (6), 1419–1426.
[19] Thilgen, C., Herrmann, A., Diederich, F. (1997) Configurational description of chiral fullerenes and fullerene derivatives with a chiral functionalization pattern. *Helv. Chim. Acta*, **80** (1), 183–199.
[20] Szumna, A. (2010) Inherently chiral concave molecules – from synthesis to applications. *Chem. Soc. Rev.*, **39**, 4274–4285.
[21] Wang, T., Zhang, Y.-F., Hou, Q.-Q., *et al.* (2013) C_3-symmetrical tribenzotriquinacene derivatives: Optical resolution through cryptophane synthesis and supramolecular self-assembly into nanotubes. *J. Org. Chem.*, **78** (3), 1062–1069.
[22] Higashibayashi, S., Sakurai, H. (2008) Asymmetric synthesis of a chiral buckybowl, trimethylsumanene. *J. Am. Chem. Soc.*, **130** (27), 8592–8593.
[23] Sánchez-Molina, I., Grimm, B., Calderon, R. M. K., *et al.* (2013) Self-assembly, host-guest chemistry, and photophysical properties of subphthalocyanin-based metallo-supramolecular capsules. *J. Am. Chem. Soc.*, **135** (28), 10503–10511.
[24] Brotin, T., Dutasta, J.-P. (2009) Cryptophanes and their complexes – present and future. *Chem. Rev.*, **109** (1), 88–130.
[25] Vysotsky, M. O., Pop, A., Broda, F., *et al.* (2001) Molecular motions within self-assembled dimeric capsules with tetraethylammonium cations as guest. *Chem. Eur. J.*, **7** (20), 4403–4410.
[26] Dalla Cort, A., Mandolini, L., Pasquini, C., Schiaffino, L. (2004) "Inherent chirality" and curvature. *New J. Chem.*, **28**, 1198–1199.
[27] Lukášek, J., Böhm, S., Dvořáková, H., *et al.* (2014) Regioselective halogenation of thiacalix[4]arenes in the cone and 1,3-alternate conformations. *Org. Lett.*, **16** (19), 5100–5103.
[28] Collet, A., Gabard, J., Jacques, J., *et al.* (1981) Synthesis and absolute configuration of chiral (C_3) cyclotriveratrylene derivatives. Crystal structure of (*M*)-(–)-2,7,12-triethoxy-3,8,13-tris-[(*R*)-1-methoxycarbonylethoxy]-10,15-dihydro-5*H*-tribenzo[a,d,g]-cyclononene. *J. Chem. Soc., Perkin Trans.* 1, 1630–1638.
[29] Guieu, S., Zaborova, E., Blériot, Y., *et al.* (2010) Can hetero-polysubstituted cyclodextrins be considered as inherently chiral concave molecules ? *Angew. Chem. Int. Ed.*, **49** (13), 2314–2318.
[30] Sendhoff, N., Weissbarth, K.-H., Vögtle, F. (1987) Synthesis and chirality of sixfold-bridged triple-decker phanes. *Angew. Chem. Int. Ed. Engl.*, **26** (8), 777–779.
[31] Hasenknopf, B., Lehn, J.-M., Kneisel, B. O., *et al.* (1996) Self-assembly of a circular double helicate. *Angew. Chem. Int. Ed.* **35** (16), 1838–1840.
[32] Yeh, R. M., Xu, J., Seeber, G., Raymond, K. N. (2005) Tetrahedral coordination cages: an extension of symmetry-based design. *Inorg. Chem.*, **44** (18), 6228–6239.
[33] Meng, W., Clegg, J. K., Thoburn, J. D., Nitschke, J. R. (2011) Controlling the transmission of stereochemical information through space in terphenyl-edged Fe_4L_6 cages. *J. Am. Chem. Soc.*, **133** (34), 13652–13660.

[34] Prince, R. B., Barnes, S. A., Moore, J. S. (2000) Foldamer-based molecular recognition. *J. Am. Chem. Soc.*, **122** (12), 2758–2762.
[35] Shi, Z.-M., Chen, S.-G., Zhao, X., *et al.* (2011) *meta*-substituted benzamide oligomers that complex mono-, di- and tricarboxylates: folding-induced selectivity and chirality. *Org. Biomol. Chem.*, **9**, 8122–8129.
[36] Mislow, K., Bolstad, R. (1955) Molecular dissymmetry and optical inactivity. *J. Am. Chem. Soc.*, **77** (24), 6712–6713.
[37] Pfeiffer, P., Quehl, K. (1931) Über einen neuen Effekt in Lösungen optisch-aktiver Substanzen. *Chem. Ber.*, **64** (10), 2667–2671.
[38] Kuhn, R. (1932) Über einen neuen Effekt in Lösungen optisch-aktiver Substanzen. *Chem. Ber.*, **65** (1), 49–51.
[39] Ferrand, Y., Kendhale, A. M., Kauffmann, B., *et al.* (2010) Diastereoselective encapsulation of tartaric acid by a helical aromatic oligoamide. *J. Am. Chem. Soc.*, **132** (23), 7858–7859.
[40] Maeda, K., Morino, K., Okamoto, Y., *et al.* (2004) Mechanism of helix induction on a stereoregular poly((4-carboxyphenyl)acetylene) with chiral amines and memory of the macromolecular helicity assisted by interaction with achiral amines. *J. Am. Chem. Soc.*, **126** (13), 4329–4342.
[41] Helmich, F., Lee, C. C., Schenning, A. P. H. J., Meijer, E. W. (2010) Chiral memory via chiral amplification and selective depolymerization of porphyrin aggregates. *J. Am. Chem. Soc.*, **132** (47), 16753–16755.
[42] Mizuno, Y., Aida, T., Yamaguchi, K. (2000) Chirality-memory molecule: Crystallographic and spectroscopic studies on dynamic molecular recognition events by fully substituted chiral porphyrins. *J. Am. Chem. Soc.*, **122** (22), 5278–5285.
[43] Green, M. M., Peterson, N. C., Sato, T., *et al.* (1995) Helical polymer with a cooperative response to chiral information. *Science*, **268** (5219), 1860–1866.
[44] Ishi-i, T., Crego-Calama, M., Timmerman, P., *et al.* (2002) Enantioselective formation of a dynamic hydrogen-bonded assembly based on the chiral memory concept. *J. Am. Chem. Soc.*, **124** (49), 14631–14641.
[45] Green, M. M., Garetz, B. A., Munoz, B., *et al.* (1995) Majority rules in the copolymerization of mirror image isomers. *J. Am. Chem. Soc.*, **117** (14), 4181–4182.
[46] van Gestel, J., Palmans, A. R. A., Titulaer, B., *et al.* (2005) "Majority-rules" operative in chiral columnar stacks of C_3-symmetrical molecules. *J. Am. Chem. Soc.*, **127** (15), 5490–5494.
[47] Huang, X., Rickman, B. H., Borhan, B., *et al.* (1998) Zinc porphyrin tweezer in host-guest complexation: Determination of absolute configurations of diamines, amino acids, and amino alcohols by circular dichroism. *J. Am. Chem. Soc.*, **120** (24), 6185–6186.
[48] Bentley, K. W., Wolf, C. (2013) Stereodynamic chemosensor with selective circular dichroism and fluorescence readout for in situ determination of absolute configuration, enantiomeric excesses, and concentration of chiral compounds. *J. Am. Chem. Soc.*, **135** (33), 12200–12203.
[49] Girard, C., Kagan, H. B. (1998) Nonlinear effects in asymmetric catalysis and stereoselective reactions: ten years of investigation. *Angew. Chem. Int. Ed.*, **37** (21), 2922–2959.
[50] Haberhauer, G., Kallweit, C. (2010) A bridged azobenzene derivative as a reversible, light-induced chirality switch. *Angew. Chem. Int. Ed.*, **49** (13), 2418–2421.
[51] Johnson, R. S., Yamazaki, T., Kovalenko, A., Fenneri, H. (2007) Molecular basis for water-promoted supramolecular chirality inversion in helical rosette nanotubes. *J. Am. Chem. Soc.*, **129** (17), 5735–5743.
[52] Eelkema, R., Feringa, B. L. (2006) Phosphoric acids as amplifiers of molecular chirality in liquid crystalline media. *Org. Lett.*, **8** (7), 1331–1334.
[53] Haino, T., Tanaka, T., Fukazawa, Y. (2008) Self-assembly of tris(phenylisoxazolyl)benzene and its asymmetric induction of supramolecular chirality. *Chem. Commun.*, 468–470.
[54] Forgan, R. S., Sauvage, J.-P., Stoddart, J. F. (2011) Chemical topology: complex molecular knots, links, and entanglements. *Chem. Rev.*, **111** (9), 5434–5464.
[55] Simon, J. (1986) Topological chirality of certain molecules. *Topology*, **25** (2), 229–235.
[56] Flapan, E. (1991) Topological techniques to detect chirality. In *New Developments in Molecular Chirality*, (ed. P. G. Mezey), pp. 209–239. Kluwer Academic Publishers.

[57] Chambron, J.-C., Sauvage, J.-P., Mislow, K., *et al.* (2001) A [2]catenane and a [2]rotaxane as prototypes of topological and Euclidean molecular "rubber gloves." *Chem. Eur. J.*, **7** (19), 4085–4096.
[58] McArdle, C. P., Jennings, M. C., Vittal, J. J., Puddephatt, R. J. (2001) Self-assembly of rings, catenanes, and a doubly braided catenane containing gold(I): the hinge-group effect in diacetylide ligands. *Chem. Eur. J.*, **7** (16), 3572–3583.
[59] Ponnuswamy, N., Cougnon, F. B. L., Clough, J. M., *et al.* (2012) Discovery of an organic trefoil knot. *Science*, **338** (6108), 783–785.
[60] Pentecost, C. D., Chichak, K. S., Peters, A., *et al.* (2007) A molecular Solomon link. *Angew. Chem. Int. Ed.*, **46** (1–2), 218–222.
[61] Peinador, C., Blanco, V., Quíntela, J. M. (2009) A new doubly interlocked [2]catenane. *J. Am. Chem. Soc.*, **131** (3), 920–921.
[62] Ponnuswamy, N., Cougnon, F. B. L., Dan Pantoş, G. Sanders, J. K. M. (2014) Homochiral and *meso* figure eight knots and a Solomon link. *J. Am. Chem. Soc.*, **136** (23), 8243–8251.
[63] Mitchell, D. K., Sauvage, J.-P. (1988) A topologically chiral [2]catenand. *Angew. Chem. Int. Ed. Engl.*, **27** (7), 930–931.
[64] Chambron, J.-C., Mitchell, D. K., Sauvage, J.-P. (1992) Synthesis, characterization, and a proton NMR study of topologically chiral copper(I) [2]-catenates and achiral analogues. *J. Am. Chem. Soc.*, **114** (12), 4625–4631.
[65] Kaida, Y., Okamoto, Y., Chambron, *et al.* (1993) The separation of optically-active copper(I) catenates. *Tetrahedron Lett.*, **34** (6), 1019–1022.
[66] Jäger, R., Vögtle, F. (1997) A new synthetic strategy towards molecules with mechanical bonds: Nonionic template synthesis of amide-linked catenanes and rotaxanes. *Angew. Chem. Int. Ed.*, **36** (9), 930–944.
[67] Ottens-Hildebrandt, S., Schmidt, T., Harren, J., Vögtle, F. (1995) Sulfonamide-based catenanes – Regioselective template synthesis. *Liebigs Ann.* (10), 1855–1860.
[68] Yamamoto, C., Okamoto, Y., Schmidt, T., *et al.* (1997) Enantiomeric resolution of cycloenantiomeric rotaxane, topologically chiral catenane, and pretzel-shaped molecules: Observation of pronounced circular dichroism. *J. Am. Chem. Soc.*, **119** (43), 10547–10548.
[69] Nierengarten, J.-F., Dietrich-Buchecker, C. O., Sauvage, J.-P. (1994) Synthesis of a doubly interlocked [2]-catenane. *J. Am. Chem. Soc.*, **116** (1), 375–376.
[70] Dietrich-Buchecker, C., Sauvage, J.-P. (1999) Lithium templated synthesis of catenanes: Efficient synthesis of doubly interlocked [2]-catenanes. *Chem. Commun.*, 615–616.
[71] Chambron, J.-C., Sauvage, J.-P. (2013) Topologically complex molecules obtained by transition metal templation: It is the presentation that determines the synthesis strategy. *New J. Chem.*, **37**, 49–57.
[72] Dietrich-Buchecker, C. O., Nierengarten, J.-F., Sauvage, J.-P., *et al.* (1993) Dicopper(I) trefoil knots and related unknotted molecular systems: Influence of ring size and structural factors on their synthesis and electrochemical and excited state properties. *J. Am. Chem. Soc.*, **115** (24), 11237–11244.
[73] Dietrich-Buchecker, C. O., Sauvage, J.-P., Kintzinger, J.-P., *et al.* (1992) A di-copper(I) trefoil knot and its parent ring compounds: Synthesis, solution studies and X-ray structures. *New J. Chem.*, **16**, 931–942.
[74] Rapenne, G., Dietrich-Buchecker, C., Sauvage, J.-P. (1999) Copper(I)- or iron(II)- templated synthesis of molecular knots containing two tetrahedral or octahedral coordination sites. *J. Am. Chem. Soc.*, **121** (5), 994–1001.
[75] Perret-Aebi, L.-E., von Zelewsky, A., Dietrich-Buchecker, C., Sauvage, J.-P. (2004) Stereoselective synthesis of a topologically chiral molecule: The trefoil knot. *Angew. Chem. Int. Ed.*, **43** (34), 4482–4485.
[76] Cao, P.-F., Mangadlao, J., Advincula, R. (2015) A trefoil knotted polymer produced through ring expansion. *Angew. Chem. Int. Ed.*, **54** (17), 5127–5131.
[77] Safarowski, O., Nieger, M., Fröhlich, R., Vögtle, F. (2000) A molecular knot with twelve amide groups–One-step synthesis, crystal structure, chirality. *Angew. Chem. Int. Ed.*, **39** (9), 1616–1618.

[78] Vögtle, F., Hünten, A., Vogel, E., *et al.* (2001) Novel amide-based molecular knots: Complete enantiomeric separation, chiroptical properties, and absolute configuration. *Angew. Chem. Int. Ed.*, **40** (13), 2468–2471.
[79] Feigel, M., Ladberg, R., Engels, S., *et al.* (2006) A trefoil knot made of amino-acids and steroids. *Angew. Chem. Int. Ed.*, **45** (34), 5698–5702.
[80] Hasenknopf, B., Lehn, J.-M., Boumediene, N., *et al.* (1997) Self-assembly of tetra- and hexanuclear circular helicates. *J. Am. Chem. Soc.*, **119** (45), 10956–10962.
[81] Ayme, J.-F., Beves, J. E., Leigh, D. A., *et al.* (2012) A synthetic molecular pentafoil knot. *Nat. Chem.*, **4** (1), 15–20.

2

Homochirogenesis and the Emergence of Lifelike Structures

Pedro Cintas

2.1 Introduction and Scope

The quest for a plausible origin of homochirality in nature has lured and preoccupied scientists dating back to Pasteur's times and his seminal discovery of molecular asymmetry. The handedness of matter manifests itself through multiple size scales, from tiny astrophysical particles (electrons or neutrinos) to discrete molecules, provided they are configurationally stable, and to large macromolecular or supramolecular entities. We cannot think of life without invoking a complete stereochemical bias that ensures error-free replication and translational processes at cellular level. L-Amino acids and D-sugars dominate life's building blocks, particularly in proteins and oligonucleotides, but their origins remain uncertain nevertheless. For some of us this has become the most interesting and intriguing question, the pursuit being quixotic too. Solving a difficult problem will take a long time, but homochirogenesis is also circumscribed to prebiotic chemistry scenarios, for which reliable evidence will be difficult to find. Fortunately, the empirical and theoretical tools have now matured to the point where abiotic mechanisms can be assessed in detail. Any abiotic theory presupposes an initial enantiomeric excess that would have further been amplified leading to the enantiomeric purity that characterizes the biomolecules present in living organisms. Even in the RNA-world hypothesis, where primitive fragments of ribonucleotides could have evolved along with peptides or hybrid peptide-nucleic acid structures, the establishment of a small enantiomeric imbalance appears to be a prerequisite. The latter can be rationalized in terms of *deterministic* or *chance* models. Well known deterministic mechanisms involve the inherent asymmetry caused by parity violation or the

Chirality in Supramolecular Assemblies: Causes and Consequences, First Edition. Edited by F. Richard Keene.

action of some physical force (in particular, asymmetric photolysis by circularly polarized light) on racemates. Chance mechanisms, by contrast, assume the occurrence of asymmetric transformations that may trigger the formation of either enantiomer with equal probability. Representative examples include the spontaneous crystallization of conglomerates as well as site-selective adsorption or catalysis on chiral minerals. In recent years there have been significant advances in enantioenrichment and deracemization processes, which have been checked on small molecules, including some of pharmaceutical interest (i.e. chiral drugs). Such concepts will not be dealt with further in this chapter. The reader is instead referred to recent reviews, chapters and monographs [1–10] illustrating the facets of abiotic mechanisms mentioned above.

Unlike abiogenesis, biotic models claim the existence of the two enantiomeric kinds of biomolecules, followed by the extinction of one of them. In other words, life emerged after the enantiodiscrimination step. This "two equal runners–one tripped" approach, pioneered by Nobelist George Wald, has been revisited recently and provides much food for thought [11]. We shall revisit these arguments later on in connection with the generation of chiral sequences at a macromolecular level. In general, biotic theories are no longer supported because clear-cut signs of the extinct chirality remain elusive. The existence of D-amino acids in bacterial walls [12, 13], for instance, or the fact that both enantiomers of certain natural products are present in plants or animals [14], have nothing to do with biotic remnants, but are associated with Darwinian mechanisms that boosted fitness and survival against predators.

In a volume devoted to supramolecular assemblies, it is pertinent to move the discussion of chirogenesis upwards and forwards, from the usual scenario of small molecules to oligomeric sequences and polymers in general. Either conjecture on these aspects is much more speculative than those based on discrete species as one invariably assumes a bottom-up approach where chiral monomers appeared first. Naturally occurring polymers (DNA, RNA, or polypeptides) exhibit a given handedness resulting from the stereogenicity of their constituents together with topological features typical of large and flexible chains. Those ingredients can also be recreated in artificial polymers and supramolecular aggregates linked by weak interactions, even starting from achiral monomers [15, 16].

Although the spontaneous crystallization and sorting out of conglomerates is thought to be a plausible and robust source of prebiotic homochirality, the issue of chiral crystallization will be treated separately in another chapter. Accordingly, and with the sole exception of terminology or contextual purposes, crystallization mechanisms will be omitted. Readers are also referred to a recent and comprehensive chapter on crystallization of chiral molecules for an in-depth analysis [17].

2.2 The Racemic State: Mirror Symmetry Breaking

Practitioners of asymmetric synthesis know well how difficult it is to achieve a high level of enantiopurity – enantiomeric excesses larger than 95% – as racemization is an ubiquitous foe. Nevertheless, there will always be an enantiomeric imbalance at the molecular level caused by stochastic fluctuations. A pure racemic state composed exactly of $N/2$ molecules of one enantiomer and $N/2$ of the other is rather an oddity, because this assumption goes against statistics. The macroscopic analogy is well exemplified by the experiment of tossing a coin with two equiprobable outcomes, which is described by a binomial

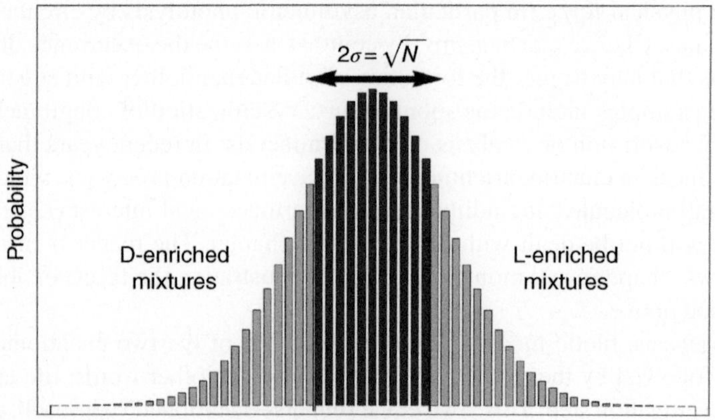

Figure 2.1 *The statistical distribution of a racemic mixture can be represented by a binomial distribution with standard deviation equal to* $\sqrt{N}/2$

(Bernoullian) distribution (Figure 2.1). Each trial occurs with equal probability (1/2). The binomial distribution is characterized by a mean value, $N/2$, together with its standard deviation, $\sigma = \sqrt{N}/2$. As shown in Figure 2.1 about 68% of racemic mixtures (dark gray) lie within one standard deviation from the mean value, whereas the deviation from $N/2$ is higher for the remaining 32% (light gray).

Ignoring an in-depth mathematical treatment, the discrete probability distribution of having exactly $N/2$ molecules of one enantiomer and $N/2$ of the other, is given approximately by:

$$P_{1/2}(N/2, N) \sim (2/\pi N)^{1/2}.$$

Thus, such a probability is inversely proportional to the number of molecules (N). If one considers one mole of a racemate, $N = N_A$ (Avogadro number $= 6.022 \times 10^{23}$ molecules), the probability of finding the racemate state, $P_{1/2}(N/2, N) \sim 10^{-12}$, is so small that it should be neglected in realistic terms. The binomial distribution tells us that there are unavoidable deviations from the racemic state, based exclusively on statistical grounds, which obviate the need to invoke an external chiral force like in deterministic mechanisms. Random fluctuations, coupled with an amplification mechanism, might then account for the enantiopurity of biomolecules. This possibility was introduced by Pearson in the late nineteenth century [18, 19], and revived later on by Mills through a quantitative analysis to understand better the magnitude of such stochastic fluctuations [20].

A thought-provoking experiment illustrating the stochastic bias again, yet being an intuitive analogy of enantiomer separation, was devised by Cairns-Smith some decades ago [21]. Consider a box containing 100 balls, 50 of one color, 50 of the other one, on which we perform sequential trials, each involving throwing away at random half of the balls and duplicating those that remain (Figure 2.2). After some time, all of the balls in the box will have the same color. This experiment, however, is not realistic enough because balls are not allowed to change their color – the model ignores racemization, which will be a driving force in nature [22].

As mentioned before, autocatalytic reactions could amplify minuscule imbalances, yielding samples with measurable enantiomeric excesses [23–25]. A synthetic model

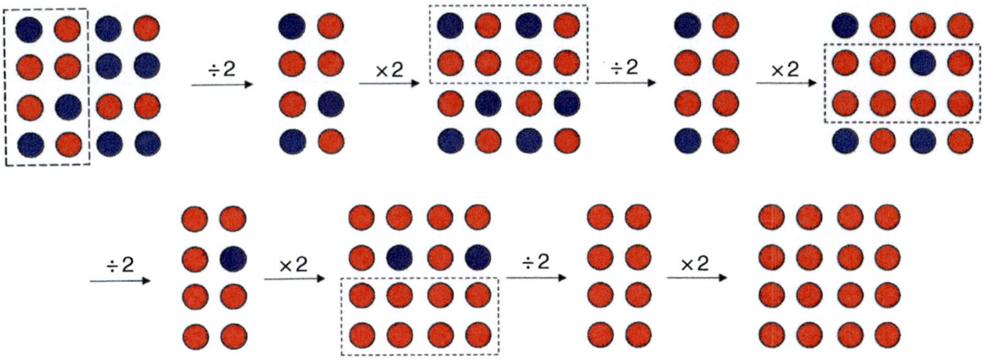

Figure 2.2 *Cairns-Smith's experiment showing the creation of random imbalances from a racemic mixture leading eventually to single handedness. Reproduced and adapted with permission from Ref. [22]*

capable of reproducing such a huge amplification is the Soai reaction involving the addition of organozinc reagents to a narrow range of pyrimidyl aldehydes [7]. The protocol is robust and the preferred chiral sense may be biased by a wide repertory of chiral molecules and crystals. Unfortunately, this transformation will hardly occur in a prebiotic scenario. A more credible amplification (actually an example of stereoisomeric disproportionation) of enantioenriched systems can take place for some chiral substances where the solubility of the racemic compound (e.g. DL-crystal) differs from that of the pure enantiomers as suggested by Morowitz in 1969 [26]. If pure enantiomers are more soluble than the racemate, an evaporative process will cause the precipitation of the latter, often leading to large enantiomeric excesses in the liquid phase. This simple physical mechanism may account for enantioenrichment in some amino acids and nucleosides [27]. Enantiomeric demixing can also be triggered by chaotic regimes that put the system far from equilibrium conditions and drive the autocatalytic step to completion, such as flow gradients [28, 29], vortex motion [30], or temperature gradients [31].

2.2.1 Is There a Chiral Ancestor?

The problems of chirogenesis and the origin of life are inextricably linked. However, when one moves to prebiotic chemical reactions like the classical formose reaction or amino acid condensations, messy and intractable mixtures are obtained. Although some hypotheses have been proposed to channel both the selection and accumulation of prebiotic reactants [32], we still miss an initial self-replicator coding chiral information and acting as template to form further copies. Valuable candidates to achieve this goal could be ribonucleotides or polypeptides but neither of them would have been easily obtained by prebiotic routes. Even if one bypasses the problematic syntheses of pentoses (ribose in particular) and the difficult coupling of free sugars and nucleobases, for instance through the intermediacy of heterocycles like 2-aminoxazole [33], the selection bottleneck demands a satisfactory explanation.

Suppose, after all, that an RNA strand (perhaps composed of a few nucleotides) could be assembled at a pre-evolutionary stage and then it was able to exert a supramolecular control – serving as template for a complementary strand linked by intermolecular hydrogen bonding among nucleobases. This a priori feasible and simple solution has, however, been called into question by some experts [34]. Firstly, a homochiral sequence is strongly required. *In vitro* experiments by Orgel and coworkers demonstrated that a chiral defect at 1-ribonucleotide level inhibits the oligomerization step [35]. As further proven by Goldanskii *et al.* [36, 37], the enantiomeric switching (L-ribose replacing D-ribose) causes chain termination as the nucleobase lies in a large angle (*ca.* 100º) relative to its position for H-bonding to occur. The situation is depicted in Figure 2.3 through a homochiral RNA fragment (···3′-GAGC-5′···) acting as template for its complementary strand (···5′-CUCG-3′···), but the incorporation of L-ribose (in red) in the guanosine residue (ent-G) disrupts replication.

The conclusive statement is that a chiral defect within an informational polymer has a detrimental effect, losing the ability of being a matrix for further copying. Obviously, homochirality is not necessarily a condition for matrix replication. One could envisage a heterochiral sequence of alternating D- and L-enantiomers encoding a message that should

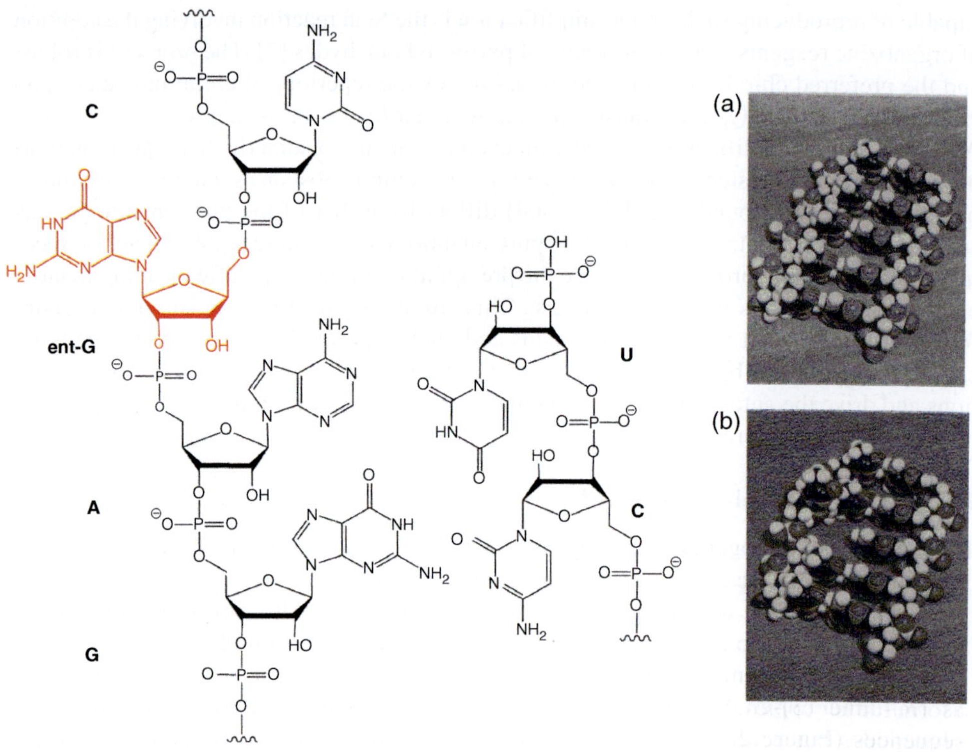

Figure 2.3 *In a polymeric chain of RNA acting as template for self-replication, the presence of L-ribose in a guanosine residue (ent-G) fully impedes the replication ability. The image shows a [poly(A)·poly(T)] DNA fragment. The double-strand matrix-replica structure with homochiral chains complement each other (a), while this does not hold for the double-strand structure with a chiral defect (b). Image reproduced with permission from Ref. [37]. Copyright 1996 National Academy of Sciences, United States*

be readable. In that sequence, however, a chiral defect would also have the same effect – the loss of chemical information to generate a complementary replica [37]. Current directions aimed at designing artificial replicators using monomer-based alphabets exploit this unequivocal strategy: polymers containing readable information should possess perfectly controlled co-monomer sequences [38].

2.3 Asymmetric Oligomerization

A significant portion of chemist's nightmare on homochirogenesis is caused by the accepted orthodox assumption that *homochirality precedes life* (and hence evolution). Under this premise, amplification is compulsory to channel the exclusive or prevalent formation of a single enantiomer; otherwise both replication and specific metabolic reactions would have been extremely difficult. Without denying the role of asymmetric autocatalysis, the puzzle of homochirality vanishes (to a large extent) if one assumes the opposite and rather iconoclastic hypothesis – that *life precedes homochirality* or, perhaps, that homochirality might have evolved with life [39]. This possibility hinges on the availability of an informational polymer existing in nonracemic form that directs its self-replication. The emergence of homochirality *after* or *with* the origin of biomolecular evolution has been dismissed in the literature [40, 41]. Since the early 2000s, however, polymerization models accounting for the generation of homochirality have gained increased interest and support. It is pertinent to quote first the Sandars model, which represents a manifestation of the bifurcation theory. Along with a detailed polymerization model, Sandars incorporated the key ingredients of enantiomeric cross-inhibition and chiral feedback in which only the largest polymers can enhance mirror symmetry breaking [42]. All the subsequent models can be regarded as variations on Sandars' proposal, taking into account issues like reversibility, open and closed systems, energy flow, the effects of chiral bias, as well as formation (and dissociation) of heteropolymers [43–51]. All models assume in general that an achiral precursor can directly afford a couple of chiral monomers (L/D or *R*/*S* according to the stereochemical nomenclature) at a slow rate, and then be consumed in processes in which homopolymers of all lengths catalyze the production of such monomers. Some premises in these models can actually be unrealistic in terms of lab-scale experiments. Thus, in open-flow systems exchanging both matter and energy with the exterior, the achiral precursor is constantly supplied, which leads to homochiral chains of infinite length. Closed systems, for which no matter flow occurs with the surrounding environment, and inclusion of reversible steps, enable the analysis of asymmetric reactions under thermodynamic conditions [49]. The situation can be extrapolated to copolymerization of two chemically distinct monomers, such as (*R*)- and (*S*)-enantiomers producing a variety of homochiral and heterochiral sequences (Figure 2.4) [51]. It is noteworthy that, in closed systems, the competing heteropolymerization reactions must be treated separately to avoid multiple counting. If this point is left unchecked, the model would cause a violation of the constant mass constraint. This also ensures that the maximum length for all oligomers produced or consumed never exceeds the upper limit. This model successfully reproduces the generation of homochiral oligopeptides from an initial nonracemic mixture of amino acids [51] (see section 2.3.2).

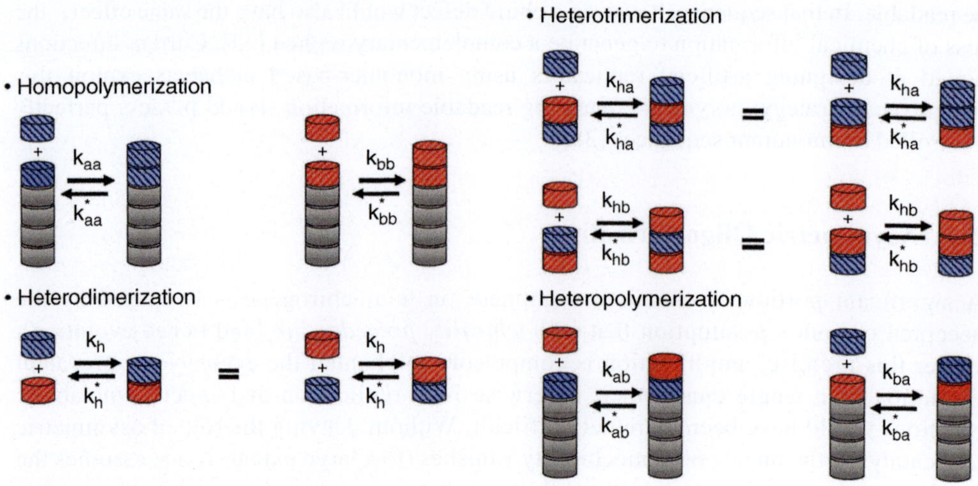

Figure 2.4 Schematic representation of copolymerization of (R)- and (S)-configured monomers (red and blue, respectively) into the growing homochiral or heterochiral chains. All steps are considered to be reversible and the system is closed. Reproduced with permission from Ref. [51]. Copyright 2012 Royal Society of Chemistry

2.3.1 Homochirality and Critical Chain Length

A general conclusion of kinetic models is that homochiral polymers can be obtained, even in the presence of significant enantiomeric cross-inhibition. In any case, these analyses are harder to grasp than structural ones. In fact, the emergence of chirality in macromolecular systems is a consequence of the statistics of the polymerization process. In this context, it is fair to recognize the seminal work of Green and Garetz, who, in their often overlooked 1984 paper, showed that an atactic vinyl polymer is actually a mixture of chiral chains, as long as the degree of polymerization is high enough [52].

Like in small molecules, in a polymer chain of length n formed by R and S monomers, the total number of possible configurational sequences is $N_{total}=2^n$, most coming in pairs of chemically equivalent chains if one ignores the effect of chain-end groups. The calculation of truly chiral diastereomers is not trivial, because depending on even/odd values for n, all the permutations with repetition of half of a chain may produce both *meso* and *palindromic* diastereomeric sequences [2]. *Meso* forms are obviously achiral, whereas a *palindromic* sequence will be chiral. However, the latter does not necessarily come in an equivalent pair (only for n odd); the sequence has not equivalent counterpart for n even. For the purpose of this article, it suffices to say that the total number of diastereomers (both chiral + *meso*, regardless of even/odd values for n) is roughly 2^{n-1}. For large chains, the number of *meso* chains will usually be small with respect to the chiral chains. Accordingly, $N_{chiral} \sim 2^{n-1}$. This figure contrasts with the actual number of available chains of length n, which is much smaller; $N_{avail}=N_A/n$, where N_A is the Avogadro number (if one considers 1 mol of monomers). Clearly, as the polymerization degree (n) increases, the number of available chains decreases monotonically (a polymer is actually a distribution of chain lengths) while the number of possible chiral chains increases exponentially. The intersection of such plots, estimated by Green and Garetz to occur approximately at $n=70$ [52], allows us to figure out the transition from racemic to nonracemic states. For n values less than that threshold, the number of all different configurational sequences will involve the number of all possible polymer sequences, including those intrinsically chiral. These samples, however, will be racemic (i.e. no optical rotation will be conventionally observed) as the probability of finding enantiomeric pairs will be roughly 1:1, thus resembling a binomial-like distribution. In stark contrast, for n values above or larger than the intersection point, the number of possible chiral chains (2^{n-1}) will be much greater than the number of available chains given by N_A/n. These samples will not be racemic, consisting of a random distribution of enantiomerically pure diastereomers. If one considers, for instance, a chain of 130 monomers, the number of polymer sequences is simply astronomical (~10^{40}). The chance of finding a single racemic chain – that is, a polymeric sequence compensated with its enantiomer – is essentially zero. In other words, any existing sequence at this level of complexity is virtually unique. It is worth noting that this breakthrough is theoretically supported by Sandars' model [42], which predicts that bifurcation to a chiral state occurs more readily for higher polymers and with less requirements on the feedback fidelity. This was attributed to "the existence of more stages at which the remaining minority enantiomer can be removed by cross-inhibition," which is actually the case.

Avetisov and Goldanskii have suggested two levels of complexity depending on the number of possible sequences of L- and D-links in the polymeric macromolecule, such as nucleotides [37]. In a *chemical level of complexity*, chains do not exceed a critical length, which also depends on the initial amount of monomers, the system will be racemic and contain the whole spectrum of possible sequences of links. Chains of 40–50 nucleotides fully satisfy the preceding statement. As n (number of nucleotides in a chain) rises, let us say 150, the range of sequences is so high that cannot be accommodated in the chemical space available. In fact, most statistical sequences do not exist at all and the existing ones will be chiral and unique as discussed above. This scenario, called *biological or biochemical complexity*, is characteristic of nucleic acids and enzymes whose functions emerge from structural restrictions, i.e. those carrying informational sequences only.

Within the chirality-at-critical-length framework, it may not be difficult to foresee the evolutionary transmission of stereochemical information, as revived recently in a concise and lucid account by Brewer and Davis [53]. This so-called *chiral encoding* merely requires an achiral monomer, or a pair of enantiomeric molecules (capable of accommodating structural variations in alkyl chain for example), capable of polymerizing leading to a stereoisomeric product. The latter acting as template will produce a duplex where the stereochemical information has been transferred to the growing chain [54, 55]. Nevertheless this scheme may be impractical unless chain separation, as the key step, works efficiently under experimental conditions, thus enabling further copying cycles.

Application of the above ideas to homochirogenesis of nucleic acid analogs has been mainly investigated by Eschenmoser and associates [56, 57]. Figure 2.5 constitutes an intuitive representation of how homochiral D- and L-libraries of higher oligomers are specifically assembled from a mixture of all possible diastereomers of the initial mixture of smaller oligomers in an enzyme-free process. The group explored the ligative self-assembly of unnatural pyranosyl tetramer 2′,3′-cyclophosphates (e.g. 4′-ATCG) under dilute conditions. Oligomerization of the all-D-tetramer occurs readily, whereas all heterochiral diastereomers are seldom detected. The only exception is the tetramer bearing the

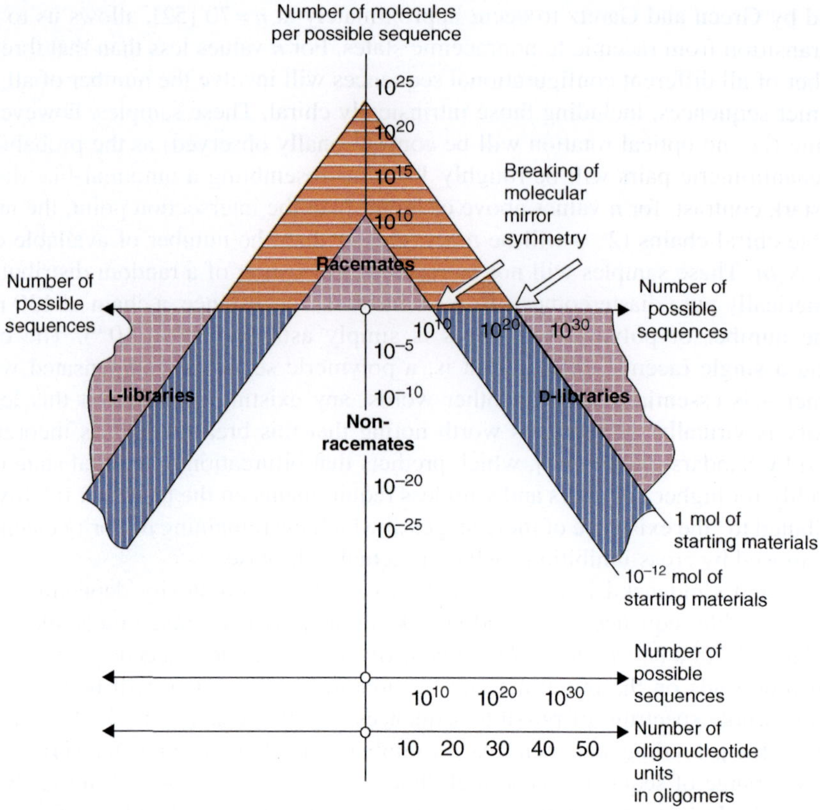

Figure 2.5 *Homochiral D- and L-libraries of oligonucleotides, even if derived from a racemic mixture of monomers, would not themselves constitute a racemate, since the constitutional complexity imposed by the sequence length exceeds a critical limit that makes deracemization unavoidable. Reproduced with permission from Ref. [56]. Copyright 1997 Elsevier Ltd*

L-ribopyranosyl unit (instead of the corresponding D-ribopyranosyl) at the 4′-end of the tetramer, whose oligomerization rate is about 10% of that shown by the homochiral tetramer. If oligomerization is allowed to occur in the presence of the heterochiral diastereomers, the process remains essentially unaffected, thus proving a high level of chiroselection. The only side reaction that hampers the ligation process is the hydrolytic deactivation of 2′,3′-cyclophosphate end groups [56].

An intriguing question in this context is how replication of long oligonucleotide sequences, which is required for the evolution of biological complexity, could ultimately be achieved. It is well known that short oligonucleotides replicate faster than longer sequences, which results in the survival of the shortest with a concomitant loss of genetic information. In a recent study, Braun and coworkers suggest a geochemical microenvironment where physical mechanisms drive the accumulation and replication of long strands at the expenses of the shorter ones [58]. Taking inspiration from hydrothermal vents as plausible prebiotic scenarios, where huge thermal and chemical gradients exist, this group shows that a heat flow across an open pore in a submerged rock enables a length-selective accumulation. Strands of 75 nucleotides

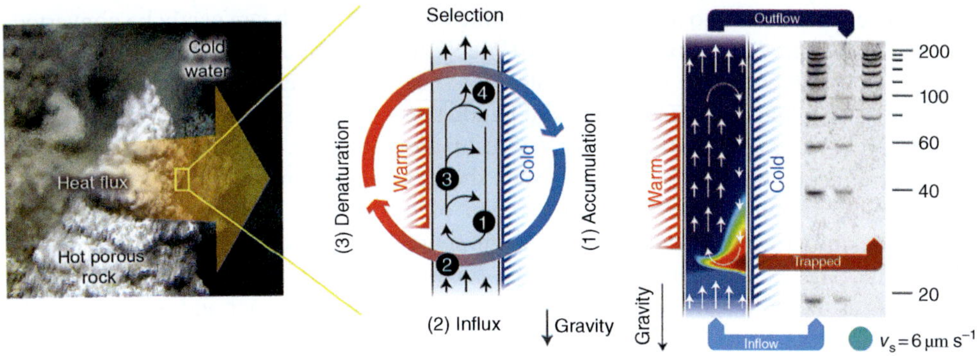

Figure 2.6 *A continuous flow of thermal energy through a porous rock gives rise to selection for long nucleotide sequences. The experimental setup (right) involves passing a flow of DNA solution upwards into a capillary where local convection shuttles the molecules between warm and cold zones, thus facilitating cyclic denaturation and replication of nucleotides. The longer sequences are trapped and pile up in the cold side, while convection along the warm side swept the shorter sequences out of the capillary. Reproduced with permission from Ref. [58]. Copyright 2015 Macmillan Ltd*

and larger are trapped and survive owing to a combination of molecular thermophoresis and laminar convection, whereas shorter sequences are flushed out of the pore (Figure 2.6). Although this microenvironment clearly selects for long sequences, whose critical length is determined by the inflow speed, this process does not guarantee the ideal nucleotide sequence. A Darwinian selection would still be required to obtain a functional genome [59]. Overall, this proof of concept is quite valuable in assessing the pivotal role of physical mechanisms to favor long oligonucleotide strands, which would presumably be inherently chiral.

2.3.2 Polymerization Models: Homochiral Peptides

The biogenesis of peptides is crucial to understand the multifaceted puzzle of the origin of life [60]. Short sequences, mostly dipeptides, have been found in meteoritic samples and numerous models simulate the dimerization step in particular [48–51, 61]. In general, peptides are poor replicators. Yet, Ghadiri and his group were able to demonstrate that certain peptides are prone to replication as exemplified by an amphiphilic peptide linked noncovalently to two smaller peptide fragments having appropriate nucleophilic and electrophilic ends, respectively. The major peptide catalyzes the amide bond formation between such fragments, hence promoting its own replication, which splits subsequently in two replicas [62]. These replicating peptides of a given chirality (e.g. all-L) can also template their formation enantioselectively from a racemic mixture of peptide fragments. The amplification process can even discriminate structures with a single stereomutation, such as the presence of one D-amino acid in a homochiral sequence [63]. It is true, however, that this sort of replicators do not show complete chiral encoding in the sense used above [53] because only homochiral libraries will undergo amplification.

Compelling examples of stereoregular polymerization are provided by the generation of isotactic oligopeptides in racemic crystals of α-amino acids in the presence of chiral initiators, as explored extensively by Lahav and coworkers [8, 64–66]. Starting from hydrophobic racemic amino acids (phenylalanine or valine) which can easily be converted into their activated *N*-carboxyanhydrides (PheNCA or ValNCA), thought to occur in prebiotic niches as well, the oligomerization of such derivatives in water results in the formation of racemic β-sheets of opposite handedness packed in either antiparallel (for oligo-PheNCA) or parallel (for oligo-ValNCA) arrangements, in agreement with the packing arrangements of the monomers. Such sheets serve as templates for further homochiral copying (Figure 2.7). Desymmetrization of the racemic oligopeptides could be achieved by initiating the reaction with enantiopure methyl esters or thioesters of the amino acids. The enantiomeric excess found for isotactic tetrapeptides favors the same handedness as the initiator. This trend is reversed beyond the tetramer and increases with

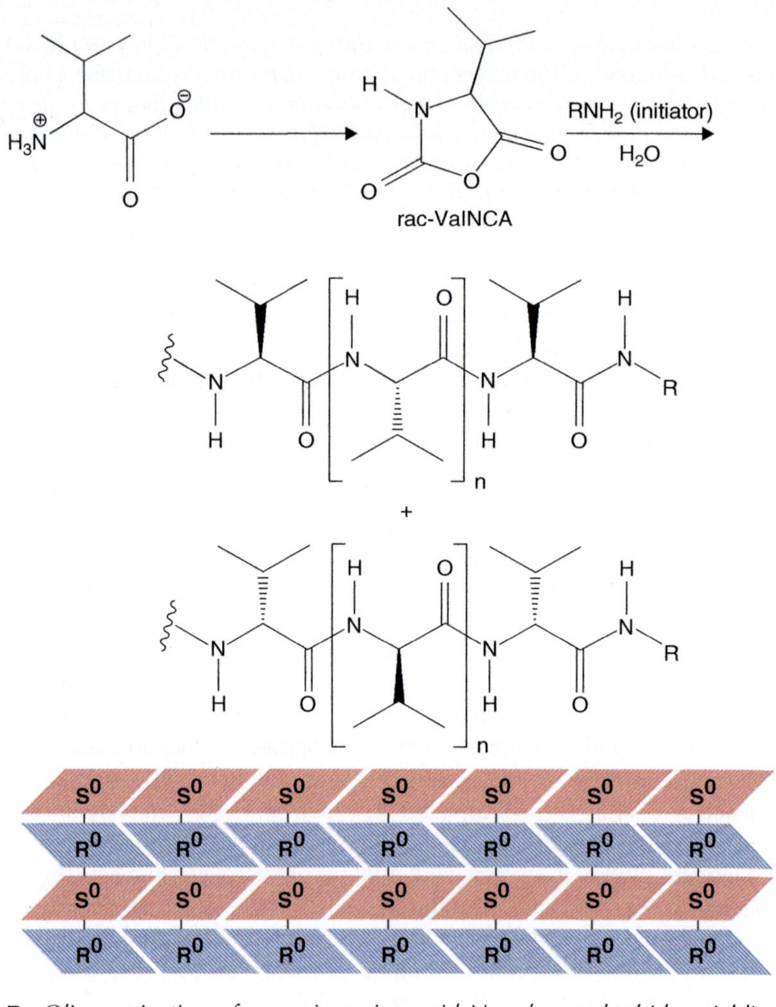

Figure 2.7 Oligomerization of racemic amino acid N-carboxyanhydrides yielding racemic β-sheets that act as templates for the synthesis of homochiral chains. Each chain is composed of amino acid residues of a single chirality. Reproduced and modified from Ref. [67]. Copyright 2012 American Chemical Society

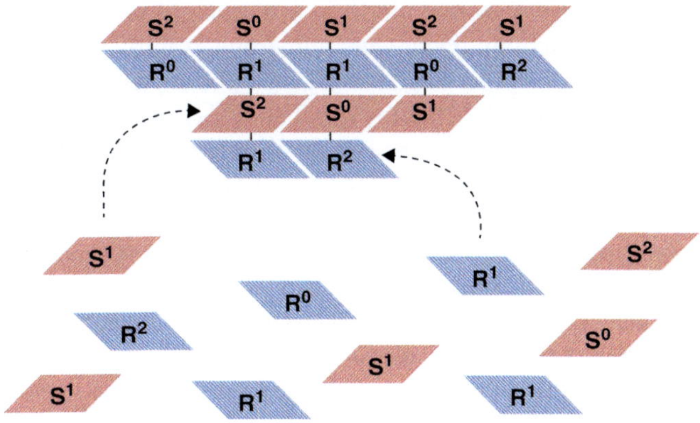

Figure 2.8 *Generation of nonenantiomeric pairs of homochiral copolymers by enantioselective occlusion of two minor guest species: (R_1S_1) and (R_2S_2) within the sites of a host species (R_0S_0), all assumed to be in ideally racemic proportions. Reproduced with permission from Ref. [67]. Copyright 2012 American Chemical Society*

chain length, a fact consistent with the formation of rippled antiparallel β-sheets [64]. Again, this templation leads to only one homochiral sequence (all-R or all-S), while a full chiral encoding should enable the templation of different *R*- and *S*-sequences within a chain; in principle not a distant goal by modifying the structures of monomers.

The desymmetrization of isotactic peptides can alternatively be carried out by a copolymerization strategy that involves the initial formation of quasi-racemic crystals of host and guest amino acids [65, 66]. Thus, when (*RS*)-PheNCA is crystallized with small amounts of an enantiomerically pure amino acid derivative, such as (*S*)-3-(2-thienyl)alanineNCA (ThieNCA), the latter occupies random sites of only host (*S*)-PheNCA in the racemic crystals. Accordingly, the subsequent polymerization generates copeptides of (*S*)-Phe and (*S*)-Thie residues as well as homochiral oligopeptides composed of (*R*)-Phe units only. This process has also been modeled mathematically and a schematic representation is shown in Figure 2.8 [67].

Unlike the situation depicted in Figure 2.7, where the β-sheet is globally racemic, the presence of racemic host (R_0S_0) and racemic guest (R_1S_1) and (R_2S_2) amino acids leads to a sequence of pairs of copeptide (*R*)- and (*S*)-chains within the growing template differing from each other. Mirror symmetry breaking occurs stochastically in the β-sheet template because a statistical departure from the racemic composition of the peptide chains is created. Such sequences also vary with chain length and the relative amounts of host and guest monomers [65]. The mechanism relies clearly on the presence of different amino acids. The species of a given handedness attach to sites of matching chirality within the growing β-sheet affording strands of uniform chirality, albeit in alternating configurations through the rows. Finally, the remaining free chiral monomers do not necessarily give rise to an optically active solution.

2.3.3 Lessons from Artificial Systems

Mirror symmetry breaking in supramolecular chemistry is not a rare phenomenon. Self-assembly of achiral monomers often leads to macro- and supramolecular architectures of single chirality, which may also be biased under the action of external physical fields (e.g. circularly polarized light or vortex motion). A recent and vast treatment of supramolecular chirality and self-assembly documents numerous examples [68]. Most, if not all, cases have nothing to do with prebiotic chemistry and, consequently, a potential link with systems related to the origin of homochirality is debatable. However, like the Soai reaction, the intellectual value of such contributions is to provide mechanistic insights that explain why biological systems are expected to be homochiral rather than heterochiral or racemic.

It is within this framework that the elegant work of Percec's group on dendritic peptides should be highlighted [69–72]. We shall concentrate on the self-assembly of a dendron composed of multiple alkyl tails, phenyl groups, and ether linkages, attached to a dipeptide of tyrosine and alanine; in short: (4-3,4-3,5)12G2-CH_2-Boc-X-Tyr-Y-Ala-OMe, where X and Y denote the configurations of the parent amino acids. Both L-L and D-D dendritic dipeptides were shown to self-assemble in an organic solvent (cyclohexane) and in the bulk state into helical pores – helical columnar arrangements that mimic the transmembrane water channel Aquaporine. In order to evaluate the role played by stereochemistry in this process, the group was able to prepare all possible configurational permutations of the dipeptides. Thus, for instance, the reaction of *rac*-Ala with L-Tyr produces the L-DL dipeptide, which contains a 50 : 50 mixture of L-L and L-D dipeptides (Figure 2.9).

As mentioned, self-assembly of homochiral and heterochiral dipeptides can be observed in solution and the solid state. While the helical sense is dictated by the stereochemistry of tyrosine residues, the configuration of alanine largely determines, in an allosteric fashion, the structural features (i.e. length and diameter) of supramolecular pores. The cooperative process in solution involves transition from nucleation to elongation, which takes place at 24 °C for homochiral dipeptides with a molar enthalpy of monomer addition (h_e) of −35 kcal/mol. The corresponding enthalpy for heterochiral Tyr-Ala sequences is h_e = −30 kcal/mol at 28 °C (Figure 2.10). However, the most remarkable differences between homochiral and heterochiral sequences are observed during solid-state self-assembly. The former undergoes fast crystallization after a very short period of annealing whereas the heterochiral pore is scarcely crystalline even after long annealing periods. This behavior is likewise observed for various racemic permutations of the dendritic dipeptides; the resulting pores do not exhibit crystalline order, nor does annealing induce spontaneous deracemization. These experiments demonstrate that the mimicry of a membrane channel (Aquaporine) via homochiral dendritic dipeptides yields highly crystalline structures, whereas heterochiral and racemic pores are essentially micellar. The corollary is that supramolecular structures assembled from achiral, racemic, and heterochiral building blocks are expected to be less ordered and less efficient in transferring functional information than homochiral assemblies [70, 72].

Optical resolution of supramolecular architectures harnesses recent advancements in living supramolecular polymerization [73], which enables a greater control of chain growth relative to other polymerization strategies. A salient illustration is provided by Japanese authors on the polymerization of corannulene molecules functionalized with amide groups containing side chains [74]. These corannulene derivatives are chiral due to their C_5-symmetric nonplanar bowl shape, but undergo racemization even at 25 °C. The monomer adopts a cagelike conformation in low-polarity solvents, although it spontaneously polymerizes into a one-dimensional assembly upon heating. The *N*-methyl derivative of such a monomer lacks the capacity for intramolecular hydrogen bonding and hence self-assembly

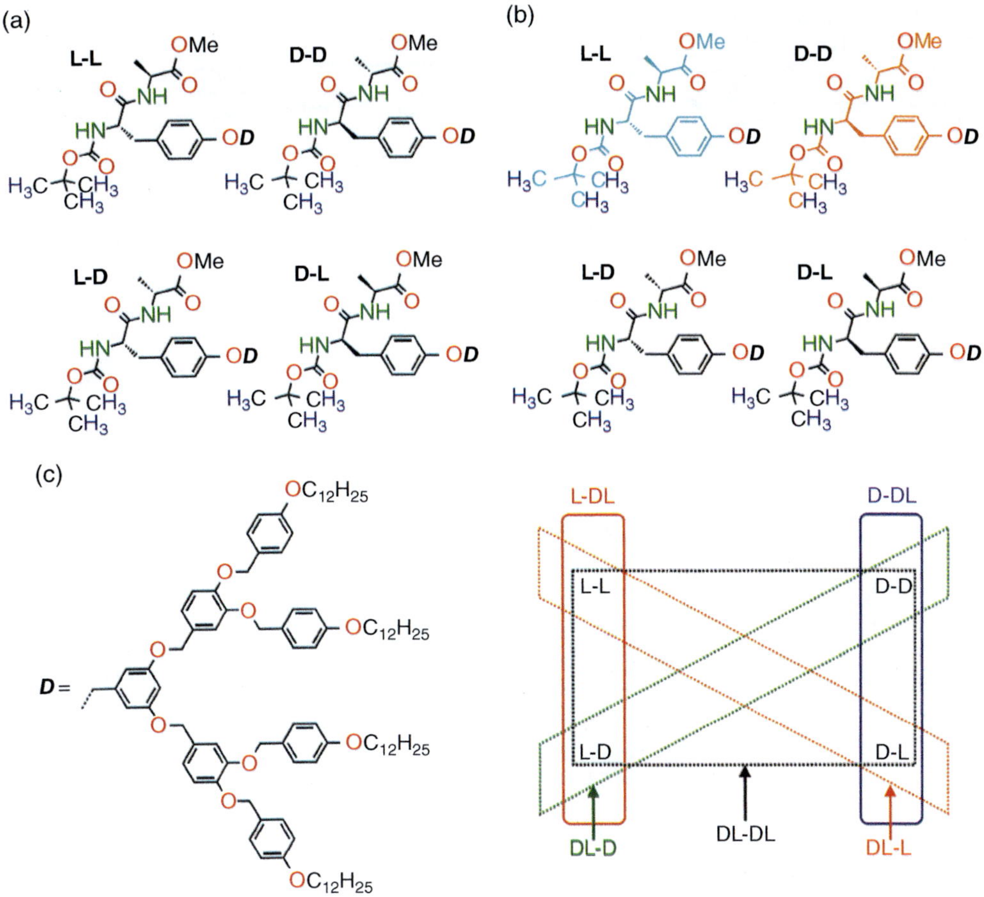

Figure 2.9 *Structures of homochiral and heterochiral dendritic dipeptides (4-3,4-3,5)12G2-CH$_2$-Boc-X-Tyr-Y-Ala-OMe composed of tyrosine and alanine. The color code shown in dipeptides serves to illustrate the cross-section of the supramolecular self-assemblies (see Figure 2.10). The rectangles depicted below indicate the enantiomeric composition of racemized dendritic dipeptides. Colored boxes are composed of equimolar proportions of the enantiopure dipeptides that they enclose. Reproduced with permission from Ref. [71]. Copyright 2011 American Chemical Society*

but it can appropriately initiate the polymerization at room temperature. When a chiral version of this initiator is employed, polymerization takes place only when the configuration of the chiral center of the monomer matches that of the initiator. Accordingly, a chiral initiator of single chirality triggers enantioselective polymerizations and resolution of a racemic mixture of monomers M_S and M_R could be achieved by using either I_R or I_S molecules as initiators (Figure 2.11). This procedure mimics, to some extent, the symmetry breaking in racemic β-sheets of oligopeptides induced by chiral initiators [64].

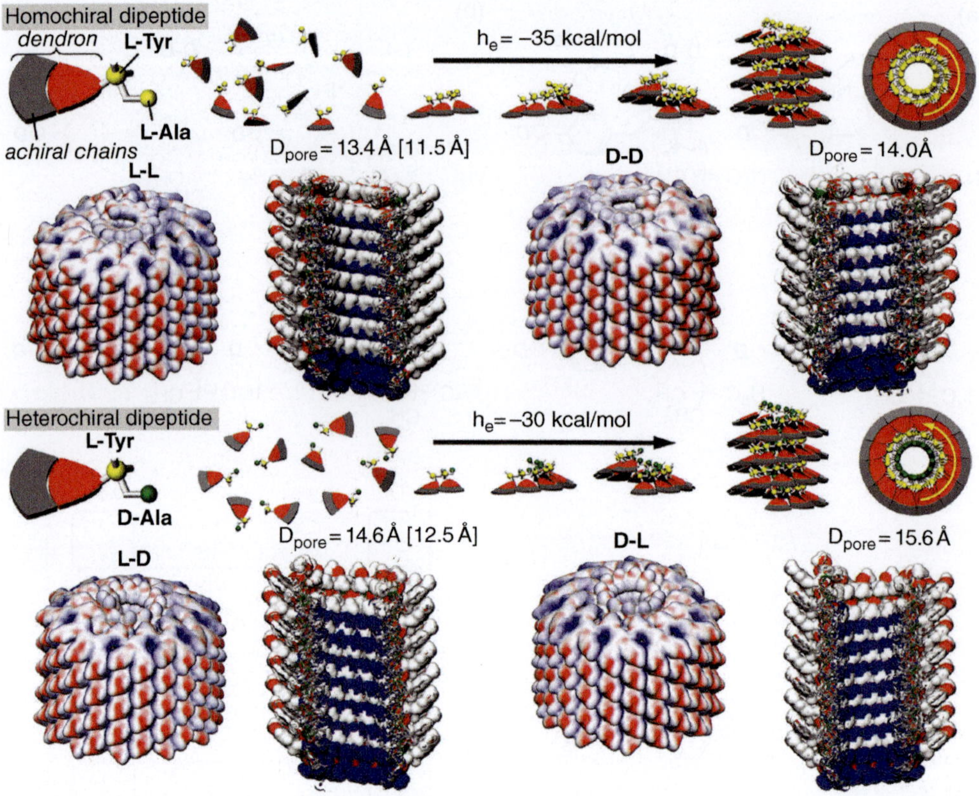

Figure 2.10 *Self-assembly of homochiral (top) and heterochiral (bottom) Tyr-Ala-containing dendritic dipeptides into supramolecular helical pores. The images show the corresponding side views and cross sections of the porous columns as determined by X-ray analysis in the solid state. Reproduced with permission from Ref. [71]. Copyright 2011 American Chemical Society*

2.4 Biochirality in Active Sites

Any epilogue to chemical dissertations on chirogenesis cannot conclude without giving a glimpse on arguments that lie beyond prebiotic chemistry and are therefore less speculative. It is precisely here, in the interplay between synthetic chemistry and molecular biology, where the enforcement and perpetuation of chirality can be understood.

Had evolution preceded homochirality by generating informational self-replicators, this primeval machinery would still need the capacity of discriminating monomers in a racemic world. Two recurring questions (among others) in context are whether there is a universal relationship between L-amino acids and D-sugars, and how the enantiopurity is retained with fidelity through translation and protometabolism [75]. Both experimental and theoretical simulations suggest that present aminoacylation and peptide-bond-forming reactions are unable to incorporate D-amino acids. Once again, this points to homochiral preferences not related to prebiotic development, but arising from the properties of

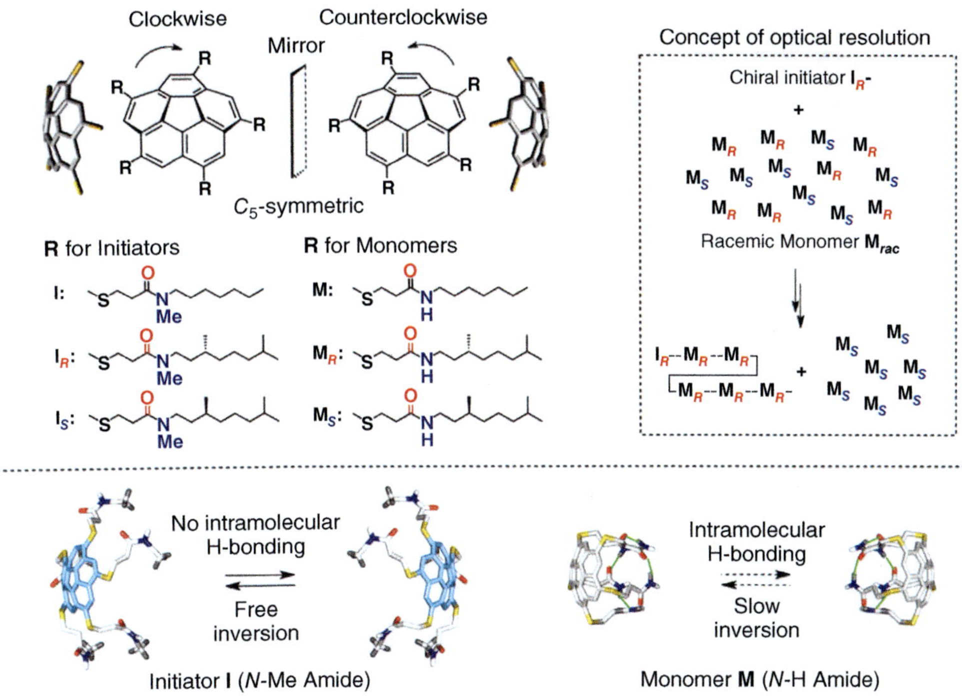

Figure 2.11 *Chain-growth polymerization of functionalized corannulene derivatives with initiators (I) and monomers (M). Bottom: schematic representation of optical resolution of a racemic mixture of monomers carried out by stereoselective polymerization with an initiator of single chirality (I_R). Reproduced with permission from Ref. [74]. Copyright 2015 American Association for the Advancement of Science*

primitive RNA fragments. The chirality of the sugar ring is decisive in peptide synthesis and configurational changes have unfavorable effects [75–77].

Peptide biosynthesis requires a complex multienzymatic cascade. Among the plethora of enzymes, D-aminoacyl-tRNA deacylase (DTD) is crucial in preventing incorporation of D-amino acids into proteins. That is, the enzyme has the ability to recognize and remove the D-configuration from starting amino acids before they are taken by transfer RNA (tRNA) molecules. This process is not completely known. With the advent of recent techniques to crystallize tRNAs and their complexes with different ligands [78], further goals can now be envisaged. Ahmad *et al.* have shown the details of the aforementioned specificity by crystallizing a DTD enzyme (from *Plasmodium falciparum*) complexed with a molecule (D-Tyr3AA) that mimics a D-amino acid attached to tRNA [79]. The 5′-OH group is free in the substrate analog instead of being attached to tRNA in the actual substrate. Likewise, the natural ester linkage between the amino acid and adenosine is replaced by an amide functionality that makes it resistant to hydrolysis (Figure 2.12).

The enantioselective mechanism that rejects L-amino acids from the active site relies critically on a dipeptide (Gly-*cis*Pro), which generates a binding pocket where the D-configuration of the enzyme-ligand complex fits accurately (Figure 2.13). Remarkably,

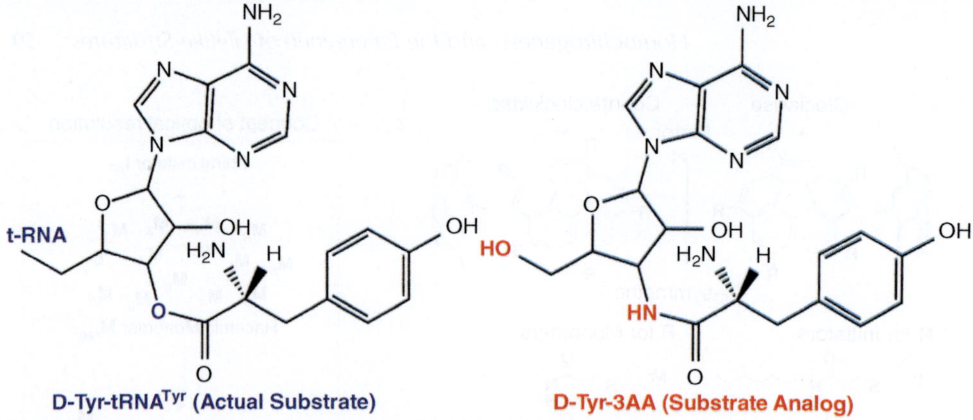

Figure 2.12 Comparative structures of natural and analogous substrates of D-aminoacyl-tRNA deacylase (DTD). See text and Ref. [79] for an explanation

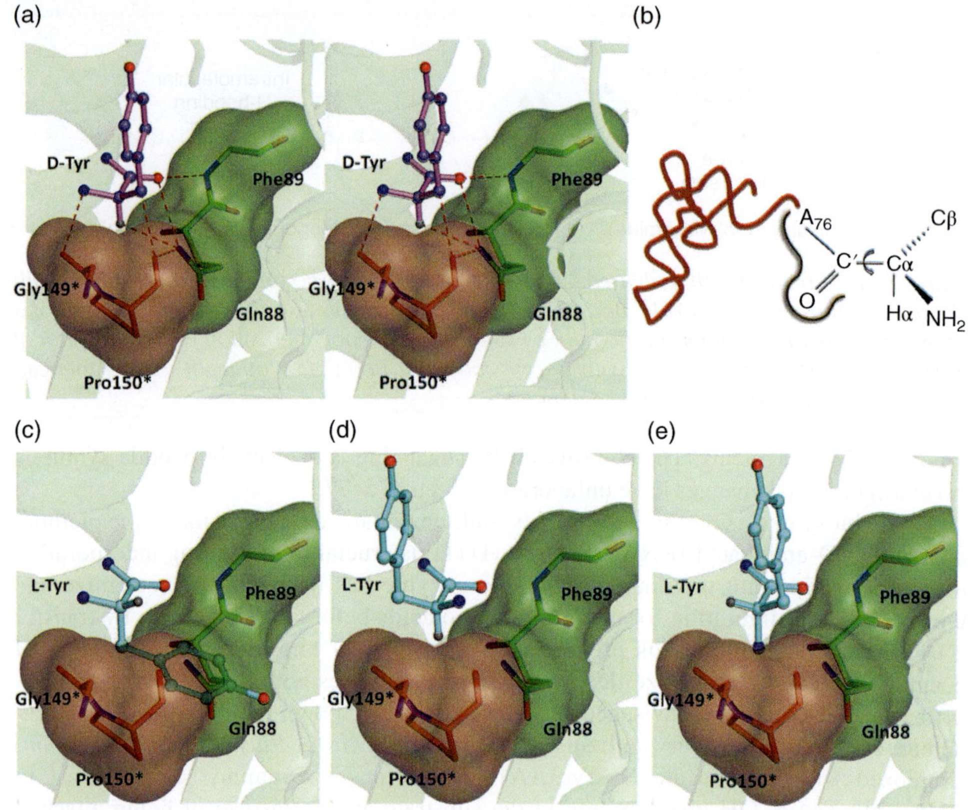

Figure 2.13 Mechanism of L-amino acid rejection; the cross-subunit Gly-cisPro is shown in brown. (a) Stereoscopic views of the conformation of D-amino acids observed in the pocket. (b) The adenosine moiety and the carbonyl oxygen are tightly fixed, and the only allowed flexibility involves torsion around the Cα-C′ bond. This rotation affords three possibilities of binding an L-amino acid. (c) Conformation I: the side chains swaps positions with Hα, severe short contacts of the side chain atoms with active site residues are observed. (d) Conformation II: the side chain swaps positions with the NH_2 group, short contact of the side chain with C75 of tRNA, Cβ is 2.56 Å from the amide nitrogen (N8) of the substrate. (e) Conformation III: the NH_2 group swaps positions with Hα, the nonpolar side chain of Pro150 is an unfavorable environment for the NH_2 group. Reproduced from Ref. [79] under the terms of the Creative Commons Attribution License. Copyright 2013 by Ahmad et al.

that binding pocket structure remains scarcely unaffected through numerous organisms, from bacteria to mammals. This suggests that DTD, at least, would have been an early ingredient of the homochiral puzzle. The functionality of primitive RNA structures could thus have evolved with the process of protein synthesis.

At this stage, one may wonder whether a prebiotic scenario favoring D-amino acids and L-sugars would have resulted in identical biosynthetic pathways mirroring the present available structures composed of L-amino acids and D-sugars. Dynamic simulations (over 500 ps) with the inverted form of histidyl-tRNA (HisRS) show, however, that the substrates (D-amino acids and ATP with L-sugar) fail to have favorable interactions with the active site residues [80]. That simulations suggest that protein biosynthesis involving mirror structures might not evolve in a mirror-symmetric manner. Obviously, this conjecture represents an open question in this epilogue and future studies simulating primeval biomachineries will be required.

2.5 Conclusions

Our current interpretations accounting for the abiotic origin of biomolecular homochiraltiy lie substantially in the amplification of stochastic imbalances and polymerizations above a critical chain length, the latter leading to inherently chiral structures out of a complex mixture of diastereomers. Moreover, these substrates may serve as intermediates or catalysts for mirror-symmetry breaking processes. It is certainly speculative to indicate that templation and self-replication of atactic polymeric chains have the last word. However, it is also quite plausible to imagine synthetic routes to long homochiral sequences of nucleotides and oligopeptides, among other credible macromolecules, endowed with enough functionality to initiate the characteristics of life. One can convince oneself that the ability for symmetry breaking is an intrinsic property of macro- and supramolecular systems and this aspect brings an element of interest in further pursuits on the origin of biochirality.

Acknowledgements

Financial support from the Spanish Ministry of Economy and Competitiveness (Grant CTQ2013-44787P on prebiotic chemistry and homochirality) and the Gobierno de Extremadura y FEDER (Grant GR15022) is gratefully acknowledged. The author has benefitted over the years from discussions with some talented colleagues, in particular Mark Green, Meir Lahav, and Cristóbal Viedma.

References

[1] U. Meierhenrich (2008) *Amino Acids and the Asymmetry of Life*, Springer, Berlin.
[2] A. Guijarro and M. Yus (2009) *The Origin of Chirality in the Molecules of Life*, Royal Society of Chemistry, Cambridge.
[3] R. M. Hazen and D. S. Sholl, Chiral selection on inorganic crystalline surfaces, *Nature Mater.*, **2**, 367–374 (2003).
[4] W. L. Noorduin, E. Vlieg, R. M. Kellogg, and B. Kaptein, From Ostwald ripening to single chirality, *Angew. Chem. Int. Ed.*, **48**, 9600–9606 (2009).

[5] C. Viedma and P. Cintas, On the chiral homogeneity of nature: from atoms to small molecules, *Isr. J. Chem.*, **51**, 997–1006 (2011).
[6] P. Cintas and C. Viedma, On the physical basis of asymmetry and homochirality, *Chirality*, **24**, 894–908 (2012).
[7] K. Soai, T. Kawasaki, and A. Matsumoto, Asymmetric autocatalysis of pyrimidyl alkanol and its application to the study on the origin of homochirality, *Acc. Chem. Res.*, **47**, 3643–3654 (2014).
[8] I. Weissbuch and M. Lahav, Crystalline architectures as templates of relevance to the origins of homochirality, *Chem. Rev.*, **111**, 3236–3267 (2011).
[9] J. E. Hein, D. Gherase, and D. G. Blackmond (2013) Chemical and Physical Models for the Emergence of Biological Homochirality, in P. Cintas (Ed.), *Biochirality. Origins, Evolution and Molecular Recognition*, Springer, Berlin, pp. 83–108.
[10] J. M. Ribó, C. Blanco, J. Crusats, *et al.*, Absolute asymmetric synthesis and enantioselective autocatalytic reaction networks: theoretical games, speculations on chemical evolution and perhaps a synthetic option, *Chem. Eur. J.*, **20**, 17250–17271 (2014).
[11] M. M. Green and V. Jain, Homochirality in life: two equal runners, one tripped, *Orig. Life Evol. Biosph.*, **40**, 111–118 (2010).
[12] S. Martínez-Rodríguez, A. I. Martínez-Gómez, F. Rodríguez-Vico, *et al.* Natural occurrence and industrial applications of D-amino acids: an overview, *Chem. Biodiv.*, **7**, 1531–1548 (2010).
[13] H. Lam, D.-C. Oh, F. Cava, *et al.*, D-Amino acids govern stationary phase cell wall remodeling in bacteria, *Science*, **325**, 1552–1555 (2009).
[14] J. M. Finefield, D. H. Sherman, M. Kreitman, and R. M. Williams, Enantiomeric natural products: occurrence and biogenesis, *Angew. Chem. Int. Ed.*, **51**, 4802–4836 (2012).
[15] T. Sierra (2009) Expression of Chirality in Polymers, in D. Amabilino (Ed.), *Chirality at the Nanoscale: Nanoparticles, Surfaces, Materials and More*, Wiley-VCH, Weinheim, pp. 115–189.
[16] F. R. Keene (2012) Chirality, in P. A. Gale and J. W. Steed (Eds.), *Supramolecular Chemistry: From Molecules to Nanomaterials*, John Wiley & Sons, Chichester, Vol. 1, pp. 217–235.
[17] C. Viedma, G. Coquerel, and P. Cintas (2015), Crystallization of Chiral Molecules, in T. Nishinaga (Ed.), *Handbook of Crystal Growth, Vol.* IA. *Fundamentals: Thermodynamics and Kinetics*, Elsevier, Amsterdam, pp. 951–1002.
[18] K. Pearson, Chance or vitalism, *Nature*, **58**, 495–496 (1898).
[19] K. Pearson, Asymmetry and vitalism, *Nature*, **59**, 30 (1898).
[20] W. H. Mills, Some aspects of stereochemistry, *Chem. Ind. (London)*, **51**, 750–759 (1932).
[21] A. G. Cairns-Smith, Chirality and the common ancestor effect, *Chem. Brit.*, **22**, 559–561 (1986).
[22] C. Rougeot, Deracemization of Active Compound Precursors by Physical Treatments. European PhD Thesis, Université Paul Sabatier Toulouse III-Université de Rouen, p. 62 (2012).
[23] G. Lente, Stochastic kinetic models of chiral autocatalysis: a general tool for the quantitative interpretation of total asymmetric synthesis, *J. Phys. Chem. A*, **109**, 11058–11063 (2005).
[24] C. Blanco, M. Stich, and D. Hochberg, Temporary mirror symmetry breaking and chiral excursions in open and closed systems, *Chem. Phys. Lett.*, **505**, 140–147 (2011).
[25] C. Blanco, J. Crusats, Z. El-Hachemi, *et al.*, Spontaneous emergence of chirality in the limited enantioselectivity model: autocatalytic cycle driven by an external reagent, *ChemPhysChem*, **14**, 2432–2440 (2013).
[26] H. J. Morowitz, A mechanism for the amplification of fluctuations in racemic mixtures, *J. Theor. Biol.*, **25**, 491–494 (1969).
[27] R. Breslow and Z. L. Cheng, On the origin of terrestrial homochirality for nucleosides and amino acids, *Proc. Natl. Acad. Sci. USA*, **106**, 9144–9146 (2009).
[28] G. Metcalfe and J. M. Ottino, Autocatalytic processes in mixing flows, *Phys. Rev. Lett.*, **72**, 2875–2878 (1994).
[29] Marcos, H. C. Fu, T. R. Powers, and R. Stocker, Separation of microscale chiral objects by shear flow, *Phys. Rev. Lett.*, **102**, 158103 (2009).
[30] T. M. Hermans, K. J. M. Bishop, P. S. Stewart, *et al.*, Vortex flows impart chirality-specific lift forces, *Nature Commun.*, **6**, 5640 (2015).
[31] C. Viedma and P. Cintas, Homochirality beyond grinding: deracemizing chiral crystals by temperature gradient under boiling, *Chem. Commun.*, **47**, 12786–12788 (2011).

[32] A. W. Schwartz, Evaluating the plausibility of prebiotic multistage syntheses, *Astrobiology*, **13**, 784–789 (2013).
[33] B. H. Patel, C. Percivalle, D. J. Ritson, *et al.*, Common origins of RNA, protein and lipid precursors in a cyanosulfidic protometabolism, *Nature Chem.*, **7**, 301–307 (2015).
[34] L. E. Orgel, Prebiotic chemistry and the origin of the RNA world, *Crit. Rev. Biochem. Mol. Biol.*, **39**, 99–123 (2004).
[35] G. F. Joyce, G. M. Visser, C. A. van Boeckel, *et al.*, Chiral selection in poly(C)-directed synthesis of oligo(G), *Nature*, **310**, 602–604 (1984).
[36] V. I. Goldanskii, V. A. Avetisov, and V. V. Kuzmin, Chiral purity of nucleotides as a necessary condition of complementarity, *FEBS Lett.*, **207**, 181–183 (1986).
[37] V. Avetisov and V. Goldanskii, Mirror symmetry breaking at the molecular level, *Proc. Natl. Acad. Sci. USA*, **93**, 11435–11442 (1996).
[38] J.-F. Lutz, Coding macromolecules: inputting information in polymers using monomer-based alphabets, *Macromolecules*, **48**, 4759–4767 (2015).
[39] M. Wu, S. I. Walker, and P. G. Higgs, Autocatalytic replication and homochirality in biopolymers: is homochirality a requirement of life or a result of it? *Astrobiology*, **12**, 818–829 (2012).
[40] W. A. Bonner, The origin and amplification of biomolecular chirality, *Orig. Life Evol. Biosph.*, **21**, 59–111 (1991).
[41] W. A. Bonner, Chirality and life, *Orig. Life Evol. Biosph.*, **25**, 175–190 (1995).
[42] P. G. H. Sandars, A toy model for the generation of homochirality during polymerization, *Orig. Life Evol. Biosph.*, **33**, 575–587 (2003).
[43] A. Brandenburg and T. Multamäki, How long can left and right handed life forms coexist? *Int. J. Astrobiol.*, **3**, 209–219 (2004).
[44] A. Brandenburg, A. C. Andersen, S. Höfner, and M. Nilsson, Homochiral growth through enantiomeric cross-inhibition, *Orig. Life Evol. Biosph.*, **35**, 225–241 (2005).
[45] J. A. Wattis and P. V. Coveney, Symmetry-breaking in chiral polymerisation, *Orig. Life Evol. Biosph.*, **35**, 243–273 (2005).
[46] M. Gleiser and S. I. Walker, An extended model for the evolution of prebiotic homochirality: a bottom-up approach to the origin of life, *Orig. Life Evol. Biosph.*, **38**, 293–315 (2008).
[47] Y. Saito and H. Hyuga, Chirality selection in open flow systems and in polymerization, *J. Phys. Soc. Jpn.*, **74**, 1629–1635 (2005).
[48] R. Plasson, H. Bersini, and A. Commeyras, Recycling Frank: spontaneous emergence of homochirality in non-catalytic systems, *Proc. Natl. Acad. Sci. USA*, **101**, 16733–16738 (2004).
[49] C. Blanco and D. Hochberg, Chiral polymerization: symmetry breaking and entropy production in closed systems, *Phys. Chem. Chem. Phys.*, **13**, 839–849 (2011).
[50] J. A. D. Wattis and P. V. Coveney, Sequence selection during copolymerization, *J. Phys. Chem. B*, **111**, 9546–9562 (2007).
[51] C. Blanco and D. Hochberg, Homochiral oligopeptides by chiral amplification: interpretation of experimental data with a copolymerization model, *Phys. Chem. Chem. Phys.*, **14**, 2301–2311 (2012).
[52] M. M. Green and B. A. Garetz, The configurational stereochemistry of atactic vinyl homopolymers, *Tetrahedron Lett.*, **25**, 2831–2834 (1984).
[53] A. Brewer and A. P. Davis, Chiral encoding may provide a simple solution to the origin of life, *Nature Chem.*, **6**, 569–574 (2014).
[54] A. L. Weber, The triose model: glyceraldehyde as a source of energy and monomers for prebiotic condensation-reactions, *Orig. Life Evol. Biosph.*, **17**, 107–119 (1987).
[55] L. E. Orgel, Unnatural selection in chemical systems, *Acc. Chem. Res.*, **28**, 109–118 (1995).
[56] M. Bolli, R. Micura, and A. Eschenmoser, Pyranosyl RNA: chiroselective self-assembly of base sequences by ligative oligomerization of tetranucleotide-2′,3′-cyclophosphates (with a commentary concerning the origin of biomolecular homochirality), *Chem. Biol.*, **4**, 309–320 (1997).
[57] A. Eschenmoser, Etiology of potentially primordial biomolecular structures: from vitamin B_{12} to the nucleic acids and an inquiry into the chemistry of life's origin: a retrospective, *Angew. Chem. Int. Ed.*, **50**, 12412–12472 (2011).
[58] M. Kreysing, L. Keil, S. Lanzmich, and D. Braun, Heat flux across an open pore enables the continuous replication and selection of oligonucleotides towards increasing length, *Nature Chem.*, **7**, 203–208 (2015).

[59] I. A. Chen, Replicating towards complexity, *Nature Chem.*, **7**, 191–192 (2015).
[60] G. Danger, R. Plasson, and R. Pascal, Pathways for the formation and evolution of peptides in prebiotic environments, *Chem. Soc. Rev.*, **41**, 5416–5429 (2012).
[61] M. Gleiser and S. I. Walker, Toward homochiral protocells in noncatalytic peptide synthesis, *Orig. Life Evol. Biosph.*, **39**, 479–493 (2009).
[62] H. L. Lee, J. R. Granja, J. A. Martinez, *et al.*, A self-replicating peptide, *Nature*, **382**, 525–528 (1996).
[63] A. Saghatelian, Y. Yokobayashi, K. Soltani, and M. R. Ghadiri, A chiroselective peptide replicator, *Nature*, **409**, 797–801 (2001).
[64] I. Weissbuch, R. A. Illos, G. Bolbach, and M. Lahav, Racemic β-sheets as templates of relevance to the origin of homochirality of peptides: lessons from crystal chemistry, *Acc. Chem. Res.*, **42**, 1128–1140 (2009).
[65] J. G. Nery, G. Bolbach, I. Weissbuch, and M. Lahav, Homochiral oligopeptides generated by induced "mirror symmetry breaking" lattice-controlled polymerizations in racemic crystals of phenylalanine *N*-carboxyanhydride, *Chem. Eur. J.*, **11**, 3039–3048 (2005).
[66] I. Weissbuch, L. Leiserowitz, and M. Lahav, Achiral organic, inorganic, and metal crystals as auxiliaries for asymmetric transformations, *Isr. J. Chem.*, **51**, 1017–1033 (2011).
[67] C. Blanco and D. Hochberg, Models for mirror symmetry breaking via β-sheet-controlled copolymerization: (i) mass balance and (ii) probabilistic treatment, *J. Phys. Chem. B*, **116**, 13953–13967 (2012).
[68] M. Liu, L. Zhang, and T. Wang, Supramolecular chirality in self-assembled systems, *Chem. Rev.*, **115**, 7304–7397 (2015).
[69] B. M. Rosen, C. J. Wilson, D. A. Wilson, *et al.*, Dendron-mediated self-assembly, disassembly, and self-organization of complex systems, *Chem. Rev.*, **109**, 6275–6540 (2009).
[70] V. Percec and P. Leowanawat, Why are biological systems homochiral?, *Isr. J. Chem.*, **51**, 1107–1117 (2011).
[71] B. M. Rosen, M. Peterca, K. Morimitsu, *et al.*, Programming the supramolecular helical polymerization of dendritic dipeptides via the stereochemical information of the dipeptide, *J. Am. Chem. Soc.*, **133**, 5135–5151 (2011).
[72] B. M. Rosen, C. Roche, and V. Percec (2013), Self-Assembly of Dendritic Dipeptides as a Model of Chiral Selection in Primitive Biological Systems, in P. Cintas (Ed.), *Biochirality. Origins, Evolution and Molecular Recognition* (Topics in Current Chemistry 333), Springer, Berlin, pp. 213–254.
[73] R. D. Mukhopadhyay and A. Ajayaghosh, Living supramolecular polymerization, *Science*, **349**, 241–242 (2015).
[74] J. Kang, D. Miyajima, T. Mori, *et al.*, A rational strategy for the realization of chain-growth supramolecular polymerization, *Science*, **347**, 646–651 (2015).
[75] N. Nandi (2012) *Chirality in Biological Nanospaces. Reactions in Active Sites*, CRC Press-Taylor & Francis Group, Boca Raton, FL, USA.
[76] K. Thirumoorthy and N. Nandi, Role of chirality of the sugar ring in the ribosomal peptide synthesis, *J. Phys. Chem. B*, **112**, 9187–9195 (2008).
[77] S. M. Hecht, Probing the synthetic capabilities of a center of biochemical catalysis, *Acc. Chem. Res.*, **25**, 545–552 (1992).
[78] R. Giegé, Crystallogenesis at the heart of the interplay between science and technology in the quest to comprehend tRNA biology, *Cryst. Growth Des.*, **13**, 405–414 (2013).
[79] S. Ahmad, S. B. Routh, V. Kamarthapu, *et al.*, Mechanism of chiral proofreading during translation of the genetic code, *eLife*, **2**, e01519 (2013).
[80] S. D. Banik and N. Nandi (2013), Chirality and Protein Biosynthesis, in P. Cintas (Ed.), *Biochirality. Origins, Evolution and Molecular Recognition*, Springer, Berlin, pp. 255–306.

3

Aspects of Crystallization and Chirality

Roger Bishop

3.1 Introduction

This chapter provides an overview of the various outcomes, especially those related to chirality, which may occur on crystallization of a chemical substance. It is intended to be introductory and informative rather than specialist in nature, mindful of a target audience of synthetic chemists interested in the solid state aspects of their products. Molecular structures of chiral compounds are usually drawn as one enantiomer only, but their racemic mixtures are generally used in the crystallizations described. For convenience, the illustrative examples have been chosen from small organic compounds but, of course, the phenomena discussed are applicable to all classes of molecules.

3.2 Crystal Space Groups

3.2.1 Space Group Listing

The space group is a unique three-dimensional description of the symmetry present in a particular crystalline solid, and the definitive account of these appears in International Tables for Crystallography [1]. For the present account, however, we only need to be cognizant of those space groups that accommodate chiral structures. Combinations of the various symmetry operations result in a total of 230 possible space groups. The majority (165) of these are achiral, while the remaining 65 (the Sohncke space groups) are chiral.

Chirality in Supramolecular Assemblies: Causes and Consequences, First Edition. Edited by F. Richard Keene.

Table 3.1 provides a full list of the 230 space groups, with the chiral cases indicated in bold type. The latter comprise two subgroups of 22 (11 pairs) and 43 space groups, respectively. The 11 pairs contain at least one screw axis that is not of the 2_1 (twofold) type. Consequently their enantiomers adopt different space group designators: $P4_1$ and $P4_3$; $P4_122$ and $P4_322$; $P4_12_12$ and $P4_32_12$; $P3_1$ and $P3_2$; $P3_112$ and $P3_212$; $P3_121$ and $P3_221$; $P6_1$ and $P6_5$; $P6_2$ and $P6_4$; $P6_122$ and $P6_522$; $P6_222$ and $P6_422$; $P4_132$ and $P4_332$. The remaining subgroup of 43 space groups contains only proper rotations and the 2_1 screw axis. Since this is a planar arrangement, enantiomeric crystals of the latter subgroup have unchanged space group designators. Thus, as examples, reflection of a $P4_1$ crystal would produce its enantiomeric $P4_3$ crystal: both contain a fourfold screw axis but with opposite senses of helicity. However, reflection of a $P2_12_12_1$ crystal results in its enantiomeric $P2_12_12_1$ crystal, containing material of the opposite handedness but with the same symmetry descriptor.

A pure chiral molecule can only crystallize in one of the 65 chiral space groups. An achiral molecule, or a chiral one present as a 1 : 1 mixture of enantiomers (racemic mixture), will normally crystallize in one of the 165 achiral space groups. However, if either of these materials forms an overall chiral assembly in its crystals then this can only be accommodated in a chiral space group. This situation occurs most frequently if the molecules form helices of the same sense during crystallization. A very familiar example of this behavior is that of the simple achiral molecule urea **1** [2]. Pure urea yields crystals in the (achiral) tetragonal space group $P\bar{4}2_1m$. However, inclusion compounds are formed when it is crystallized from long *n*-alkane solvents. A 1 : 1 mixture of (chiral) hexagonal $P6_122$ and $P6_522$ inclusion crystals is then produced, containing right- or left-handed helical ribbon urea assemblies, respectively. These chiral properties are only present in the solid state, of course, and are lost on dissolution of the crystals.

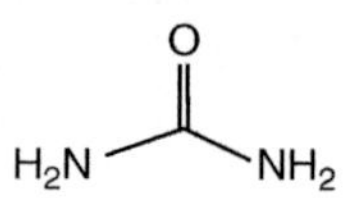

1

3.2.2 Data and Statistics

Crystal structures are archived in, and may be retrieved from, several databases. The three major ones specialize in inorganic compounds [3], organic and organometallic compounds [4, 5], and proteins [6].

The Cambridge Structural Database (CSD) [4, 5], as of 16 February 2015 [7], contained 754 897 structures: 589 932 (78.1%) of these formed in centrosymmetric space groups and 164 965 (21.9%) adopted noncentrosymmetric space groups. The number of Sohncke space group examples was 124 574 (16.5%). The frequency of obtaining a particular space group varies considerably. Around 83% of the structures crystallized in the following six space groups: $P2_1/c$ (including $P2_1/n$) 34.6%, $P\bar{1}$ 24.4%, $C2/c$ 8.3%, $P2_12_12_1$ 7.3%, $P2_1$ 5.2% and $Pbca$ 3.4%.

Since fewer chiral structures have been determined, their statistics are less mature: $P2_12_12_1$ 7.3%, $P2_1$ 5.2%, $P1$ 0.9%, $C2$ 0.8%, $P2_12_12$ 0.4%. All the remaining chiral space groups are under 0.3% frequency.

Table 3.1 *The 230 crystal space groups, with the 65 Sohncke space groups indicated in bold type*

Point groups	Space groups					
			Triclinic crystal system			
1	$\mathbf{P1}$					
$\bar{1}$	$P\bar{1}$					
			Monoclinic crystal system			
2	$\mathbf{P2}$	$\mathbf{P2_1}$	$\mathbf{C2}$			
m	Pm	Pc	Cm	Cc		
$2/m$	$P2/m$	$P2_1/m$	$C2/m$	$P2/c$	$P2_1/c$	$C2/c$
			Orthorhombic crystal system			
222	$\mathbf{P222}$	$\mathbf{P222_1}$	$\mathbf{P2_12_12}$	$\mathbf{P2_12_12_1}$	$\mathbf{C222_1}$	$\mathbf{C222}$
	$\mathbf{F222}$	$\mathbf{I222}$	$\mathbf{I2_12_12_1}$			
$mm2$	$Pmm2$	$Pmc2_1$	$Pcc2$	$Pma2$	$Pca2_1$	$Pnc2$
	$Pmn2_1$	$Pba2$	$Pna2_1$	$Pnn2$	$Cmm2$	$Cmc2_1$
	$Ccc2$	$Amm2$	$Aem2$	$Ama2$	$Aea2$	$Fmm2$
	$Fdd2$	$Imm2$	$Iba2$	$Ima2$		
mmm	$Pmmm$	$Pnnn$	$Pccm$	$Pban$	$Pmma$	$Pnna$
	$Pmna$	$Pcca$	$Pbam$	$Pccn$	$Pbcm$	$Pnnm$
	$Pmmn$	$Pbcn$	$Pbca$	$Pnma$	$Cmcm$	$Cmce$
	$Cmmm$	$Cccm$	$Cmme$	$Ccce$	$Fmmm$	$Fddd$
	$Immm$	$Ibam$	$Ibca$	$Imma$		
			Tetragonal crystal system			
4	$\mathbf{P4}$	$\mathbf{P4_1}$	$\mathbf{P4_2}$	$\mathbf{P4_3}$	$\mathbf{I4}$	$\mathbf{I4_1}$
$\bar{4}$	$P\bar{4}$	$I\bar{4}$				
$4/m$	$P4/m$	$P4_2/m$	$P4/n$	$P4_2/n$	$I4/m$	$I4_1/a$
422	$\mathbf{P422}$	$\mathbf{P42_12}$	$\mathbf{P4_122}$	$\mathbf{P4_12_12}$	$\mathbf{P4_222}$	$\mathbf{P4_22_12}$
	$\mathbf{P4_322}$	$\mathbf{P4_32_12}$	$\mathbf{I422}$	$\mathbf{I4_122}$		
$4mm$	$P4mm$	$P4bm$	$P4_2cm$	$P4_2nm$	$P4cc$	$P4nc$
	$P4_2mc$	$P4_2bc$	$I4mm$	$I4cm$	$I4_1md$	$I4_1cd$
$\bar{4}2m$	$P\bar{4}2m$	$P\bar{4}2c$	$P\bar{4}2_1m$	$P\bar{4}2_1c$	$P\bar{4}m2$	$P\bar{4}c2$

(Continued)

Table 3.1 (*Continued*)

Point groups	Space groups					
	$P\bar{4}b2$	$P\bar{4}n2$	$I\bar{4}m2$	$I\bar{4}c2$	$I\bar{4}2m$	$I\bar{4}2d$
$4/mmm$	$P4/mmm$	$P4/mcc$	$P4/nbm$	$P4/nnc$	$P4/mbm$	$P4/mnc$
	$P4/nmm$	$P4/ncc$	$P4_2/mmc$	$P4_2/mcm$	$P4_2/nbc$	$P4_2/nnm$
	$P4_2/mbc$	$P4_2/mnm$	$P4_2/nmc$	$P4_2/ncm$	$I4/mmm$	$I4/mcm$
	$I4_1/amd$	$I4_1/acd$				
			Trigonal crystal system			
3	$\boldsymbol{P3}$	$\boldsymbol{P3_1}$	$\boldsymbol{P3_2}$	$\boldsymbol{R3}$		
$\bar{3}$	$P\bar{3}$	$R\bar{3}$				
32	$\boldsymbol{P312}$	$\boldsymbol{P321}$	$\boldsymbol{P3_112}$	$\boldsymbol{P3_121}$	$\boldsymbol{P3_212}$	$\boldsymbol{P3_221}$
	$\boldsymbol{R32}$					
$3m$	$P3m1$	$P31m$	$P3c1$	$P31c$	$R3m$	$R3c$
$\bar{3}m$	$P\bar{3}1m$	$P\bar{3}1c$	$P\bar{3}m1$	$P\bar{3}c1$	$R\bar{3}m$	$R\bar{3}c$
			Hexagonal crystal system			
6	$\boldsymbol{P6}$	$\boldsymbol{P6_1}$	$\boldsymbol{P6_5}$	$\boldsymbol{P6_2}$	$\boldsymbol{P6_4}$	$\boldsymbol{P6_3}$
$\bar{6}$	$P\bar{6}$					
$6/m$	$P6/m$	$P6_3/m$				
622	$\boldsymbol{P622}$	$\boldsymbol{P6_122}$	$\boldsymbol{P6_522}$	$\boldsymbol{P6_222}$	$\boldsymbol{P6_422}$	$\boldsymbol{P6_322}$
$6mm$	$P6mm$	$P6cc$	$P6_3cm$	$P6_3mc$		
$\bar{6}m2$	$P\bar{6}m2$	$P\bar{6}c2$	$P\bar{6}2m$	$P\bar{6}2c$		
$6/mmm$	$P6/mmm$	$P6/mcc$	$P6_3/mcm$	$P6_3/mmc$		
			Cubic crystal system			
23	$\boldsymbol{P23}$	$\boldsymbol{F23}$	$\boldsymbol{I23}$	$\boldsymbol{P2_13}$	$\boldsymbol{I2_13}$	
$m\bar{3}$	$Pm\bar{3}$	$Pn\bar{3}$	$Fm\bar{3}$	$Fd\bar{3}$	$Im\bar{3}$	$Pa\bar{3}$
	$Ia\bar{3}$					
432	$\boldsymbol{P432}$	$\boldsymbol{P4_232}$	$\boldsymbol{F432}$	$\boldsymbol{F4_132}$	$\boldsymbol{I432}$	$\boldsymbol{P4_332}$
	$\boldsymbol{P4_132}$	$\boldsymbol{I4_132}$				
$\bar{4}3m$	$P\bar{4}3m$	$F\bar{4}3m$	$I\bar{4}3m$	$P\bar{4}3n$	$F\bar{4}3c$	$I\bar{4}3d$
$m\bar{3}m$	$Pm\bar{3}m$	$Pn\bar{3}n$	$Pm\bar{3}n$	$Pn\bar{3}m$	$Fm\bar{3}m$	$Fm\bar{3}c$
	$Fd\bar{3}m$	$Fd\bar{3}c$	$Im\bar{3}m$	$Ia\bar{3}d$		

3.2.3 Space Group Prediction

Since molecules pack regularly within a crystal, it might be thought easy to predict space groups and in detail packing arrangements for a new case. This situation is far from reality and we shall see that crystal packing is much more subtle than simplistic views suggest. Nonetheless, considerable progress towards prediction has been made and success can be achieved in some cases [8–10]. This extremely important area is likely to advance significantly in the near future.

3.3 Fundamentals of Crystallization for a Racemic Mixture

A racemic mixture contains a 1 : 1 mixture of (+) and (−) enantiomers and various outcomes can occur on its crystallization. Here the enantiomers are designated A and A*, and no other chemical species are present in the crystals discussed. More complex situations will be described later. The four fundamental crystal outcomes are shown diagrammatically in Figure 3.1, with racemic crystals indicated by rectangles and chiral crystals shown as a pair of parallelograms. Particular attention should be directed to the differences in enantiomer association within these crystal outcomes. The black spheres indicate sites occupied randomly by either A or A*.

3.3.1 Racemic Compound

Formation of a racemic compound (or true racemate) is what most would regard as the typical behavior expected on crystallization of a racemic mixture. The opposite enantiomers (A and A*) have a strong affinity for each other and assemble into a close, intimate and regular arrangement. Addition of a small excess of one pure enantiomer will result in a lowering of the melting point.

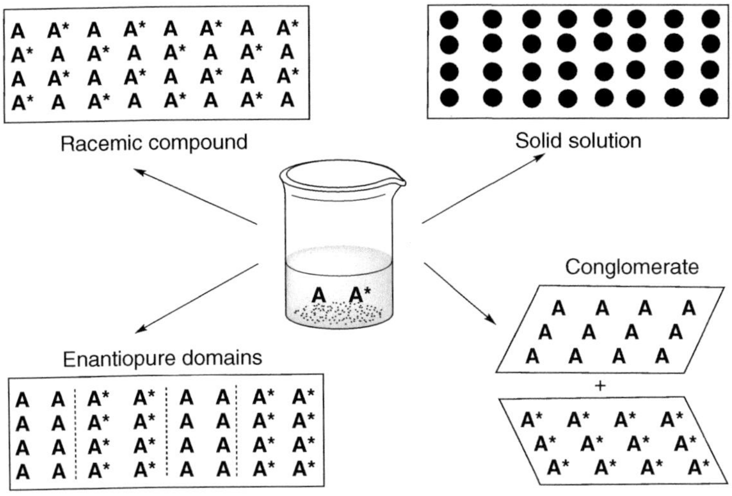

Figure 3.1 *Diagrammatic representation of the fundamental outcomes observed on crystallization of a racemic mixture containing equal numbers of the opposite enantiomers A and A**

3.3.2 Solid Solution

A solid solution can occur in cases where there is reduced discrimination between the components. The concept is nicely illustrated by the behavior of the Werner complexes [$NiCl_2$(4-phenylpyridine)$_4$] **2** and [$CoCl_2$(4-phenylpyridine)$_4$] **3**. If $NiCl_2$, $CoCl_2$ and 4-phenylpyridine (ratio 0.5 : 0.5 : 4.0) are ground together, then the solid compound $Ni_{0.5}Co_{0.5}Cl_2$(4-phenylpyridine)$_4$] **4** is produced. This new material has the same X-ray powder diffraction pattern as **2**. In other words **2** and **4** are isostructural. The latter, however, is a solid solution with its Ni and Co ions randomly occupying the metal sites in the crystalline solid [11].

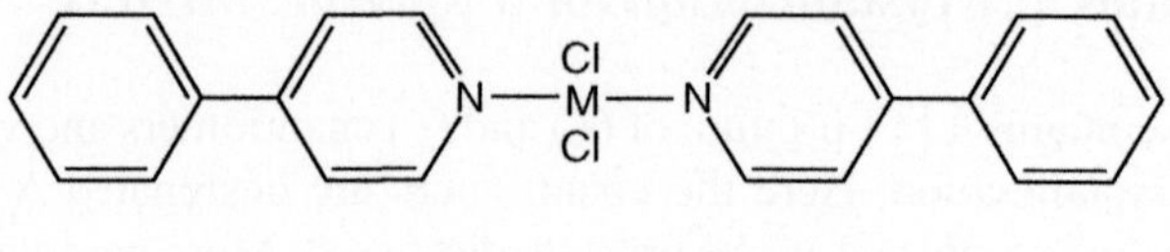

2 M = Ni **3** M = Co **4** M = Ni or Co

In the situation illustrated in Figure 3.1 (a pseudoracemate), the attraction energies between identical enantiomers (A and A, or A* and A*) are close to those between the opposite enantiomers (A and A*). Discrimination now is poor and random occupancy occurs in the solid, as indicated diagrammatically using the black spheres. Addition of a small excess of one pure enantiomer will result in little change to the melting point. The experimental crystal structure data is that of an average A plus A* combination and attempts at its refinement will remain unsatisfactory.

3.3.3 Enantiopure Domains

In cases where attractions are stronger between their own type (A only; or A* only), rather than between A and A*, the enantiomers can form enantiopure regions within the crystal. These domains are indicated diagrammatically using dashed lines in Figure 3.1. The assemblies are most commonly chains, helices, and layers. Having exhausted their favorable attractions in forming these enantiopure regions, they can no longer propagate further and must pack alongside identical assemblies of the opposite handedness. Therefore, although a considerable amount of enantiomer separation has taken place, the complete crystal is still racemic.

A typical example is the racemic keto alcohol **5** [12]. These molecules, linked by C=O···H-O hydrogen bonds, form single enantiomer chains along the *b* direction. The chains then assemble parallel to each other creating enantiopure layers in the *ab* plane. Adjacent planes have opposite handedness as shown in Figure 3.2.

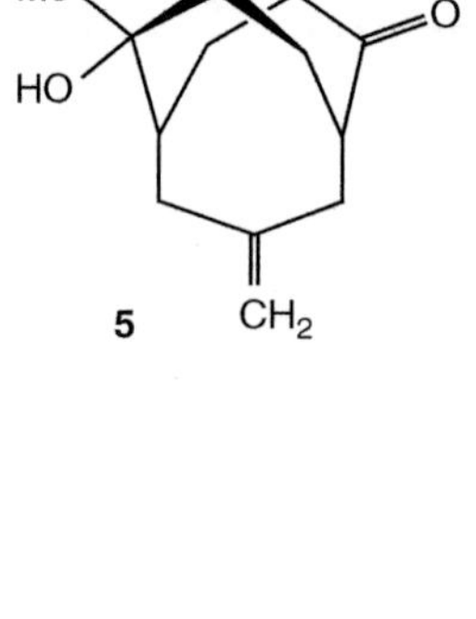

5

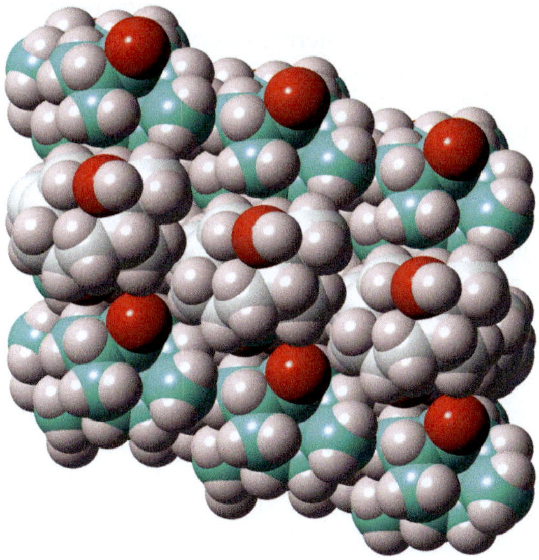

Figure 3.2 *Edge-on view of the enantiopure chains and layers in the crystal structure of* ***3*** *(projected on the* ac *plane). Color code: C opposite enantiomers light blue or off white, and O red. Refcode SIYWON [12]*

3.3.4 Conglomerates

If the assembly process described in section 3.3.3 is able to continue over three dimensions, then self-resolution takes place to yield a conglomerate. This is a physical mixture of enantiomorphous pure (+) and pure (−) crystals, and its melting point is lower than that of the pure enantiomer. Addition of a small excess of one pure enantiomer will result in elevation of the melting point. Since a conglomerate is a mixture, then in principle the enantiomeric crystals can be separated physically to yield pure samples of both A and A*. Hence every conglomerate is potentially a bulk source of both pure enantiomers. The implications of this important property will be explored later in greater detail.

3.4 More Complex Crystallization Behavior

It would be expected generally that molecules will assemble into a crystal following the targets of close-packing and highest density [13]. These ideals may be in conflict, however, with awkwardly shaped molecules or intermolecular attractions that have a strong directional requirement. The latter is best illustrated by the formation of ice from liquid water at 0 °C, which is accompanied by an amazing density decrease from 0.9998 to 0.9168 g cm^{-3}. Formation of the final set of hydrogen bonds results in a rather open packing structure and causes ice to float. Hence crystal structures tend to be a compromise between factors that may be opposing each other. Sometimes the most suitable outcome can be achieved relatively simply. In other instances, much more complex behavior has to be adopted. The fundamentals (section 3.3) therefore require extension to take account of additional solid state properties. These will allow the best compromise to be reached, and include crystallographically independent molecules, alternative crystal structures, molecular inclusion, and combinations of these phenomena.

3.4.1 Crystallographically Independent Molecules

A solute in solution has all its molecules occupying an identical environment. This is often also the case in the solid state. However, many crystals contain molecules that occupy different local environments and so these are crystallographically independent. As seen above for water, there is also no necessary correlation between a simple molecular structure and the complexity of the crystal structure that is adopted. A particularly striking case is the thermodynamically favored α-crystal form of hydroquinone **6**. This highly symmetrical molecule crystallizes in space group $R\bar{3}$ using three independent molecules (A–C). This induced crystal complexity overcomes the major packing difficulties evidently encountered by **6** and results in a unit cell containing an astonishing 54 molecules [14]. If the crystal is dissolved, thereby eliminating the packing problems, the solute of course reverts to being just identical achiral molecules of **6**. Hence the characteristics of molecules in the solid state can be unexpected and surprisingly complex.

6

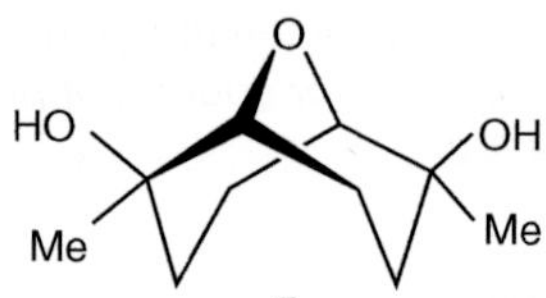

7

The relatively simple molecule **7** is another that encounters severe packing difficulties [15]. Crystallization of racemic **7** gives a structure in the chiral space group $P2_12_12$ that contains six independent molecules (A–F) and their enantiomers (A*–F*). The framework of **7** has a certain amount of conformational freedom, so it is easy to generate these different independent molecules that contain slightly different bond lengths and torsion angles. The descriptor Z' is used to indicate the number of independent molecules in the asymmetric unit of a crystal structure. Its value also corrects for such molecules situated on symmetry sites. Hence the crystal structure of **7** has $Z' = 5$, rather than six, because two of its independent molecules surround a twofold axis and each contributes only 0.5 to the total value.

The implications and significance of crystal structures containing multiple molecules in the asymmetric unit ($Z' > 1$) have been the subject of robust debate since the mid-2000s and this is ongoing [16]. Steed has estimated that the frequency of the phenomenon is 11.5% for organic structures and 8.8% for the entire CSD. It is well known to vary considerably across different classes of compound and is higher (14.6%) for chiral space groups [17].

3.4.2 Kryptoracemates

Crystallographically independent molecules can also result in kryptoracemate formation [18, 19]. The 1 : 1 mixture of enantiomorphous crystals produced is very easily mistaken for a conglomerate. The fundamental difference is that if one crystal of a conglomerate is

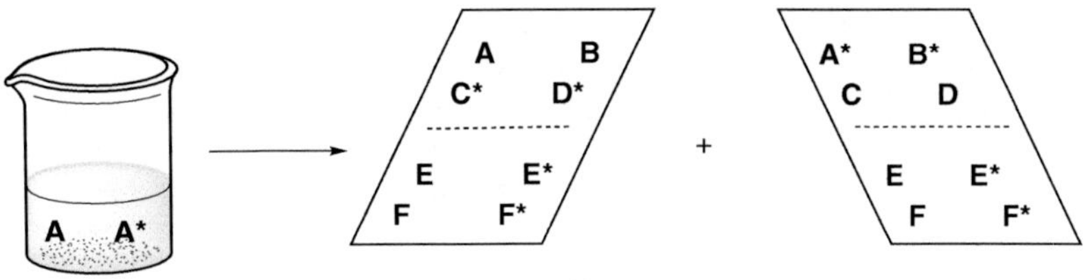

Figure 3.3 Diagrammatic representation of the kryptoracemate formed on crystallization of diol 7

dissolved, then an optically active solution is obtained. In contrast, dissolution of one crystal of a kryptoracemate (or false conglomerate) yields an optically inactive solution. The kryptoracemate phenomenon is explained here using the example of diol **7** shown in Figure 3.3.

Crystallization of racemic **7** yields a 1 : 1 mixture of enantiomorphous crystals in the chiral space group $P2_12_12$. Single crystal X-ray determination reveals a three-dimensional network constructed from two very different subunits [15]. Molecules of **7** assembled into pillars afford a host-like framework, which surrounds guest-like columns of **7**. Hydrogen bonds are used to build both the pillars and columns, and these subunits are cross-linked by additional hydrogen bonding. In one enantiomeric crystal the chiral pillars comprise A, B, C* and D* independent molecules, and the columns contain E, E*, F and F* molecules. The enantiomorphous crystal contains A*, B*, C and D pillars, and E, E*, F and F* achiral columns. Of course, this complex chiral construction is only a solid-state property. Dissolution causes reversion to the original racemic solution of just A and A* molecules.

Fábián and Brock have recently devised elegant methods for the identification of kryptoracemates deposited in the CSD, and have described 181 confirmed organic examples [20]. Our own work suggests that the frequency of occurrence is around 1% for slightly twisted C_2-symmetric organic molecules with limited conformational flexibility [15].

3.4.3 Quasiracemates

A quasiracemate is an extension of a racemic compound in which one of the enantiomers (say A*) has been replaced by the same enantiomer of a different chemical species (say X*). The pure enantiomer X* is sterically similar to the A* molecule it has replaced and would normally use similar interaction forces. The equimolar mixture of A and X* enantiomers for crystallization can be prepared deliberately, as well as being discovered by accident. Its ordered crystal outcome is illustrated in Figure 3.4.

Quasiracemates are often based on racemic compound crystal structures in which A and A* interact around an inversion center. Replacement of A* by X* then often leads to association of A and X* around a pseudo-inversion center and formation of the quasiracemate structure [21, 22]. Representative examples are the deliberate combinations **8** and **9**, or **10** and **11**. Recent research into quasiracemate compounds has been reviewed by Zhang and Curran [23].

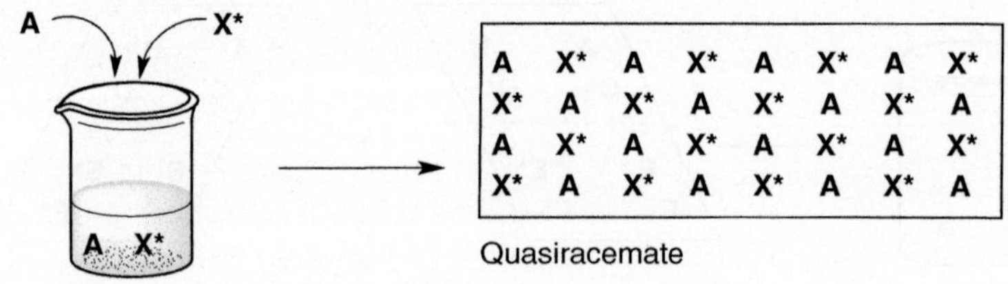

Figure 3.4 *Diagrammatic representation of the formation of a quasiracemate crystal*

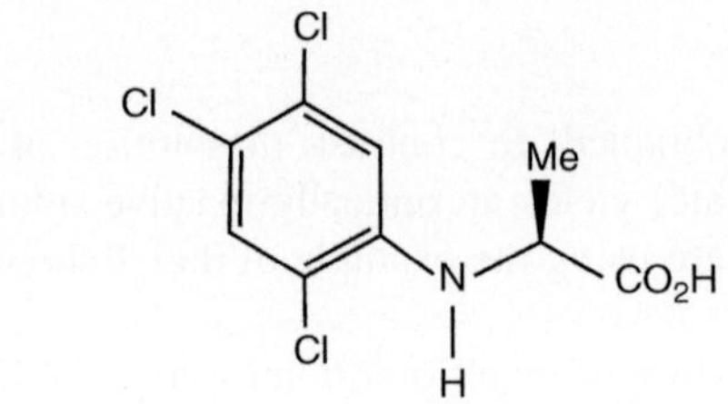

8

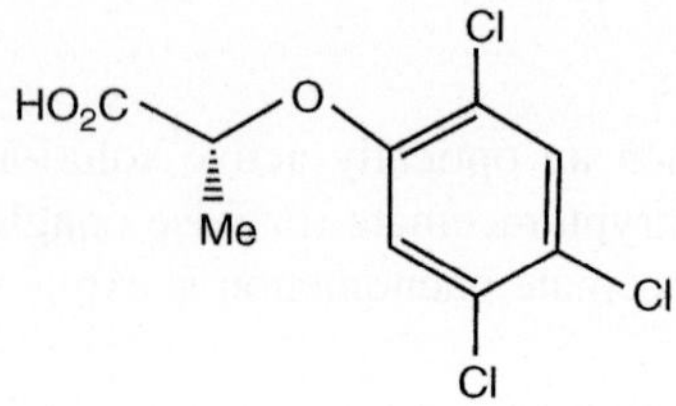

9

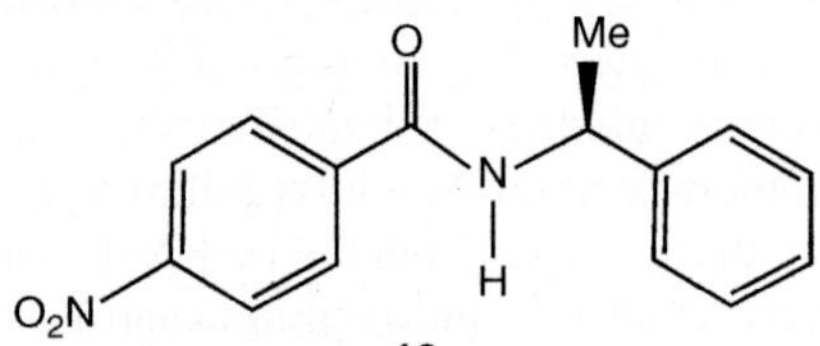

10

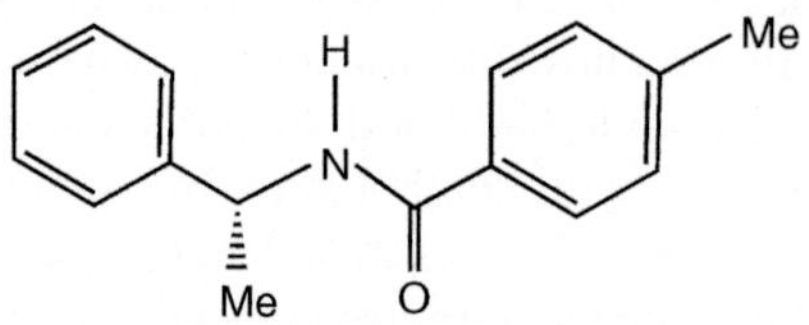

11

Quasiracemates are also often found by chance (see section 3.5.4). One of the most striking examples is the case of ammonium (2*R*,3*R*)-(+)-bitartrate and ammonium (*S*)-(-)-bimalate. This 1:2 combination **12** was originally reported by Pasteur [24] in 1853, and has been recently reinvestigated in greater detail by Wheeler *et al.* [25]. Remarkable dumbbell-shaped crystals are produced from this pair of components (Figure 3.5). The "handgrip" contains the tartrate salt only, whereas the two "weights" are quasiracemate laths constructed from the tartrate and malate enantiomers.

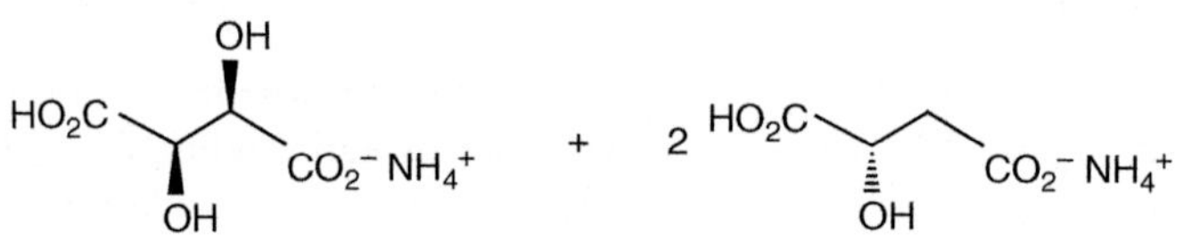

12

3.5 Multiple Crystal Forms

Considerable time and effort was required in the early days of crystallography to complete each structural determination. There was a general view that nearly all molecules would give just one result – *the* crystal structure. However, it is now apparent that many substances can yield multiple answers. Their proportion is increasing rapidly as X-ray

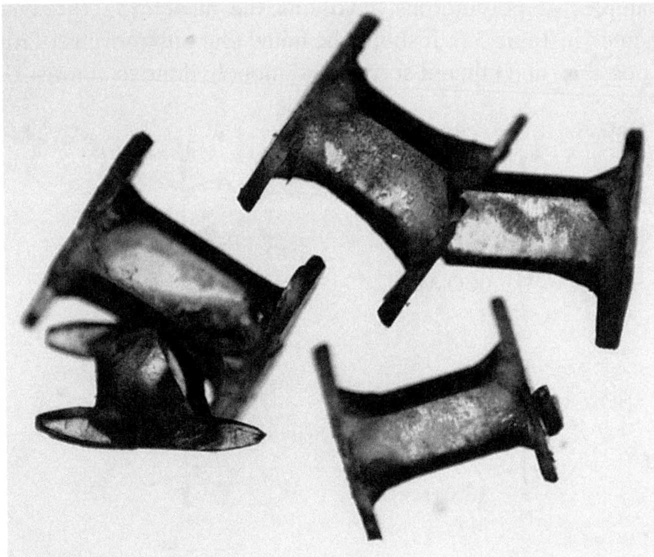

Figure 3.5 *The dumbbell-like crystals of **12** produced from ammonium (2R,3R)-(+)-bitartrate and ammonium (S)-(−)-bimalate [25]. Photograph kindly supplied by Professor Kraig Wheeler of Eastern Illinois University*

determinations become easier to perform and as more thorough screening is carried out. Most of these multiple crystal forms can be subdivided into the polymorph, solvate, hydrate, and cocrystal structural categories. At least 50% of organic compounds have been claimed to exhibit polymorphism [26]. With the additional opportunities available for solvate, hydrate, and cocrystal formation, it appears that that the majority of compounds will yield multiple crystal forms if they are investigated thoroughly.

3.5.1 Polymorphs

Polymorphs are crystals of identical chemical composition that contain different spatial arrangements of their component molecules [27, 28]. These alternative packing modes mean that polymorphs will have different physical properties such as melting point, solubility, and biouptake.

Some years ago, Desiraju observed that such alternative structures represented a significant complication for crystal engineering [29]. They have proved to be considerably less welcome in the pharmaceutical industry. Specific case histories of these substances are described by Bernstein [27]. Polymorph formation is a common phenomenon and thorough screening is now routine for new pharmaceutical products to avoid potential problems relating to patents, formulations, and earnings.

Calculation of the crystal landscape for a given molecule reveals that it is common for several potential polymorphs to be within 10 kJ mol^{-1} or so of the computed crystal structure energy minimum [8]. Factors such as nucleation and crystal growth are difficult to control and therefore which, if any, of these crystal arrangements will be readily accessible by experiment is still problematic [30].

Illustrative examples of polymorphs, involving the bicyclo[3.3.0]octane diols **13–16** [31–35], are presented in Table 3.2. It should be noted that polymorphs of multicomponent crystals also are possible, and ethanol solvate and monohydrate examples are given here.

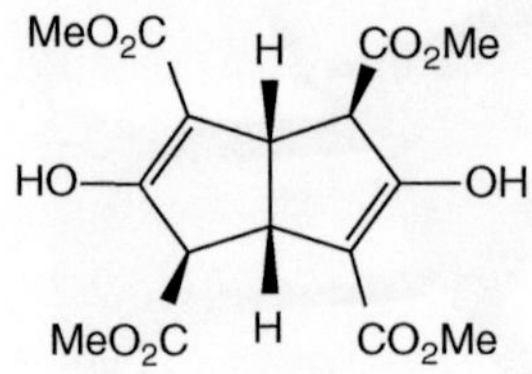

13

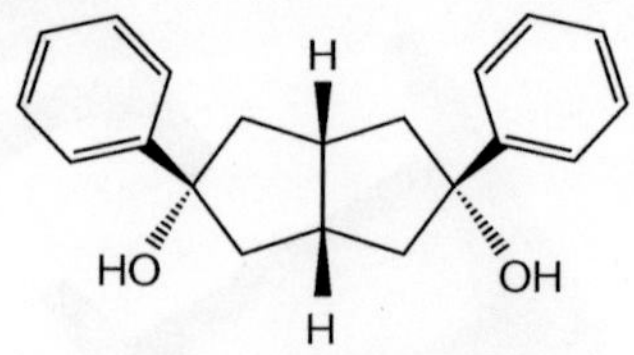

14

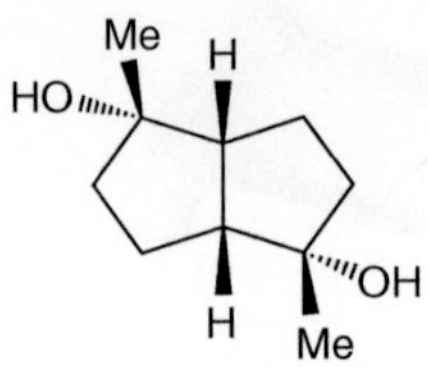

15

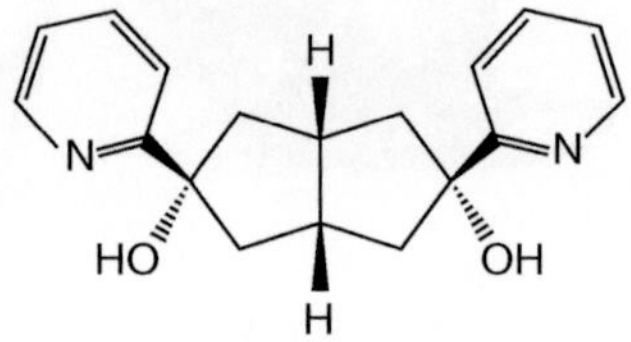

16

The polymorph structures of tetraester **13** are described briefly to illustrate how this phenomenon can occur. Molecule **13** has two hydroxy groups that act as hydrogen bond donors, and in both structures there is intramolecular hydrogen bonding O-H···O=C (ester carbonyl group). This is the only use made of the hydroxy group in polymorph 1. The tetraester molecules then assemble into centrosymmetric pairs utilizing C-H···O weak hydrogen bonding (Figure 3.6). Note that this hydrogen is donated to O2, which is the oxygen atom of the hydroxy group. Repetition of the dimeric unit then generates the complete crystal structure [31, 32].

Additional hydroxy group hydrogen bonding occurs in the polymorph 2 structure [32]. Here the hydroxy hydrogen atoms function as donors to both intra- and intermolecular carbonyl oxygen atoms. Thus adjacent molecules of the tetraester **13** now assemble into enantiopure chains along the *b* direction (Figure 3.7).

A switch from monodirectional to bifurcated hydrogen bonding is the reason that these two polymorphs are produced. The same solvent (methanol) was used but the crystallization temperature was lowered from 295 K to 273 K. A less important factor here is that the molecules in these two crystal structures have differing conformations: one of the ester groups rotates through about 90°. Alternative molecular conformations, however, are the major cause of polymorphic behavior for many other molecules.

Crystallization of the racemic diol **15** is a particularly interesting case [34]. Its solvent-free polymorph 1 is formed in achiral space group $P2_1/c$ from either cyclohexane or ethanol. However, a different solvent-free structure, polymorph 2, is produced in the chiral space group $P4_1$ (and $P4_3$) if acetone, benzene, 1,4-dioxane or tetrahydrofuran solutions are used. Note that there is no obvious correlation between the nature of these solvents and the observed crystalline outcomes. All these crystals were grown at 295 K. The crystallization solution contains the (+)- and (−) enantiomers of **3** (symbolized here as A and A*), but the crystals produced contain two crystallographically independent molecules (A, B) and their enantiomers (A*, B*), as illustrated in Figure 3.8.

Table 3.2 *Polymorphs formed by the bicyclo[3.3.0]octane diol derivatives* ***13–16***

Compound	Crystallization solvent	T (K)	Designator	Crystal system	Space group	Refcode	Ref.
13	Acetone	295	Polymorph 1	Monoclinic	$P2_1/n$	WIZHUJ	31, 32
13	Methanol	273	Polymorph 2	Triclinic	$P\bar{1}$	WIZHUJ01	32
(**14**)·(ethanol)	Ethanol	295	Ethanol polymorph 1	Monoclinic	$C2/c$	–	33
(**14**)·(ethanol)	Ethanol	295	Ethanol polymorph 2	Triclinic	$P\bar{1}$	–	33
15	Cyclohexane	295	Polymorph 1	Monoclinic	$P2_1/c$	ENITUR	34
15	1,4-Dioxane	295	Polymorph 2	Tetragonal	$P4_1$ and $P4_3$	ENIVAZ	34
16	Diethyl ether	295	Polymorph 1	Monoclinic	$C2/c$	PEKJEW01	35
16	Acetone and water	295	Polymorph 2	Orthorhombic	$Pna2_1$	PEKJEW	35
(**16**)·(water)	Tetrahydrofuran and water	295	Hydrate polymorph 1	Monoclinic	$P2_1/c$	PEKJUM01	35
(**16**)·(water)	Acetonitrile and water	295	Hydrate polymorph 2	Monoclinic	$P2_1/n$	PEKJUM	35

Notes: Compounds **13** and **15** are chiral but their racemates were used here. Compounds **14** and **16** are achiral. T (K) is the temperature of crystal growth.

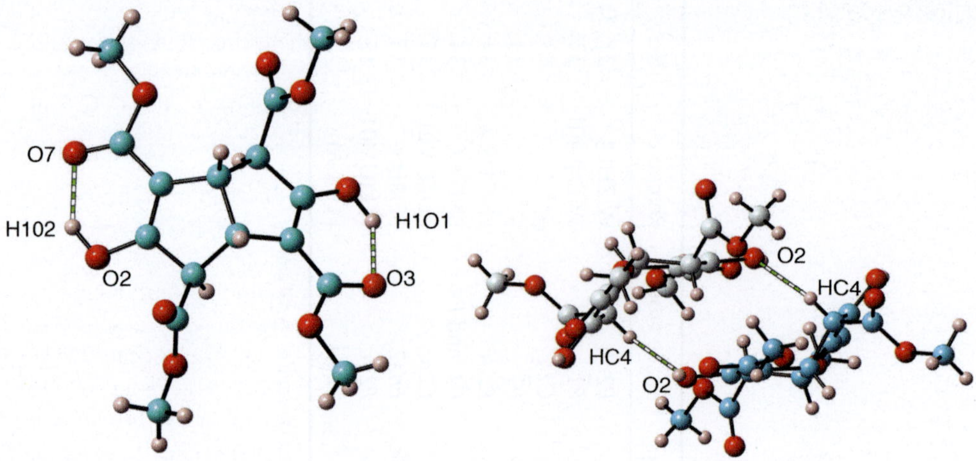

Figure 3.6 *Polymorph 1 structure of the tetraester* ***13****. Left: One molecule of* **13** *showing the intramolecular O-H···O=C hydrogen bonds (green dashed lines). Right: The centrosymmetric dimer, assembled using C-H···O weak hydrogen bonds (green dashed lines), which acts as the crystal repeat unit. Color code: C opposite enantiomers blue or off white, and O red. Refcode WIZHUJ [31, 32]*

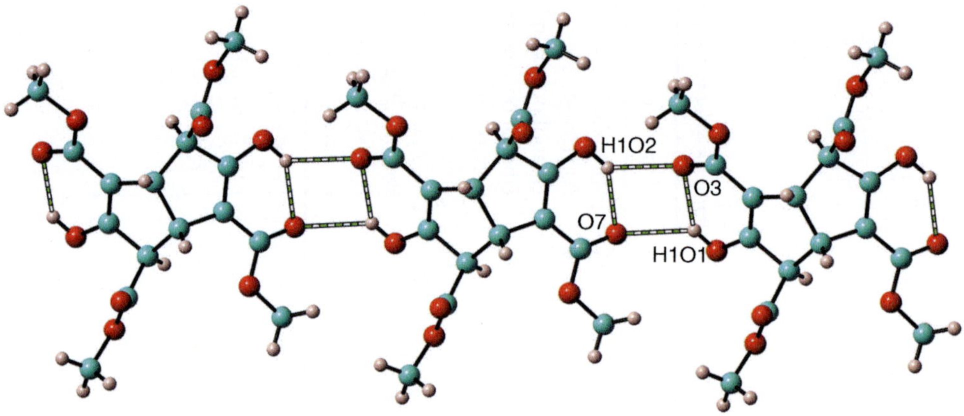

Figure 3.7 *Polymorph 2 structure of the tetraester* ***13****. The intra- and intermolecular O-H···O=C hydrogen bonding between three molecules of the same chirality is shown (green dashed lines) with* b *horizontal. Refcode WIZHUJ01 [32]*

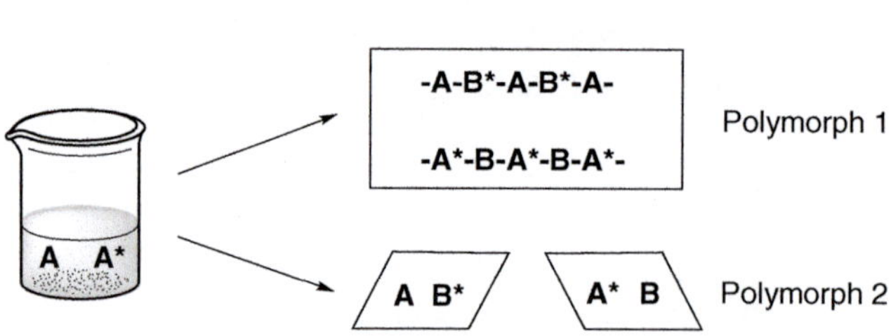

Figure 3.8 *Diagrammatic representation of the diol* ***15*** *properties in solution and the solid state. The racemic polymorph 1 and kryptoracemate polymorph 2 structures are contrasted. Both of these polymorphs grow from solution at 295 K*

The structure of polymorph 1 uses enantiomer ordering to form infinite chains of A-B*-A-B*-A- and A*-B-A*-B-A*-molecules. Each crystal is therefore racemic, as indicated by its space group. At first sight it appears that polymorph 2 is a conglomerate, as the product is a 1 : 1 mixture of enantiomeric crystals in space groups $P4_1$ and $P4_3$. Closer examination, however, shows that these are AB* crystals or A*B crystals, respectively. Hence polymorph 2 is a further example of a kryptoracemate.

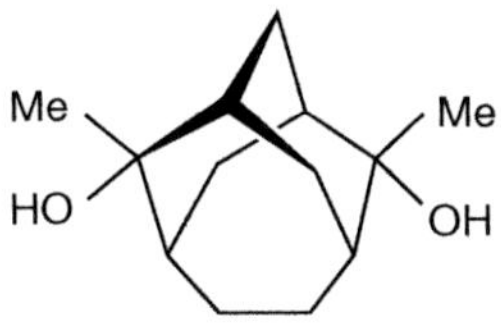

17

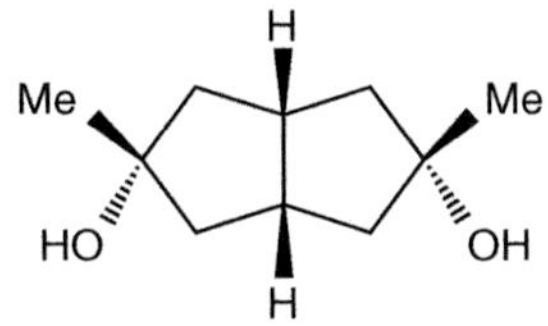

18

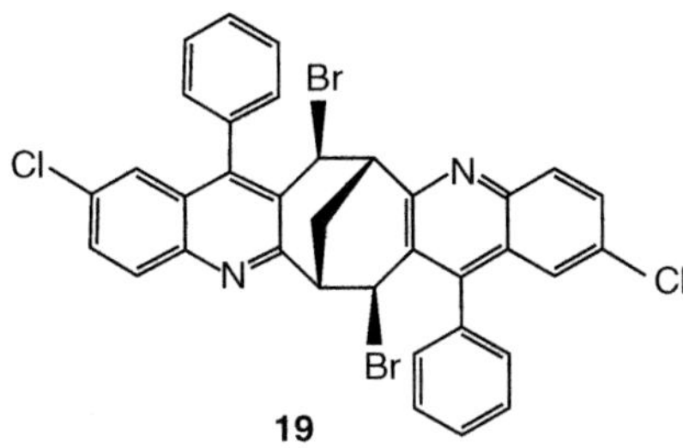

19

It should be noted that each of the compounds **17–19** produces different crystalline substances (Table 3.3) containing the same two molecular components [34, 36, 37]. These materials should not be described as polymorphs as different ratios of host to guest are present and so the pairs of compounds have different formulae. They are usually described as solvates or hydrates, depending on the nature of the guest.

3.5.2 Solvates

The ideal outcome is close packing of molecules in a crystal [13]. However, many pure compounds with irregular shapes have difficulty assembling in an efficient complementary manner. Others interact with each other through molecular attractions like hydrogen bonds that have strong directional requirements. Both situations tend to result in a lower packing coefficient, and hence amorphous solids could result. To avoid this situation, the substance may instead include guest solvent molecules to maintain an efficient packing coefficient and higher crystal density.

There is a continuum of multicomponent inclusion interactions ranging from clathrates, via coordinato-clathrates, through to complexes [38]. The position on this scale is largely dependent on the strength of the host-guest interaction. Clathrate crystals have strong host networks with the guest molecules physically enclosed but otherwise with weak host-guest interactions. At the other extreme, inclusion complexes frequently are mutually hydrogen bonded and the distinction between host and guest is now largely a case of their relative size. Considerable work has been carried out on solvate compounds during recent decades and this has been reviewed [39–43].

Table 3.3 *Crystalline materials formed by the compounds **17–19***

Compound	Crystallization solvent	T (K)	Crystal system	Space group	Refcode	Ref.
$(\mathbf{17})_4$·(1,2-dichlorobenzene)	1,2-Dichlorobenzene	295	Tetragonal	$I4_1/acd$	VUSYEN	36
$(\mathbf{17})_3$·(1,2-dichlorobenzene)	1,2-Dichlorobenzene	273	Trigonal	$P3_121$ and $P3_221$	VUSYIR	36
$(\mathbf{18})_2$·(water)	Toluene and water	295	Monoclinic	$P2/c$	ENIVED	34
(**18**)·(water)	Isooctane and water	295	Monoclinic	$P2_1/c$	ENIVON	34
(**19**)·(*d*-chloroform)	*d*-Chloroform	295	Orthorhombic	$Pna2_1$	AJEQAH	37
(**19**)·$(d\text{-chloroform})_2$	*d*-Chloroform	295	Triclinic	$P\bar{1}$	AJEQEL	37

Notes: Compound **17** is chiral but its racemate was used here. The tetragonal product contains two interpenetrating host sublattices, whereas the trigonal product is a conglomerate. Compound **18** is achiral. Compound **19** is chiral but its racemate was used here. Its two types of crystals were produced concomitantly. T (K) is the temperature of crystal growth.

A huge range of different topologies can be employed for inclusion and these have been classified in detail [38]. The more common types are clathrates, tubulates, intercalates, hosts containing hollow spaces and zeolite networks. A good level of understanding of solvate compounds has been reached but accurate forecast of new examples (and particularly their crystal structures) is still problematic.

Some hosts give a series of compounds in which the host:guest ratios, space groups, and molecular construction are entirely different from each other. Others yield an essentially isostructural series of inclusion compounds when crystallized from a range of solvents. The tetraester **13** behaves in the latter manner (in addition to forming the solvent-free polymorphs described in section 3.5.1). A series of closely related structures (tubulates) containing parallel guest-filled tubes is produced from many solvents [32, 44]. Figure 3.9 shows the crystal structure of $(\mathbf{13})_2$·(*p*-xylene), with one of the guest tubes shown running horizontally.

The numerical ratio of components in a binary solvate is usually 1 : 1, 1 : 2, 1 : 3, 1 : 4, etc. Less common ratios, such as 2 : 3 or 4 : 5, indicate that other factors are in play. Solvate $(\mathbf{20})_4$·(benzene)$_5$, for example, contains three crystallographically independent benzene molecules [45]. One of these (light green) surrounds an inversion center and therefore contributes only half a molecule per two molecules of **20** (Figure 3.10).

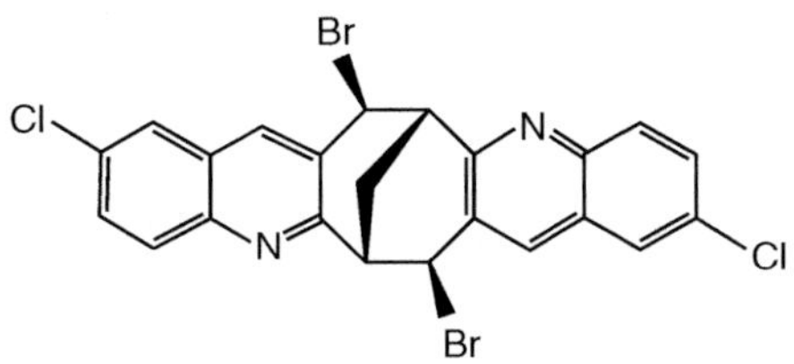

20

3.5.3 Hydrates

Water is often included in crystals as a hydrate compound: indeed it has been suggested that around 30% of pharmaceutical molecules are capable of this behavior [46]. The frequency of hydrate formation is also very high when compared to inclusion of specific organic recrystallization solvents [47]. This is due, in part, to water being an extremely effective hydrogen bonding agent that does not occupy much space. However, another practical reason is that most organic substances are deliberately exposed to water at some stage during their extraction or experimental workup.

The CSD has been used extensively to investigate the frequency of hydrate formation and to probe the factors underpinning this phenomenon [48, 49]. Water is highly versatile in the manner by which it is included. It may be present as an individual molecule, or in aggregates such as rings, clusters, cages, chains, and tapes [50]. Prediction of new examples of hydrate formation, and their potential structures, remains problematic. This is a very important area of structural chemistry that would reward deeper investigation.

The monohydrate of racemic diol **21** is shown here as a representative of the simple hydrates. Water molecules act as hydrogen bond linkers between the hydroxy groups of four adjacent molecules and thus generate a layered crystal structure (Figure 3.11). This hydrate is noteworthy for its almost instantaneous melting point at 150 °C, a considerably higher temperature than the water boiling point [51].

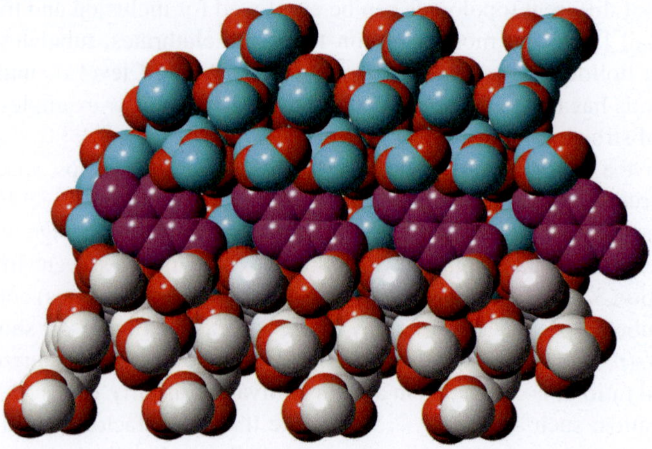

Figure 3.9 *The (**13**)$_2$·(p-xylene) crystal structure showing one of the guest-containing tubes (b horizontal). Four columns of host molecules surround each tube, but one column (off-white enantiomer) has been omitted here. No hydrogen atoms are shown for clarity. Color code: host C opposite enantiomers light blue or off-white, O red, and p-xylene magenta. Refcode FAHLAD [32]*

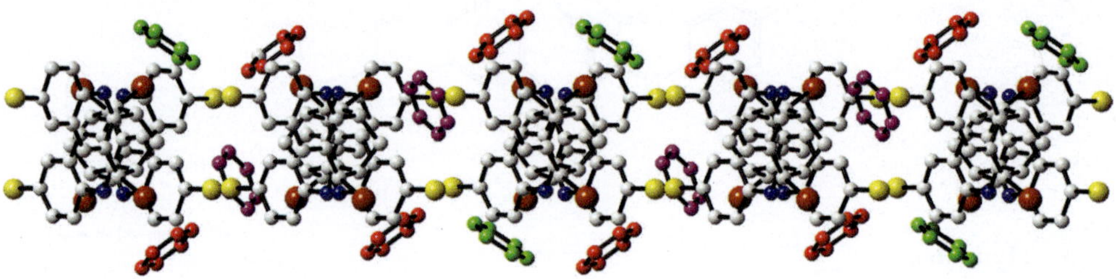

Figure 3.10 *Projection view (a horizontal) of compound (**20**)$_4$·(benzene)$_5$ and showing the three independent benzene guests. Its empirical formula arises from the combination of benzene guests (magenta 1.0, red 1.0, and light green 0.5) with each pair of host molecules. Color code for both host enantiomers: C off white, N dark blue, Br brown and Cl yellow. All hydrogen atoms are omitted for clarity. Refcode XOGKUZ [45]*

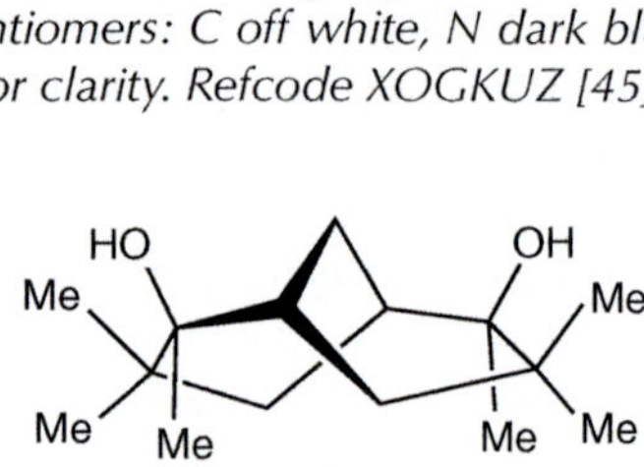

21

3.5.4 Cocrystals

Different types of molecules sometimes crystallize together, yielding stoichiometric multi-component crystals. A common definition of cocrystal in this context is that the individual components should be solids at room temperature. They are often comparable in size and

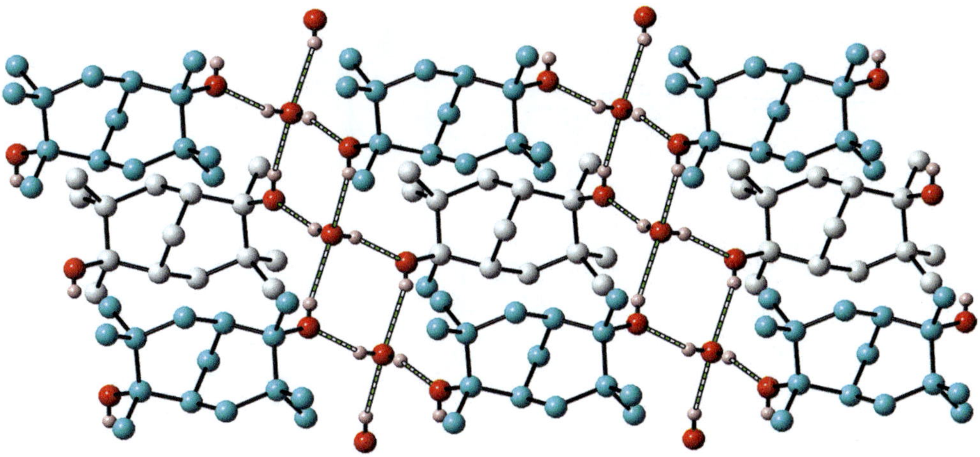

Figure 3.11 *Projection view in the* ab *plane of the (**21**)·(H_2O) structure. The only H atoms shown are those involved in O-H···O hydrogen bonds (which are indicated by green dashed lines). Refcode JAPZAB [51]*

usually mutually attracted by means of hydrogen bonding or other strong interactions. Hence the host-guest description used previously is no longer meaningful.

Cocrystals of this type are often discovered during failed fractional crystallization purification of a reaction product. Their components are therefore closely related in structure – for example they are isomers, have comparable molecular structures, or can be quasiracemates (section 3.4.3). This structurally diverse group of compounds has been analyzed recently by Brock [52]. One example, the 1 : 1 complex of hydroquinone **22** and *p*-benzoquinone **23** is so well known that it has earned its own name, quinhydrone **24**. The structure of this representative material (Figure 3.12) is stabilized by both strong donor–acceptor complexation and O-H···O=C hydrogen bonding [53].

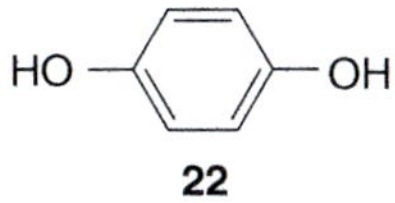

22

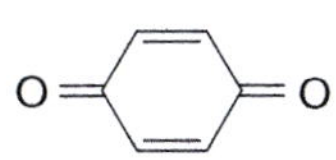

23

(22)·(23) = 24

There is no compulsion, however, for the components to share closely related molecular structures. These cases have usually been discovered whilst carrying out synthetic work, but the key requirement for successful combination is really complementary supramolecular interaction. This can involve very different partners. Recognition of this has led to cocrystal design becoming a highly active area of contemporary research. Many pharmaceutical drugs have low aqueous solubility (hence poor biouptake) while others have poor

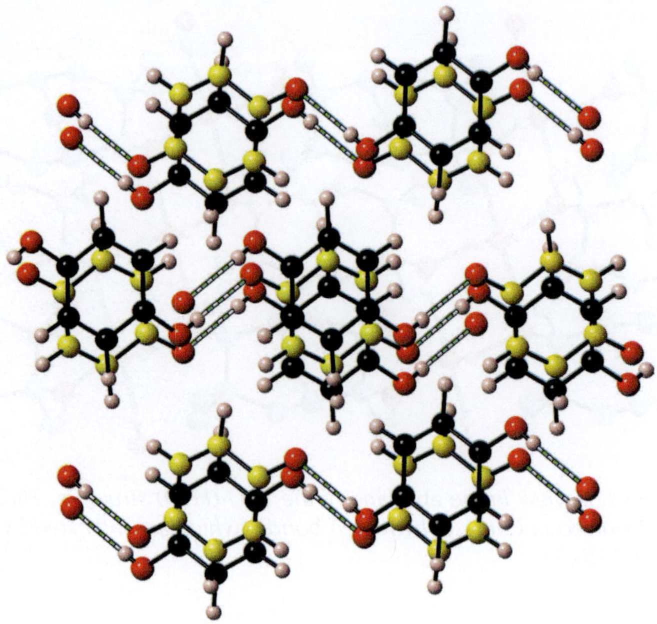

Figure 3.12 The (22)·(23) or quinhydrone **24** crystal structure. Color code: hydroquinone **22** C black, p-benzoquinone **23** C yellow, and O red. The O-H···O=C hydrogen bonds are indicated by green dashed lines. Refcode QUIDON01 [53]

formulation characteristics or stability problems. A cocrystal form of the drug may improve these properties, permit lower dosages, or circumvent existing patent protection [54, 55]. The approach used is to combine the active pharmaceutical ingredient (API) with a generally recognized as safe (GRAS) coformer molecule [56].

A good example of this principle is carbamazepine **25**, an anticonvulsant drug with poor aqueous solubility and low bioavailability, that does however, form cocrystals with a wide range of co-former molecules [57]. Its 1 : 1 complex with saccharine **26** is illustrated in Figure 3.13.

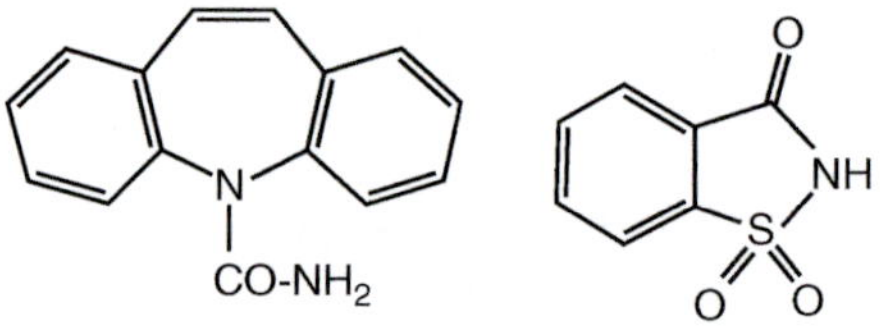

25 **26**

Efavirenz **27** is a chirally pure substance that is effective against HIV by acting as a non-nucleoside reverse transcriptase inhibitor. It can exist in several polymorphic arrangements, but it also yields several cocrystalline complexes [58]. One of these is its 2 : 1 combination with 4,4′-bipyridine **28** and it is noteworthy that the opposite ends of this simple coformer subtend different interaction modes to efavirenz (as shown in Figure 3.14). It appears that API molecules capable of forming cocrystals can often do so with different GRAS components by employing a variety of different molecular interactions.

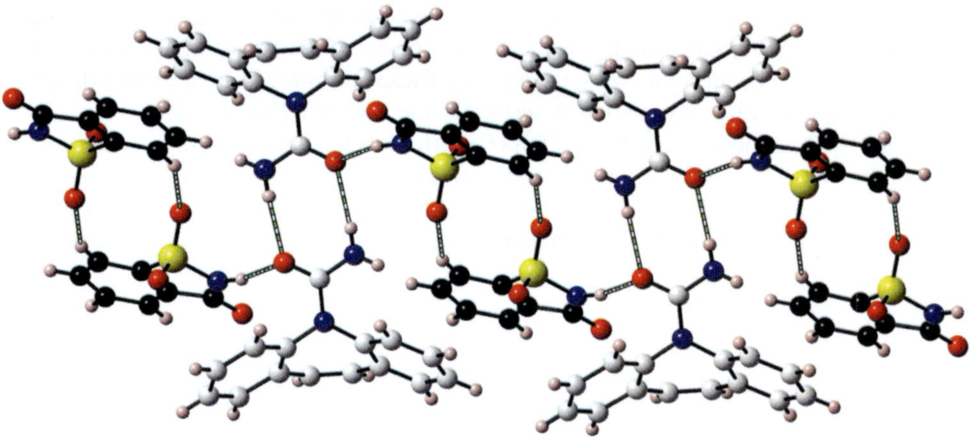

Figure 3.13 The intermolecular attractions under 2.40 Å present in the (carbamazepine)·(saccharine) cocrystal (**25**)·(**26**). These N-H···O=C, S=O···H-C and C=O···H-NH interactions are indicated by green dashed lines. Color code: C of **25** off white and of **26** black, N blue, O red and S yellow. Refcode UNEZAO [57]

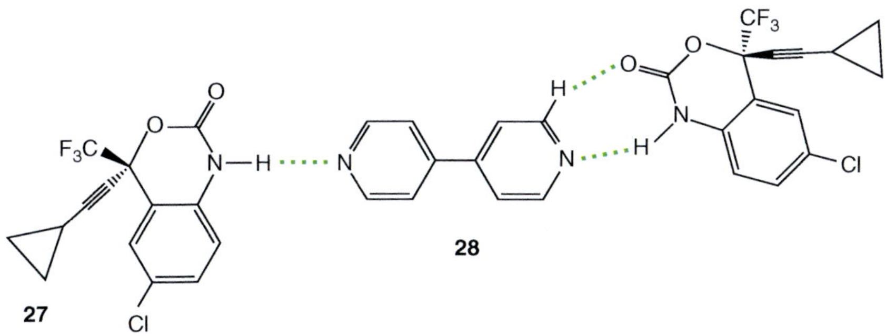

Figure 3.14 The interaction modes of efavirenz **27** with its crystal coformer 4,4′-bipyridine **28** are indicated using green dashed lines [58]

3.6 Conglomerates Revisited

The phenomenon of conglomerate formation (section 3.3.4) is arguably the most significant crystallization behavior presented in this chapter. Its implications are therefore worth examining in greater detail.

3.6.1 Frequency of Conglomerate Formation

How many racemic compounds yield a conglomerate on crystallization? Nobody really knows. Misconceptions occasionally still appear in print based on early literature sources. In popular review articles, Secor recorded only some 30 examples in 1963 [59], while Collet, Brienne and Jacques had located 248 examples by 1980 [60]. This can lead the unwary into believing that conglomerate formation is a rare phenomenon. Correction must be made for the later account by Jacques *et alia*, who proposed a frequency of around 5–10% for uncharged organic molecules and about two or three times higher for charged ones (both ionic and zwitterionic substances) [61]. These percentages arose from a detailed study of a sample of 1308 compounds described in the Beilstein Handbook.

Since the CSD has now reached over 750 000 entries it is likely to contain information on tens of thousands of conglomerate structures. Unfortunately, it is not yet possible to separate electronically these cases from the bulk data on Sohncke space groups. Very many of the latter crystals, of course, will have originated from sources other than racemic mixtures.

3.6.2 Enantiomer Resolution

A conglomerate is a self-resolved mixture of pure (+)- and pure (−) enantiomeric crystals and therefore these can be mechanically separated. The first such example was the classic separation of sodium ammonium (2*R*,3*R*)-(+)-tartrate **29** and its (2*S*,3*S*)-(−)-enantiomer **30** reported by Pasteur in 1848 [62]. A 2003 poll chose this as the most beautiful experiment in the history of chemistry [63].

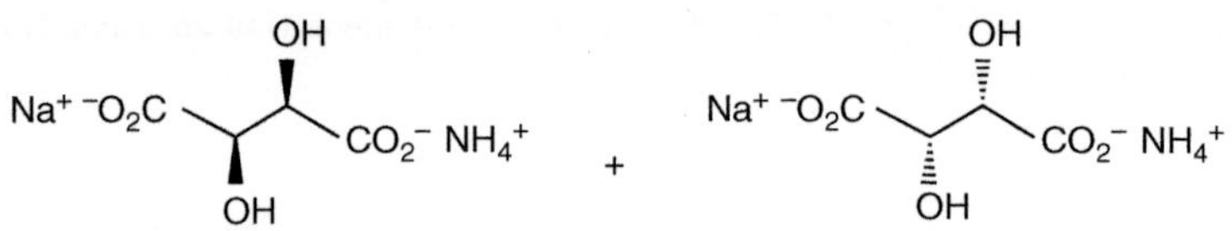

29 **30**

Crystallization of the racemic mixture produced a conglomerate in the form of the tetrahydrate crystals. Segregation was possible since these crystals have enantiomorphic hemihedral facets (visible under a microscope) that permit enantiomer separation using tweezers. Unfortunately this is a rare property. Alternatively, growth of sufficiently large crystals can allow their sign of optical rotation to be measured and, if so, then mechanical separation is still possible. However, this technique is exceptionally tedious, and has little practical value other than providing small amounts of chirally pure material for use in seeding experiments.

The favored separation technique uses an enantiomerically pure seed crystal and a supersaturated solution of a racemate (which is already known to crystallize as a conglomerate). This technique was first devised by Gernez, a student of Pasteur, in 1866 [64]. Thus the supersaturated solution of **29/30** deposited the (+) enantiomer (**29**)$\cdot(H_2O)_4$ when in contact with a (+) seed crystal, or (**30**)$\cdot(H_2O)_4$ if a (−)- seed were employed. If such a resolution is only partially successful, then a second or third pass will usually readily afford the pure enantiomer. The seeding method is widely applicable to conglomerates, although there are instances of it failing completely.

Jungfleish [65] realized that both enantiomers could be deposited simultaneously if both (+) and (−) seeds were added to remote parts of the same crystallization solution. Enantiomer crystals of up to 200 g mass could be obtained for the **29/30** system. This concept has been applied to a number of separations of industrial value. Zaugg has described methodology that affords around 13 g of each methadone **31** enantiomer per run [66]. Considerably larger scale industrial applications have included the antihypertensive L-α-methyldopa **32**, broadspectrum antibiotic chloramphenicol **33**, and food additive L-glutamic acid **34**. For many years racemic glutamic acid was produced from acrylonitrile using a nonstereospecific process, and then the L-isomer separated as above (13 000 tons per year).

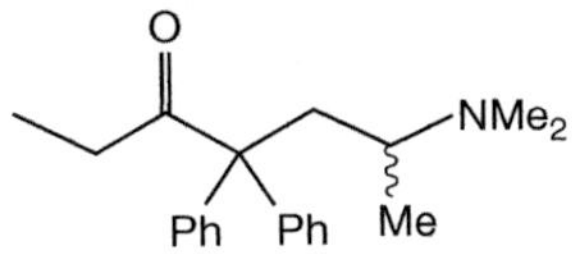

31

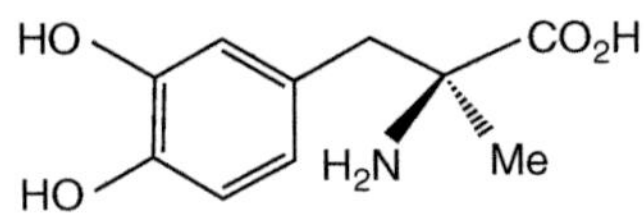

32

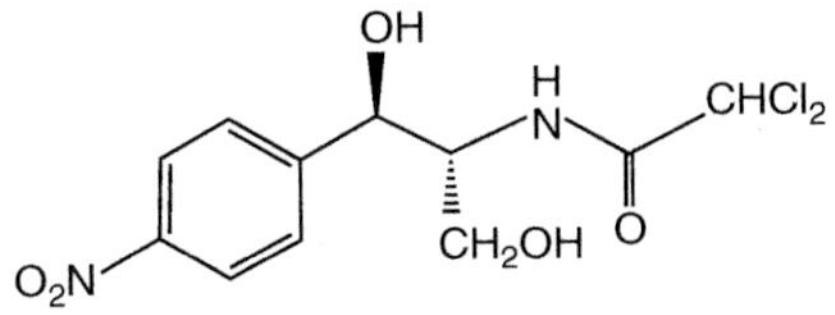

33

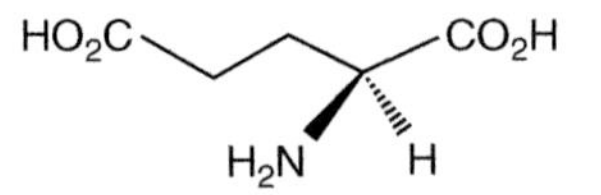

34

The chief drawback of the seeding method is that the principle is very simple but its practice can be extremely difficult to accomplish. Devising the appropriate experimental conditions can represent a major exercise in chemical engineering. Collet *et al.* have described several of these specialized cases in detail [60].

3.6.3 Increasing the Chiral Pool

The fundamental significance of conglomerate formation is that thermodynamics has already resolved the opposite enantiomers for you. Therefore the discovery of new conglomerate systems is highly important. Each one represents a formal enantiomer resolution, and they offer a major resource of enantiopure substances from outside the traditional chiral pool. It is therefore crucial to research and understand the characteristics of crystallization that result in conglomerate formation [67, 68].

It is now also realized that most molecules are potentially capable of yielding more than one crystal form. Hence merely changing the crystallization solvent may be sufficient to cause a switch from a racemic crystal to a conglomerate. One such case is the dialcohol **35**. Crystallization from benzene yields racemic crystals in $P2_1/c$, wherein partial enantiomer separation into layers has taken place (Figure 3.15). If dimethyl sulfoxide (DMSO) is used, however, then the inclusion compound $(\mathbf{35})_2$·(DMSO) is produced as a conglomerate in space group $P2_12_12_1$ (Figure 3.16). In this example, the hydrogen bonded $(\text{-O-H})_4$ cycles within the racemic crystal have been replaced by O-H···O=S interactions, thereby allowing complete enantiomer separation to occur [69].

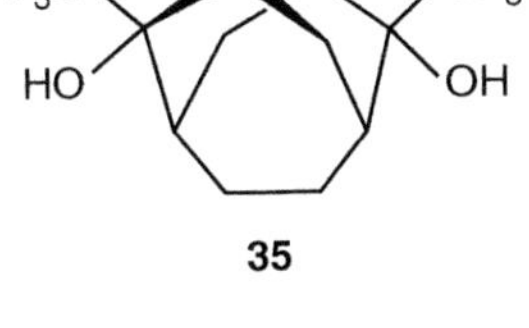

35

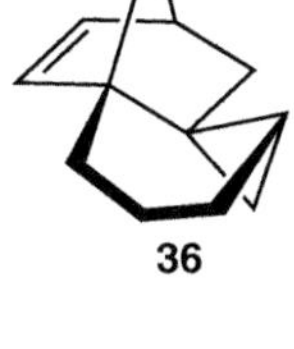

36

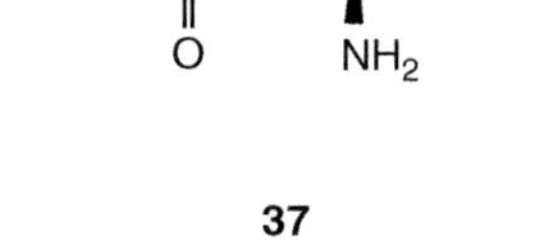

37

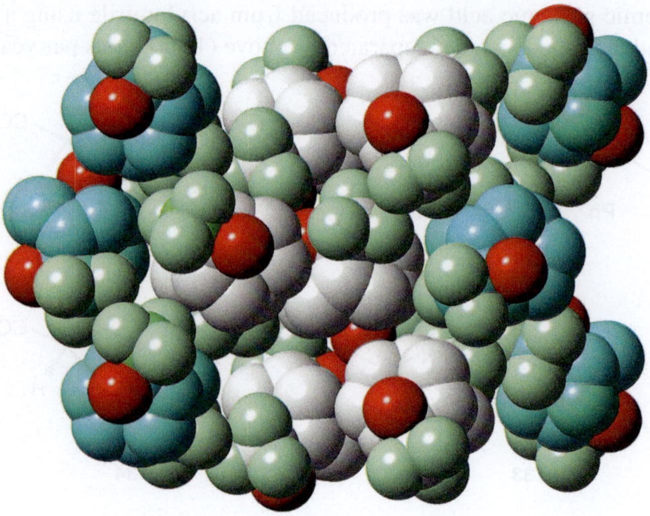

Figure 3.15 *Part of the crystal structure of racemic **35** projected on the* bc *plane and showing the layers edge on. Color code: C atoms of the opposite enantiomers are colored light blue or off white, O red and F light green. Hydrogen atoms are omitted for clarity. Refcode HITPEF [69]*

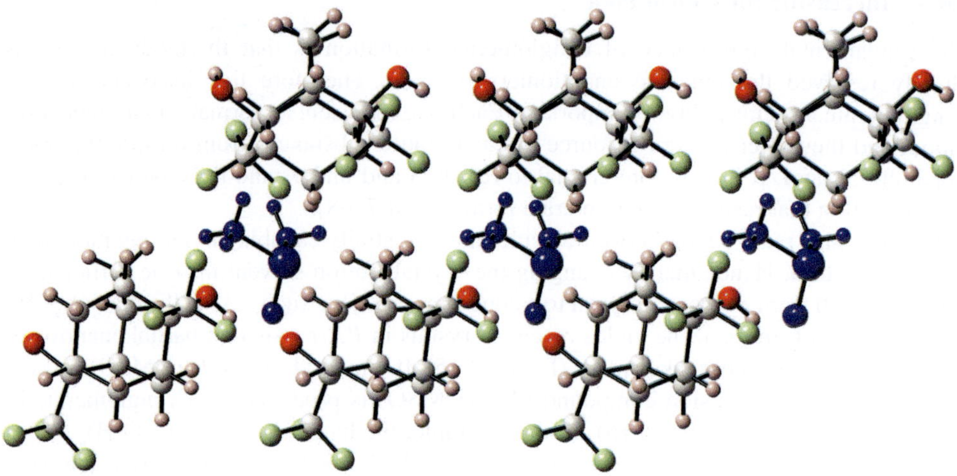

Figure 3.16 *Part of the $(\mathbf{35})_2\cdot$(DMSO) structure projected on the* ab *plane. Color code: the enantiomerically pure host C off white, O red, F light green and all dimethyl sulfoxide guest atoms dark blue. Refcode HITPIJ [69]*

Further variables are the temperature and pressure conditions under which crystallization occurs. If crystals of the hexagonal $P6_122$ and $P6_522$ urea inclusion crystals (section 3.2.1) are cooled, then the reduced thermal motion of the guests can result in adoption of the $P2_12_12_1$ space group [70]. This reversible process is a reminder that alternative crystal arrangements may be formed at different temperatures, and that these will include Sohncke space groups. In particular, the Pasteur sodium ammonium tartrate tetrahydrate conglomerate is only produced at temperatures below 300 K. Above this value, racemic monohydrate crystals (Scacchi's salt) are obtained in an achiral space group. Similarly, sodium potassium tartrate (Seignette's salt) crystallizes as a tetrahydrate conglomerate only below 267 K [71]. Relatively small changes in temperature can therefore cause major changes in solid state behavior. The effects of pressure have mainly been studied during solvothermal preparations of metallic complexes and metal organic framework compounds. These typically have achiral network structures.

Other techniques for growing crystals can also result in conglomerate formation. Seeding of an appropriate supercooled racemic melt can also cause precipitation of one specific enantiomer [60]. At least one case of conglomerate formation by sublimation is known. The racemic norbornenol **36** at 293 K and 20 Torr resulted in enantiomerically pure crystals in space group $P3_1$ [72].

The use of chiral additives, such as glucose or fructose, has been claimed to encourage conglomerate nucleation. However, inoculation by enantiopure particles from the laboratory atmosphere is difficult to eliminate as an alternative explanation [59]. More convincingly, Ostromisslensky seeded a supersaturated solution of sodium ammonium tartrate **29/30** with L-(−)-asparagine **37** and precipitated (+)-(**29**)·$(H_2O)_4$ [73]. The seed crystal that facilitates resolution is believed to be isomorphous with the enantiomer induced to crystallize.

Employment of enantiopure solvents during crystallization, and the use of nonsoluble chiral surfaces, to encourage single enantiomer growth also have much potential.

3.6.4 Chemical Modification

A significant obstacle to conglomerate formation is the strong tendency for opposite enantiomers of many racemic compounds to associate around inversion centers during their crystallization. If the techniques of section 3.6.3 are ineffective, then minor chemical modification of the substance may provide an alternative strategy.

For example, carboxylic acids commonly associate in the solid state by hydrogen bonding as infinite chains (catemers) or as centrosymmetric dimers (Figure 3.17). The ratio of catemer : dimer formation is around 7 : 93 for simple carboxylic acids that have no other strong competing interactions [74, 75]. Therefore it may be preferable to use a carboxylic acid salt or ester derivative in the search for new conglomerate crystals.

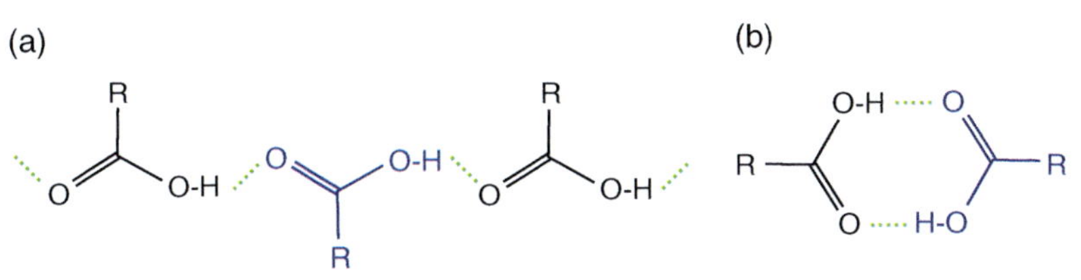

Figure 3.17 *Common hydrogen bonded motifs for racemic carboxylic acids, with the opposite enantiomers colored black or blue: (a) catemer, and (b) centrosymmetric dimer*

Other convenient minor modifications are converting an amine into its hydrochloride or hydrobromide salt, forming an ester from an alcohol, and so on. Reliable procedures for adding (and later removing) organic protecting groups are well documented [76]. The most systematic modification studies have been made on the amino acids. Some (for example threonine, glutamic acid, and asparagine) yield conglomerates directly. Most of the others follow suit if their salts or acyl derivatives are used instead [60]. The higher frequency of conglomerate formation amongst ionic and zwitterionic substances (section 3.6.1) results from the isotropic electrostatic charge surrounding an ion, as opposed to the anisotropic nature of hydrogen bonds and many other intermolecular attractions. The directionality imposed by these latter forces encourages formation of the centrosymmetric assemblies so frequently encountered in racemic crystals of neutral organic molecules.

Collet, Brienne and Jacques remarked that "it is nearly always possible to find a conglomerate among derivatives of a given compound to be resolved" [60]. It should be noted, however, that this optimism related specifically to their amino acid research. The situation for all the other classes of organic compounds is simply unknown at present. It is probable, however, that the frequencies for conglomerate formation found in section 3.6.1 are considerable underestimates of the true situation.

References

[1] International Tables for Crystallography, International Union of Crystallography, http://it.iucr.org (accessed April 9, 2016).

[2] M. D. Hollingsworth and K. D. M. Harris, Urea, Thiourea, and Selenourea, in D. D. MacNicol, F. Toda and R. Bishop (Eds.), *Comprehensive Supramolecular Chemistry*, Vol. 6, Pergamon, Oxford, Ch. 7, pp. 177–237 (1996).

[3] Inorganic Crystal Structure Database, FIZ Karlsruhe, www.fiz-karlsruhe.de/icsd.html (accessed April 3, 2016).

[4] Cambridge Structural Database (CSD), Cambridge Crystallographic Data Centre, www.ccdc.cam.ac.uk (accessed April 3, 2016).

[5] E. Pidcock, J. A. Chisholm, P. A. Wood, *et al.* (2012), The Cambridge Structural Database System and its Applications in Supramolecular Chemistry and Materials Design, in J. W. Steed and P. A. Gale (Eds.), *Supramolecular Chemistry: From Molecules to Nanomaterials*, Vol. 6, John Wiley & Sons, Ltd., Chichester, pp. 2927–2946.

[6] Worldwide Protein Data Bank: www.wwpdb.org (accessed April 3, 2016).

[7] Cambridge Structural Database (Statistics as at 16 February 2015), http://www.ccdc.cam.ac.uk/products/csd/statistics (accessed April 3, 2016).

[8] S. L. Price, Computed crystal energy landscapes for understanding and predicting organic crystal structures and polymorphism, *Acc. Chem. Res.*, **42**, 117–126 (2009).

[9] M. A. Neumann, F. J. J. Leusen and J. Kendrick, A major advance in crystal structure prediction, *Angew. Chem. Int. Ed.*, **47**, 2427–2430 (2008).

[10] A. J. Cruz-Cabeza, S. Karki, L. Fábián, *et al.*, Predicting stoichiometry and structure of hydrates, *Chem. Commun.*, **46**, 2224–2226 (2010).

[11] E. Batisai, M. Lusi, T. Jacobs and L. J. Barbour, A mechanochemically synthesised solid solution enables engineering of the sorption properties of a Werner clathrate, *Chem. Commun.*, **48**, 12171–12173 (2012).

[12] I. Y. H. Chan, R. Bishop, D. C. Craig, M. L. Scudder and W. Yue, 8-Methyl-5-methylene-2-oxotricyclo[5.3.1.$1^{3,9}$]dodecan-*endo*-8-ol, *Acta Crystallogr., Sect. E*, **63**, o841 (2008).

[13] A. I. Kitaigorodskii (1973) *Molecular Crystals and Molecules*, Academic Press, New York.

[14] S. C. Wallwork and H. M. Powell, The crystal structure of the α form of quinol, *J. Chem. Soc., Perkin Trans. II*, 641–646 (1980).

[15] R. Bishop and M. L. Scudder, Multiple molecules in the asymmetric unit ($Z'>1$) and the formation of false conglomerate crystal structures, *Cryst. Growth Des.*, **9**, 2890–2894 (2009).
[16] J. W. Steed, University of Durham, UK, Z' web site, http://www.zprime.co.uk (accessed April 3, 2016).
[17] K.M. Anderson, K. Afarinkia, H. Yu, *et al.*, When $Z'=2$ is better than $Z'=1$ – supramolecular centrosymmetric hydrogen-bonded dimers in chiral systems, *Cryst. Growth Des.*, **6**, 2109–2113 (2006).
[18] I. Bernal, J. Cai, S. S. Massoud, *et al.*, The phenomenon of kryptoracemic crystallization. Part 1. Counterion control of crystallization pathway selection. *J. Coord. Chem.*, **38**, 165–181 (1996).
[19] G. A. Morales and F. R. Fronczek, A kryptoracemic hydroperoxide, *Acta Crystallogr., Sect. C*, **52**, 1266–1268 (1996).
[20] L. Fábián and C. P. Brock, A list of organic kryptoracemates, *Acta Crystallogr., Sect. B*, **66**, 94–103 (2010).
[21] S. L. Fomulu, M. S. Hendi, R. E. Davis and K. A. Wheeler, Structural studies of enantiomers, racemates, and quasiracemates. 2(2,4,5-Trichloroanilino)propanoic acid and 2-(2,4,5-trichloropenoxy)propanoic acid, *Cryst. Growth Des.*, **2**, 637–644 (2002).
[22] M. S. Hendi, P. Hooter, R. E. Davis, *et al.*, Structural studies of enantiomers, racemates, and quasiracemates: *N*-(4-methylbenzoyl)methylbenzylamine and *N*-(4-nitrobenzoyl)methylbenzylamine, *Cryst. Growth Des.*, **4**, 95–101 (2004).
[23] Q. Zhang and D. P. Curran, Quasienantiomers and quasiracemates: new tools for identification, analysis, separation, and synthesis of enantiomers, *Chem. Eur. J.*, **11**, 4866–4880 (2005).
[24] L. Pasteur, Nouvelles recherches sur les relations qui peuvent exister entre la forme cristalline, la composition chimique et le Phénomène rotatoire moléculaire, *Ann. Chim. Phys.*, **38**, 437–483 (1853).
[25] K. A. Wheeler, R. C. Grove, R. E. Davis and W. S. Kassel, Rediscovering Pasteur's quasiracemates, *Angew. Chem. Int. Ed.*, **47**, 78–81 (2008).
[26] G. P. Stahly, Diversity in single- and multiple-component crystals. The search for and prevalence of polymorphs and cocrystals, *Cryst. Growth Des.*, **7**, 1007–1026 (2007).
[27] J. Bernstein (2002) *Polymorphism in Molecular Crystals*, Oxford University Press, Oxford.
[28] J. Bernstein, Polymorphism – a perspective, *Cryst. Growth Des.*, **11**, 632–650 (2011).
[29] G. R. Desiraju (1989) *Crystal Engineering: The Design of Molecular Solids*, Elsevier, Amsterdam, Ch. 10, pp. 285–301.
[30] S. L. Price, Why don't we find more polymorphs? *Acta Crystallogr., Sect. B*, **69**, 313–328 (2013).
[31] A. Vega, O. Donoso-Tauda, A. Ibañez and C. A. Escobar, Five bicyclo[3.3.0]octa-2,6-dienes, *Acta Crystallogr., Sect. C*, **64**, o199 (2008).
[32] J. Gao, M. M. Bhadbhade and R. Bishop, Different crystal structures of a rich hydrogen bond acceptor compound resulting from alternative C–H···O and orthogonal C=O···C=O molecular interaction patterns, *CrystEngComm*, **14**, 138–146 (2012).
[33] I. Y. H. Chan, Solvent Control of Enantiomeric Ordering, Self-Resolution and Resolution, Ph. D. thesis, University of New South Wales, Australia (2010).
[34] I. Y. H. Chan, V. T. Nguyen, R. Bishop, *et al.*, The design of "awkward" molecules expected to yield multiple crystal forms, *Cryst. Growth Des.*, **10**, 4582–4589 (2010).
[35] J. Gao, M. M. Bhadbhade and R. Bishop, Polymorphic crystals formed by an achiral diol under ambient conditions, *Cryst. Growth Des.*, **12**, 5746–5756 (2012).
[36] A. T. Ung, R. Bishop, D. C. Craig, *et al.*, Prediction and structure of polymorphic lattice inclusion compounds of 2,7-dimethyltricyclo[4.3.1.0^{3,8}]undecane-*syn*-2,*syn*-7-diol, *Tetrahedron*, **49**, 639–648 (1993).
[37] J. Ashmore, R. Bishop, D. C. Craig and M. L. Scudder, Comparison of the crystal structures of concomitant pseudodimorphs formed between a diquinoline host and *d*-chloroform guest, *Mendeleev Commun.*, **13**, 144–146 (2003).
[38] E. Weber and H.-P. Josel, A proposal for the classification and nomenclature of host-guest-type compounds, *J. Inclusion Phenom.*, **1**, 79–85 (1983).
[39] J. L. Atwood, J. E. D. Davies and D. D. MacNicol (Eds.), *Inclusion Compounds*, Vols. 1–3, Academic Press, London (1984).

[40] J. L. Atwood, J. E. D. Davies and D. D. MacNicol (Eds.), *Inclusion Compounds*, Vols. 4–5, Oxford University Press, Oxford (1991).
[41] D. D. MacNicol, F. Toda and R. Bishop (Eds.), *Comprehensive Supramolecular Chemistry*, Vol. **6**, Pergamon, Oxford (1996).
[42] F. H. Herbstein (2005) *Crystalline Molecular Complexes and Compounds: Structures and Principles*, Oxford University Press, Oxford.
[43] R. Bishop, Synthetic Clathrate Systems, in J. W. Steed and P. A. Gale (Eds.), *Supramolecular Chemistry: From Molecules to Nanomaterials*, Vol. 6, John Wiley & Sons, Ltd., Chichester, pp. 3033–3056 (2012).
[44] J. Gao, M. M. Bhadbhade and R. Bishop, Solvent–guest control of two extremely similar tetrahydrofuran inclusion structures, *Acta Crystallogr., Sect. B*, **70**, 126–131 (2014).
[45] J. Ashmore, R. Bishop, D. C. Craig and M. L. Scudder, The effect of chlorine substitution on the inclusion properties of a diquinoline host molecule, *CrystEngComm*, **4**, 194–198 (2002).
[46] H. P. Stahl, The Problem of Drug Interactions with Incipients, in D. D. Breimer (Ed.), *Towards Better Safety of Drugs and Pharmaceutical Products*, Elsevier – North Holland Biomedical Press, Amsterdam, pp. 265–280 (1980).
[47] G. R. Desiraju, Hydration in organic crystals: prediction from molecular structure, *J. Chem. Soc., Chem. Commun.*, 426–428 (1991).
[48] A. L. Gillon, N. Feeder, R. J. Davey and R. Storey, Hydration in molecular crystals – a Cambridge Structural Database analysis, *Cryst. Growth Des.*, **3**, 663–673 (2003).
[49] L. Infantes, L. Fábián and W. D. S. Motherwell, Organic crystal hydrates: what are the important factors for formation, *CrystEngComm*, **9**, 65–71 (2007).
[50] M. Mascal, L. Infantes and J. Chisholm, Water oligomers in crystal hydrates – what's news and what isn't?, *Angew. Chem. Int. Ed.*, **45**, 32–36 (2006).
[51] R. Bishop, D. C. Craig and M. L. Scudder, Preparation and crystal structure of a twin twist-boat bicyclo[3.3.1]nonane monohydrate, *J. Chem. Soc., Perkin Trans. 1*, 1473–1477 (1989).
[52] S. P. Kelley, L. Fábián and C. P. Brock, Failures of fractional crystallization: ordered co-crystals of isomers and near isomers, *Acta Crystallogr., Sect. B*, **67**, 79–93 (2011).
[53] T. Sakurai, On the refinement of the crystal structures of phenoquinone and monoclinic quinhydrone, *Acta Crystallogr., Sect. B*, **24**, 403–412 (1968).
[54] Ö. Almarsson and M. J. Zawarotko, Crystal engineering of the composition of pharmaceutical phases. Do pharmaceutical co-crystals represent a new path to improved medicines? *Chem. Commun.*, 1889–1896 (2004).
[55] C. B. Aakeröy and P. D. Chopade, Cocrystals: Synthesis, Structure and Applications, in P. A. Gale and J. W. Steed (Eds.), *Supramolecular Chemistry: From Molecules to Nanomaterials*, Vol. 6, John Wiley & Sons, Ltd., Chichester, pp. 2975–2992 (2012).
[56] The Select Committee on GRAS Substances (SCOGS) database provides details of the current 378 GRAS substances – see http://www.accessdata.fda.gov/scripts/fdcc/?set=SCOGS (accessed April 9, 2016). Further information on GRAS and the Federal Food, Drug, and Cosmetic Act is available at http://www.fda.gov (accessed April 9, 2016).
[57] S. G. Fleischman, S. S. Kuduva, J. A. McMahon, *et al.*, Crystal engineering of the composition of pharmaceutical phases: multiple-component crystalline solids involving carbamazepine, *Cryst. Growth Des.*, **3**, 909–919 (2003).
[58] S. Mahapatra, T. S. Thakur, S. Joseph, *et al.*, New solid state forms of the anti-HIV drug efavirenz. Conformational flexibility and high *Z'* issues, *Cryst. Growth Des.*, **10**, 3191–3202 (2010).
[59] R. M. Secor, Resolution of optical isomers by crystallization procedures, *Chem. Rev.*, **63**, 297–309 (1963).
[60] A. Collet, M.-J. Brienne and J. Jacques, Optical resolution by direct crystallization of enantiomer mixtures, *Chem. Rev.*, **80**, 215–230 (1980).
[61] J. Jacques, A. Collet and S. H. Wilen (1981) *Enantiomers, Racemates, and Resolutions*, John Wiley & Sons, Inc., New York.
[62] L. Pasteur, Recherches sur les relations qui peuvent exister entre la forme cristalline, la composition chimique et les sens de la polarisation rotatoire, *Ann. Chim. Phys.*, [3] **24**, 442–459 (1848).

[63] M. Freemantle, Chemistry at its most beautiful, *Chem. Eng. News*, **81**(34), 27–30 (2003).
[64] D. Gernez, Séparation des tartrates droits et des tartrates gauche à l'aide des solutions saturées, *Compt. Rend. Acad. Sci.*, **63**, 843–888 (1866).
[65] E. Jungfleisch, Decomposition of racemic acid, *J. Pharm. Chim.*, [5], **5**, 346–352 (1882); *J. Chem. Soc. Abstr.*, 1882, **42**, 602–604.
[66] H. E. Zaugg, A mechanical resolution of *dl*-methadone base, *J. Am. Chem. Soc.*, **77**, 2910 (1955).
[67] K. Kinbara, Y. Hashimoto, M. Sukegawa, *et al.*, Crystal structures of the salts of chiral primary amines with achiral carboxylic acids: recognition of the commonly-occurring supramolecular assemblies of hydrogen-bond networks and their role in the formation of conglomerates, *J. Am. Chem. Soc.*, **118**, 3441–3449 (1996).
[68] G. Levilain and G. Coquerel, Pitfalls and rewards of preferential crystallization, *Cryst. Eng. Comm.*, **12**, 1983–1992 (2010).
[69] R. Bishop, G. A. Downing, D. C. Craig and M. L. Scudder, Enantiomeric self-resolution through dimethylsulfoxide complexation, *J. Incl. Phenom. Mol. Rec. Chem.*, **31**, 145–160 (1998).
[70] Y. Chatani, Y. Taki and H. Tadokoro, Low temperature form of urea adducts with *n*-paraffins, *Acta Crystallogr., Sect. B*, **33**, 309–311 (1977).
[71] E. L. Eliel, S. H. Wilen and L. N. Mander (1994) *Stereochemistry of Organic Compounds*, John Wiley & Sons, Inc., New York.
[72] L. A. Paquette and C. J. Lau, An example of spontaneous resolution by sublimation, *J. Org. Chem.*, **52**, 1634–1635 (1987).
[73] I. Ostromisslensky, Untersuchungen im gebiete der spiegelbildisomerie, *Ber. Deutsch. Chem. Ges.*, **41**, 3035–3046 (1908).
[74] L. Leiserowitz, Molecular packing modes. Carboxylic acids., *Acta Crystallogr., Sect. B*, **32**, 775–802 (1976).
[75] T. R. Shattock, K. K. Arora, P. Vishweshwar and M. J. Zaworotko, Hierarchy of supramolecular synthons: persistent carboxylic acid····pyridine hydrogen bonds in cocrystals that also contain a hydroxyl moiety, *Cryst. Growth Des.*, **8**, 4533–4545 (2008).
[76] P. G. M. Wuts and T. W. Greene (2006) *Greene's Protective Groups in Organic Synthesis*, 4th edn., John Wiley & Sons, Inc., New York.

4

Complexity of Supramolecular Assemblies

Jonathan A. Kitchen and Philip A. Gale

4.1 Introduction

4.1.1 Supramolecular Chirality

Supramolecular chirality has gained significant attention since the mid-2000s [1–6]. The phenomenon arises from the nonsymmetric spatial arrangement of constituents in self-assembling architectures including the assembly of both chiral and achiral components through noncovalent interactions. Elaborate architectures have been constructed using a variety of noncovalent interactions – including hydrogen bonding [7], π stacking [8], halogen bonding [9], coordination complexation [10] and combinations thereof [11–16]. This has resulted in supramolecular self-assembling systems with wide-ranging functionalities and hence applications from the biological arena to advanced nanotechnology and beyond.

The aim of this chapter is to introduce, through appropriate examples, some of the concepts of chiral metallo-supramolecular assemblies illustrating the depth and breadth that this research area has to offer. Examples will focus on inorganic aspects of supramolecular self-assembly where dynamic coordination chemistry approaches are used to generate larger more complex chiral architectures. Both transition-metal-based assemblies and *f*-block-directed assemblies of a broad range of structures will be discussed including mononuclear systems, di- and trinuclear helicates, tetrahedral helicates (cages), and other higher nuclearity architectures/clusters. By introducing these examples we will highlight particularly important aspects of supramolecular chirality such as self-assembly processes in general, transfer of chirality through chirogenesis and the amplification of chirality, and

Chirality in Supramolecular Assemblies: Causes and Consequences, First Edition. Edited by F. Richard Keene.

discuss the consequences this has for developing new functional and chiral materials. Initially, brief introductions to self-assembly, and supramolecular chirogenesis will be presented before select examples of assembly of achiral components into metallo-supramolecular architectures will be discussed, before moving on to enantioselective synthesis of supramolecular architectures.

A number of review articles covering the field of supramolecular chirality have been published in recent years and the reader is directed to these for more specific and in-depth discussion [11–25].

4.1.2 Self-Assembly

Self-assembly [26] is a fundamental process on the molecular scale responsible for the formation of systems of molecules and ultimately living systems. From the coalescence of the lipid bilayer cell membrane to the assembly of the DNA double helix there is perhaps no more important principle for molecular organization. However, self-assembly can operate on many different levels of scale [27] extending even into the world of design and architectural production [28]. Self-assembly has been defined in the context of this chapter as the process in which a supramolecular species forms spontaneously from its components [29]. Whitesides and Boncheva have highlighted that successful self-assembly is determined by five characteristics of the system [27]. These are the self-assembling components themselves, the interactions between them, reversibility (or adjustability) allowing for error correction in the formation of an assembly, environment – in solution or at an interface and finally mass transport and agitation – where molecules must be able to interact and therefore must be mobile.

4.1.3 Supramolecular Chirogenesis

Supramolecular chirogenesis combines supramolecular self-assembly and molecular chirality and in essence is the generation or transfer of asymmetry through noncovalent interactions [2, 30, 31]. Given its prevalence in naturally occurring systems (e.g. the DNA double helix and the α-helix structures of some proteins) there is, understandably, significant research effort devoted to understanding the factors that control the phenomenon in synthetic systems. Given that noncovalent interactions (e.g. hydrogen bonding, metal coordination, π-based interactions) underpin chirogenesis, it is not surprising that factors influencing these interactions go on to effect the overall chirality of induction/transfer processes. Factors such as interaction strength (e.g. metal-ligand bond strength), electronic properties (ligand substituents), stoichiometry, steric effects, geometry preferences (from ligands, metals and anions) as well as external factors (e.g. polarity and temperature) all play a pivotal role. Perhaps the most studied mode of self-assembly from the perspective of inorganic chemistry has been coordination chemistry where the assembly between ligand and metal can result in the formation of a chiral metal-containing complex. With metal-based systems being the focal point of this chapter, various attempts to prepare chiral metallo-supramolecular self-assemblies will be discussed in which the primary interaction under discussion is the M-L coordination bond.

4.2 Generating Supramolecular Chirality through Assembly of Achiral Components

Metal-ligand coordination has played an important role in the development of self-assembled architectures over the last few decades. The dynamic nature of the coordination bond means that metal-ligand interactions are ideal building blocks for the construction of many interesting, and often potentially chiral, architectures exhibiting a range of functionalities.

Throughout this chapter we will attempt to demonstrate the breadth of metal-based supramolecular chiral systems through choice of appropriate examples to highlight some of the key concepts in this vibrant field as well as introduce some of the design paradigms used by various researchers.

4.2.1 Supramolecular Chirality – Metallo-Helicates

Helical structures have captured the attention of supramolecular chemists for many years [10, 32–39]. The similarity of these complexes to self-assembled biological systems (e.g. DNA) has resulted in significant research efforts into the targeted formation of helical architectures through self-assembly via a coordination chemistry approach. The first report of such metallo-helicates was by Lehn in 1987, where he described the spontaneous assembly of double-stranded helicates from covalently linked 2,2′-bipyridyl ligands and Cu^{+} [40]. In these systems the specific combination of ligand design and the tetrahedral coordination geometry of the copper ion resulted in the formation of dinuclear and trinuclear double-stranded helicates where the organic ligands wrapped around the ions defining the helical axis. Since this discovery there has been an explosion of interest in the development of metallo-helicates with a variety of ligand systems, metal ions, nulcearities and types of helicate (e.g. linear versus circular) [18, 19]. Chirality is an inherent characteristic of helicates that exist in either a right-handed (*P*, or *plus*) or left-handed (*M*, or *minus*) form depending on the rotation of the helix (clockwise or anticlockwise – Scheme 4.1). Helical structures can be single-, double-, triple- or quadruple-stranded and can exist as linear or circular assembles. In the following section we will discuss a broad miscellany of metallo-helicates from single stranded through to larger self-assemblies, demonstrating the depth and breadth of this growing research field.

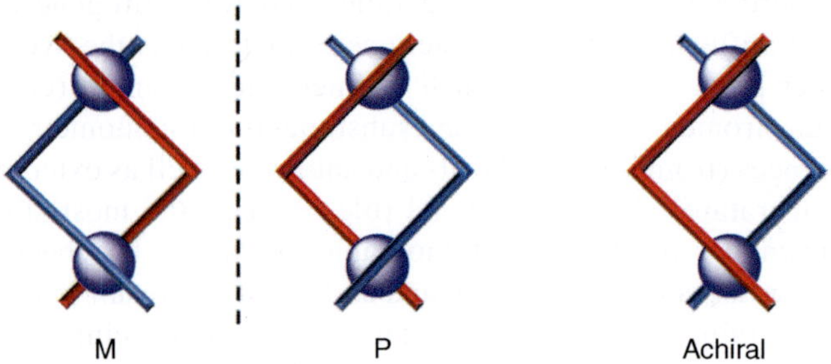

Scheme 4.1 *Diagramatic representation of M and P helicates (related by mirror plane) and the achiral mesocate*

4.2.1.1 Single Stranded Helicates

Reports of single-stranded helicates are not as frequent as those for double- and triple-stranded helicates, but nonetheless they result in constrained systems with possible chiral arrangements. Frequently, but certainly not always, dinuclear single-stranded helicates are formed when there are significantly short contacts between the metal centres. A particularly elegant example of this was reported by Ward and coworkers using pyridyl/pyrazolyl ligand systems (Figure 4.1) [41].

In this study these researchers developed three ligands (**1–3**) and studied their coordination behaviour with Ni^{2+}, Cu^{2+}, Cd^{2+} and Ag^{+}. They demonstrated that the Cu and Ni systems gave dinuclear double-stranded helicates whilst the Ag^{+} complexes were single-stranded helicates with short Ag···Ag interactions (Figure 4.1). The solid-state complex exists as a racemic mixture (a commonly observed phenomenon when using achiral building blocks to construct helicates).

Circular single-stranded helicates were developed by Piguet and coworkers when examining the disassembly of dinuclear triple-stranded lanthanide helicates using amide-linked

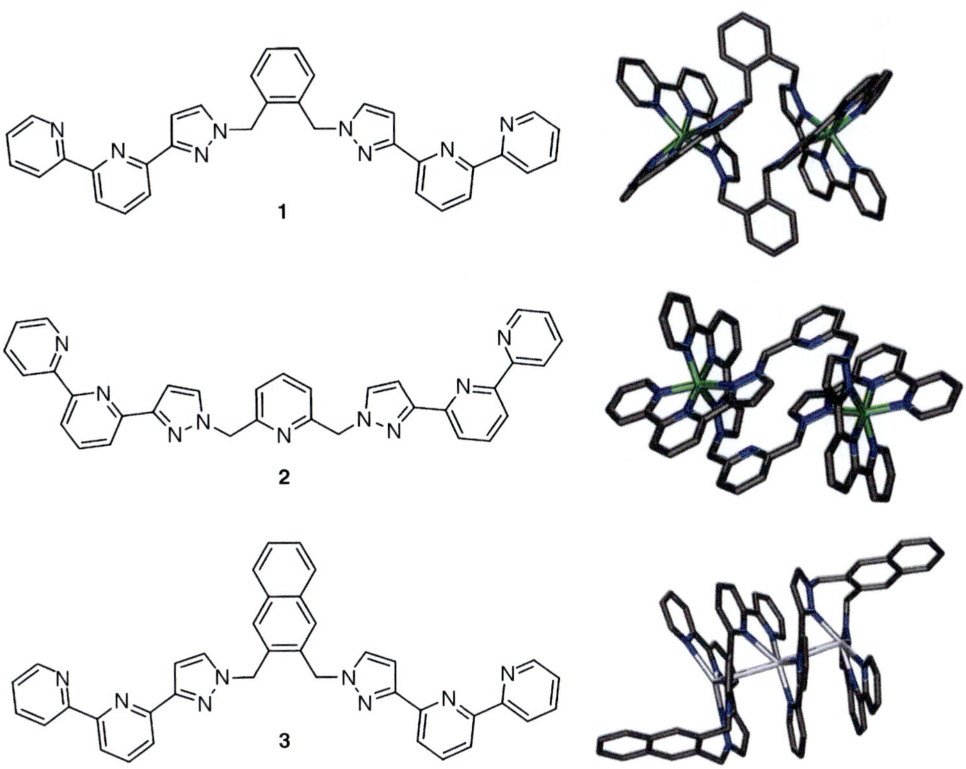

Figure 4.1 *Ligands **1–3** used by Ward and coworkers to construct helicates and mesocates. Ni^{2+} helicate $[Ni_2(\mathbf{1})_2]$ (top), Ni^{2+} mesocate of $[Ni_2(\mathbf{2})_2]$ (centre) and Ag^{+} single-stranded helicate of $[Ag_3(\mathbf{3})_2]$ (bottom) [41]*

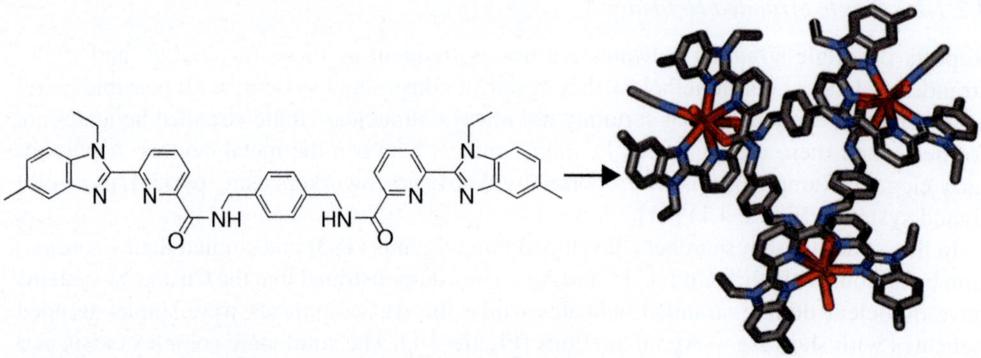

Figure 4.2 *Ligand system used by Piguet and coworkers to generate lanthanide circular single-stranded helicates. Structure of the Eu^{3+} helicate [42]*

benzimidazole ligands (Figure 4.2) [42]. These helicates were trinuclear Eu^{3+} complexes with three ligands assembled around the Eu^{3+} centres and the remainder of the coordination sphere completed by solvent molecules and anions. Lanthanide-based helicates will be discussed in more detail in the triple-stranded helicate section.

4.2.1.2 Double Stranded Helicates

Double-stranded helicates hold a particular fascination within the metallo-supramolecular research community, and since the first report of dinuclear double-stranded helicates by Lehn[40] there have been many examples of such metallo-helicates formed from a multitude of different ligands, and using a variety of different metal ions. Double-stranded helicates have been the subject of reviews [33, 43–47] since the mid-2000s, which highlight the breadth of research within this field. In addition to further developing the complexity of double-stranded helicates, there has recently been a move towards exploring a variety of applications for double-stranded helical architectures, including asymmetric catalysis, medical applications, anion binding, magnetically interesting compounds and luminescent systems [48–51].

Double-stranded dinuclear helicates are some of the most commonly studied helical supramolecular systems. For a system truly to be a helical structure the ligand must coordinate two different metal centres whilst adopting a twisted conformation to give the homochiral helicate ($\Delta\Delta$ or $\Lambda\Lambda$). If the ligand coordinates two metal centres but fails to adopt the twisted conformation, the system is achiral and said to be a meso-helicate ($\Delta\Lambda$ or $\Lambda\Delta$). Control over which of the two coordination modes is adopted is not well understood and it is only recently that Rice and coworkers have developed a ligand system that selectively adopts either a helicate or a meso-helicate structure dependent upon the size of the metal ion or the steric bulk of the ligands [52]. These researchers developed three bipy-thiazole ligands (**4–6**, Figure 4.3) with different substituents appended to the linking group.

When complexed with Cd^{2+}, Co^{2+} and Fe^{2+} they found that the steric bulk of the substituent groups influenced the nature of the dinuclear species formed. Ligand **4**, with the least steric bulk, formed helicate structures with both Cd^{2+} and Fe^{2+}, whilst **5**, which is more bulky than **4**,

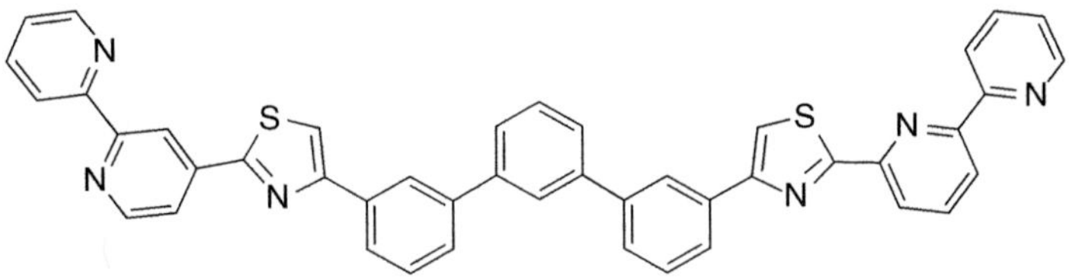

4

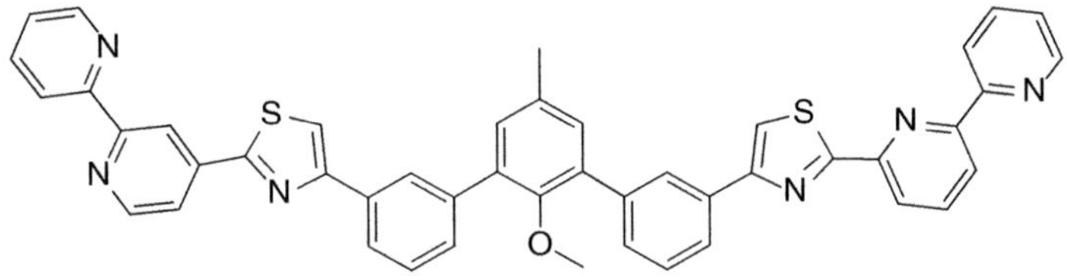

5

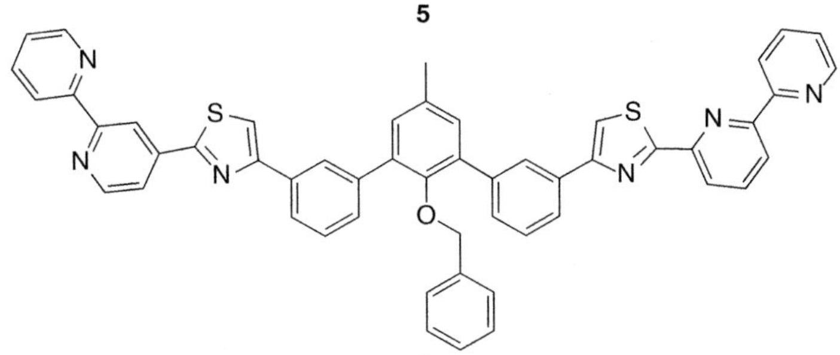

6

Figure 4.3 *Thiazole containing ligands **4–6** used by Rice and coworkers for the formation of dinuclear helicates [52]*

forms a double-stranded helicate with Cd^{2+} but forms a dinuclear double meso-helicate with the smaller Co^{2+} ion, postulated to be a result of the steric interactions increasing as the smaller cation was used. To test steric control they complexed the bulky phenyl-substituted ligand **6** with Cd^{2+} and found that the system adopted the meso-helicate structure, despite $[Cd_2(\mathbf{4})_2]$ and$[Cd_2(\mathbf{5})_2]$ giving double-stranded helicates (Figure 4.4). From this, the authors were able to suggest that by including sterically demanding interactions into the systems they were able to influence the resultant metallo-supramolecular assembly.

Another case of metal-controlled helicate formation was described by Maeda and coworkers who used rigid tetrapyrrole-based ligands (**7–10**, Figure 4.5) [53].

In order for a helicate to form, the two coordination pockets of the ligand need to be capable of linking two metal centres, rather than both pockets binding to only one metal. In this case the ability of the tetrapyrrole ligands to do both was exploited to give metal-selective control of helicate formation. The principle built on the work of Dolphin and coworkers who found that ligands of this type can form dinuclear double-stranded helicates when reacted with tetrahedral metals [54–58]. Maeda *et al.* used substituted tetrapyrroles with aromatic groups appended to the backbone of the ligand system. When reacted with Zn^{2+} the result was a [2+2] double-stranded helical system where the zinc ions adopted tetrahedral geometries (Figure 4.5). Further expanding the ligands such that two tetrapyrrole units were linked

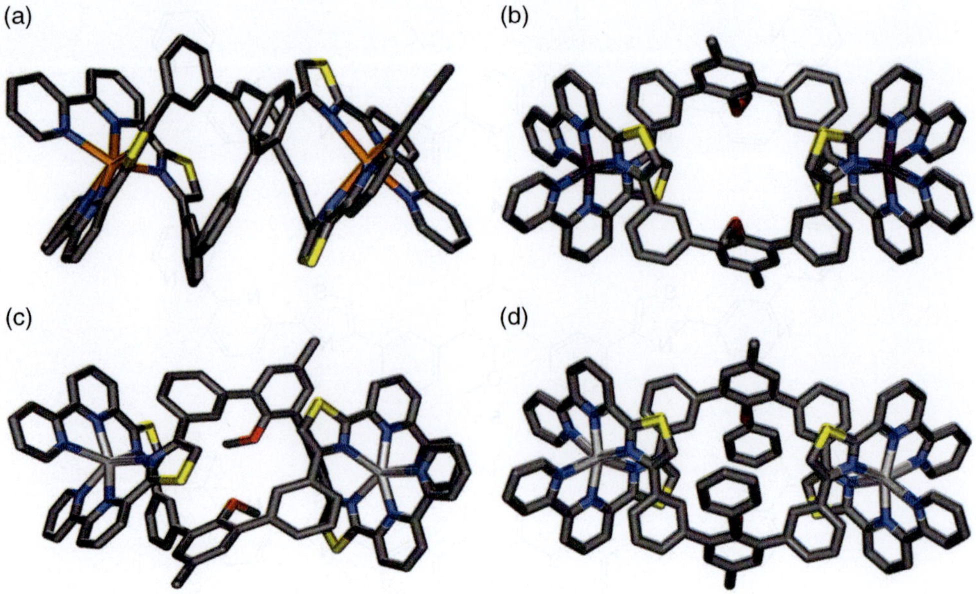

Figure 4.4 *Helicates (**a** and **c**) and mesocates (**b** and **d**) formed using ligands **4–6** [52]*

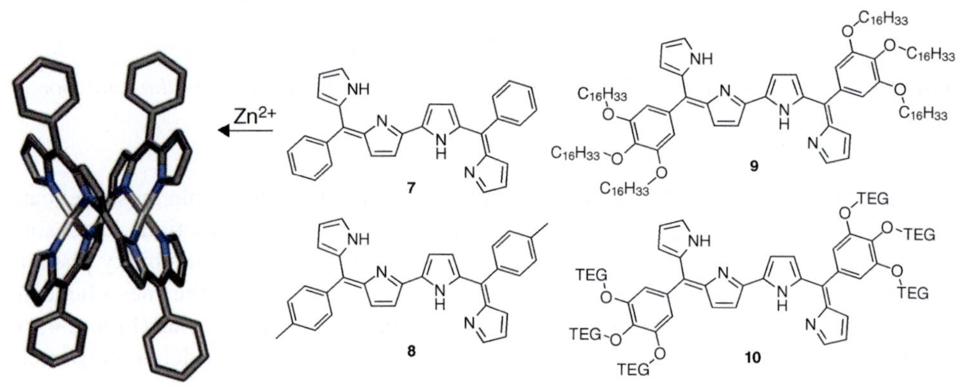

Figure 4.5 *Tetrapyrrole based ligands **7–10** used by Maeda and coworkers to generate Zn^{2+} helicates (left) [53]*

(through the aromatic substituents) into a large macrocycle gave a system that responded differently to different metals. When Zn^{2+} was used a Zn-bridged [2+4]-type dimer resulted whereas when metals that adopt square planar geometries (Ni^{2+} and Cu^{2+}) were used the structures adopted were [1+2] type complexes. In these studies it was revealed that the Zn-bridged [2+4]-type dimer exhibited thermally responsive synchronized spring motions, highlighting that helical motifs may be used in new advanced, switchable materials.

Metal ion-dependent interconversion between a tetranuclear grid and a dinuclear double-stranded helicate was shown by Lehn and coworkers [59]. They reported a system that formed a tetra-copper(II) grid with Cu^{2+} ions and a di-copper(I) double-stranded helicate

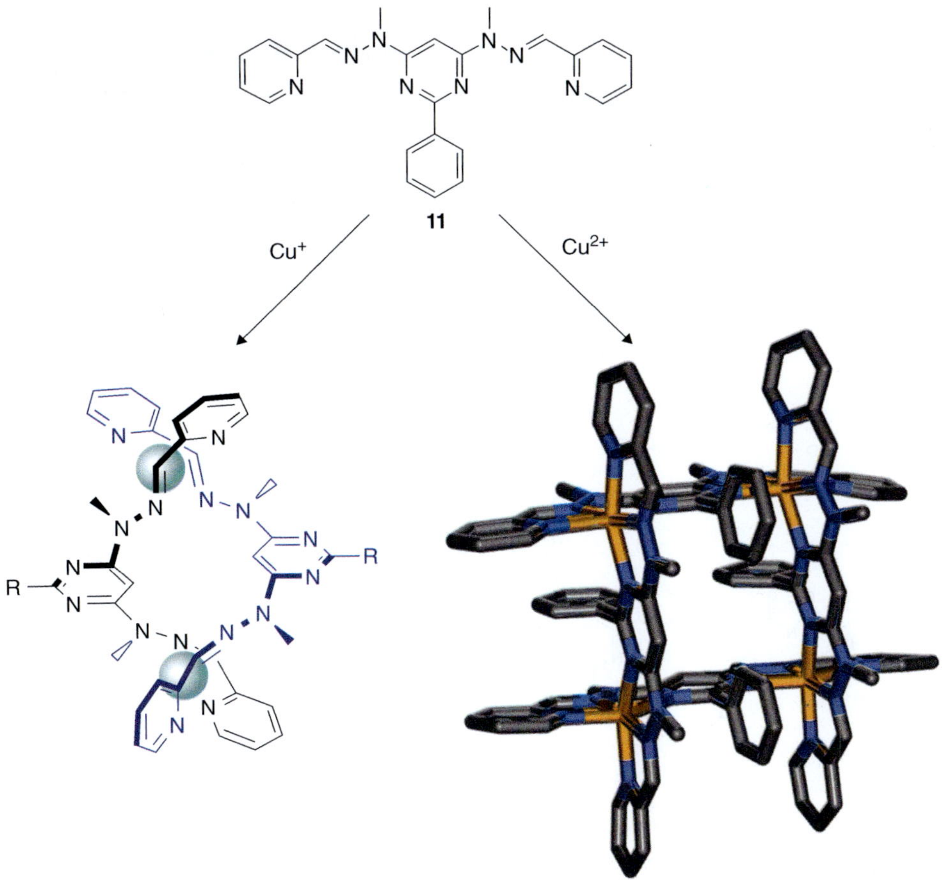

Figure 4.6 *Oxidation state control over product formation demonstrated by Lehn and coworkers [59]*

with Cu^{+} using ligand **11**. The system was dynamic and switchable between the grid and helicate on reduction or oxidation of the copper metal centres. The oxidation state switching was attributed to both the coordination geometry preferences of Cu^{2+} vs. Cu^{+} and the ability of the ligand to adopt either a bis(bidentate) (for helicate) or a bis(tridentate) (grid) coordination mode (Figure 4.6).

Metal ion helicate conformational control is not the only mechanism by which to manipulate the structure of dinuclear double-stranded helicates. Whilst undertaking anion binding research for anion extraction, Plieger and coworkers found that anions altered the structures of di-copper(II) double-stranded helicates constructed from salicylaldoxime ligand (**12**) [60–62].

These ligands formed helical structures with anions in a central cavity that bridged the Cu^{2+} centres (for ClO_4^-, NO_3^-, SO_4^{2-} and BF_4^-) and a metal-only complex when OAc^- was used. The encapsulated anion dictates the Cu···Cu distance in these structures and hence alters the helical twist angle (Figure 4.7) giving the structures a more helical arrangement.

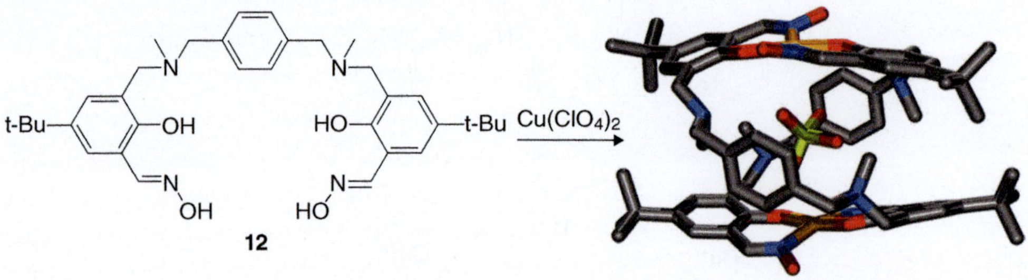

Figure 4.7 *Anion influenced helicate formation demonstrated by Pleiger and coworkers [60]*

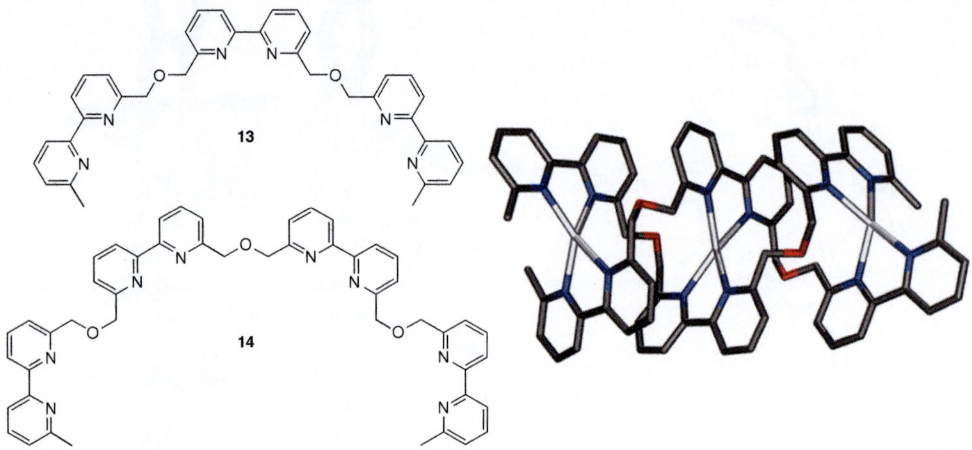

Figure 4.8 *Ligands used by Lehn and coworkers to generate oligo-helicates with Cu^+ and Ag^+ (left). Structure of a trinuclear silver helicate formed from **13** [64] (right)*

The above examples have exclusively focused on dinuclear double-stranded helical systems but examples of larger polynuclear double-stranded helicates exist. Lehn and coworkers initially prepared oligo-bipyridyl ligand systems that gave rise to a series of tri-, tetra- and pentanuclear linear double-stranded helicates (Figure 4.8) [63]. More recently Lehn and coworkers developed ether-linked oligo-polypyridyl ligands with bidentate and tridentate binding sites alternating along the ligand [64]. The result was heterometallic double-stranded helicates that displayed self-sorting.

There are many examples of dinuclear double-stranded helicates that have appeared in the literature recently and this chapter presents only a very small selection of double-stranded helical structures derived from achiral building blocks. The reader is directed to recent reviews in this area for more focused discussions on specific types of helicates.

4.2.1.3 Triple Stranded Helicates

Triple Stranded helicates form when a self-recognition event occurs to give a homochiral helicate structure. As is typically the case, the system design plays a significant role in the

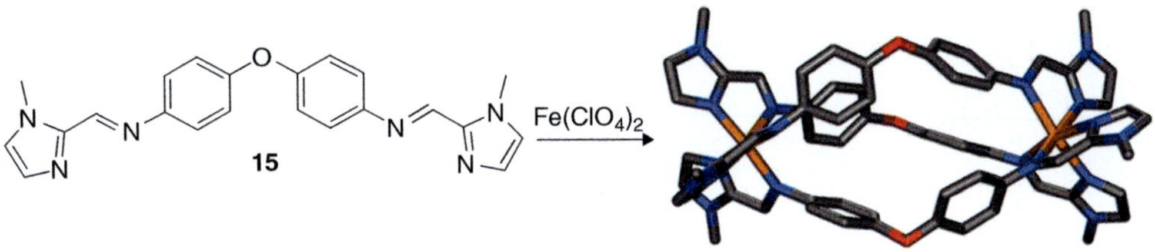

Figure 4.9 *Imidazole based ligand **15** used by Kruger and coworkers to generate dinuclear SCO active Fe^{2+} triple-stranded helicate [65]*

formation of the helicates and is crucial for the nature of the assembly formed – homochiral helicate (ΔΔ or ΛΛ) versus meso-helicate (ΔΛ). A variety of different ligand systems have been studied over the last decade, investigating control over formation, stability of helicates and more recently investigating the use of such helicates for advanced material and medical applications [49].

A magnetically interesting system has been developed by Kruger and coworkers. Who reported the synthesis of a di-iron(II) triple-stranded helicate that exhibited partial Spin-Crossover (SCO) behaviour [65] was reported in which the ligand system consists of two bidentate imidazole-imine binding pockets linked via an oxyphenyl bridge (**15**). When complexed with $Fe(ClO_4)_2 \cdot 6H_2O$ a racemic mixture of homochiral dimetallic helicates (ΔΔ and ΛΛ) resulted (Figure 4.9). The complex displayed partial temperature dependent SCO to give a mixed spin state where each of the iron(II) centres adopts an opposite spin state (i.e. HS-LS) showing that helical structures are useful in the development of advanced materials with switchable properties.

Nitschke and coworkers have also developed di-iron(II) triple-stranded helicates; however, the aim of their research was not to develop SCO-active compounds but rather to assess the postsynthetic modification of metallo-supramolecular assemblies [66]. The system under investigation was a dynamic subcomponent self-assembly of the C_3 symmetric ligands **16–18**, 2-pyridine carboxaldehyde and Fe^{2+} giving rise to either a $[Fe_2L_3]^{4+}$ helicate (the kinetic product) or a $[Fe_4L_4]^{8+}$ tetrahedron (thermodynamic product) – Figure 4.10.

The research carried out aimed to covalently postsynthetically modify the ligand strand in order to trap the kinetically metastable helicate. The self-assembly system was treated with a variety of acylating groups (Figure 4.10) that, by effectively blocking the third site on the ligand, prevented structural reorganization and hence trapped the triple-stranded helicate. The helicate existed as a racemic mixture of homochiral helicates. This study showed that the nature of the ligand can significantly influence the formation of the helical structure and that postsynthetic modification is a useful tool for controlling helicate formation, allowing for kinetic products to be isolated preferentially to thermodynamic products.

Typically ligands that possess relatively flexible linkers are used to prepare helicates; however, Fitchett and coworkers have shown the utility of much more rigid systems for the selective formation of dinuclear triple-stranded helicates. In one study by this group they used a series of five symmetrically substituted dipyridylpyridazine ligands (**19–23**) that differed in the nature of the steric bulk of the substituents [67]. They found that di-zinc(II)

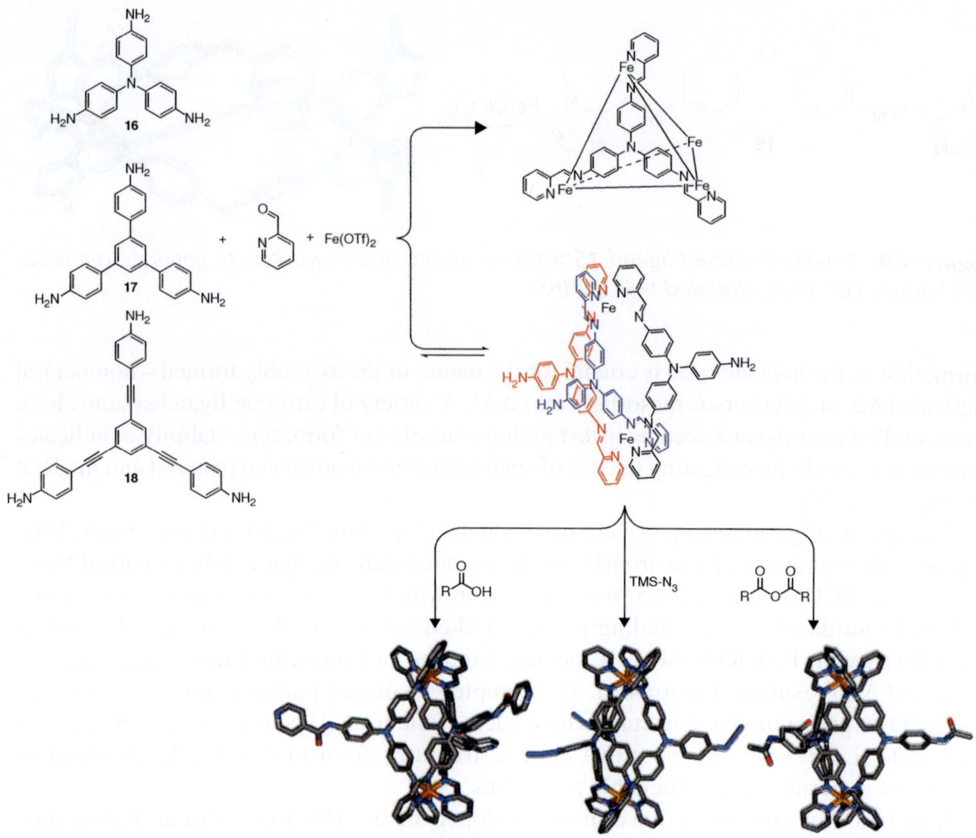

Figure 4.10 Schematic showing approach used by Nitschke and coworkers for controlling di-iron triple-stranded helicates formation over tetrahedral cage through postsynthetic modification [66]

triple-stranded helicates [M_2L_3] were isolated in the solid state from reactions between $Zn(ClO_4)_2$ and the less sterically hindered ligands **19** and **21**, whilst the more hindered ligands **20**, **22** and **23** gave rise to [2×2] grid structures (Figure 4.11).

In a second study the same group investigated the highly constrained bridged imidazole **24** (Figure 4.12) as a carboxylate mimic [68].

These compounds showed a preference for forming [Cu_2L_3] helicates when Cu^+ and non-coordinating anions were used. When **24** was complexed with Cu^{2+} salts, it formed a M_2L_2 dimer with $Cu(OAc)_2$ giving bridging and terminal acetate counteranions, and a quadruple helix with $Cu(ClO_4)_2$. Only with Cu^+ and noncoordinating anions (BF_4^-, ClO_4^- and NO_3^-) did Cu_2L_3 helicates form (CuI gave a coordination polymer). The ethyl bridge in the ligand strands twist in the same direction giving rise to the helical twist (albeit very small, ~8°) and the systems crystallize in chiral space groups. The studies by Fitchett showed that relatively rigid achiral ligands can form chiral triple-stranded helicates that are highly dependent on the metal salt used as they only formed when the correct combination of metal ion and anion was chosen.

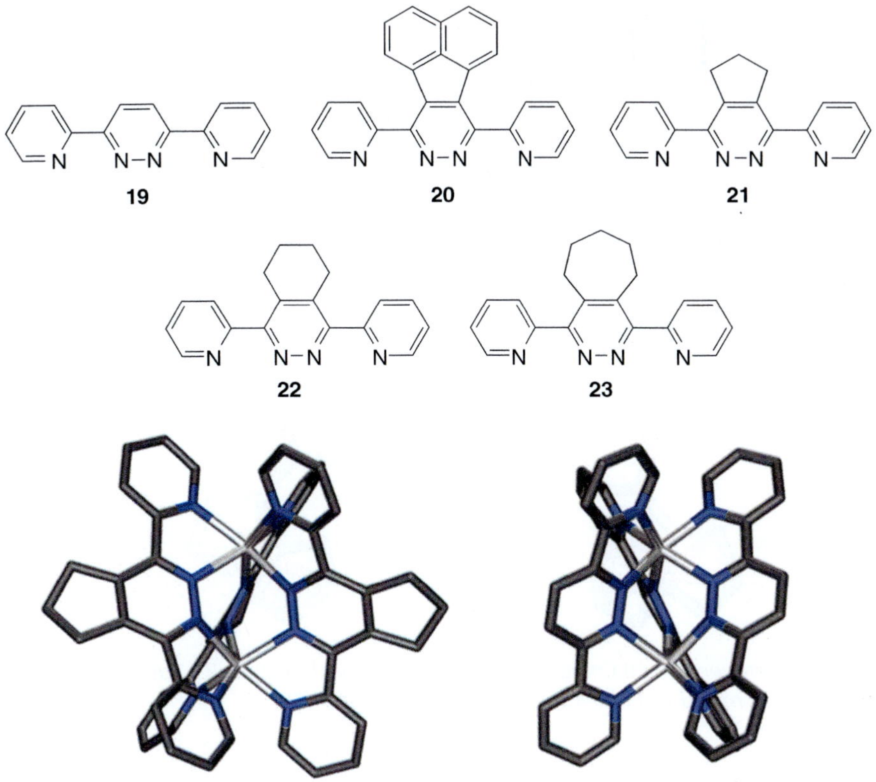

Figure 4.11 *Rigid pyridazine ligands **(19–23)** used by Fitchett and coworkers for the formation of Zn^{2+} triple-stranded helicates (bottom) [67]*

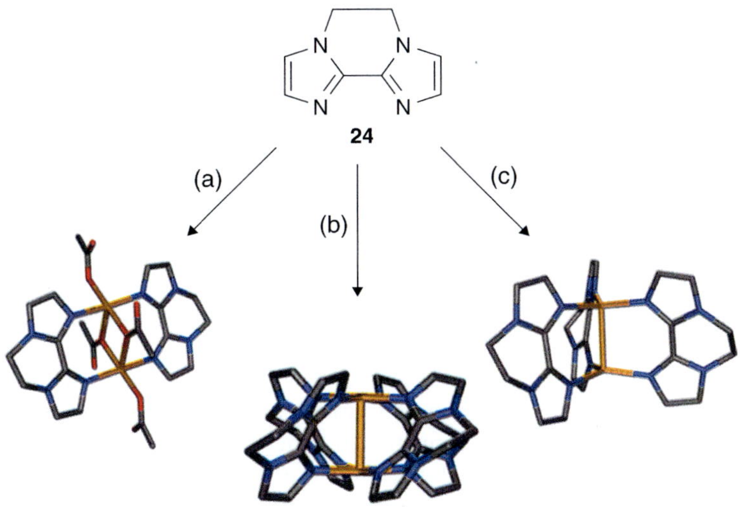

Figure 4.12 *Constrained bridged imidazole used by Fitchett and coworkers for the formation of copper complexes. **(a)** $Cu(OAc)_2$; **(b)** $Cu(ClO_4)_2$; **(c)** $CuClO_4$ [68]*

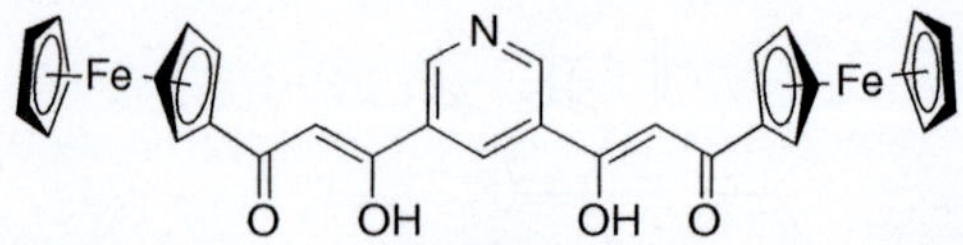

25

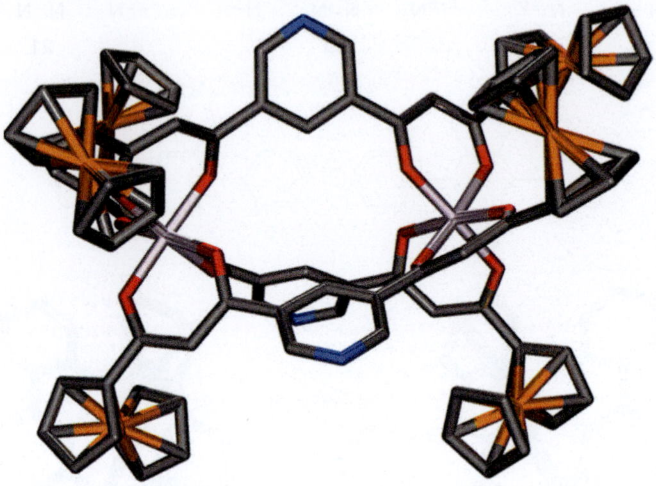

Figure 4.13 *Ferrocene containing ditopic-enolate based ligand* ***25****, used by Raja and coworkers to form mixed p-d block triple-stranded helicates [69]*

With the considerable interest in the development of dinuclear triple-stranded helicates, and to a small extent a degree of control being established over their formation, chemists have developed ever more elaborate ligand systems capable of forming such self-assembly structures. Recently, Raja and coworkers developed organoiron containing ditopic ligands where the two enolate binding pockets were capped by ferrocenyl groups (**25**) (Figure 4.13) [69].

By using In^{3+} and Ga^{3+} the group prepared the first examples of mixed *d*-block and *p*-block containing triple-stranded helicates. In the solid state the helicates exist as a racemic mixture of ΔΔ and ΛΛ, an observation that was further verified by nuclear magnetic resonance (NMR) studies in solution and variable-temperature NMR (VT-NMR) studies. This work is particularly interesting as it shows that interesting functional groups can be appended to the periphery of helicates without affecting the overall supramolecular architecture. In this particular case the introduction of organometallic species has the potential to open up routes to novel helically chiral catalysts.

Whilst metal-only helicates have dominated the literature, other structure-directing components can be used to orient organic ligands in a helical manner. Given the correct binding sites, anions are ideal for the formation of helicates, either alone or in combination with metal cations (ion-pairs) [70–75]. Such ion-pair triple-stranded helicates were studied by Custelcean and coworkers who used an asymmetric achiral ligand (**26**) that incorporated a metal-binding site (2,2′-bpy) and known hydrogen bonding anion binding sites (urea groups) [73].

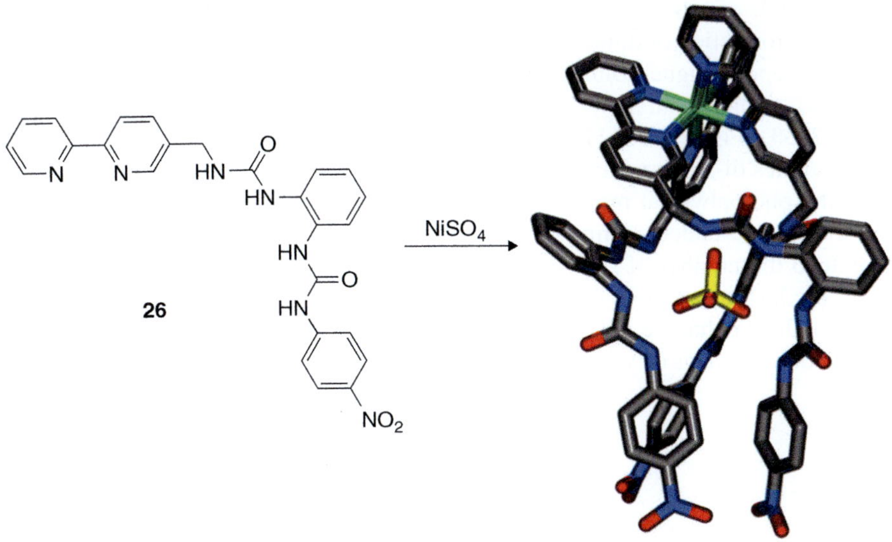

Figure 4.14 *System used by Custelcean and coworkers for the formation of triple-stranded ion-pair helicates [73]*

The group used ditopic ligand **26** to bind Fe^{2+} or Ni^{2+} in the bpy chelate pocket and SO_4^{2-} in the urea pocket giving rise to [ML_3A] triple-stranded helicates (Figure 4.14). In an interesting extension of this work the same group have recently used computational methods to design ligands suitable for forming ion-pair helicates [70]. The study, which screened a vast number groups to link the metal-binding site and the anion-binding site, identified eight synthetically feasible ligands for complexation of $FeSO_4$ as well as four targets to generate *f*-block $LnPO_4$ ion-pair helicates. Work is now underway to synthesize the identified targets, and if these do indeed form ion-pair helicates it will be an excellent example of the role computation methods are now playing in aiding the design and development of complex chiral supramolecular architectures.

Whilst first row transition metals are commonly employed for the formation of metallo-helicates, trivalent lanthanide ions are also able to form chiral self-assemblies with ligands containing hard donor atoms and there is currently a thrust towards the design and preparation of multinuclear systems. Interest in multinuclear systems in part stems from the potential enhancement of the interesting photophysical and/or magnetic properties of the Ln^{3+} series when in close proximity with other 4*f* metal centres. Subsequently there has been significant interest in recent years in the design of ligands for the self-assembly of multiple Ln^{3+} centres. Hamacek, Piguet and Bünzli have been instrumental in the development of lanthanide-containing multinuclear systems including dinuclear triple-stranded helicates [76–78] as well as higher order assemblies (tetranuclear supramolecular cages [79–81] and pentanuclear Ln^{3+} helicates [82]). Piguet reported the first dinuclear lanthanide triple-stranded helicate in 1992 [83] and, just like the transition metal-based helicates, much research has gone into understanding the self-assembly process as well as developing a variety of ligands for formation of *f*-block containing helical structures [34, 42, 84–101].

Given the high coordination number for the Ln^{3+} series (typically 8–9) many ditopic ligands for the formation of dinuclear triple-stranded helicates use tridentate chelate pockets. In several reports Piguet, Bünzli and co-workers reported the formation of polynuclear helicates from ligands containing pyridyl amides and pyridyl benzimidazoles in varying numbers (**27–30**) (Figure 4.15) [102].

In these studies di-, tri- and tetra-nuclear triple-stranded helicates were studied in detail including the photophysical properties and thermodynamics of assembly. Through their detailed studies and understanding of the formation of helicates, Bünzli and Piguet were able to extend the number of Ln^{3+} ions incorporated in their self-assembled systems and move towards larger and more complex assemblies, including heterobimetallic helicates. Heterobimetallic lanthanide systems (**31–35**, Figure 4.16) are extremely attractive targets for the development of versatile multi-emission bioprobes [90, 103–106]. By taking

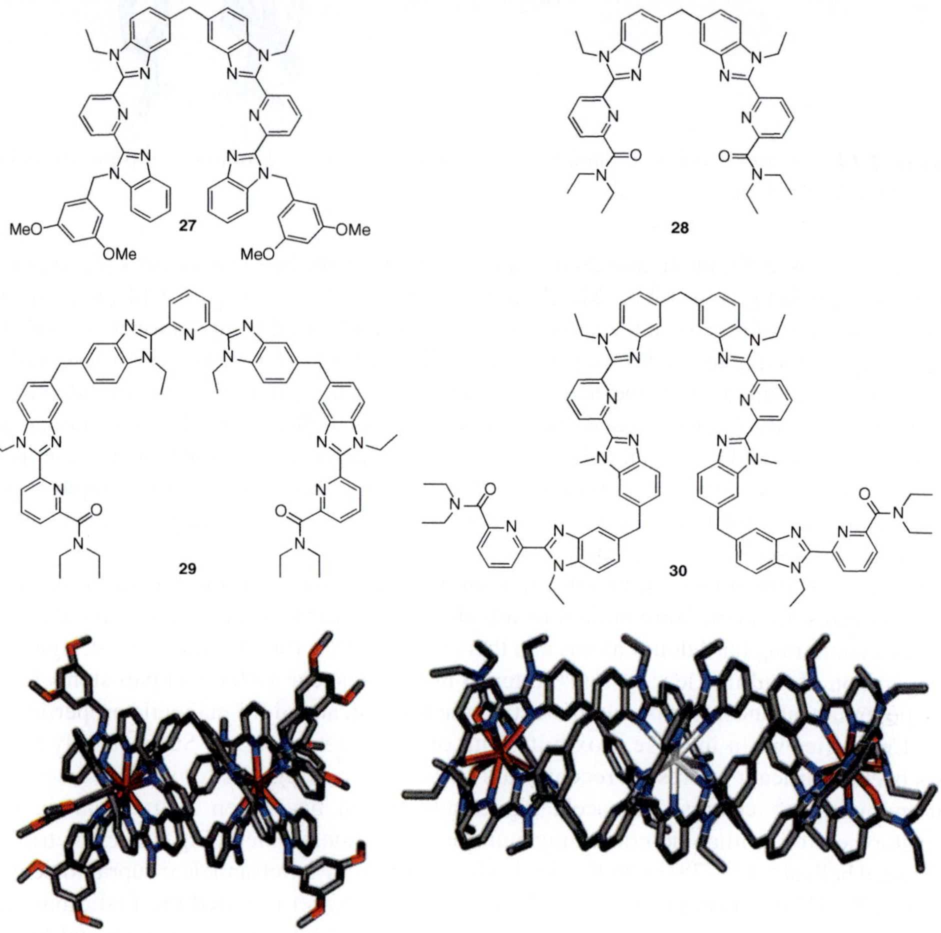

Figure 4.15 *Ligand strands generated by Piguet et al. for the preparation of di-, tri- and tetranuclear linear triple-stranded lanthanide containing helicates. Structures of [Eu_2(**27**)$_3$] and [Eu_2La(**29**)$_3$] (Eu = red, La = white) [102]*

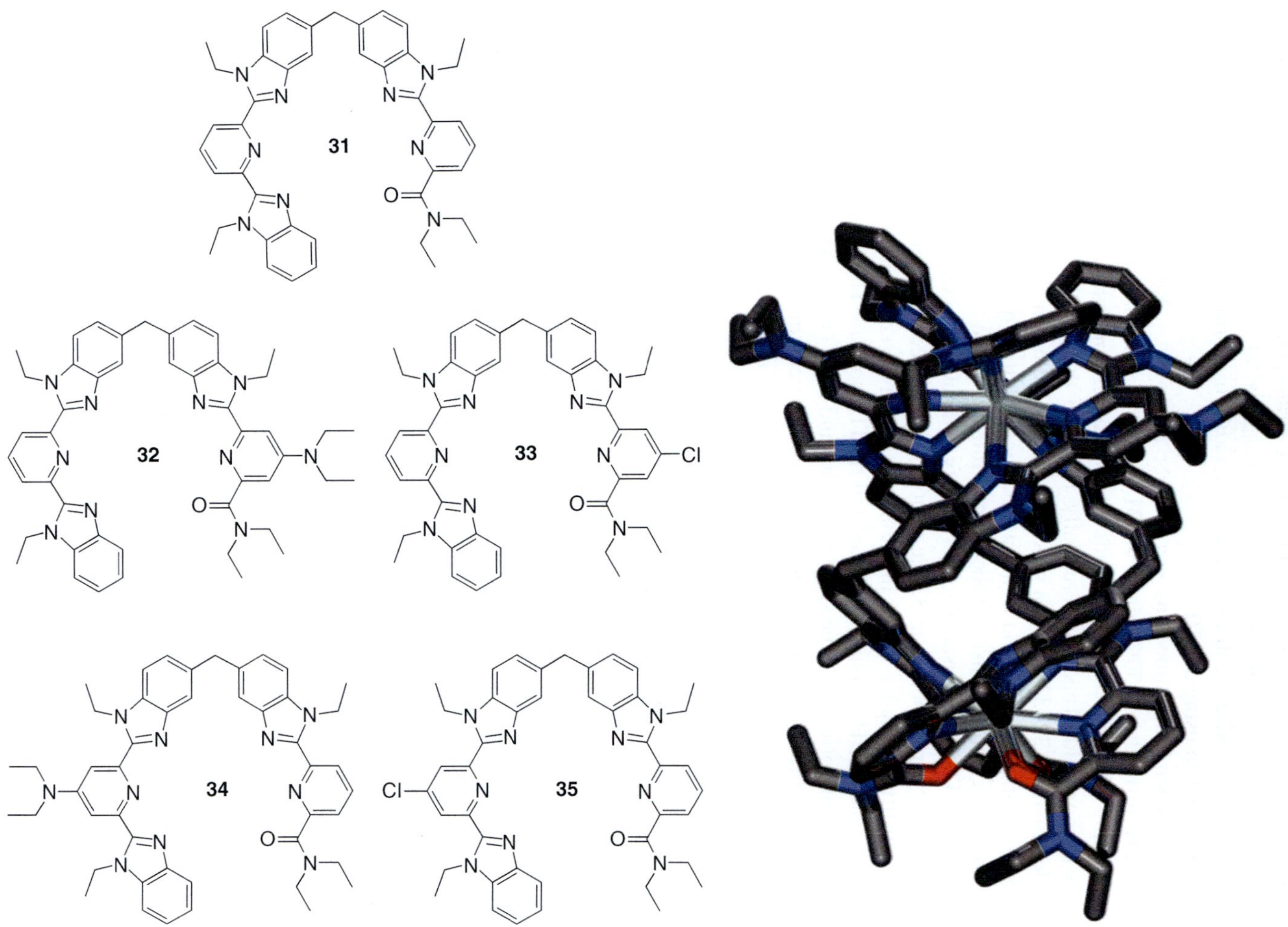

Figure 4.16 *Heteroditopic ligands developed by Bünzli and coworkers for the preparation of dilanthanide triple-stranded helicates and structure of the Pr_2 system. The heteroditopic nature of these systems allowed for the isolation of heterometallic mixed lanthanide helicates [103–107]*

advantage of the different ionic radii of lanthanides, careful ligand design and the thermodynamically controlled self-assembly of Ln^{3+} triple-stranded helicates, Bünzli and coworkers were able to form heterobimetallic systems using La^{3+}/Lu^{3+} ions (this heteropair was chosen as they maximize the difference in ionic radii) that formed the $[LuLa(\mathbf{L})_3]$ helicates in >90% yield when using **31** [107].

However, when lanthanide ions were closer in size (e.g. Eu^{3+} and Tb^{3+}) the hetero system only formed in ~50% yield. In **31** the benzimidazole-pyridine-benzimidazole (bpb) binding site preferentially binds to the larger ion whilst the benzimidazole-pyridine-carboxamide (bpa) preferentially binds to the smaller ions. In the second-generation ligands (**32–33**) the bpa was modified by inclusion of NEt_2 (**32**) or Cl (**33**) at the *para* position where electron-donating (**32**) or -withdrawing (**33**) ability was expected to influence the electron density at the pyridyl nitrogen atom [108]. The rationale behind this was that in **32** the increased electron density at the nitrogen would favour the smaller, and harder, Ln^{3+} ions whereas in **33** the opposite was expected. Interestingly, they found that the substituents had a major influence on the preference for the system to adopt head-head-head (HHH) versus head-head-tail (HHT) isomers. The HHH (where all three ligand strands are oriented in the same direction) and HHT (where one ligand strand runs in the opposite direction to the other two) isomers are common when developing heteroditopic ligands with binding sites that have similar affinities for the metal ions used. The substituents on the bpa caused preferential formation of the HHT isomer and hence led to lower selectivity for heteropairs of lanthanide ions. In the third generation of ligands (**34** and **35**) the substituents were moved to the *para* position of the bpb binding pocket where the electron withdrawing Cl substituent on **35** was expected to give preferential binding to the larger and softer Ln^{3+} ions [109]. Ligand **34** was not expected to have any increased selectivity and was synthesized as a control ligand. When complexed with Pr^{3+}, homobimetallic helicates were isolated as single crystals and the molecular structure revealed the HHH isomer (Figure 4.16). These crystallized as a racemic mixture of *M* and *P* helicates and the stability of the helicate was partially attributed to weak intramolecular π-stacking interactions. Solution speciation studies were carried out in acetonitrile and revealed that there is NMR evidence for the formation of the desired heterobimetallic systems. The study by Bünzli highlighted how the assembly of triple-stranded helicates is highly ligand dependent and can be fine tuned through appropriate choice of ligand substituents.

Given the high coordination requirement for lanthanides compared to first-row transition metals, it is also possible to design heteroditopic ligands for triple-stranded helicate formation containing one binding pocket for Ln^{3+} ions (usually a tridentate chelate) and one chelate pocket for transition metals (typically bidentate). Using this approach, Piguet and coworkers were able to develop mixed *f-d* triple-stranded helicates with ligands **36–38** [103–106]. A variety of lanthanides and transition metal combinations were used (Figure 4.17) to form heterometallic helicates as dimetallic (**36** and **37**) and trimetallic systems (**38**). These studies again show how careful ligand design can generate interesting chiral, multimetallic species.

Albrecht and coworkers also developed heteroditopic ligands capable of forming mixed *f-d* triple-stranded helicates. Using a mixed bidentate/tridentate ligand (**39**) they were able to isolate heterodinuclear $[LaTi(\mathbf{39})_3]$ complexes based on the geometric preferences of the metal species (Figure 4.18) [110].

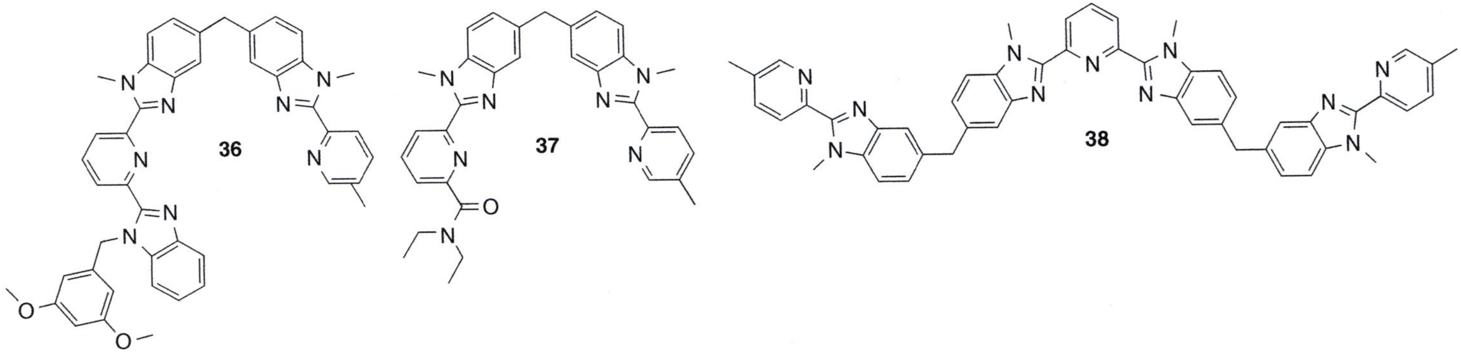

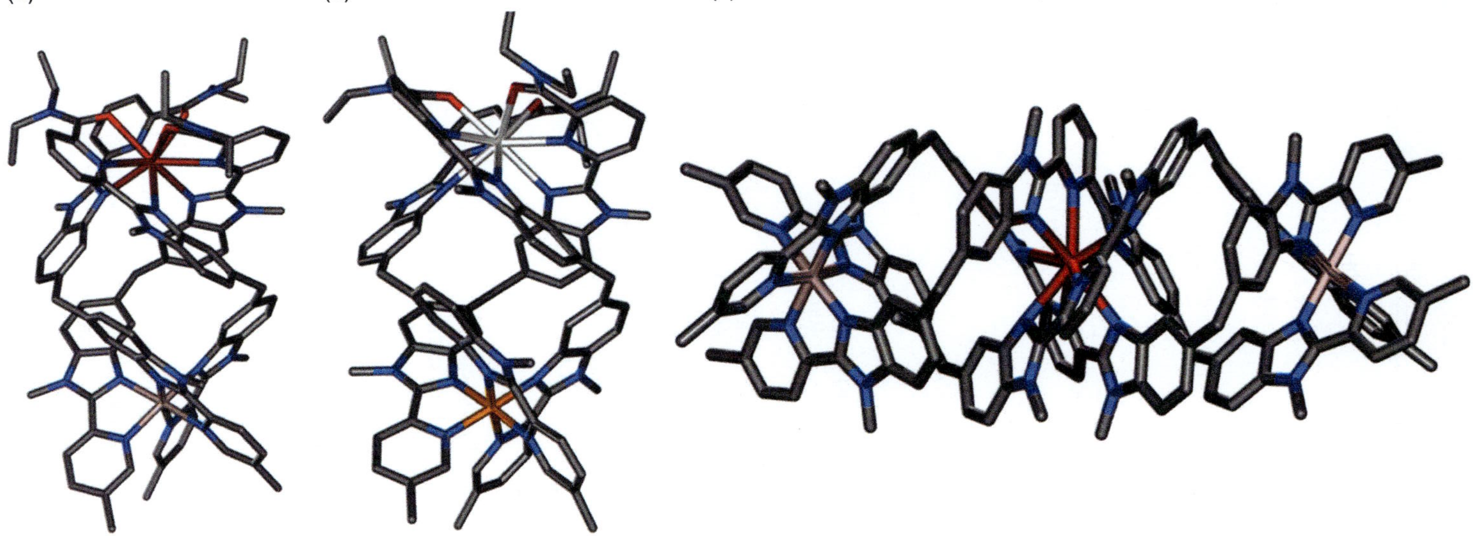

Figure 4.17 *Designer heteroleptic ligands developed by Piguet et al. for the preparation of mixed metal f-d triple-stranded helicates. (a) = [Eu(**37**)$_3$Ni]; (b) = [La(**37**)$_3$Fe] and (c) = [Ni$_2$(**38**)$_3$Eu] [103–106]*

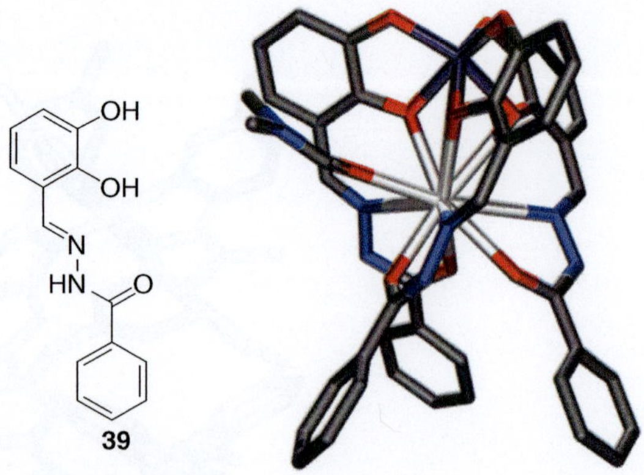

Figure 4.18 Mixed f-d triple-stranded helicate developed by Albrecht and coworkers and the molecular structure of [Ti(**39**)$_3$La] [110]

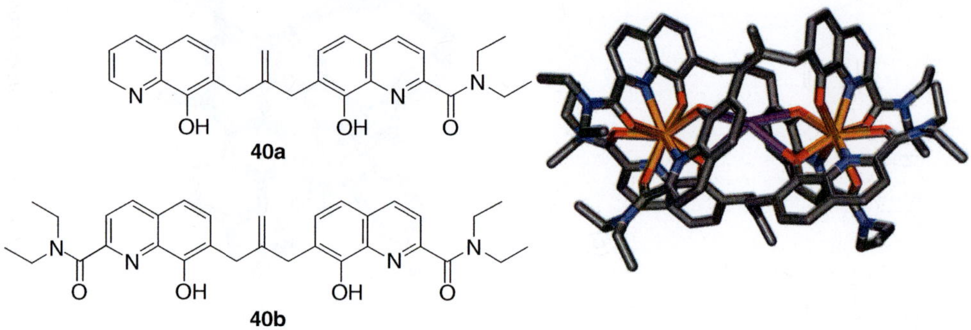

Figure 4.19 Ligands developed by Albrecht for the preparation of mixed f-p triple-stranded helicates Structure of the homodinuclear [Yb$_2$(**40b**)$_3$] system [111]

Similarly *f-p* based helicates can be prepared using the same designer ligand approach [32, 111, 112]. One notable example of this was again the work of Albrecht who used the 8-hydroxyquinoline-based ligand (**40**) where the tridentate pocket binds to Yb^{3+}, Er^{3+} and Nd^{3+} and the bidentate hydroxyquinoline ligand to Al^{3+} [111]. In these systems, alkali metal carbonates were added to deprotonate the ligand and also provide a cation (K^+) to template the formation of helicates of composition [LnKAl(**40a**)]$^+$ (Figure 4.19).

4.2.1.4 Higher Order Assemblies

Whilst double- and triple-stranded helicates are two somewhat commonly observed supramolecular architectures, larger and more complex chiral self-assemblies have been isolated, and much like the smaller helicates many factors influence their formation. Examples of such higher order assemblies include quadruple-stranded helicates, tetrahedral cages, expanded helicates, circular helicates and knots.

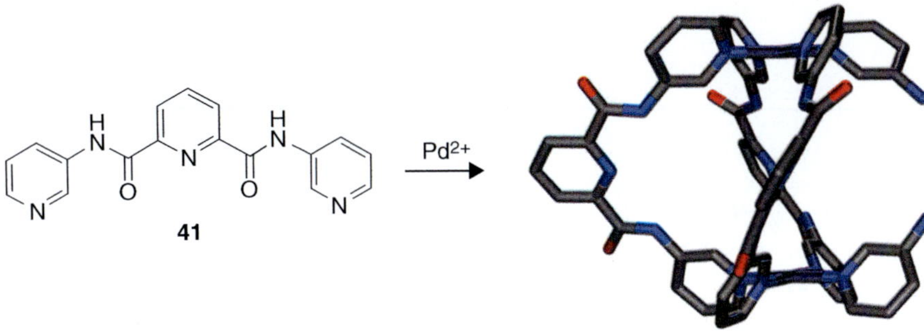

Figure 4.20 *Pyridyl amide based ligands used by Hanan and coworkers to yield the quadruple-stranded di-palladium helicate [Pd_2(**41**)$_4$] [113]*

Quadruple-stranded helicates have been successfully formed from both *d*- and *f*-block metals. Hanan and coworkers used the pyridyl diamide ligand **41** to prepare self-assembled Pd^{2+} helicate [Pd_2(**41**)$_4$]$(NO_3)_4$ [113].

The solid state structure (Figure 4.20) shows the azimuthal angle (45°) of the helicate formed from the twist between the two square-planar Pd^{2+} centres. The asymmetric unit contained both *M* and *P* helicates and the large central cage formed in these helicates encapsulates three nitrate counterions. Given the chiral nature of the system and the large central cavity, research is now ongoing to use these systems for selective guest recognition and use in host-guest chemistry.

Given the high coordination requirement of the lanthanide ions, they can form quadruple-stranded helicates when lower denticity ligands are used (i.e. ditopic bidentate systems). A particularly attractive example of this came from Murugesu and coworkers who used ligands **42–44** to form helicates or mesocates with Dy^{3+} of general composition [Dy_2L_4] and studied not only the formation of the supramolecular Dy^{3+} helicates but also the relationship between helicate structure/topology and single molecule magnetism (SMM) behaviour [114]. The solid-state molecular structures were obtained and complexes [Dy_2(**42**)$_4$] and [Dy_2(**43**)$_4$] were shown to form racemic mixtures of quadruple-stranded helicates whilst [Dy_2(**44**)$_4$] formed the mesocate structure (Figure 4.21). This study resulted in the first quadruple lanthanide helicate and mesocate SMM's and was able to highlight ligand dependent nature of the self-assembly process (i.e. helicate versus mesocate), which, in turn, altered the coordination sphere/geometry around the Dy^{3+} and hence the SMM behaviour. Thus control over helicate formation is an important method by which to alter other physical properties of advanced materials.

Much like the case of the smaller helicate assemblies anions can alter – and indeed be the major driving force for – the structure of the supramolecular assembly obtained. For example, Liu and coworkers studied the effect of anions on the formation of large La^{3+} cluster helicates using the multidentate ligand **45** [115].

When $La(ClO_4)_3$ was complexed with **45** the result was the first example of a lanthanide hexanuclear circular helicate (Figure 4.22). Two of the large tetrahedral perchlorate anions were encapsulated in the central cavity of the helicate. When a smaller, trigonal planar nitrate anion was used in the complexation, the result was a tetranuclear quadruple helical cluster (Figure 4.22) of composition [La(**45**)$_4$$(NO_3)$]$^{3+}$ with a nitrate counteranion

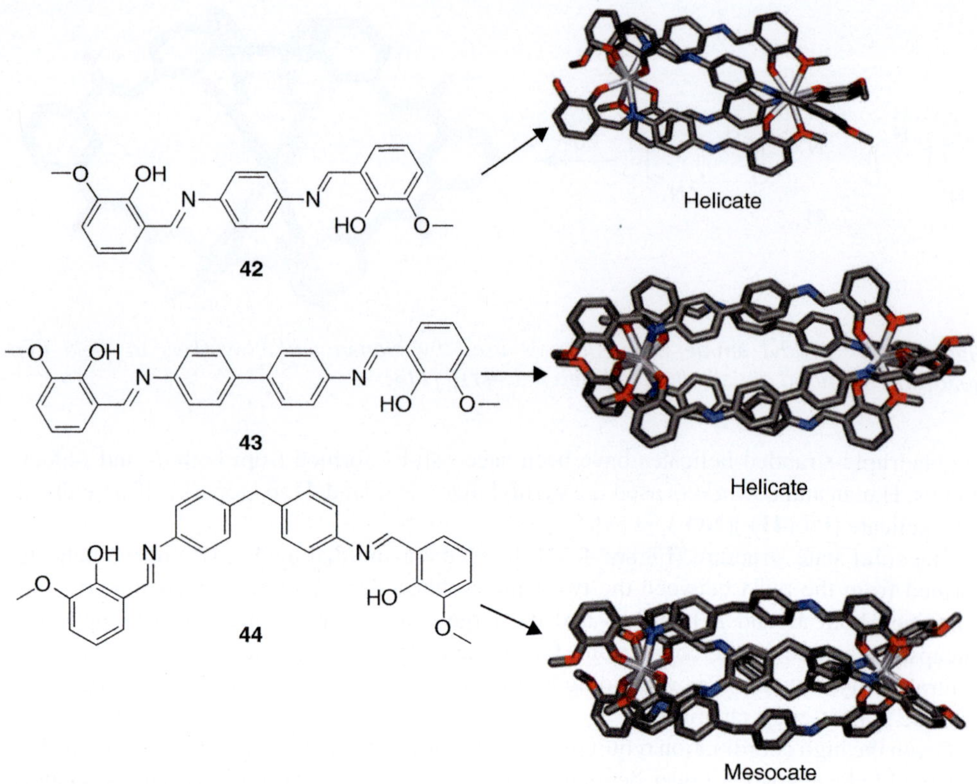

Figure 4.21 *Ligands developed by Murugesu and coworkers for the preparation of Dy^{3+} helicates and mesocates [114]*

encapsulated in the centre. The anion-dependent formation of the assemblies was attributed to the templating ability of the two different shaped anions. The role of anion templation was further confirmed when conversion between the circular helicate and the quadruple helicate was observed on addition of NO_3^- to the circular helicate (Figure 4.22). To test this further, the group added a variety of anions to the quadruple helicate ($H_2PO_4^-$, OAc^-, HCO_3^-, HSO_4^-, F^-, Cl^-, Br^- and I^-) and only observed the nitrate-templated quadruple-stranded helicate.

Hamacek and coworkers have used multitopic ligands based on pyridine-2,6-dicarbonyl motifs to effect the formation of larger Ln^{3+} based self-assemblies. For example tripodal ligands **46** and **47** formed self-assembled Ln^{3+} 3D tetranuclear helicates (i.e. tetrahedral cages) with large central cavities (Ln = La, Lu, Nd and Eu) (Figure 4.23) [79, 81].

The motivation behind this was to use the central cavities for sensing via host-guest chemistry. The cages were prepared by the reaction of one equivalent of **46** or **47** with one equivalent of $Ln(ClO_4)_3$. Single crystals of $[Tb_4(\mathbf{46})_4](ClO_4)_{12}$ and $[Eu_4(\mathbf{47})_4](OH)(ClO_4)_{11}$ revealed the tetrahedral nature of the complexes. Much like the case for the previous linear helicates, the ligand design influenced the structure of the resulting 3D helicates. The amide (**46**) versus ester (**47**) linkages in the ligands resulted in significantly different cage structures (Figure 4.23). In the Tb^{3+} complex of **46** the exo-CH_3 is orientated into the central

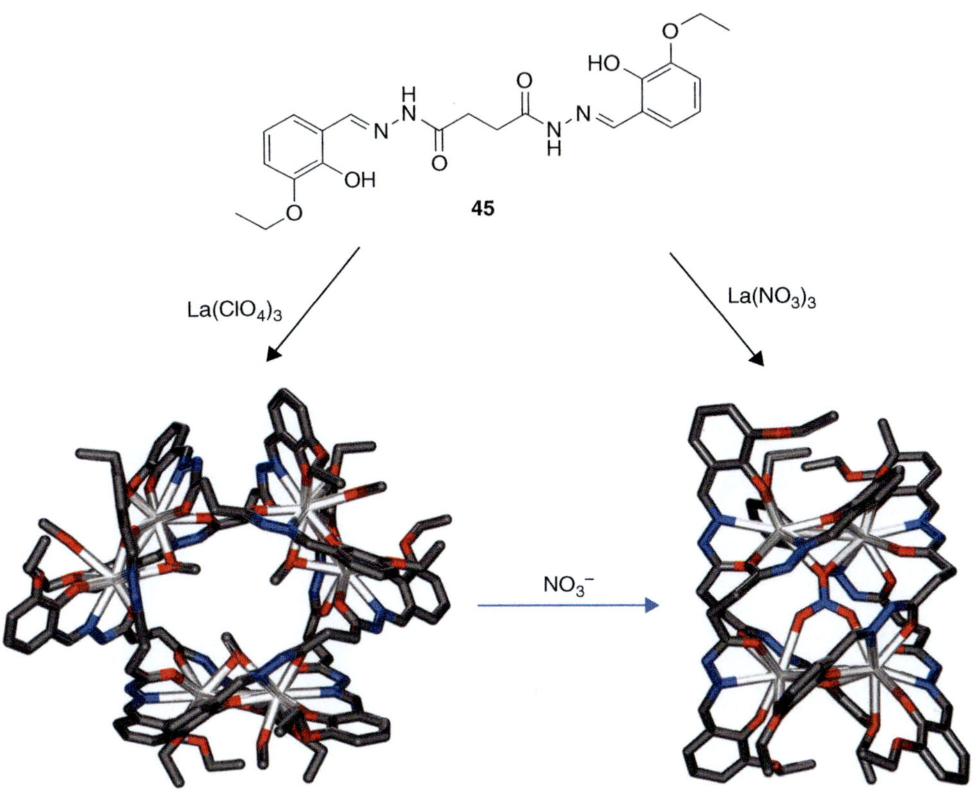

Figure 4.22 Dynamic assembly of hexa-nuclear circular La^{3+} helicate and tetranuclear quadruple helicate studied by Liu and coworkers. The system was highly dependent on the shape of the anion present and could convert from the circular helicate to the linear [115]

cavity whereas in the Eu^{3+} complex of **47** the exo-CH_3 is oriented away from the cavity. The result of this is a smaller central cavity in $[Tb_4(\mathbf{46})_4]^{12+}$ compared to $[Eu_4(\mathbf{47})_4]^{12+}$ which further results in the central cavity in the latter being occupied by a ClO_4^- anion. The thermodynamic stability differs for each of these systems, as the 3D helicates formed from **46** are not stable in the presence of excess metal. The 3D helicates formed from **47** were found to be thermodynamically stable in excess metal (with one exception – $[La_4(\mathbf{47})_4]^{12+}$). The encapsulated ClO_4^- anion in cages constructed using ligand **47** was observed to undergo slow exchange with other anions, highlighting the stability of these self-assemblies and also their applicability for host-guest chemistry based applications.

Three-dimensional helicates have also been constructed using achiral ligands and first-row transition metals. Nitschke and coworkers have been instrumental in the development of complex supramolecular architectures through subcomponent self-assembly for a range of applications. They have developed many different assemblies where the nature of the ligands dictated the properties of the resulting assembly, which have included systems for use in catalysis, encapsulation of reactive species, transition state characterization, separation science and magnetism [116–120]. One particularly elegant example of

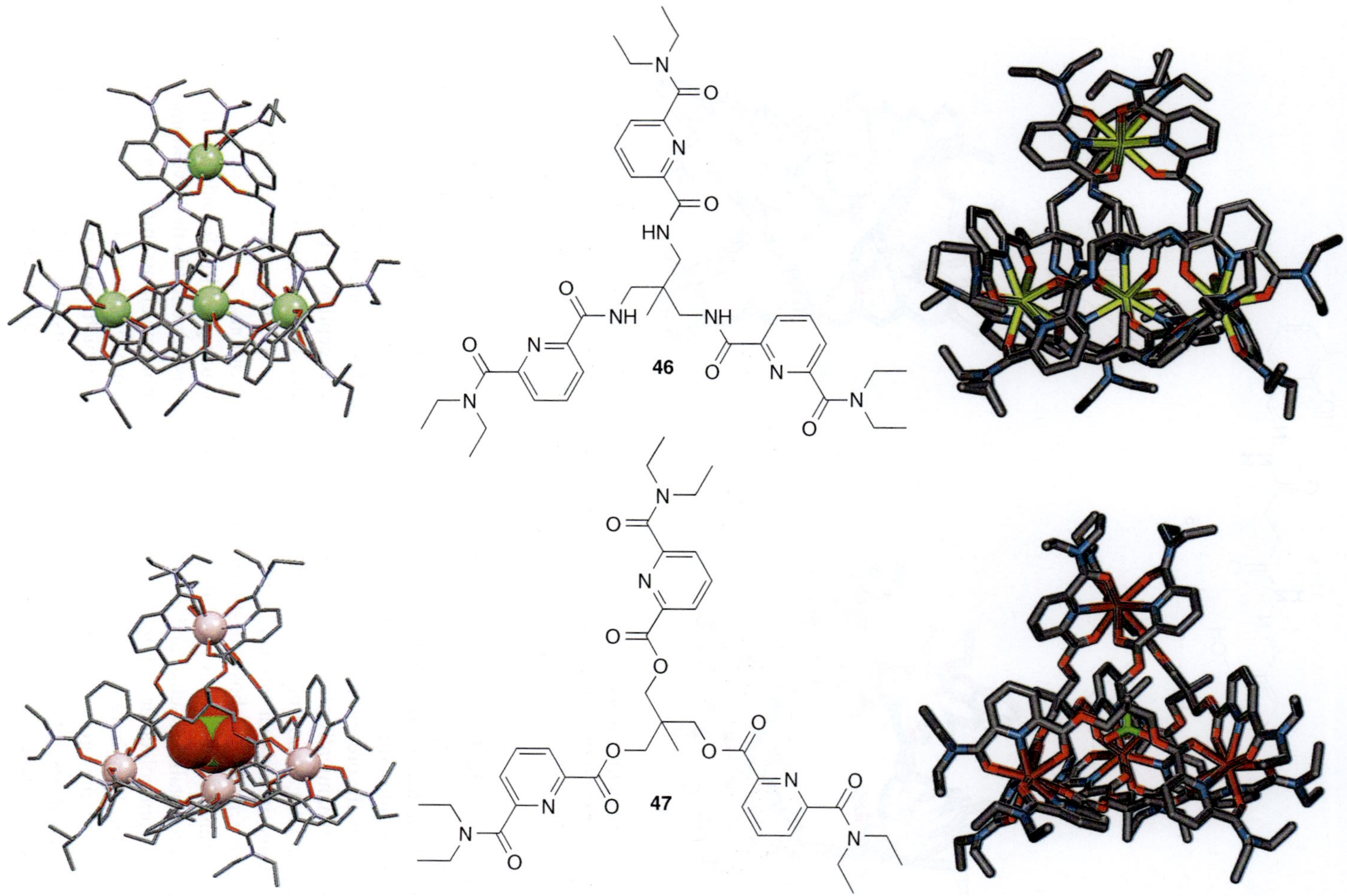

Figure 4.23 *Tetrahedral cages developed by Hamacek et al.* Tb^{3+} *cage (top) and* Eu^{3+} *cage (bottom) where the nature of the central cavity was highly dependent on the structure of the ligands* ***46*** *or* ***47*** *[79, 81]*

encapsulation and stabilization of reactive species came from a self-assembled tetrahedral capsule that stabilized the highly pyrophoric white phosphorus in air and also allowed it to be solubilized in water [120]. Using a bis(sulfonated) diamine (**48**), 2-formylpyridine and Fe^{2+} (Figure 4.24) the group were able to self-assemble a tetrahedral cage in water where P_4 was encapsulated within the central cavity.

In a separate encapsulation study the same group used larger pyrene-based linkers (**49a** and **49b**) to construct tetrahedral cages (Figure 4.25) [121]. By changing the isomer of the central pyrene groups (1,6-pyrene versus 2,7-pyrene) the system showed selectivity for encapsulation of a variety of large guest molecules (including: C_{60}, C_{70}, pyrene, cholesterol). [Fe_4(**49a**)$_6$] showed encapsulation of guests, while [Fe_4(**49b**)$_6$], which was more porous because of the larger separation between imine units, did not bind guest molecules. In solution these systems assembled into complex mixtures of stereoisomers consisting of homochiral (ΔΔΔΔ/ΛΛΛΛ), heterochiral (ΔΔΔΛ/ΛΛΛΔ) and achiral (ΔΔΛΛ) species, all in equilibrium. The solid state structure of [Fe_4(**49b**)$_6$] crystallized as a racemic mixture of ΔΔΔΛ/ΛΛΛΔ (Figure 4.25). The studies by Nitschke have been pivotal in shifting the focus of higher order metallo self-assemblies from interesting structural entities to application based systems and have also shown that rationale design of ligands is essential in developing diversity in complex supramolecular systems.

In another study of tetrahedral cage systems, Kruger and coworkers extended their SCO active triple-stranded helicates to a face capped [Fe_4L_4] tetrahedral cage using the triazine-imidazole-based ligand **50** (Figure 4.26) [122]. The system formed the expected tetrahedral cage structure and was the first example of a SCO-active tetrahedral cage, a property brought about through careful ligand design.

Moving beyond tetrahedral cages, and in an extension of their 3D tetranuclear Ln^{3+} helicates, Hamacek and coworkers generated large self-assembled 'towerlike' pentanuclear assemblies (Figure 4.27) using the unsymmetrical multimodal ligand **51** [82]. The results of combining **51** and **46** with Eu^{3+} or Nd^{3+} (in a 3:1:5 ratio) were complexes [Ln_5(**46**)(**51**)$_3$]$(ClO_4)_{15}$ in which **51** arranged five metal centres into a towerlike configuration and **46** capped the bottom of the assembly (Figure 4.27). The system reported by Hamacek is another excellent example of rational ligand design being used to generate self-assemblies in a predictable manner. This system involved an extra level of supramolecular complexity in that it used inclusion of two different ligands into one complex through self-assembly highlighting how multicomponent self-assembly can now be exploited for the generation of higher order architectures.

Molecular knots, and related architectures, are the final higher order self-assembly structures that will be discussed. Knots represent some of the most elaborate structural motifs and, as such, have been the target of much research in recent years. Leigh and coworkers reported a particularly elegant example of such research where they studied the self-assembly of a large series of tri-topic pyridyl based imine ligands derived from di-aldehyde **52** with Fe^{2+} [123, 124]. In this study pyridyl-dialdehyde (**52**) was reacted with amine and Fe^{2+} to give pentameric circular double-stranded helicates where each Fe^{2+} ion was coordinated by two 2-pyridyl-imine groups (from two different ligand strands) and one 2,2′-bpy group from a third ligand strand (Figure 4.28).

The formation of these circular helicates was highly dependent on the amine choice, metal-salt choice (only $FeCl_2$ resulted in circular helicates), stoichiometry and reaction conditions (solvent choice and concentration). Given the spacial position that the imine

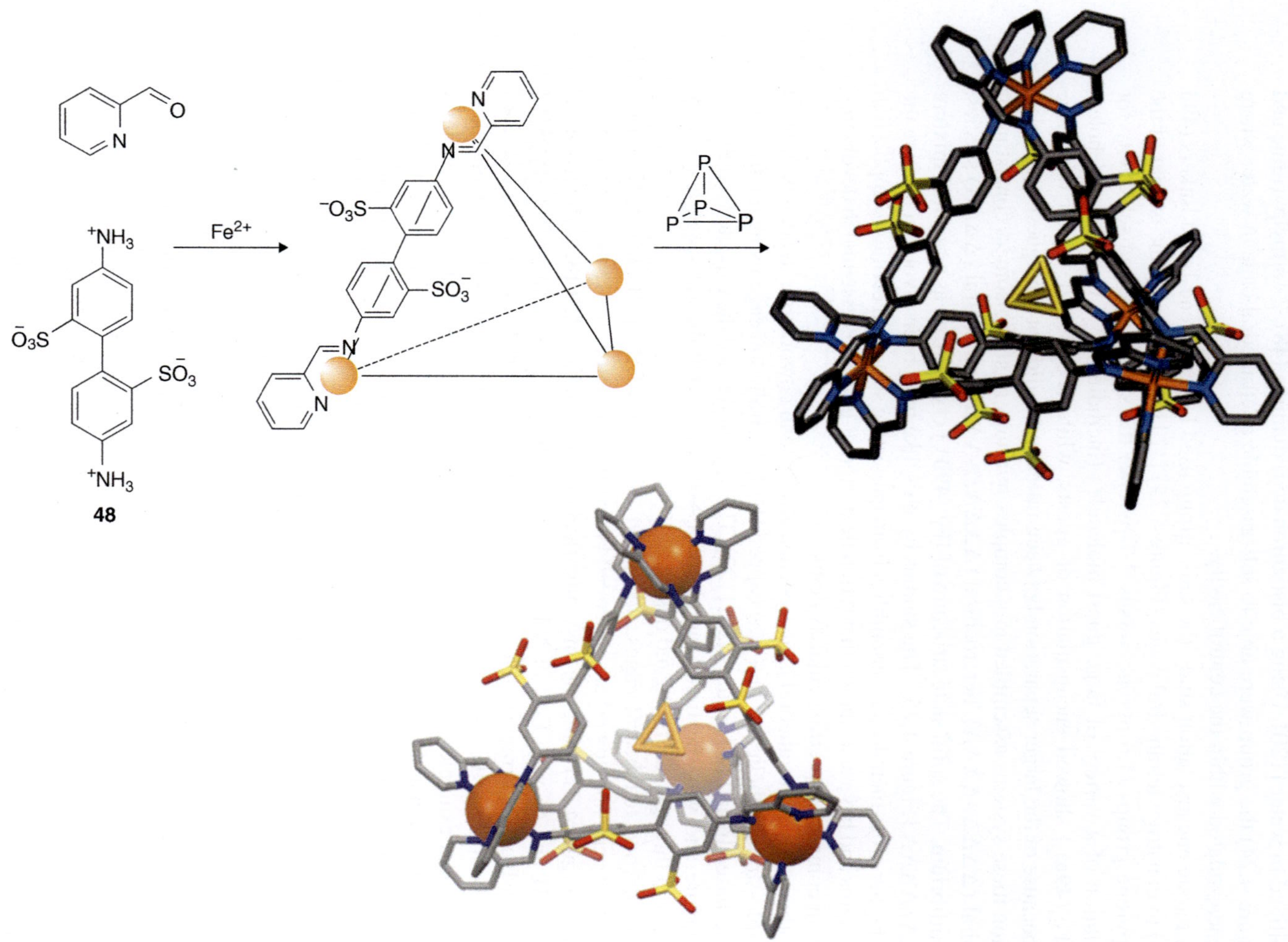

Figure 4.24 *Self-assembly system developed by Nitschke et al. In this case the tetra-iron cage was able to encapsulate and stabilize P_4 in aqueous environments [120]*

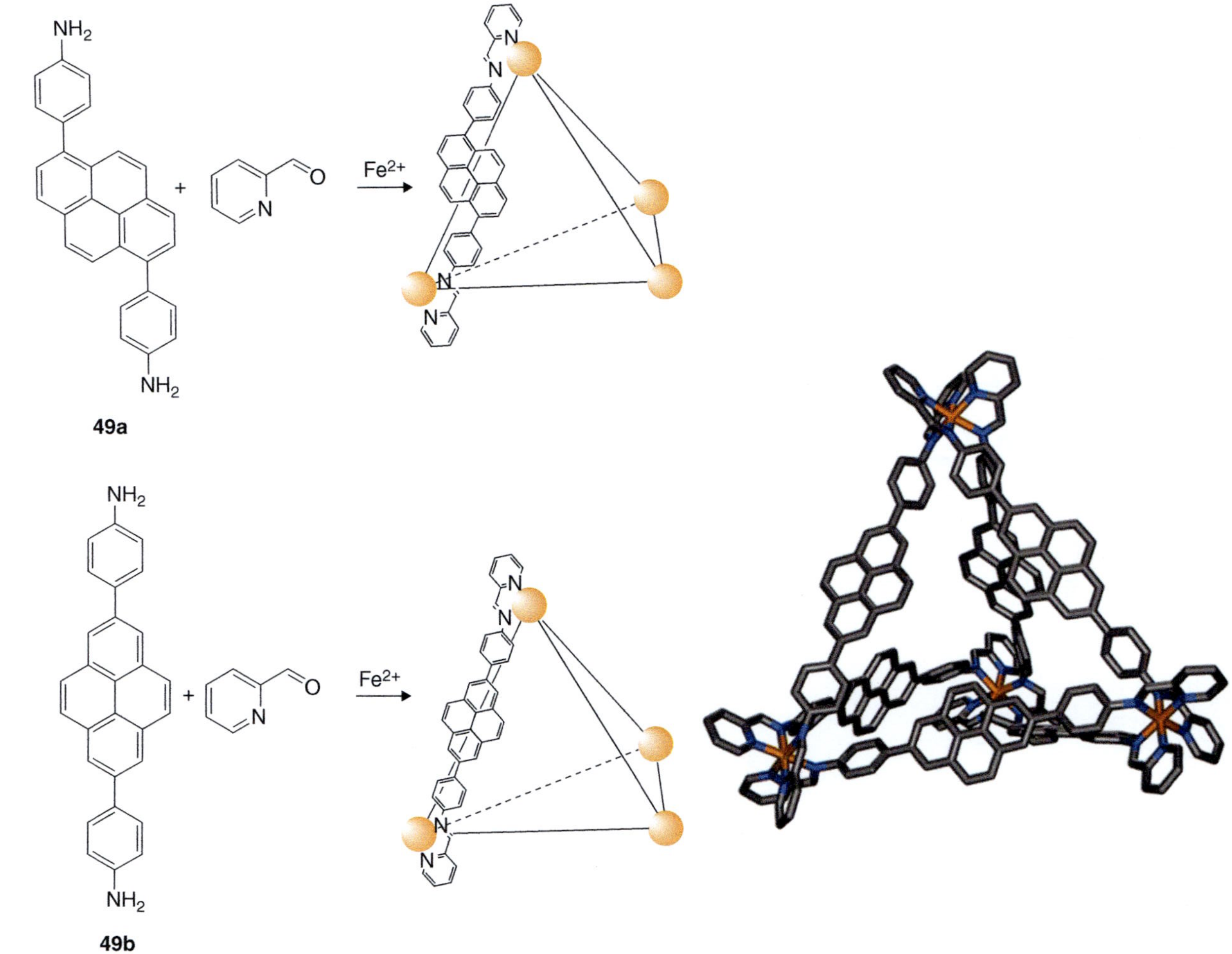

Figure 4.25 Pyrene based cages developed by Nitschke et al. where different positional isomers gave different central cavity sizes and hence different encapsulation properties [121]

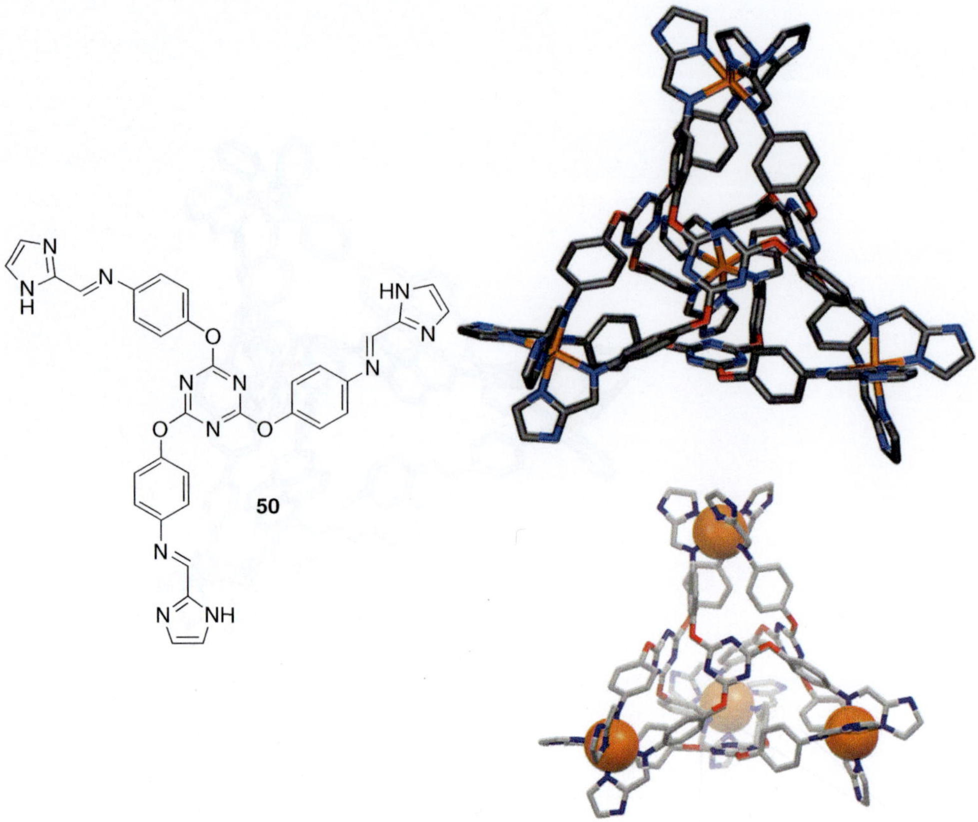

50

Figure 4.26 *Face capped iron(II) tetrahedral cage developed by Kruger et al. that displayed SCO behaviour [122]*

groups adopted in the circular helicates – in relatively close proximity to an imine on a neighbouring ligand strand – the researchers then extended these systems into molecular pentafoil knots. In order to prepare these linked systems they used diamines in the synthesis in order to connect the closely spaced pendant aldehydes (Figure 4.28). Much like in the formation of the circular double-stranded helicates, the nature of the diamine significantly influenced the formation of the knots. When using aliphatic chains, the knots did not form; however, when glycol linkers were used the knots formed. This observation was attributed to the ability of the glycol linker to introduce a low-energy turn through the gauche effect (the preference for O-C-C-O chains to adopt torsion angles of 60° instead of 180° as is the case for all carbon chains).

In a separate study, Leigh and coworkers modified the aforementioned ligands by altering the linker between the central 2,2′-bpy unit and the terminal 2-pyridylimine binding sites to give **53**. Using the same approach as above – cyclic double-stranded helicate formation to prepare knots – they were able to isolate a tetra Fe^{2+} supramolecular Solomon knot using a one-pot procedure in 75% yield (Figure 4.29). These two studies by Leigh highlighted the

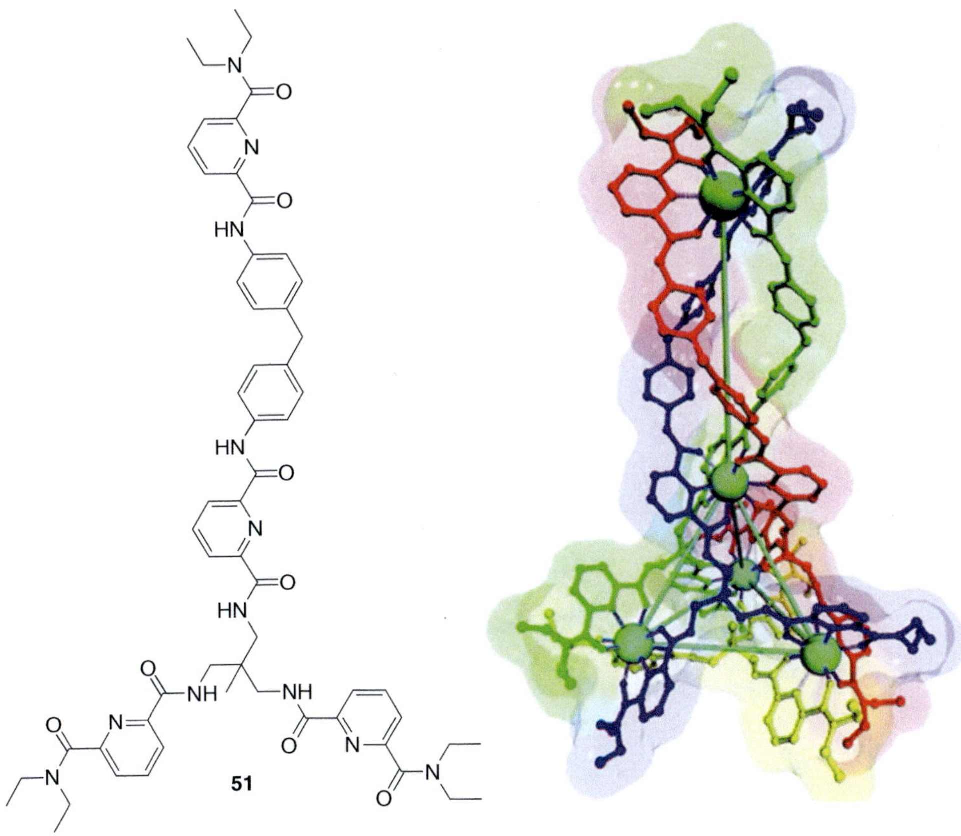

51

Figure 4.27 Ligand **51** developed by Hamacek et al. and the resulting geometry optimized structure of the pentanuclear tower like structure [Eu_5(**51**)$_3$(**46**)]. Reproduced with permission from Ref. [82]. Copyright (2011) American Chemical Society

utility of circular helicates to act as scaffolds for the development of intricate chiral supramolecular architectures such as knots [125].

4.3 Enantioselective Supramolecular Assemblies

The previous section has briefly detailed some of the most recent research into using achiral ligands to construct chiral supramolecular architectures. However, a reoccurring theme throughout has been that the systems usually exist as mixtures of enantiomers. Given the importance of chirality and chiral recognition in nature as well as in science and technology, significant research has gone into the development of artificial chiral systems, including enantioselective supramolecular assemblies. The most common approach to enantioselective synthesis of self-assembled architectures is to include a known chiral group into one of the components; most often the ligand as introducing asymmetric carbon atoms into ligand backbones is well known and relatively trivial. The remainder of this chapter will

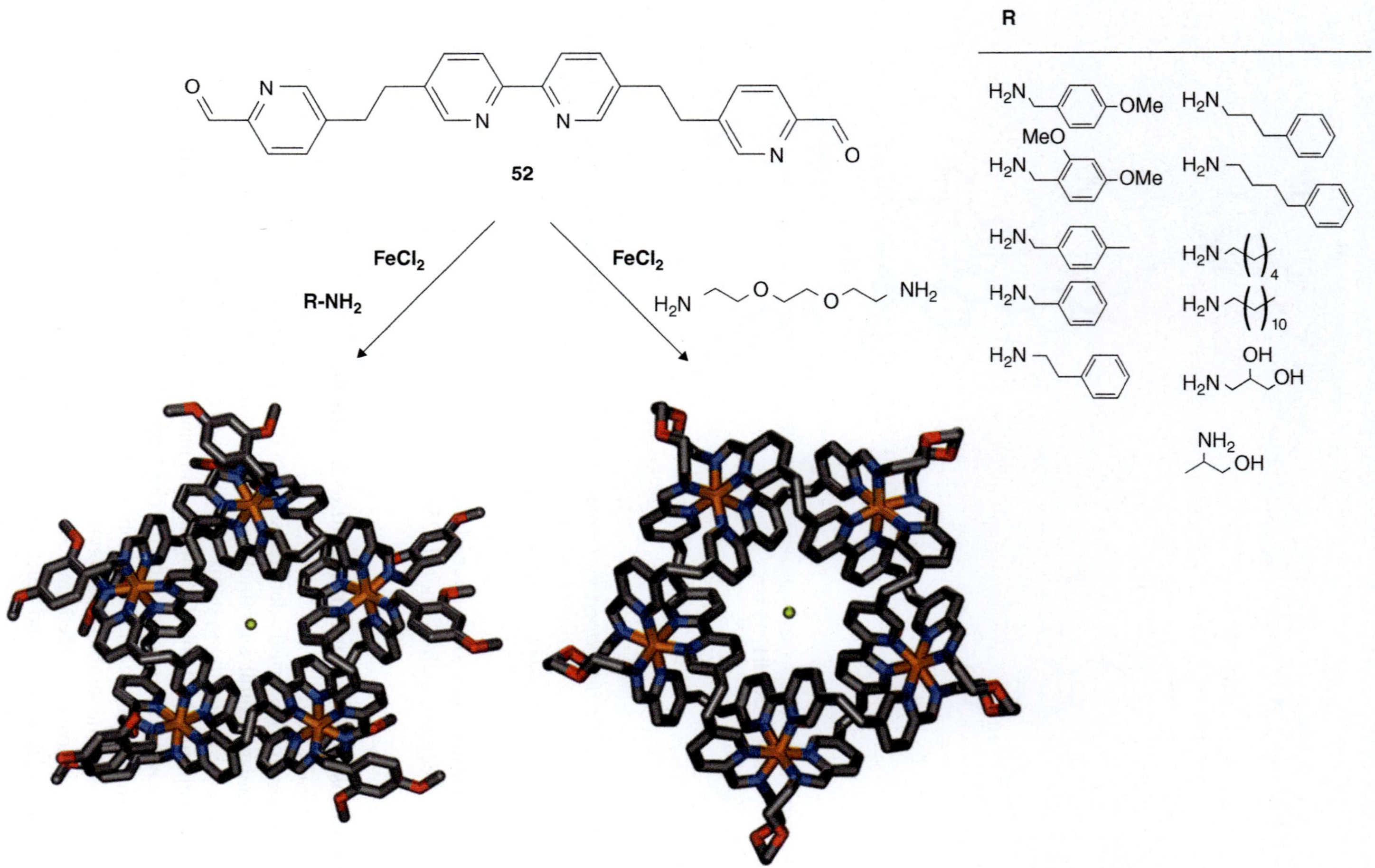

Figure 4.28 Self-assembly system developed by Leigh and coworkers. The nature of the R group dictated the end structure as either a circular helicate or a molecular knot. In each case a Cl^- is bound into the central cavity (green atom) [123, 124]

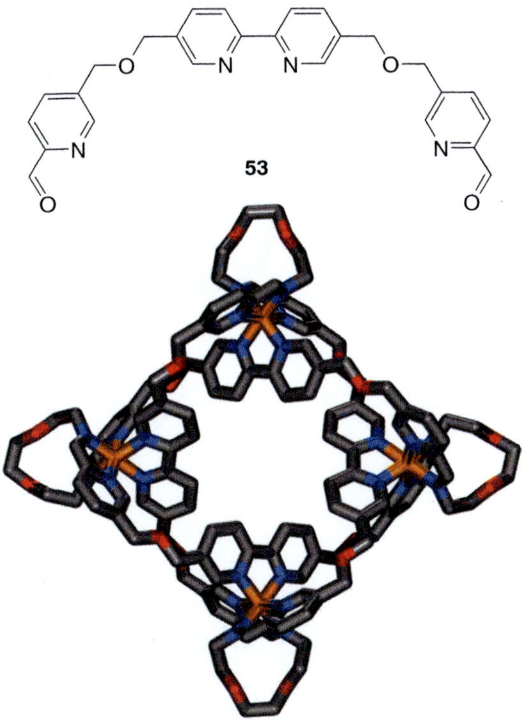

Figure 4.29 *Tetra-iron(II) molecular Solomon knot developed by Leigh and coworkers from di-aldehyde* ***53*** *[125]*

introduce, again through appropriate *recent* examples, research efforts into developing systems that exhibit enantioselective supramolecular synthesis and will focus on some of the supramolecular topologies mentioned above including mononuclear bundles, helicates, cages, and interlocked systems.

4.3.1 Mononuclear Bundles

We begin our discussion on enantioselective synthesis by looking at the simplest systems – mononuclear complexes.

In recent years there has been significant interest in developing chiral luminescent probes for selective chiral sensing and cellular imaging applications, and whilst there are different approaches to realizing materials of this type, Gunnlaugsson and coworkers have focused on the development of lanthanide luminescent self-assemblies [51]. As such they have carried out significant research into the development of *enantioselective* mononuclear and dinuclear lanthanide self-assemblies. Their initial study incorporated chiral 2-naphthalene groups into 2,6-pyridyldiamine based ligands to give the tridentate ligands **56** and **57** [126].

Complexation with Eu^{3+}, Tb^{3+} and Sm^{3+} resulted in the selective formation of the Λ complexes with **56** and the Δ complex when **57** was used. Single crystals of composition $[Ln(\mathbf{L})_3]^{3+}$ were grown and all crystallized in chiral spacegroups with only one enantiomer present (Figure 4.30). The large naphthalene rings resulted in significant intramolecular

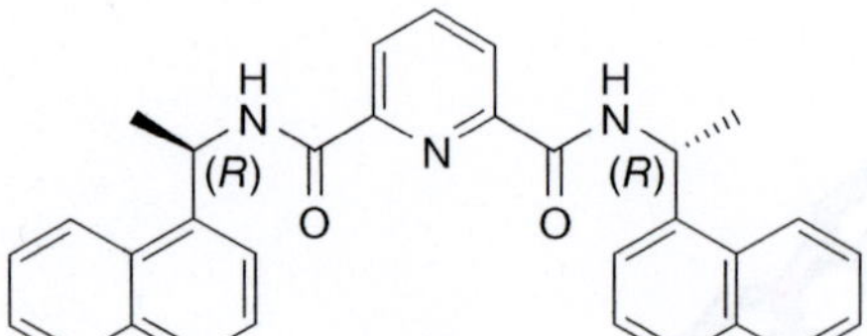

56

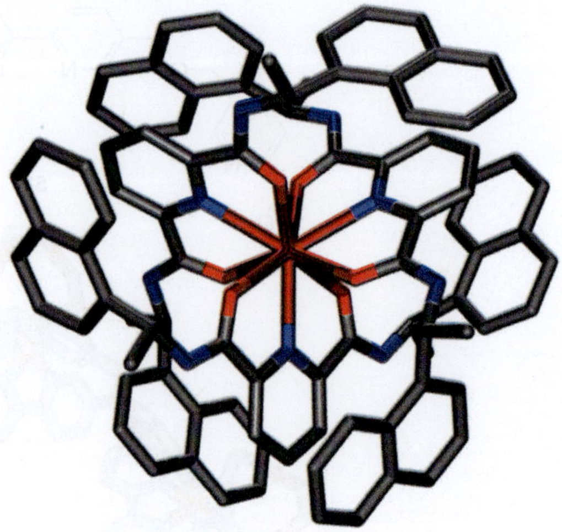

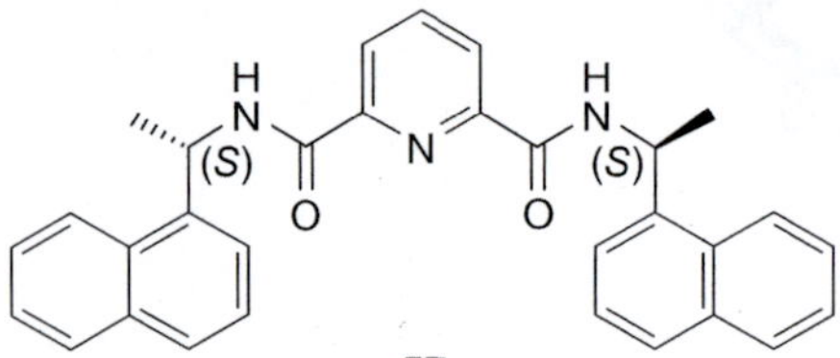

57

Figure 4.30 *Molecular bundles developed by Gunnlaugsson and coworkers where the chiral ligands **56** and **57** transfer their chirality into the [Eu(**L**)$_3$] bundles that form [126]*

π-stacking between the naphthalene arms and the central pyridine giving a tightly packed molecular bundle. These chiral bundles further packed into helical tubes through hydrogen bonding interactions between amide NH protons and triflate counterions (Figure 4.31). With applications as potential bio-probes these systems must be stable to racemization, and given the optical properties (i.e. the naphthalene group is a good antenna for sensitizing Eu^{3+}, and to a lesser extent Tb^{3+} emission) the solution assembly/stability can be monitored using absorption and emission spectroscopy.

In solution the 3:1 L:M species was observed using UV/vis and luminescent titrations, and CD spectroscopy revealed solutions of enantiomers to have equal and opposite spectra. Circularly polarized luminescence (CPL) spectroscopy was used to assess the chirality of

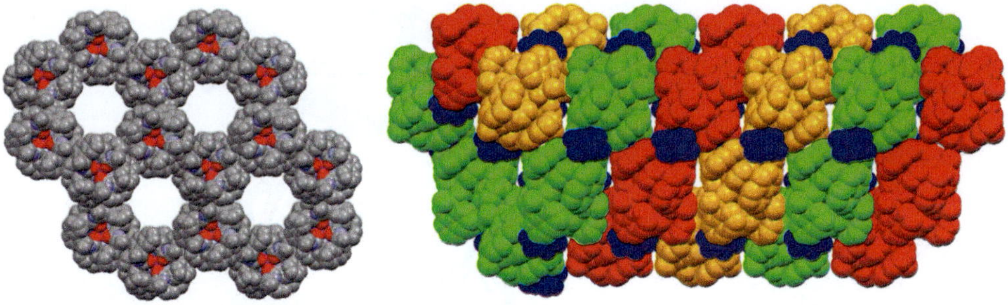

Figure 4.31 Packing of the mononuclear bundles developed by Gunnlaugsson et al. [126]

the bundles and revealed that the chirality of the ligand is transferred to the metal excited state resulting in chiral emission. In further studies, the same group investigated the chiroptical spectroscopy further and found that the complexes are highly stable with respect to inversion of chirality – a feature attributed to the packing of the chiral ligands about the lanthanide ions [127]. The optical purity of these systems, brought about by self-assembly processes, led Gunnlaugsson *et al.* to develop asymmetric chiral ligands for the formation of luminescent molecular bundles for incorporation into Langmuir–Blodgett films [128]. Whilst these will not be discussed in detail it is worth noting that despite one less chiral site per ligand the system still forms the 3 : 1 assembly with Ln^{3+} ions and displays chiral emission. Additionally, when the 3:1 assemblies were deposited onto solid supports the chiral emission was retained. The retention of chirality processing highlighted the potential for chiral self-assemblies to be used for other soft-materials-based applications.

Other groups have also investigated ligands for the formation of mononuclear Ln^{3+} molecular bundles. For example the group of Mazzanti developed a tetradentate ligand with a chiral oxazoline incorporated into the pyridyl-carboxylate structure (**58**) [129]. The ligand was expected to generate chiral Ln^{3+} complexes of general composition $[Ln(L)_2]^+$. Interestingly, whilst chiral mononuclear species were isolated, it was found that, when carrying out the self-assembly studies at high concentrations, instead of observing the mononuclear species, they isolated a self-assembled Eu^{3+} enantiopure ΔΔΔ triangle where the ligand chirality (*S*) was transferred to the metals (Figure 4.32).

In this system the transfer of chirality was found to be highly concentration dependent. At low concentrations, diastereomeric mononuclear complexes (Δ)[Eu(**58**)$_2$]$^+$ and (Λ)[Eu(**58**)$_2$]$^+$ were observed; however, high concentrations resulted in the triangle mentioned above via self-assembly (through the free carboxylate group bridging metal centres) of three mononuclear Δ complexes. Single crystal diffraction studies confirmed the chirality of the trinuclear species as Δ and also showed that the Eu^{3+} triangle cocrystallized with a mononuclear (Λ) [Eu.(**58**)2]$^+$ complex (Figure 4.32). Nuclear magnetic resonance studies revealed that the self-assembly for the trinuclear species *only* occurred for (Δ)[Eu(**58**)$_2$]$^+$ and not for (Λ)[Eu(**58**)$_2$]$^+$ thereby showcasing a relatively rare instance of selective homochiral recognition. In contrast to the previously discussed system of Gunnlaugsson *et al.*, where chirality was selectively transferred to the mononuclear bundles depending on the chirality of the ligand (*R* versus *S*), the inclusion of a chiral entity into the tetradentate ligand species did not result in selective formation of Δ or Λ complexes. This is most likely caused by the less constrained

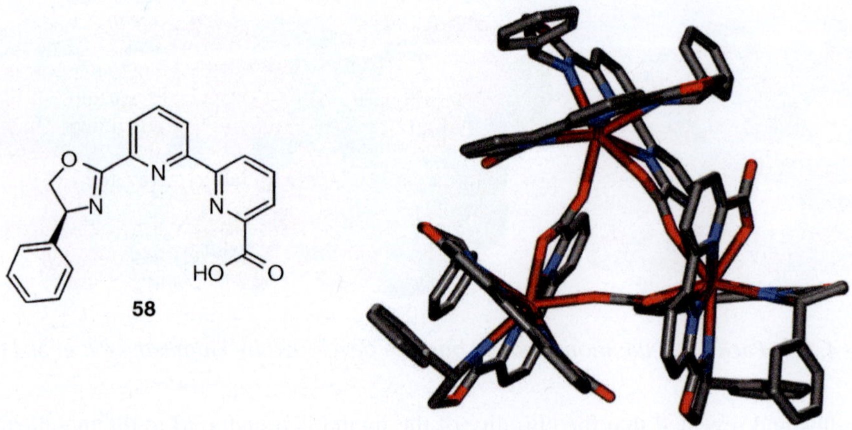

Figure 4.32 *Chiral oxazole ligand (58) developed by Mazzanti et al. and the resulting chiral Eu^{3+} triangle formed when complexation was carried out in high concentrations [129]*

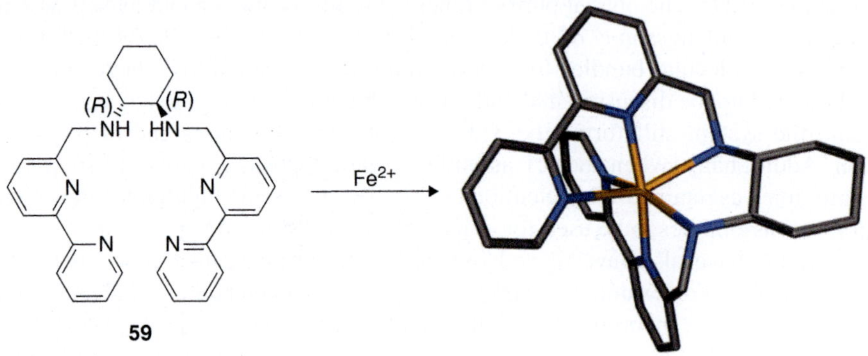

Figure 4.33 *Mononuclear chiral system developed by Constable et al. where the flexible ligand wraps around a single metal centre and induces chirality at the metal ion [131]*

2:1 complex (versus 3:1 for **56** and **57**) allowing for more flexibility with respect to the coordination sphere, and thus the chiral ligand does not constrain the ligands into predefined configurations around the metal. This work was further extended into a large heptanuclear Eu^{3+} wheel on addition of excess $Eu(CF_3SO_3)_3$ [130].

Constable and coworkers used enantiopure (*R*,*R*) hexadentate diamine based ligand **59** to prepare mononuclear *d*-block complexes (Figure 4.33) [131].

Given the relatively flexible nature of **59**, through inversion at the nitrogen atoms both *M* and *P* mononuclear helical structures should be possible; however, when **59** was complexed with Cu^{2+}, Zn^{2+} and Fe^{2+} only the *M*-[M(**59**)] complex was observed. This observation was attributed to the more favourable *anti*-configuration adopted between the NH and the adjacent CH on the cyclohexane ring for the *M* helicate as opposed to the *syn* configuration, which would be less energetically favoured, in the case of the *P* helicate. This study showed that enantiopure ligands can be used to induce chirality at *d*-metal centres, an area of research that has been well studied over the years.

4.3.2 Helicates

Enantioselective synthesis of multinuclear systems has also been the subject of extensive research in recent years, again targeting their potential application in biomaterials, biological systems and chiral smart materials. Given that helicates are formed from achiral ligands as mixtures of *M* and *P* helicates, research has emerged where chiral ligands have been developed to induce absolute helicity (i.e. selectively isolate either *M* or *P* helicates) in the resultant self-assembly.

Setsune and coworkers recently reported the isolation of a di-nickel single-stranded helicate from the benzenehexapyrrole-α,ω-dialdehyde **60** (Figure 4.34) [132].

Reaction of **60** with $Ni(OAc)_2 \cdot 4H_2O$ resulted in isolation of the helical assembly [Ni_2(**60**)] in 92% yield; however, it was isolated as a mixture of *M* and *P* helicates (**60** is achiral). To generate diastereoselective formation of helicates the group reacted [Ni_2(**60**)] with chiral amines to generate chiral imines *in situ*. By monitoring the CD spectra of these complexes they found significant diastereoselectivity depending on the type of chiral auxiliary used, opening up access to a library of enantiopure helicates with different functionalities (potentially introduced via the chiral amine).

The enantioselective formation of dinuclear double-stranded helicates has also been the subject of much investigation. In the same study detailed above (i.e. ligand **59**) Constable and coworkers used the chiral hexa-dentate imine ligand **61** to selectively form *M* helicates from Zn^{2+} and Ag^{+} of general conformation [M_2(**L**)$_2$] (Figure 4.35). In these systems the rigid ligand results in stereochemical control of helicate formation [131].

Enantioselective triple-stranded helicate formation has also been investigated extensively over the past few years. Some particularly elegant examples have been reported by Gunnlaugsson and coworkers, where they effectively extended the molecular bundle type systems described above by linking two of them together. Using ligands **62–64**, which contained chiral naphthalene groups appended to the pyridine diamide moiety, which in turn is joined to a second biding unit via linking groups to give the overall C_2 symmetric *R,R* or *S,S* enantiopure ligands (Figure 4.36) [36, 133–137].

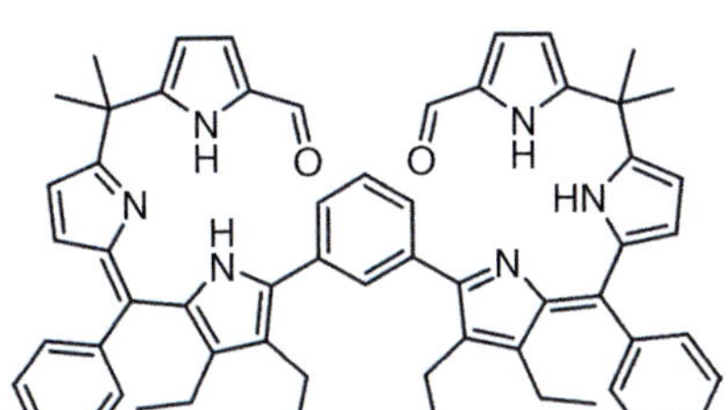

60

Figure 4.34 *Di-nickel(II) single-stranded helicate isolated from the chiral hexapyrrole based ligand **60** developed by Setsune et al. [132]*

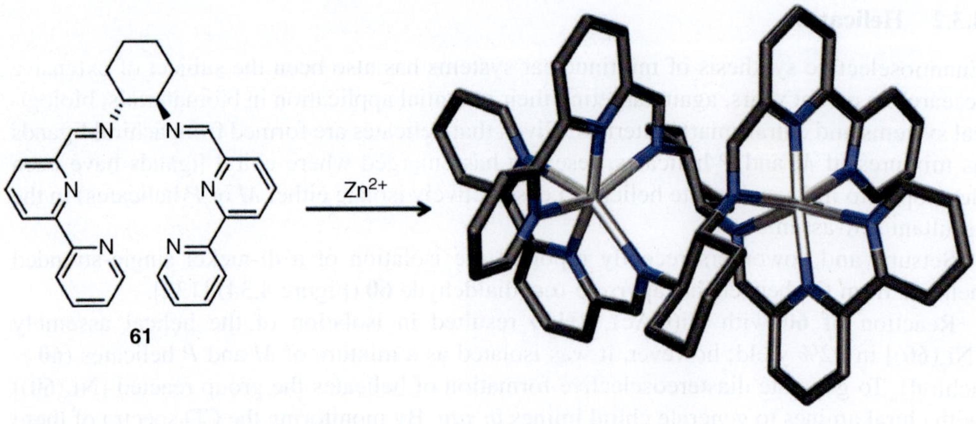

61

Figure 4.35 *Preparation of enantiopure Zn^{2+} dinuclear double-stranded helicates using ligand **61** [131]*

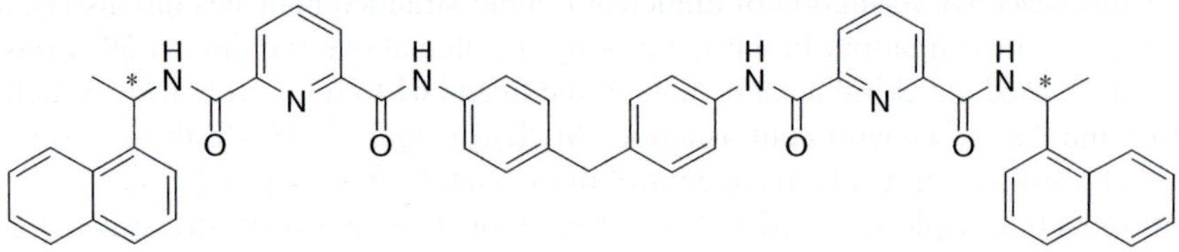

62

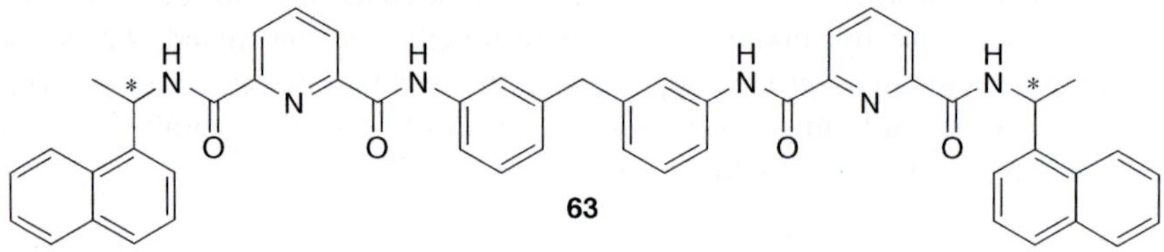

63

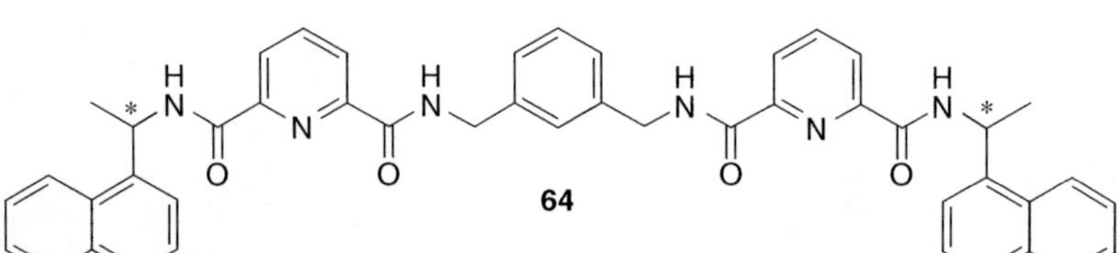

64

Figure 4.36 *Ditopic ligands developed by Gunnlaugsson et al. for the preparation of enantiopure di-lanthanide triple-stranded helicates $[Ln_2(\mathbf{L})_3]$ [134–136]*

The interaction of these chiral ligands with a variety of Ln^{3+} ions gave rise to the expected dimetallic triple-stranded helicates with control over the helicity of the self-assembly. Whilst crystallographic evidence was not obtained for these systems, molecular modelling (Figure 4.37) and detailed spectroscopic (NMR, UV/vis, luminescence, CD and CPL spectroscopies) studies were carried out to confirm the adoption of

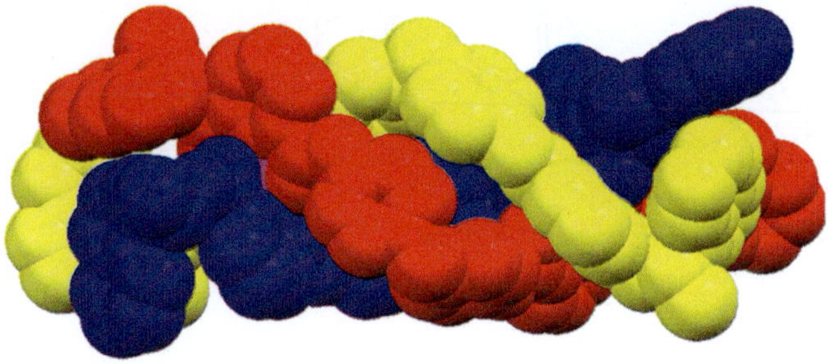

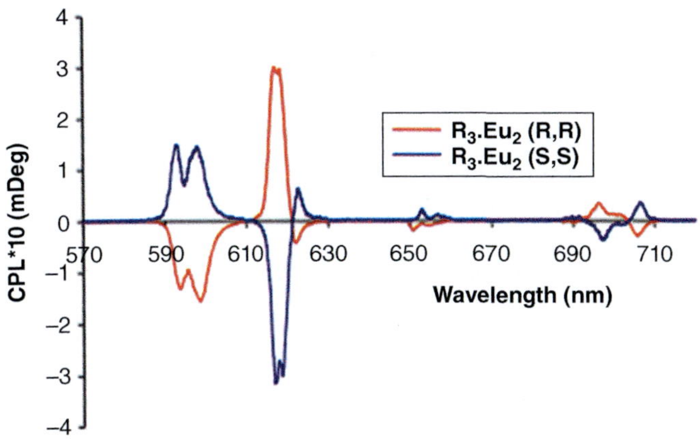

Figure 4.37 *Energy minimized structure of [Eu_2(**63**)$_3$] (left) and CPL spectra of enantiomers of Eu helicates (right). Reproduced with permission from Ref. [136]. Copyright (2009) American Chemical Society*

enantioselective helicates. The CD and CPL spectra of these systems showed the adoption of only *M* or *P* helicates (ΛΛ for *R,R* and ΔΔ for *S,S*) but not mixtures. The selective assembly of *M* or *P* helicates was put down to the favourable packing interactions from the naphthalene units at the ends of the helicates.

Solution spectroscopic titrations indicated that the formation of the dimetallic triple-stranded helicates depends on the stoichiometric amount of Ln^{3+} in solution (Figure 4.37). The same research group is currently investigating the use of helicates for the development of bioprobes and chiral sensors by attempting to make water-soluble versions for aqueous self-assembly of enantiopure helicates through rational ligand design. This study highlighted how *f*-directed assembly using enantiopure ligands can result in enantioselective triple-stranded helicate formation in which the helicates have interesting emission properties.

Hannon and coworkers also developed enantioselective triple-stranded helicates by appending chiral arginine conjugates on the periphery of metallo-supramolecular 'cylinders' (Figure 4.38). [138] These helical structures were prepared using ligand **65** and complexing with Fe^{2+} to give [Fe_2(**65**)$_3$]$^{4+}$ where they found that ligand **L-65** selectively gave the *P*-helicate whereas **R-65** exclusively gave the *M*-helicate as assessed by CD spectroscopy. The helicates were also assessed for their ability to interact with DNA and they found that

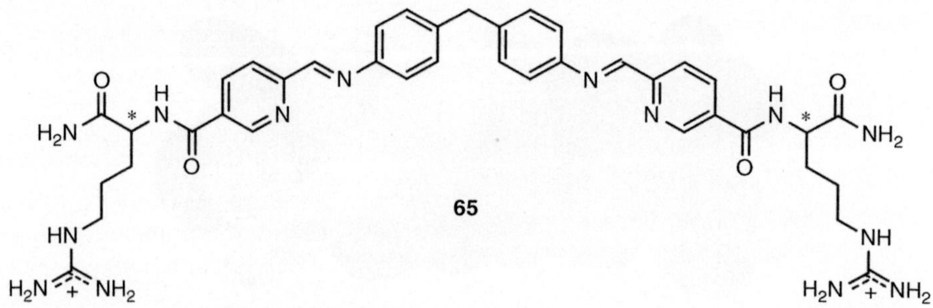

Figure 4.38 *Arginine appended ligand developed by Hannon and coworkers for the enantioselective synthesis of* Fe^{2+} *triple-stranded helicates [138]*

the *M*-helicate stabilized DNA three-way junctions more effectively than the *P*-helicate. This study showed that by controlling the helicity of self-assemblies it is possible to develop more targeted biologically relevant materials.

4.3.3 Higher Order Enantioselective Assemblies

Quadruple-stranded helicates formed from enantiopure ligands have also been targeted; for example, the work of Cui and coworkers investigated the homochiral salan-based ligand **66** and its assembly with Zn^{2+} [139].

In this system the result is a large chiral helical cage structure (Figure 4.39). The major building unit is a Zn_2L_2 dimer where each Zn^{2+} ion is coordinated in a N_2O_2 salan coordination pocket with one O donor atom from each ligand bridging the neighbouring Zn^{2+} to give overall five-coordinate Zn^{2+} and the dimer structure. The four pendant pyridyl groups in the dimer each coordinate to another Zn^{2+} ion such that two sets of dimers are linked (via py-Zn coordination) into an overall octanuclear complex [Zn_8(**66**)$_4Cl_8$], best described as a *P*-configured quadruple helix. The self-assembly described by Cui *et al.* was reported to display chiral amplification as the molar optical rotation (φ) for **66** was measured as 674.9 deg cm^3 dm^{-1} whereas for [Zn_8(**66**)$_4Cl_8$] it was 31208.7 deg cm^3 dm^{-1}. In this system, the chirality of the salan ligand is essential for helicate formation. Under identical conditions but using *rac*-**66** only a racemic dimer was isolated. The researchers went on to use the chiral cage structure to selectively recognize enantiomers of amino acids through fluorescence enhancement in solution. They were also able to separate small racemic organic compounds by adsorption in the solid (crystalline) state.

Another study that prepared enantioselective quadruple-stranded helicates was carried out by Williams and coworkers who used ridged chiral benzimidazole based ligands **67–69** (Figure 4.40) [140].

They found that the chiral ligands arrange in a helical fashion about a tetranuclear cluster of Co^{2+} ions [Co_4[(*R*/*S*)-**L**]$_4$(dpa)$_2$] (dpa = diphenylphosphinic acid, Figure 4.40) where the helicity of the tetranuclear cluster is dictated by the chirality of the ligand. The selectivity of these ligands to form the enantiopure helicates was investigated by carrying out the complexation using a mixture of (*R*)**68** and (*S*)**69**. Using ESI-MS they found that, initially, a statistical mixture of all possible combinations of ligand : cubane {i.e. five compositions of [Co_4[(*R*)-**68**]$_{4-n}$[(*S*)-**69**]$_n$(dpa)]} were present however on leaving the

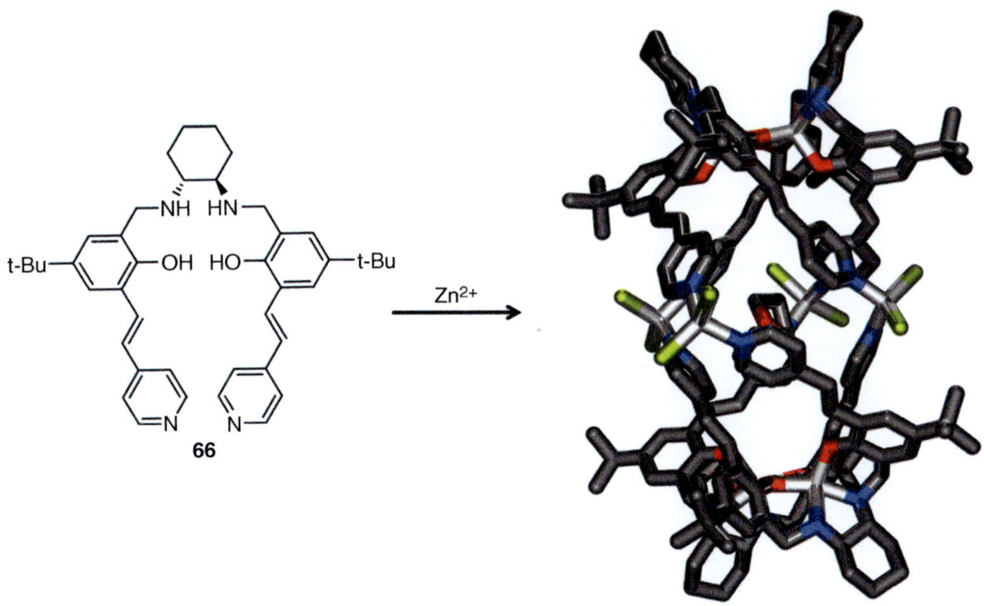

Figure 4.39 P-configured quadruple-helicate formed on complexation with Zn^{2+} [139]

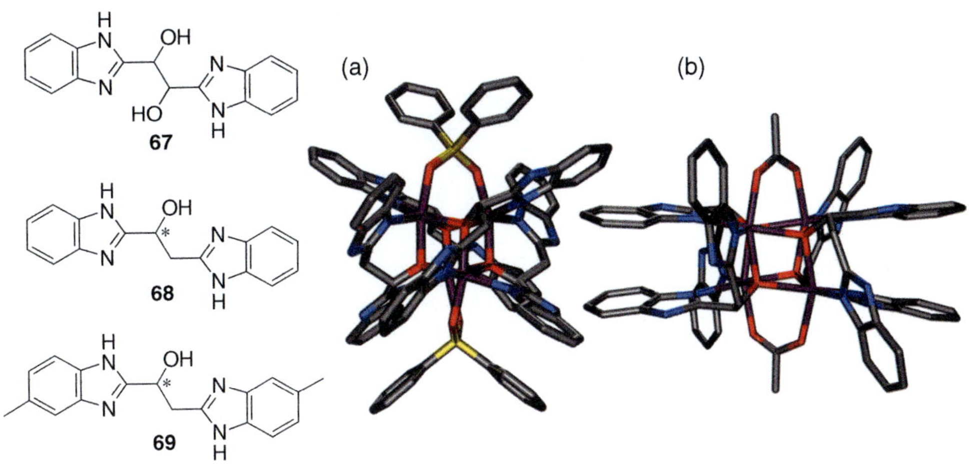

Figure 4.40 Co^{2+} cluster helicates formed using ligands **67–69** [140]

mixture to equilibrate for 3 days it was able to self-assemble into only the *RRRR*, *SSSS* and *RRSS* components (the *RRSS* component was eventually isolated and found to be a meso system).

Nitschke *et al.* were able to extend the aforementioned edged-capped tetrahedral cages into enantiopure cages again through fortuitous ligand choice. In one study they appended

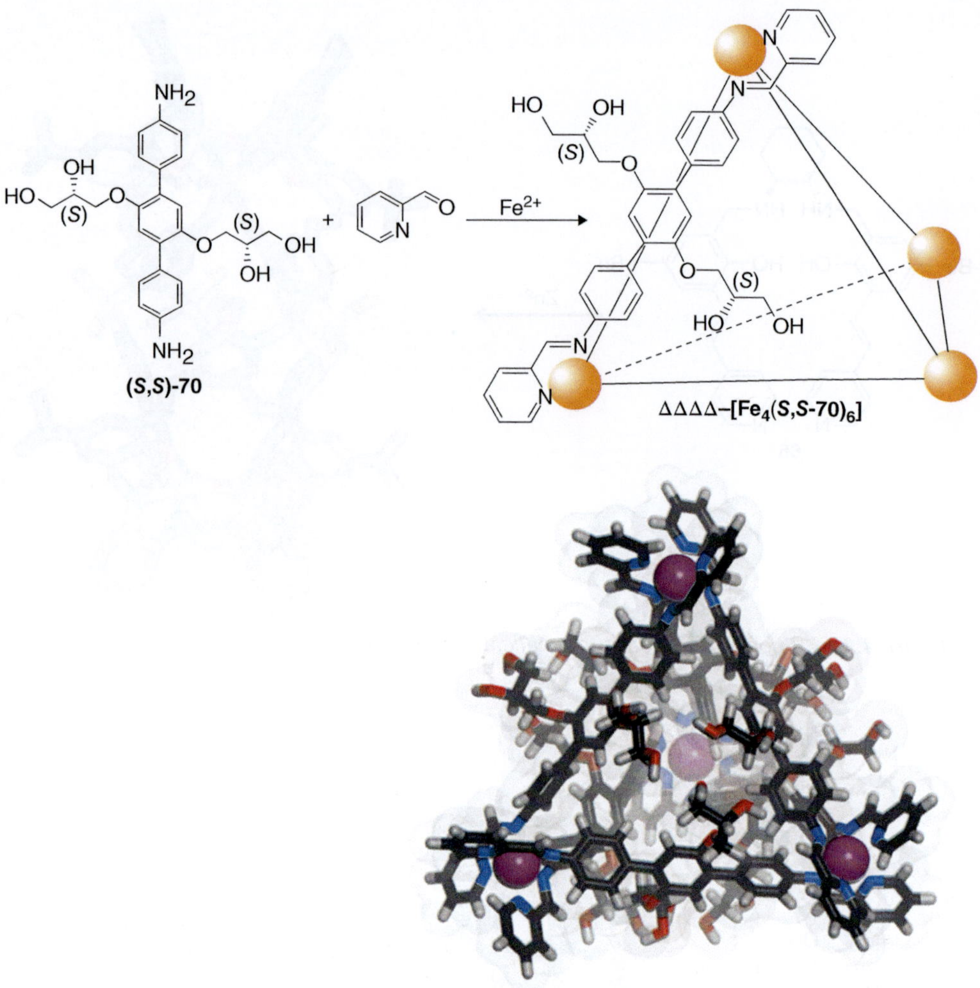

Figure 4.41 *Enantioselective cage formation using an enantiopure ligand from **70**. Geometry optimized structure (right) reproduced with permission from Ref. [141]. Copyright 2013 Wiley-VCH*

chiral 1,2-propanediol groups to the periphery of some extended diamines (***R,R*-70** and ***S,S*-70**) and used them in a subcomponent self-assembly with Fe^{2+} and 2-formylpyridine (Figure 4.41) [141]. Unlike similar self-assembled systems constructed from achiral components that gave mixtures of stereoisomers, the assemblies formed from ***R,R*-70** and ***S,S*-70** were homochiral $[Fe_4(\mathbf{70})_6]$ tetrahedron of ΔΔΔΔ and ΛΛΛΛ respectively (Figure 4.41). $[Fe_4(\boldsymbol{S,S}\text{-}\mathbf{70})_6]$ was tested for its encapsulation properties and displayed host binding for a range of small- to medium-sized molecules. This system was also able to catalyse the hydrolysis of the toxic organophosphate derivative dichlorvos. The chiral centres of the ligands in this system were significantly removed from the metal centres so it was somewhat surprising that a homochiral cage formed; however, such 'remote' chiral

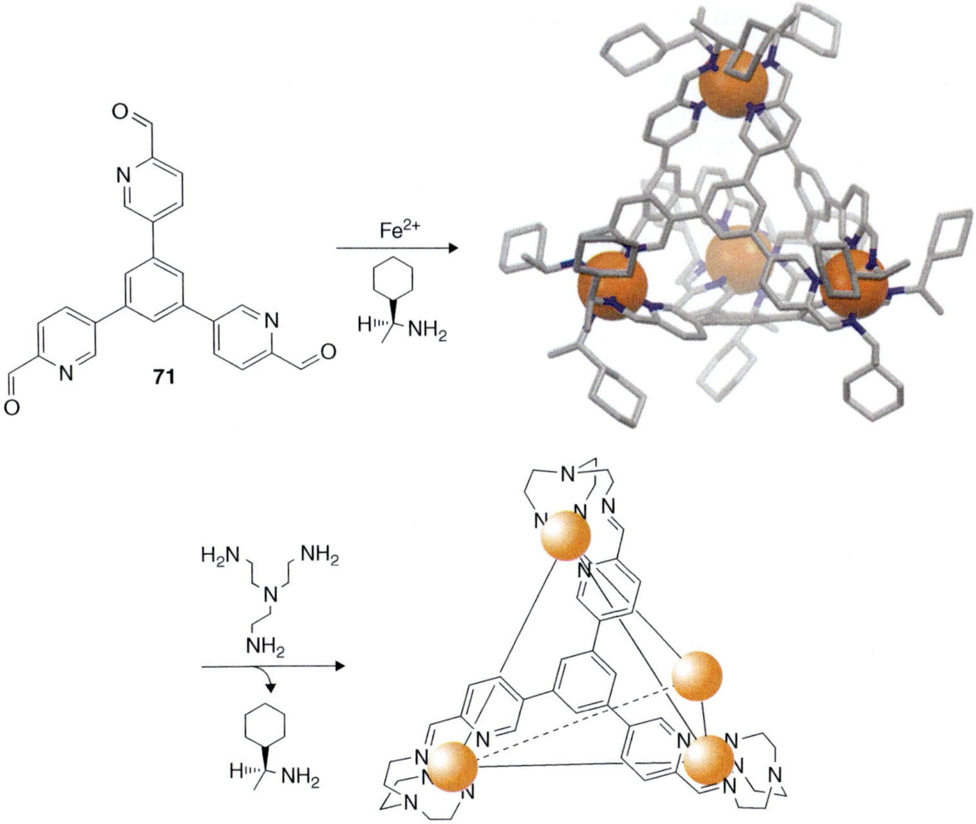

Figure 4.42 *Enantioselective cage formation (first step) and retainment of chirality on replacement of the capping group (second step) [142]*

induction highlights that only subtle changes to the system's design can result in significant changes to the overall architecture (i.e. enantiopure versus mixtures of chirality).

Stereochemical memory was also displayed by a face-capped, M_4L_4 tetrahedral cage that Nitschke *et al.* developed. In this system a homochiral tetrahedron was assembled using a tri-aldehyde (**71**), Fe^{2+} and a chiral amine (Figure 4.42) [142]. The result of using the *S*-amine was an enantiopure ΔΔΔΔ cage and using *R*-amine was an enantiopure ΛΛΛΛ cage; the researchers then undertook a subcomponent substitution by exchanging the chiral amine components (now imines) with an achiral capping group (TREN) and the result was retention of stereochemistry (99% ee). To confirm this memory/chiral templating effect, when the tetrahedron formation was carried out *without* first using the chiral amines (i.e. directly using TREN), only a racemic mixture was obtained. The long-lasting chiral transfer and durable storage of chiral information is important for the development of new chiral cages as there is potential for reusable sources of chiral information. This is an interesting case of a chiral pre-system inducing chirality to give an enantiopure chiral cage where the ligand constituents in the final species are achiral.

4.4 Conclusions and Future Outlook

In this chapter we have covered some of the many examples of chiral metallo-supramolecular assemblies. We have attempted to give a short overview of the field, focusing on the most recent examples of chirogenesis brought about through metal-ligand coordination involving *p*-, *d*- and *f*- block metal ions. The area of metallo-supramolecular chirality is fast growing and central to the further development of supramolecular chemistry in general. It is also clear from the literature that has emerged recently that there is now a drive to use chiral self-assembly for the production of application-based systems. A brief snapshot of applications was discussed within this chapter for example magnetically active helicates and cages, DNA-binding chiral-helicates, chiral-separation and sensing, and stabilization of reactive intermediates/species. This fast-moving research field will no doubt lead to more interesting and intricate architectures in the very near future. This, in turn, will allow researchers to develop our understanding of the complexity of supramolecular assemblies further and will lead to new advanced supramolecular materials with wide-ranging applications.

References

[1] T. Kreickmann, C. Diedrich, T. Pape, *et al.*, Metallosupramolecular Chemistry with Bis(benzene-o-dithiolato) Ligands, *Journal of the American Chemical Society*, **128**(36), 11808–11819 (2006).

[2] V. Borovkov and Y. Inoue, Supramolecular Chirogenesis in Host–Guest Systems Containing Porphyrinoids, in M. Crego-Calama and D. Reinhoudt (Eds), *Supramolecular Chirality*, 265, Springer, Berlin, pp. 89–146 (2006).

[3] A. Scarso and J. Rebek Jr., Chiral Spaces in Supramolecular Assemblies, in M. Crego-Calama and D. Reinhoudt (Eds), *Supramolecular Chirality*, Springer, Berlin, pp. 1–46 (2006).

[4] K. Maeda and E. Yashima, Dynamic Helical Structures: Detection and Amplification of Chirality, in M. Crego-Calama and D. Reinhoudt (Eds), *Supramolecular Chirality*, Springer, Berlin, pp. 47–88 (2006).

[5] G. Seeber, B.F. Tiedemann and K. Raymond, Supramolecular Chirality in Coordination Chemistry, in M. Crego-Calama and D. Reinhoudt (Eds), *Supramolecular Chirality*, Springer, Berlin, pp. 147–183 (2006).

[6] D. Amabilino and J. Veciana, Supramolecular Chiral Functional Materials, in M. Crego-Calama and D. Reinhoudt (Eds), *Supramolecular Chirality*, Springer, Berlin, pp. 253–302 (2006).

[7] T. Sasaki, I. Hisaki, T. Miyano, *et al.*, Linkage Control between Molecular and Supramolecular Chirality in 21-Helical Hydrogen-Bonded Networks using Achiral Components, *Nature Communications*, **4**, 1787 (2013).

[8] A.G. de Bruin, M.E. Barbour and W.H. Briscoe, Macromolecular and Supramolecular Chirality: A Twist in the Polymer Tales, *Polymer International*, **63**(2), 165–171 (2014).

[9] A. Priimagi, G. Cavallo, P. Metrangolo and G. Resnati, The Halogen Bond in the Design of Functional Supramolecular Materials: Recent Advances, *Accounts of Chemical Research*, **46**(11), 2686–2695 (2013).

[10] H. Miyake and H. Tsukube, Coordination Chemistry Strategies for Dynamic Helicates: Time-Programmable Chirality Switching with Labile and Inert Metal Helicates, *Chemical Society Reviews*, **41**(21), 6977–6991 (2012).

[11] L. Zhang, L. Qin, X. Wang *et al.*, Supramolecular Chirality in Self-Assembled Soft Materials: Regulation of Chiral Nanostructures and Chiral Functions, *Advanced Materials*, **26**(40), 6959–6964 (2014).

[12] V. Borovkov, Supramolecular Chirality in Porphyrin Chemistry, *Symmetry*, **6**(2), 256–294 (2014).

[13] M. Fujiki, Supramolecular Chirality: Solvent Chirality Transfer in Molecular Chemistry and Polymer Chemistry, *Symmetry*, **6**(3), 677–703 (2014).
[14] M. Senge, A. Ryan, K. Letchford, S. MacGowan and T. Mielke, Chlorophylls, Symmetry, Chirality, and Photosynthesis, *Symmetry*, **6**(3), 781–843 (2014).
[15] H. Miyake, Supramolecular Chirality in Dynamic Coordination Chemistry, *Symmetry*, **6**(4), 880–895 (2014).
[16] M. Famiano, R. Boyd, T. Kajino, *et al.*, Determining Amino Acid Chirality in the Supernova Neutrino Processing Model, *Symmetry*, **6**(4), 909–925 (2014).
[17] M.A. Mateos-Timoneda, M. Crego-Calama and D.N. Reinhoudt, Supramolecular Chirality of Self-Assembled Systems in Solution, *Chemical Society Reviews*, **33**(6), 363–372 (2004).
[18] C. He, Y. Zhao, D. Guo, *et al.*, Chirality Transfer through Helical Motifs in Coordination Compounds, *European Journal of Inorganic Chemistry*, **22**, 3451–3463 (2007).
[19] E.C. Constable, Stereogenic metal centres – from Werner to Supramolecular Chemistry, *Chemical Society Reviews*, **42**(4), 1637–1651 (2013).
[20] D. Chambers and E. Flapan, Topological Symmetry Groups of Small Complete Graphs, *Symmetry*, **6**(2), 189–209 (2014).
[21] G. Simonneaux, H. Srour, P. Maux, *et al.*, Metalloporphyrin Symmetry in Chiral Recognition and Enantioselective Catalysis, *Symmetry*, **6**(2), 210–221 (2014).
[22] J. Labuta, S. Ishihara, K. Ariga and J. Hill, Dynamic Processes in Prochiral Solvating Agents (pro-CSAs) Studied by NMR Spectroscopy, *Symmetry*, **6**(2), 345–367 (2014).
[23] A. Tsuda, Hydrodynamic Helical Orientations of Nanofibers in a Vortex, *Symmetry*, **6**(2), 383–395 (2014).
[24] I. Dierking, Chiral Liquid Crystals: Structures, Phases, Effects, *Symmetry*, **6**(2), 444–472 (2014).
[25] R. Kurdi, A. Táborosi, C. Zucchi and G. Pályi, Autosolvation: Architecture and Selection of Chiral Conformers in Alkylcobalt Carbonyl Molecular Clocks, *Symmetry*, **6**(3), 551–565 (2014).
[26] D. Philp and J.F. Stoddart, Self-Assembly in Natural and Unnatural Systems, *Angewandte Chemie International Edition*, **35**(11), 1155–1196 (1996).
[27] G.M. Whitesides and M. Boncheva, Beyond Molecules: Self-Assembly of Mesoscopic and Macroscopic Components, *Proceedings of the National Academy of Sciences of the United States of America*, **99**(8), 4769–4774 (2002).
[28] S. Tibbits, Design to Self-Assembly, *Architectural Design*, **82**, 68–73 (2012).
[29] L.F. Lindoy and I.M. Atkinson, Self-Assembly in Supramolecular Systems, in J.F. Stoddart (Ed.), *RSC Monographs in Supramolecular Chemistry*, Royal Society of Chemistry, Cambridge (2000).
[30] M.V. Escarcega-Bobadilla and A.W. Kleij, Artificial Chirogenesis: A Gateway to New Opportunities in Material Science and Catalysis, *Chemical Science*, **3**(8), 2421–2428 (2012).
[31] V.V. Borovkov, G.A. Hembury and Y. Inoue, Origin, Control, and Application of Supramolecular Chirogenesis in Bisporphyrin-Based Systems, *Accounts of Chemical Research*, **37**(7), 449–459 (2004).
[32] M. Albrecht, Helicates Making Head-Head-Tails of It, *Nature Chemistry*, **6**(9), 761–762 (2014).
[33] M. Boiocchi and L. Fabbrizzi, Double-Stranded Dimetallic Helicates: Assembling-Disassembling Driven by the Cu^{I}/Cu^{II} Redox Change and the Principle of Homochiral Recognition, *Chemical Society Reviews*, **43**(6), 1835–1847 (2014).
[34] J.-C. G. Bünzli, Photophysical Properties and Applications of Lanthanoid Helicates, in G. Maayan and M. Albrecht (Eds) *Metallofoldamers: Supramolecular Architectures from Helicates to Biomimetics*, John Wiley & Sons, Ltd, Chichester (2013).
[35] M. Albrecht, Supramolecular Templating in the Formation of Helicates, in C. A. Schalley, F. Vögtle and K. H. Dötz (Eds) *Templates in Chemistry I*, Springer, Berlin, pp. 105–139 (2004).
[36] C. Lincheneau, F. Stomeo, S. Comby and T. Gunnlaugsson, Recent Highlights in the use of Lanthanide-directed Synthesis of Novel Supramolecular (Luminescent) Self-Assembly Structures such as Coordination Bundles, Helicates and Sensors, *Australian Journal of Chemistry*, **64**(10), 1315–1326 (2011).
[37] M. Albrecht, Homodinuclear *f-f* and Heterodinuclear *f-p* Lanthanide Helicates, *Zeitschrift für anorganische und allgemeine Chemie*, **636**(12), 2198–2204 (2010).

[38] P.J. Lusby, Supramolecular Coordination Chemistry, *Annual Reports Section 'A' (Inorganic Chemistry)*, **108**, 292–314 (2012).
[39] C. Piguet, G. Bernardinelli and G. Hopfgartner, Helicates as Versatile Supramolecular Complexes, *Chemical Reviews*, **97**, 2005–2062 (1997).
[40] J.M. Lehn, A. Rigault, J. Siegel, *et al.*, Spontaneous Assembly of Double-Stranded Helicates from Oligobipyridine Ligands and Copper(I) Cations: Structure of an Inorganic Double Helix, *Proceedings of the National Academy of Sciences of the United States of America*, **84**(9), 2565–2569 (1987).
[41] N.K. Al-Rasbi, H. Adams, L.P. Harding and M.D. Ward, Coordination Behaviour of Bis-Terdentate N-donor Ligands: Double- and Single-Stranded Helicates, Mesocates, and Cyclic Oligomers, *European Journal of Inorganic Chemistry*, **30**, 4770–4780 (2007).
[42] J.-M. Senegas, S. Koeller, G. Bernardinelli and C. Piguet, Isolation and Characterization of the First Circular Single-Stranded Polymetallic Lanthanide-Containing Helicate, *Chemical Communications*, **17**, 2235–2237 (2005).
[43] Y. Furusho and E. Yashima, Synthesis and Function of Double-Stranded Helical Polymers and Oligomers, *Macromolecular Rapid Communications*, **32**(2), 136–146 (2011).
[44] Y. Furusho and E. Yashima, Synthesis and Function of Artificial Double Helices, *Journal of Synthetic Organic Chemistry, Japan*, **65**(11), 1121–1133 (2007).
[45] Y. Furusho and E. Yashima, Molecular Design and Synthesis of Artificial Double Helices, *Chemical Record*, **7**(1), 1–11 (2007).
[46] M. Albrecht, Artifical Molecular Double-Stranded Helices, *Angewandte Chemie International Edition*, **44**(40), 6448–6451 (2005).
[47] M.J. Hannon and L.J. Childs, Helices and Helicates: Beautiful Supramolecular Motifs with Emerging Applications, *Supramolecular Chemistry*, **16**(1), 7–22 (2004).
[48.] P.M. Selvakumar, P.Y. Jebaraj, J. Sahoo, *et al.*, The First Bromide Ion Directed Double Helicate and its Role in Catalysis, *RSC Advances*, **2**(20), 7689–7692 (2012).
[49] J.-C.G. Bünzli. Lighting Up Cells with Lanthanide Self-Assembled Helicates, *Interface Focus*, **3**(5), 1–17 (2013).
[50] S. Goetz and P.E. Kruger, A New Twist in Anion Binding: Metallo-Helicate Hosts for Anionic Guests, *Dalton Transactions*, **10**, 1277–1284 (2006).
[51] S.J. Bradberry, A.J. Savyasachi, M. Martinez-Calvo and T. Gunnlaugsson, Development of Responsive Visibly and NIR Luminescent and Supramolecular Coordination Self-Assemblies using Lanthanide Ion Directed Synthesis, *Coordination Chemistry Reviews*, **273**, 226–241 (2014).
[52] D.J. Cooke, J.M. Cross, R.V. Fennessy, *et al.*, Steric Sontrol of the Formation of Dinuclear Double Helicate and Dinuclear Meso-Helicate Assemblies, *Chemical Communications*, **49**(71), 7785–7787 (2013).
[53] T. Hashimoto, T. Nishimura, J.M. Lim, *et al.*, Formation of Metal-Assisted Stable Double Helices in Dimers of Cyclic Bis-tetrapyrroles that Exhibit Spring-Like Motion, *Chemistry – A European Journal*, **16**(38), 11653–11661 (2010).
[54] Z. Zhang and D. Dolphin, Synthesis of Triple-Stranded Complexes Using Bis(dipyrromethene) Ligands, *Inorganic Chemistry*, **49**(24), 11550–11555 (2010).
[55] Q. Miao, J.-Y. Shin, B.O. Patrick and D. Dolphin, Self-Assembly of Oligomeric Linear Dipyrromethene Metal Complexes, **18**, *Chemical Communications*, 2451–2453 (2009).
[56] Z. Zhang and D. Dolphin, A Triple-Stranded Helicate and Mesocate from the Same Metal and Ligand, *Chemical Commmunication*, **45**, 6931–6933 (2009).
[57] Z. Zhang, A. Thompson, S. J. Rettig and D. Dolphin, The Use of Dipyrromethene Ligands in Supramolecular Chemistry, *Journal of the American Chemical Society*, **120**, 13537–13538 (1998).
[58] C. Brückner, S.J. Rettig and D. Dolphin, 2-Pyrrolylthiones as Monoanionic Bidentate N,S-Chelators: Synthesis and Molecular Structure of 2-Pyrrolylthionato Complexes of Nickel(II), Cobalt(III), and Mercury(II), *Inorganic Chemistry*, **39**, 6100–6106 (2000).
[59] A.-M. Stadler, C. Burg, J. Ramirez and J.-M. Lehn, Grid-Double-Helicate Interconversion, *Chemical Communications*, **49**(51), 5733–5735 (2013).

[60] A.P.S. Pannu, J.R. Stevens and P.G. Plieger, Aryl-Linked Salicylaldoxime-Based Copper(II) Helicates and 'Boxes': Synthesis, X-ray Analysis, and Anion Influence on Complex Structure, *Inorganic Chemistry*, **52**(16), 9327–9337 (2013).

[61] J.R. Stevens and P.G. Plieger, Anion-Driven Conformation Control and Enhanced Sulfate Binding Utilising Aryl Linked Salicylaldoxime Dicopper Helicates, *Dalton Transactions*, **40**(45), 12235–12241 (2011).

[62] M. Wenzel, S.R. Bruere, Q.W. Knapp, *et al.*, Zwitterionic Dicopper Helicates: Anion Encapsulation and Binding Studies, *Dalton Transactions*, **39**(11), 2936–2941 (2010).

[63] J.-M. Lehn and A. Rigault, Helicates: Tetra- and Pentanuclear Double Helix Complexes of CuI and Poly(bipyridine) Strands, *Angewandte Chemie International Edition*, **27**(8), 1095–1097 (1988).

[64] A. Marquis, V. Smith, J. Harrowfield, *et al.*, Messages in Molecules: Ligand/Cation Coding and Self-Recognition in a Constitutionally Dynamic System of Heterometallic Double Helicates, *Chemistry – A European Journal*, **12**(21), 5632–5641 (2006).

[65] R.J. Archer, C.S. Hawes, G.N.L. Jameson, *et al.*, Partial Spin Crossover Behaviour in a Dinuclear Iron(II) Triple Helicate, *Dalton Transactions*, **40**(45), 12368–12373 (2011).

[66] D.A. Roberts, A.M. Castilla, T.K. Ronson and J.R. Nitschke, Post-assembly Modification of Kinetically Metastable $Fe^{II}_2L_3$ Triple Helicates, *Journal of the American Chemical Society*, **136**(23), 8201–8204 (2014).

[67] S.E. Bodman and C.M. Fitchett, Conformational Control of the Self-Assembly of Triple Helicates and [2×2]-Grids from Zinc(II) and 3,6-di(2-pyridyl)pyridazine based Ligands, *Dalton Transactions*, **43**(33), 12606–12613 (2014).

[68] J.L. Ferguson and C.M. Fitchett, N,N′-Dimethylene-2,2′-biimidazole-A Mimic of Carboxylate for the Formation of Complexes with Copper(I) And the Anion Directed Formation of a Solvent Pocket, *Crystal Growth and Design*, **15**, 1280 (2015).

[69] M. Raja, R.G. Iyer, C. Gwengo, *et al.*, Design, Synthesis, and Structural Characterization of a New Class of Ferrocene-Containing Heterometallic Triple-Stranded Helicates, *Organometallics*, **32**(1), 95–103 (2013).

[70] C. Jia, B.P. Hay and R. Custelcean, De Novo Structure-Based Design of Ion-Pair Triple-Stranded Helicates, *Inorganic Chemistry*, **53**(7), 3893–3898 (2014).

[71] R. Custelcean, Urea-Functionalized Crystalline Capsules for Recognition and Separation of Tetrahedral Oxoanions, *Chemical Communications*, **49**(22), 2173–2182 (2013).

[72] R. Custelcean, P.V. Bonnesen, N.C. Duncan, *et al.*, Urea-Functionalized M_4L_6 Cage Receptors: Anion-Templated Self-Assembly and Selective Guest Exchange in Aqueous Solutions, *Journal of the American Chemical Society*, **134**(20), 8525 (2012).

[73] R. Custelcean, P.V. Bonnesen, B.D. Roach and N.C. Duncan, Ion-Pair Triple Helicates and Mesocates Self Assembled from Ditopic 2,2′-bipyridine-bis(urea) ligands and Ni(II) and Fe(II) Sulfate Salts, *Chemical Communications*, **48**, 7438–7440 (2012).

[74] R. Custelcean, B.A. Moyer and B.P. Hay, A Coordinatively Saturated Sulfate Encapsulated in a Metal–Organic Framework Functionalized with Urea Hydrogen-Bonding Groups, *Chemical Communications*, **48**, 5971–5973 (2005).

[75] B.P. Hay, T.K. Firman and B.A. Moyer, Structural Design Criteria for Anion Hosts: Strategies for Achieving Anion Shape Recognition through the Complementary Placement of Urea Donor Groups, *Journal of the American Chemical Society*, **127**(6), 1810–1819 (2005).

[76] N. Dalla-Favera, J. Hamacek, M. Borkovec, *et al.*, Linear Polynuclear Helicates as a Link between Discrete Supramolecular Complexes and Programmed Infinite Polymetallic Chains, *Chemistry – A European Journal*, **14**(10), 2994–3005 (2008).

[77] J. Hamacek, Unravelling Self-Assembly of Lanthanide Helicates: Switching from Deduction to Induction, *Journal of Alloys and Compounds*, **451**(1–2), 347–351 (2008).

[78] K. Zeckert, J. Hamacek, J.-M. Senegas, *et al.*, Predictions, Synthetic Strategy, and Isolation of a Linear Tetrametallic Triple-Stranded Lanthanide Helicate, *Angewandte Chemie International Edition*, **44**(48), 7954–7958 (2005).

[79] J. Hamacek, G. Bernardinelli and Y. Filinchuk, Tetrahedral Assembly with Lanthanides: Toward Discrete Polynuclear Complexes, *European Journal of Inorganic Chemistry*, **22**, 3419–3422 (2008).

[80] J. Hamacek, C. Besnard, T. Penhouet and P.-Y. Morgantini, Thermodynamics, Structure and Properties of Polynuclear Lanthanide Complexes with a Tripodal Ligand: Insight into their Self-Assembly, *Chemistry – A European Journal*, **17**(24), 6753–6764 (2011).
[81] B.E. Aroussi, L. Guénée, P. Pal and J. Hamacek, Lanthanide-Mediated Supramolecular Cages and Host–Guest Interactions, *Inorganic Chemistry*, **50**(17), 8588–8597 (2011).
[82] B.E. Aroussi, S. Zebret, C. Besnard, *et al.*, Rational Design of a Ternary Supramolecular System: Self-Assembly of Pentanuclear Lanthanide Helicates, *Journal of the American Chemical Society*, **133**(28), 10764–10767 (2011).
[83] G. Bernardinelli, C. Piguet and A.F. Williams, The First Self-Assembled Dinuclear Triple-Helical Lanthanide Complex: Synthesis and Structure, *Angewandte Chemie International Edition*, **31**(12), 1622–1624 (1992).
[84] T. Riis-Johannessen, G. Bernardinelli, Y. Filinchuk, S. Clifford, N. Dalla Favera and C. Piguet, Self-Assembly of the First Discrete *3d-4f-4f* Triple-Stranded Helicate, *Inorganic Chemistry*, **48**(12), 5512–5525 (2009).
[85] T. Riis-Johannessen, N. Dupont, G. Canard, G. Bernardinelli, A. Hauser and C. Piguet, Towards Inert and Preorganized *d*-Block-Containing Receptors for Trivalent Lanthanides: The Synthesis and Characterization of Triple-Helical Monometallic Os^{II} and Bimetallic Os^{II}-Ln^{III} Complexes, *Dalton Transactions*, **28**, 3661–3677 (2008).
[86] J.-C.G. Bünzli and C. Piguet, Taking Advantage of Luminescent Lanthanide Ions, *Chemical Society Reviews*, **34**, 1048–1077 (2005).
[87] K. Zeckert, H. Hamacek, J.-M. Senegas, *et al.*, Predictions, Synthetic Strategy, and Isolation of a Linear Tetrametallic Triple-Stranded Lanthanide Helicate, *Angewandte Chemie International Edition*, **44**(48), 7954–7958 (2005).
[88] J.-C.G. Bünzli and C. Piguet, Lanthanide-Containing Molecular and Supramolecular Polymetallic Functional Assemblies, *Chemical Reviews*, **102**(6), 1897–1928 (2002).
[89] L.J. Charbonnière, A.F. Williams, C. Piguet, *et al.*, Structural, Magnetic, and Electrochemical Properties of Dinuclear Triple Helices: Comparison with their Mononuclear Analogues, *Chemistry – A European Journal*, **4**(3), 485–493 (1998).
[90] N. Martin, J.-C.G. Bünzli, V. McKee, C. Piguet and G. Hopfgartner, Self-Assembled Dinuclear Lanthanide Helicates: Substantial Luminescence Enhancement upon Replacing Terminal Benzimidazole Groups by Carboxamide Binding Units, *Inorganic Chemistry*, **37**(3), 577–589 (1998).
[91] F. Renaud, C. Piguet, G. Bernardinelli, J.-C.G. Bünzli and G. Hopfgartner, In Search for Mononuclear Helical Lanthanide Building Blocks with Predetermined Properties: Lanthanide Complexes with Diethyl Pyridine-2,6-Dicarboxylate, *Chemistry – A European Journal*, **3**(10), 1660–1667 (1997).
[92] C. Piguet, G. Bernadinelli and G. Hopfgarter, Helicates as Versatile Supramolecular Complexes, *Chemical Reviews*, **97**(6), 2005–2062 (1997).
[93] C. Piguet, J.-C.G. Bünzli, G. Bernardinelli, G. Hopfgartner and A.F. Williams, Self-Assembly and Photophysical Properties of Lanthanide Dinuclear Triple-Helical Complexes, *Journal of the American Chemical Society*, **115**(18), 8197–8206 (1993).
[94] S.V. Eliseeva and J.-C.G. Bünzli, Lanthanide Luminescence for Functional Materials and Bio-sciences, *Chemical Society Reviews*, **39**, 189–227 (2010).
[95] E. Deiters, B. Song, A.-S. Chauvin, *et al.*, Luminescent Bimetallic Lanthanide Bioprobes for Cellular Imaging with Excitation in the Visible-Light Range, *Chemistry – A European Journal*, **15**(4), 885–900 (2009).
[96] J.-C.G. Bünzli, S. Comby, A.-S. Chauvin and C.D.B. Vandevyver, New Opportunities for Lanthanide Luminescence, *Journal of Rare Earths*, **25**(3), 257–274 (2007).
[97] J.-C.G. Bünzli, Benefiting from the Unique Properties of Lanthanide Ions, *Accounts of Chemical Research*, **39**(1), 53–61 (2006).
[98] C. Piguet, J.-C.G. Bünzli, B. Donnio and D. Guillon, Thermotropic Lanthanidomesogens, *Chemical Communications*, **35**, 3755–3768 (2006).
[99] N. Andre, R. Scopelliti, G. Hopfgarter, *et al.*, Discriminating between Lanthanide Ions: Self-Assembly of Heterodimetallic Triple-Stranded Helicates, *Chemical Communications*, **3**, 214–215 (2002).

[100] J.A. Kitchen, R. Parkesh, E.B. Veale and T. Gunnlaugsson, Luminescent Sensing, in P.A. Gale and J.W. Steed (Eds), *Supramolecular Chemistry: From Molecules to Nanomaterials*, Vol. 5, John Wiley & Sons, Ltd, Chichester, pp. 2611–2641 (2012).

[101] J.A. Kitchen and T. Gunnlaugsson, Supramolecular Chemistry: from Sensors and Imaging Agents to Functional Mononuclear and Polynuclear Self-Assembly Lanthanide Complexes, in D.A. Atwood (Ed.), *The Rare Earth Elements*, John Wiley & Sons, Ltd, Chichester, pp. 481–495 (2012).

[102] N. Dalla-Favera, J. Hamacek, M. Borkovec, *et al.*, Linear Polynuclear Helicates as a Link between Discrete Supramolecular Complexes and Programmed Infinite Polymetallic Chains, *Chemistry – A European Journal*, **14**(10), 2994–3005 (2008).

[103] M. Cantuel, F. Gumy, J.-C.G. Bünzli and C. Piguet, Encapsulation of Labile Trivalent Lanthanides into a Homobimetallic Chromium(III)-Containing Triple-Stranded Helicate. Synthesis, Characterization, and Divergent Intramolecular Energy Transfers, *Dalton Transactions*, **22**, 2647–2660 (2006).

[104] S. Torelli, D. Imbert, M. Cantuel, *et al.*, Tuning the Decay Time of Lanthanide-Based Near Infrared Luminescence from Micro- to Milliseconds through *d*→*f* Energy Transfer in Discrete Heterobimetallic Complexes, *Chemistry – A European Journal*, **11**(11), 3228–3242 (2005).

[105] M. Cantuel, G. Bernardinelli, D. Imbert, *et al.*, A Kinetically Inert and Optically Active CrIII Partner in Thermodynamically Self-Assembled Heterodimetallic Non-Covalent d-f Podates, *Journal of the Chemical Society, Dalton Transactions*, **9**, 1929–1940 (2002).

[106] C. Piguet, E. Rivara-Minten, G. Bernardinelli, *et al.*, Noncovalent Lanthanide Podates with Predetermined Physicochemical Properties: Iron(II) Spin-State Equilibria in Self-Assembled Heterodinuclear *d-f* Supramolecular Complexes, *Journal of the Chemical Society, Dalton Transactions*, **3**, 421–433 (1997).

[107] N. Andre, T.B. Jensen, R. Scopelliti, *et al.*, Supramolecular Recognition of Heteropairs of Lanthanide Ions: A Step toward Self-Assembled Bifunctional Probes, *Inorganic Chemistry*, **43**(2), 515–529 (2004).

[108] T.B. Jensen, R. Scopelliti and J.-C.G. Bünzli, Lanthanide Triple-Stranded Helicates: Controlling the Yield of the Heterobimetallic Species, *Inorganic Chemistry*, **46**(4), 1507 (2007).

[109] T.B. Jensen, R. Scopelliti and J.-C.G. Bünzli, Tuning the Self-Assembly of Lanthanide Triple Stranded Heterobimetallic Helicates by Ligand Design, *Dalton Transactions*, **8**, 1027–1036 (2008).

[110] M. Albrecht, Y. Liu, S.S. Zhu, *et al.*, Self-Assembly of Heterodinuclear Triple-Stranded Helicates: Control by Coordination Number and Charge, *Chemical Communications*, **10**, 1195–1197 (2009).

[111] M. Albrecht, O. Osetska, J.-C.G. Bünzli, *et al.*, Homo- and Heterodinuclear Helicates of Lanthanide(III), Zinc(II) and Aluminium(III) Based on 8-Hydroxyquinoline Ligands, *Chemistry – A European Journal*, **15**(35), 8791–8799 (2009).

[112] M. Albrecht, O. Osetska, R. Froehlich, *et al.*, Highly Efficient Near-IR Emitting Yb/Yb and Yb/Al Helicates, *Journal of the American Chemical Society*, **129**(46), 14178–14179 (2007).

[113] D. Tripathy, A.K. Pal, G.S. Hanan and D.K. Chand, Palladium(II) Driven Self-Assembly of a Saturated Quadruple-Stranded Metallo Helicate, *Dalton Transactions*, **41**(37), 11273–11275 (2012).

[114] F. Habib, J. Long, P.-H. Lin, *et al.*, Supramolecular Architectures for Controlling Slow Magnetic Relaxation in Field-Induced Single-Molecule Magnets, *Chemical Science*, **3**(6), 2158–2164 (2012).

[115] B. Wang, Z. Zang, H. Wang, *et al.*, Multiple Lanthanide Helicate Clusters and the Effects of Anions on their Configuration, *Angewandte Chemie International Edition*, **52**(13), 3756–3759 (2013).

[116] S. Zarra, D.M. Wood, D.A. Roberts and J.R. Nitschke, Molecular Containers in Complex Chemical Systems, *Chemical Society Reviews*, **44**(2), 419–432 (2015).
[117] P. Ballester, M. Fujita and J. Rebek, Molecular Containers, *Chemical Society Reviews*, **44**(2), 392–393 (2015).
[118] S.H.A.M. Leenders, R. Gramage-Doria, *et al.*, Transition Metal Catalysis in Confined Sspaces, *Chemical Society Reviews*, **44**(2), 433–448 (2015).
[119] M.M.J. Smulders, S. Zarra and J.R. Nitschke, Quantitative Understanding of Guest Binding Enables the Design of Complex Host–Guest Behavior, *Journal of the American Chemical Society*, **135**(18), 7039–7046 (2013).
[120] P. Mal, B. Breiner, K. Rissanen and J.R. Nitschke, White Phosphorus is Air-Stable Within a Self-Assembled Tetrahedral Capsule, *Science*, **324**, 1697–1699 (2009).
[121] T.K. Ronson, A.B. League, L. Gagliardi, *et al.*, Pyrene-Edged $Fe^{II}_4L_6$ Cages Adaptively Reconfigure During Guest Binding, *Journal of the American Chemical Society*, **136**(44), 15615–15624 (2014).
[122] A. Ferguson, M.A. Squire, D. Siretanu, *et al.*, A Face-Capped $[Fe_4L_4]^{8+}$ Spin Crossover Tetrahedral Cage, *Chemical Communications*, **49**(16), 1597–1599 (2013).
[123] J.-F. Ayme, J.E. Beves, D.A. Leigh, *et al.*, A Synthetic Molecular Pentafoil Knot, *Nature Chemistry*, **4**(1), 15–20 (2012).
[124] J.-F. Ayme, J.E. Beves, D.A. Leigh, *et al.*, Pentameric Circular Iron(II) Double Helicates and a Molecular Pentafoil Knot, *Journal of the American Chemical Society*, **134**(22), 9488–9497 (2012).
[125] J.E. Beves, C.J. Campbell, D.A. Leigh and R.G. Pritchard, Tetrameric Cyclic Double Helicates as a Scaffold for a Molecular Solomon Link, *Angewandte Chemie International Edition*, **52**(25), 6464–6467 (2013).
[126] J.P. Leonard, P. Jensen, T. McCabe, *et al.*, Self-Assembly of Chiral Luminescent Lanthanide Coordination Bundles, *Journal of the American Chemical Society*, **129**(36), 10986–10987 (2007).
[127] O. Kotova, J.A. Kitchen, C. Lincheneau, *et al.*, Probing the Effects of Ligand Isomerism in Chiral Luminescent Lanthanide Supramolecular Self-Assemblies: A Europium 'Trinity Sliotar' Study, *Chemistry – A European Journal*, **19**(48), 16181–16186 (2013).
[128] J.A. Kitchen, D.E. Barry, L. Mercs, *et al.*, Circularly Polarized Lanthanide Luminescence from Langmuir-Blodgett Films Formed from Optically Active and Amphiphilic EuIII-Based Self-Assembly Complexes, *Angewandte Chemie-International Edition*, **51**(3), 704–708 (2012).
[129] G. Bozoklu, C. Marchal, C. Gateau, *et al.*, Diastereoselective Self-Assembly of a Homochiral Europium Triangle from a Bipyoxazoline-Carboxylate Ligand, *Chemistry – A European Journal*, **16**(21), 6159–6163 (2010).
[130] G. Bozoklu, C. Gateau, D. Imbert, *et al.*, Metal-Controlled Diastereoselective Self-Assembly and Circularly Polarized Luminescence of a Chiral Heptanuclear Europium Wheel, *Journal of the American Chemical Society*, **134**(20), 8372–8375 (2012).
[131] E.C. Constable, G. Zhang, C.E. Housecroft, *et al.*, Diastereoselective Assembly of Helicates Incorporating a Hexadentate Chiral Scaffold, *European Journal of Inorganic Chemistry*, **13**, 2000–2011 (2010).
[132] C. Eerdun, S. Hisanaga and J.-I. Setsune, One-Handed Single Helicates of Dinickel(II) Benzenehexapyrrole-α,ω-diimine with an Amine Chiral Source, *Chemistry – A European Journal*, **21**(1), 239–246 (2015).
[133] C. Lincheneau, C. Destribats, D.E. Barry, *et al.*, Lanthanide Directed Self-Assembly Synthesis and Photophysical Evaluation of Chiral Eu(III) Luminescent 'Half-Helicates', *Dalton Transactions*, **40**(45), 12056–12059 (2011).

[134] C. Lincheneau, R.D. Peacock and T. Gunnlaugsson, Europium Directed Synthesis of Enantiomerically Pure Dimetallic Luminescent 'Squeezed' Triple-Stranded Helicates; Solution Studies, *Chemistry – An Asian Journal*, **5**(3), 500–504 (2010).

[135] S. Comby, F. Stomeo, C.P. McCoy and T. Gunnlaugsson, Formation of Novel Dinuclear Lanthanide Luminescent Samarium(III), Europium(III), and Terbium(III) Triple-Stranded Helicates from a C_2-Symmetrical Pyridine-2,6-dicarboxamide-Based 1,3-Xylenediyl-Linked Ligand in MeCN, *Helvetica Chimica Acta*, **92**(11), 2461–2473 (2009).

[136] F. Stomeo, C. Lincheneau, J.P. Leonard, *et al.*, Metal-Directed Synthesis of Enantiomerially Pure Dimetallic Lanthanide Luminescent Triple-Stranded Helicates, *Journal of the American Chemical Society*, **131**(28), 9636–9637 (2009).

[137] C. Lincheneau, J.P. Leonard, T. McCabe and T Gunnlaugsson, Lanthanide Directed Self-Assembly Formations of Tb(III) and Eu(III) Luminescent Complexes from Tryptophan Based Pyridyl Amide Ligands, *Chemical Communications*, **47**(25), 7119–7121 (2011).

[138] L. Cardo, V. Sadovnikova, S. Phongtongpasuk, *et al.*, Arginine Conjugates of Metallo-Supramolecular Cylinders Prescribe Helicity and Enhance DNA Junction Binding and Cellular Activity, *Chemical Communications*, **47**(23), 6575–6577 (2011).

[139] W. Xuan, M. Zhang, Y. Liu, *et al.*, A Chiral Quadruple-Stranded Helicate Cage for Enantioselective Recognition and Separation, *Journal of the American Chemical Society*, **134**(16), 6904–6907 (2012).

[140] C. Deville, A. Spyratou, P. Aguirre-Etcheverry, *et al.*, Enantioselective Synthesis of Tetranuclear Quadruple Helicates, *Inorganic Chemistry*, **51**(16), 8667–8669 (2012).

[141] J.L. Bolliger, A.M. Belenguer and J.R. Nitschke, Enantiopure Water-Soluble $[Fe_4L_6]$ Cages: Host–Guest Chemistry and Catalytic Activity, *Angewandte Chemie International Edition*, **52**(31), 7958–7962 (2013).

[142] A.M. Castilla, N. Ousaka, R.A. Bilbeisi, *et al.*, High-Fidelity Stereochemical Memory in a $Fe^{II}_4L_4$ Tetrahedral Capsule, *Journal of the American Chemical Society*, **135**(47), 17999–18006 (2013).

5

Chirality in the Host-Guest Behaviour of Supramolecular Systems

Nicholas H. Evans and Paul D. Beer

5.1 An Introduction to Chiral Recognition and its Importance

Chiral recognition occurs when an enantiopure host (H) preferentially binds one enantiomer (E_1) of a chiral guest over the other enantiomer (E_2), due to the difference in energies of the diastereoisomeric complexes $H.E_1$ and $H.E_2$. The formation of these host-guest complexes may rely on one or more of a number of the established supramolecular, noncovalent interactions, which include hydrogen bonding, π-π stacking and dipole-dipole interactions.

The natural world's reliance on chiral recognition processes provides ample inspiration for the synthetic supramolecular chemist, with a myriad of examples including crucial enzyme substrate selectivity and reactivity, as well as our sense of taste. The fact that so many biological processes have a chiral consideration means that the most effective and safest medications are increasingly based on enantiopure chiral molecules.

A range of applications for synthetic chiral recognition have already been established. Chiral HPLC, which uses immobilized chiral molecules, is regularly used to ascertain the enantiopurity of analytical chiral samples [1]. Alternatively, chiral solvating agents can enable the determination of enantiomeric composition of samples by NMR spectroscopy [2]. Chiral sensing – typically, but not exclusively limited to, fluorescence – is a promising analytical alternative to these more established methods [3]. Furthermore, chiral recognition may be employed in the separation of enantiomers by, for example, preparative chiral HPLC or enantioselective liquid-liquid extraction, 'ELLE' [4].

The focus of this chapter will be on the role of chirality in the host-guest behaviour of enantio-discriminating host-guest systems. While the field of chiral host-guest recognition

Chirality in Supramolecular Assemblies: Causes and Consequences, First Edition. Edited by F. Richard Keene.

is still underdeveloped, there are too many systems to allow for a comprehensive discussion here. This chapter therefore covers a range of seminal, significant and illustratively useful examples.

5.2 Chiral Hosts for Chiral Guests

5.2.1 Theory of Chiral Recognition

There is some debate over the issue but it is usually stated that for chiral recognition to occur, at least three simultaneous interactions must be present between a chiral host and its chiral guest, with at least one of these interactions being stereochemically dependent [5]. The rationalization of this argument is by consideration of a geometric tetrahedral point model. If a chiral receptor has three sites of interaction with a specific enantiomer of a chiral guest, the other enantiomer of that guest will only be able interact with two of these sites (Figure 5.1). While at least one interaction must be attractive, other(s) may be repulsive – for example, steric hindrance. The tetrahedral point model is somewhat limited; while certain interactions are single point (e.g. hydrogen bonding and end-to-end dipoles), many interactions are considered to be multipoint (e.g. dipole-dipole stacking and π-π interactions).

Supramolecular chemists have studied the enantioselective recognition of a variety of chiral guest molecules, typically focusing on those that are seen as biologically relevant, and where both enantiomers are readily available. As well as being probably the most investigated, substituted α-amino acids provide a possible complication when discussing chiral recognition, for these molecules possess at least two functional groups that can exist in neutral or charged forms in the underivatized molecule. In this chapter we have chosen primarily to classify chiral hosts according to the charge of the chiral guest species.

5.2.2 Chiral Crown Ethers for Chiral Ammonium Cations

Single atom metal cations are achiral. However, many polyatomic ammonium salts – including those of substituted α-amino acid derivatives – are chiral. Donald Cram reported the enantioselective recognition of chiral ammonium salts using chiral binaphthyl crown ethers

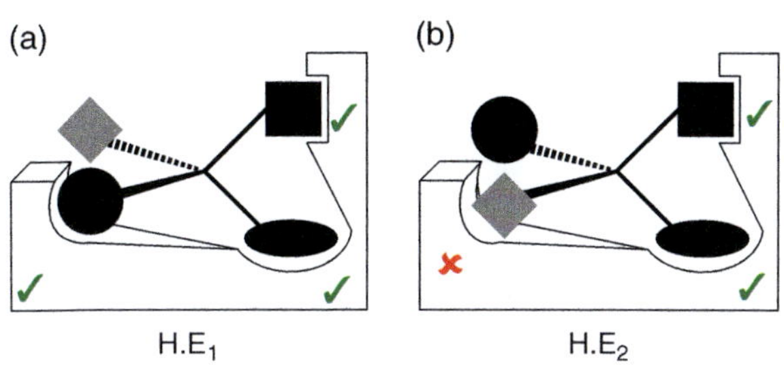

Figure 5.1 *Schematic representation of the classical tetrahedral point model for chiral recognition: (a) Enantiomer 1 (E_1) interacts with three sites of chiral host H; (b) Enantiomer 2 (E_2) interacts only at two sites of chiral host H. Hence, Enantiomer 2 will be bound less strongly than Enantiomer 1*

in the 1970s (Figure 5.2) [6, 7]. In his first work, enantiopure samples of **(*R*, *R*)-1** and **(*S*, *S*)-1** were prepared and used to extract chiral α-phenylethylammonium salts from D_2O into $CDCl_3$ [6]. In one such NMR experiment, 0.93 mol of salt was extracted per mol of enantiopure **(*S*, *S*)-1** host from a racemic sample of salt. Enantioselectivity was modest, with 62% of the salt extracted being (*R*)-α-phenylethylammonium and 38% being (*S*)-α-phenylethylammonium. The observed selectivity was in agreement with CPK molecular models of the receptors binding the chiral cation guest. In subsequent work, further binaphthyl crown ethers were prepared, including the tetra-methyl host **2** (see Figure 5.2a) that could achieve extraction into $CDCl_3$ with an enantioselectivity of 31 : 1 (in the case of the methyl ester of protonated phenylglycine) [7]. It should be noted that for the binding of the methyl ester of glycine, **(*R*, *R*)-2** binds the *R* enantiomer of the guest preferentially, once again in agreement with CPK molecular models of this particular system.

Cram then went on to demonstrate useful applications of these receptors [8, 9]. For example, by covalently attaching an appropriately functionalized analogue of receptor **(*R*, *R*)-1** to silica gel, it was possible to generate a chiral stationary phase capable of chromatographically resolving several chiral ammonium salts [8]. In other work, a 'catalytic resolving machine' was constructed (Figure 5.3) [9]. Here, separate chloroform solutions of enantiopure **(*R*, *R*)-2** and **(*S*, *S*)-2** crown ether host were placed at the bottom of the loops of a W-tube. From an aqueous reservoir at the top centre of the W-tube, a racemate of the methyl ester of phenylglycine was resolved into separate aqueous layers: the L-enantiomer being bound and transported by **(*S*, *S*)-2** through the chloroform at the bottom of one loop of the W-tube,

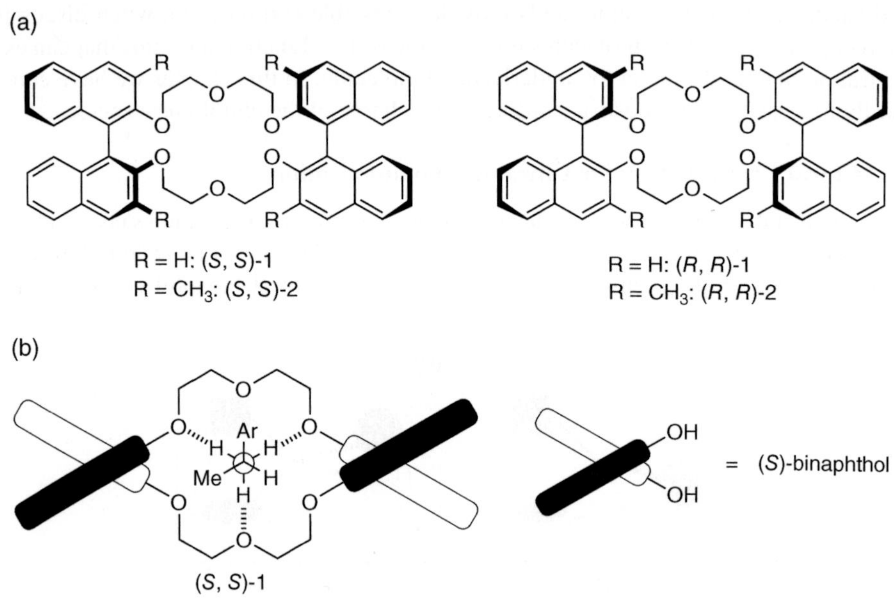

*Figure 5.2 Cram's chiral binaphthyl crown ether hosts **1** and **2** for the enantioselective binding of chiral ammonium cations: (a) structures of **1** and **2** and (b) illustration of the favourable binding mode of the R enantiomer of the phenylethylammonium cation with **(S, S)-1***

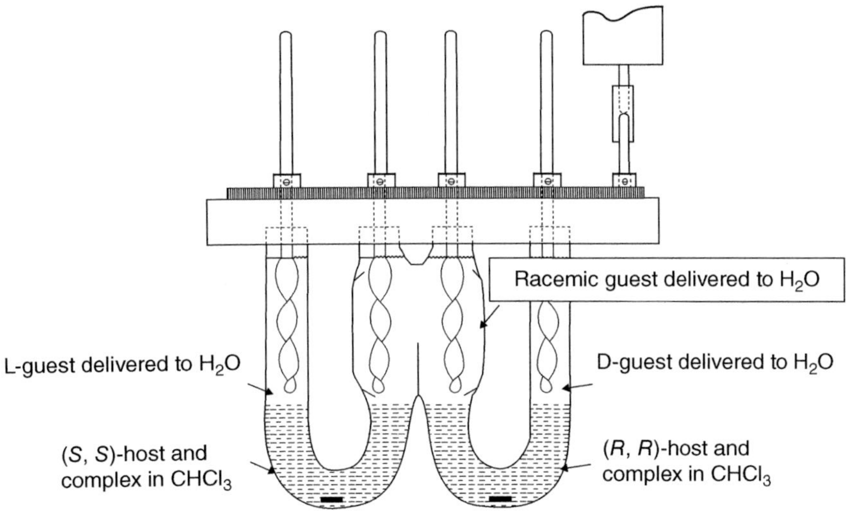

Figure 5.3 *Cram's 'catalytic resolving machine'. Reprinted with permission from M. Newcomb, J. L. Toner, R. C. Helgeson and D. J. Cram, Host-guest complexation. 20. Chiral recognition in transport as a molecular basis for a catalytic resolving machine, Journal of the American Chemical Society,* ***101****(17), 4941–4947 (1979). Copyright 1979 American Chemical Society*

while the D-enantiomer was transported by the **(*R*, *R*)-2** stereoisomer through the chloroform at the bottom of the other loop. In these studies, enantiomeric excesses (*e.e.*) of up to 90% were achieved.

5.2.3 Hosts for Chiral Anions

Naturally occurring carboxylate anions represent interesting targets for chiral recognition. However, anion recognition is generally considered challenging, particularly in water, where anions typically exhibit much higher free energies of hydration than cations. Despite this, a range of enantioselective anion receptors have been prepared and studied.

Gotor has reported a number of azamacrocycles, derived from *trans*-cyclohexane-1,2-diamine, that can operate as enantioselective receptors of chiral carboxylate anions in water (Figure 5.4) [10, 11]. For example, the hexaazamacrocycles **(*R*, *R*)-3** and **(*R*, *R*, *R*, *R*)-4** when tetra-, penta- or hexa-protonated, were able to bind a range of chiral dicarboxylate anions by a combination of electrostatic and hydrogen bonding interactions, as determined by potentiometric titration experiments [10]. Macrocycle **(*R*, *R*, *R*, *R*)-4** exhibited higher anion association constants than **(*R*, *R*)-3**, attributed by the authors to **(*R*, *R*, *R*, *R*)-4** possessing a higher charge density and being less solvated than **(*R*, *R*)-3**. Enantiomeric discrimination depended on the degree of protonation, as well as the structure of the macrocycle and anion studied. The anion bound most strongly by both macrocycles, *N*-Ac-glutamate, exhibited large differences in enantiomer association constants, for example for tetra-protonated **(*R*, *R*, *R*, *R*)-4**, log K_a (D)=4.14 and log K_a (L)=2.17 (Δ log K_a=1.97). In a later communication, a related macrocycle **(*R*, *R*, *R*, *R*)-5** displayed high levels of enantioselective recognition for the malate dianion in aqueous solution (K_a (*S*)/K_a (*R*)=11.75 at

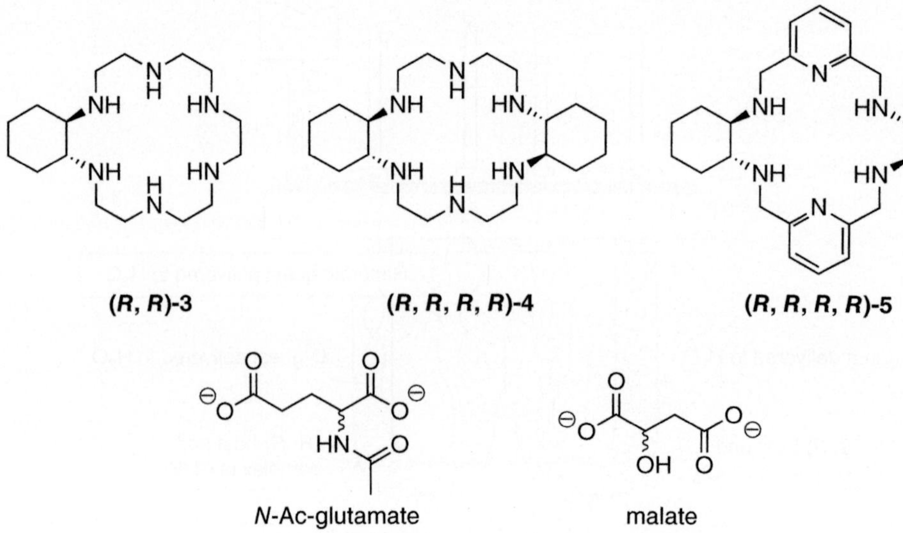

Figure 5.4 *Structures of three azamacrocycles **3-5** investigated for the enantioselective binding of chiral carboxylate anions, such as* N*-Ac-glutamate and malate*

pH 10, down to 3.89 at pH 2), unlike **(*R*, *R*)-3** and **(*R*, *R*, *R*, *R*)-4** which had shown little enantioselectivity for malate [11].

Kilburn and coworkers have studied a macrocycle that relies on strong hydrogen bond donating groups to operate as a neutral anion receptor in polar non-aqueous solvents (Figure 5.5a) [12]. The macrocycle **(*S*, *S*, *S*, *S*)-6** contains two thioureas flanked by carboxypyridines, that are separated by chiral diamines. This creates a chiral pocket that could form up to eight hydrogen bonding interactions with a single dicarboxylate anion guest. In acetonitrile, however, 1 : 1 binding stoichiometry was only observed for one enantiomer of *N*-Boc-glutamate. Measured by isothermal calorimetry, *N*-Boc-L-glutamate was bound strongly ($K_{11}=2.83\times10^4\,M^{-1}$) in a 1 : 1 stoichiometry, but *N*-Boc-D-glutamate was bound strongly in a 1 : 2 (host : guest) stoichiometry ($K_{12}=4.92\times10^4\,M^{-1}$), with only a very weak 1 : 1 binding event ($K_{11}=38\,M^{-1}$) being detected, attributed to this enantiomer not fitting into the chiral cavity of the macrocycle. Considering only the 1 : 1 stoichiometric binding events, the enantioselectivity exhibited by the macrocycle is >700 : 1.

By preparation of the bis-guanidinium macrocycle **7**, now containing additional positive charge (Figure 5.5b), the same group were able to demonstrate analogous enantioselective 1 : 1 binding of *N*-Boc-L-glutamate in an aqueous solvent system (50 : 50 DMSO-H_2O). In this system, association constants for the L-enantiomer were $K_{11}=3.8\times10^4\,M^{-1}$, $K_{12}=5.3\times10^3\,M^{-1}$, whereas the D-enantiomer was bound with association constants: $K_{11}=2.9\times10^3\,M^{-1}$, $K_{12}=1.4\times10^4\,M^{-1}$ [13].

Functionalization of the readily available chiral pool scaffold cholic acid has enabled a range of enantioselective anion receptors and transporters to be constructed, as demonstrated by Davis *et al.* (Figure 5.6) [14, 15]. In early work, guanidinium and carbamate functional groups were attached and the resulting receptors demonstrated to extract chiral carboxylate anions from aqueous buffer into chloroform by use of charged assisted

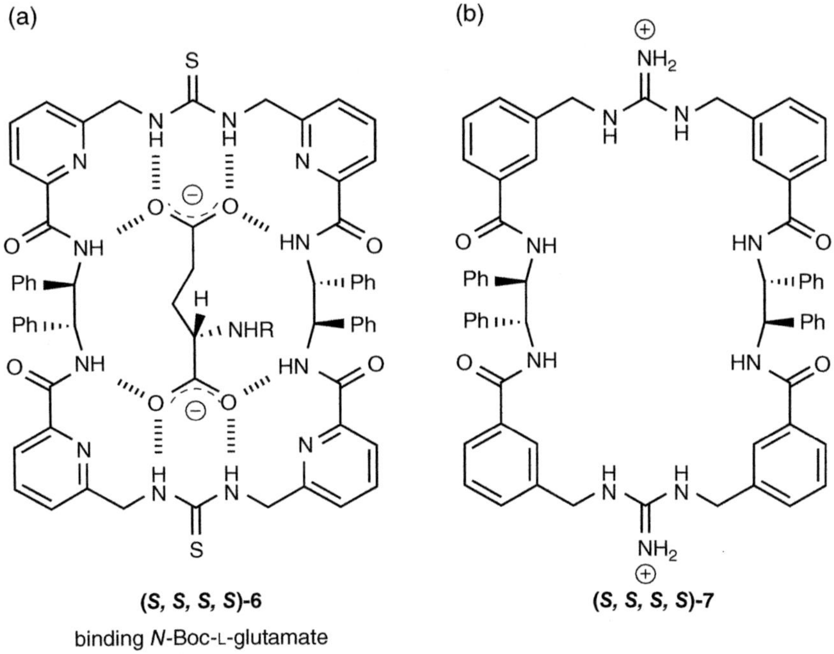

Figure 5.5 *Kilburn's chiral macrocycles for the binding of Boc-protected amino acid carboxylate anions: (a) neutral bis-thiourea* **6** *and (b) charged bis-guanidinium* **7**

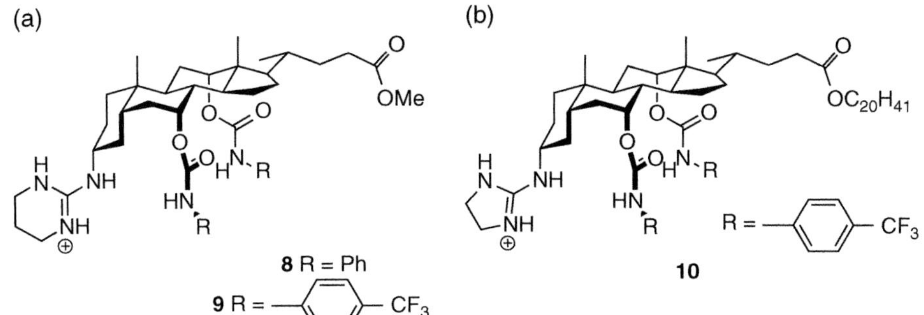

Figure 5.6 *Chiral anion receptors derived from cholic acid, used in (a) extraction and (b) transportation experiments*

hydrogen bonding. The most successful receptors **8** and **9** (Figure 5.6a), achieved enantioselective ratios of 7–10 : 1 (L : D) for a range of *N*-acetyl amino acids [14]. A lipophilic variant **10** was subsequently employed in the enantioselective transport of *N*-acetylphenylalanine (Figure 5.6b). Significant enantioselectivity (approaching that of the previous extraction experiments) and multiple turnover of receptor were observed in bulk liquid and hollow fibre membrane experiments [15].

Jurczak has reported receptor **11**, a multidentate hydrogen bond-donating diindolylmethane functionalized with chiral glucuronic acid (Figure 5.7) [16]. The receptor exhibited enantioselective binding of a selection of Boc-protected amino acid carboxylates in DMSO-d_6 (with 0.5% H_2O) with ratios of association constants K_a (D)/K_a (L) ranging from 2.40 to 2.57. In a subsequent study, the same receptor was employed in the creation of an artificial neural network. Analysis of the changes in chemical shift of proton resonances of **11** upon the addition of enantiopure chiral anion guest allowed for the 'unknown' chirality of 14 carboxylate anions to be accurately predicted on the basis of a 'training set' of 12 different carboxylate anions [17].

It is notable that a large number of chiral anion receptors have been prepared by simply appending chiral hydrogen bond donating groups to various achiral macrocycles. For example, Starnes *et al.* prepared the metallated L-nipecotic acid porphyrin derivative **12** (Figure 5.8) [18]. In dichloromethane this molecule exhibited modest enantioselectivity for (*S*)-mandelate (K_a (*S*)/K_a (*R*) = 2.1), with the authors proposing a binding model (as depicted in Figure 5.8) to rationalize this selectivity. In more recent work, a family of similar receptors has been generated, with one receptor exhibiting enantioselective ratios for the enantiomers of mandelate of 3 : 1 in dichloromethane [19].

Sessler and coworkers have studied receptors constructed as dimers of the related sapphyrin macrocycle (Figure 5.9) [20]. While the cyclic receptor **13** bound anions less strongly (in CH_2Cl_2 containing 5% CH_3OH) than two acyclic receptors also studied, it was observed in the case of *N*-Cbz-glutamate to display a higher degree of enantioselectivity (K_a (D)/K_a (L) = 4.3). It was postulated that this enhanced enantioselectivity arose from the binding demands of a preorganized cyclic receptor, being a superior host-guest geometric match than with a flexible acyclic receptor.

Lee and co-workers have reported the strapped calix[4]pyrrole receptor **14**, where chirality is derived from a BINOL (1,1′-bi-2-naphthol) incorporated into the strap (Figure 5.10) [21]. By use of isothermal titration calorimetry in anhydrous acetonitrile, the authors determined (*S*)-2-phenylbutyrate being bound by **(*S*)-14** with a $K_a = 1.0 \times 10^5\,M^{-1}$, while (*R*)-2-phenylbutyrate was bound with a $K_a = 9.8 \times 10^3\,M^{-1}$, i.e. K_a (*S*)/K_a (*R*) = 10. The proposed binding model (supported by computational calculations) suggests, in addition to the carboxylate hydrogen bonding to the convergent N-H array from the four pyrroles, that for

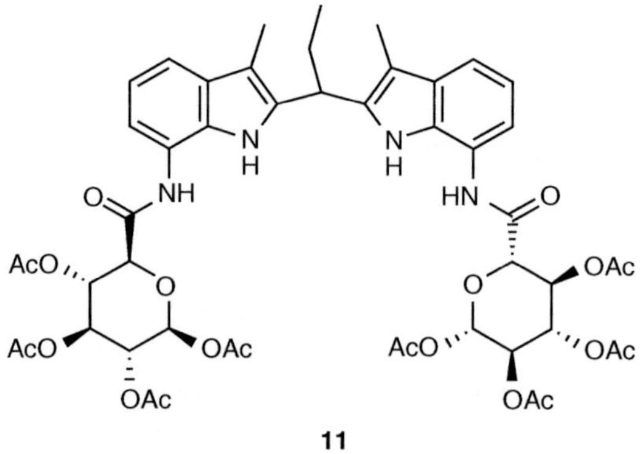

11

Figure 5.7 *Jurczak's chiral diindolylmethane anion receptor* ***11***

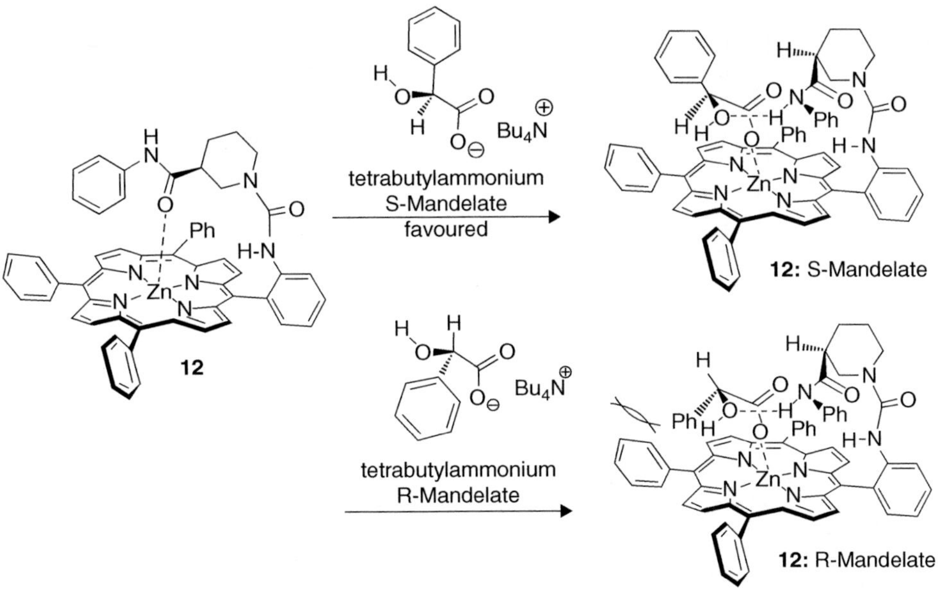

Figure 5.8 *Starnes' L-nipecotic acid-porphyrin anion receptor **12** and proposed binding to (S)- and (R)-mandelate – Reprinted with permission from X. Wu and S. D. Starnes, L-Nipecotic Acid-Porphyrin Derivative: A Chiral Host with Introverted Functionality for Chiral Recognition, Organic Letters, **14**(14), 3652–3655 (2012). Copyright 2012 American Chemical Society*

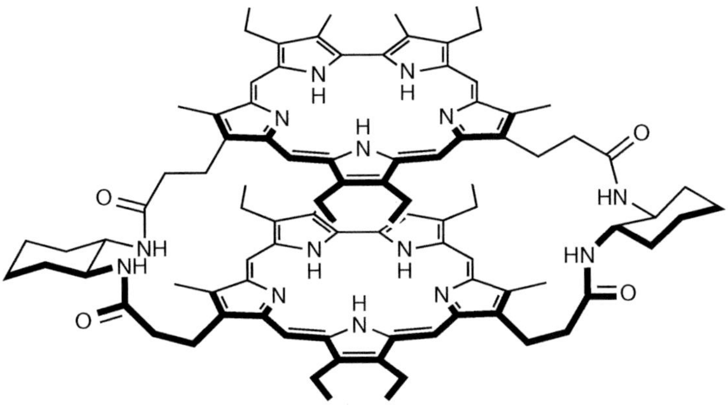

13

Figure 5.9 *Sessler's cyclic sapphyrin based dimer **13** for the chiral recognition of dicarboxylate anions*

(*S*)-14 a favourable π-π interaction between one of the naphthyl groups of the receptor and the phenyl ring of the bound substrate can occur for the *S* enantiomer of substrate, but not the *R* isomer.

A representative example of a chiral anion receptor based on a functionalized calix[4]arene scaffold, is provided by the bis-urea receptor **(*S*, *S*)-15** of Kalchenko and

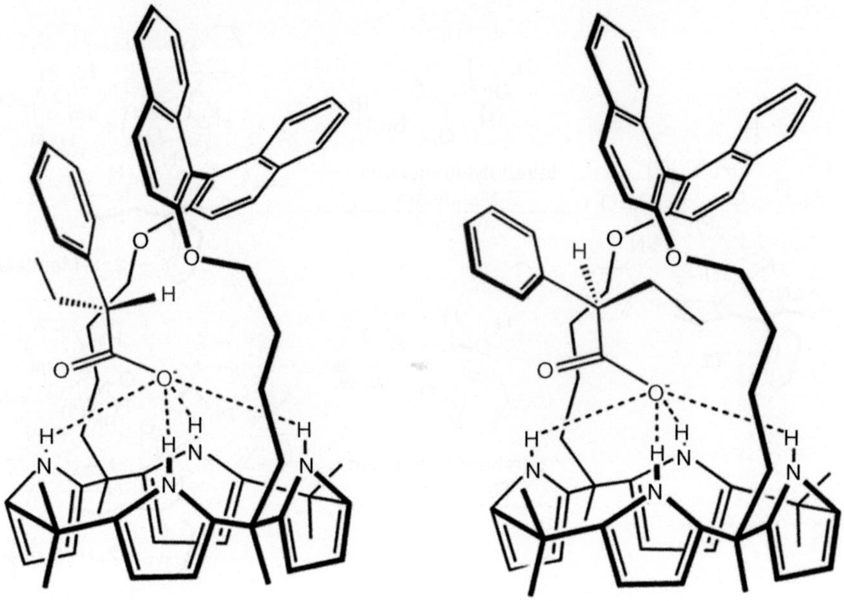

Figure 5.10 *Lee's BINOL-strapped calix[4]pyrrole receptor* ***14*** *and proposed binding to (S)- and (R)-2-phenylbutyrate. Adapted from H. Miyaji, S.-J. Hong, S.-D. Jeong et al., A Binol-Strapped Calix[4]pyrrole as a Model Chirogenic Receptor for the Enantioselective Recognition of Carboxylate Anions, Angewandte Chemie International Edition,* ***46****(14), 2508–2511 (2007). Copyright 2007 John Wiley & Sons, Inc. Reprinted by permission of John Wiley & Sons, Inc.*

Casnati (Figure 5.11) [22]. In d_6-acetone, this particular receptor binds the enantiomers of *N*-Ac-phenylalanine by hydrogen bonding, with a ratio of association constants K_a (D)/K_a (L)= 4.14. Based on computational studies, the authors propose enantioselectivity potentially arises from a repulsive interaction that occurs in the host-guest complex for the L enantiomer of *N*-Ac-phenylalanine between the phenyl ring of the guest and a carboxymethyl group of the host.

The anion binding behaviours of various chiral rhenium (I) and ruthenium (II) amide bipyridyl complexes were reported by Beer [23]. The rhenium (I) complex **(*S*, *S*)-16** in Figure 5.12 showed modest selectivity in binding the enantiomers of *N*-Cbz-glutamate (K_a (L)/K_a (D)= 1.25) in DMSO-d_6, while no discernible enantioselectivity could be observed for the rest of the complexes investigated, including those based on enantiopure samples of *Λ*- and *Δ*- tris(bipyridyl)ruthenium (II) complexes. The limited success of these systems was attributed to the conformational flexibility of the hosts.

Moody and Tucker have targeted the enantioselective electrochemical detection of chiral anions by appending the redox-active ferrocene moiety with chiral hydrogen bonding donating urea groups (Figure 5.13) [24, 25]. In their first report, modest enantioselectivity in the binding of 2-phenylbutyrate by receptor **(*S*)-17** was measurable by UV-vis spectroscopy in DMSO (K_a (*R*)= 2910 M^{-1}, K_a (*S*)= 2350 M^{-1}, K_a (*R*)/K_a (*S*)= 1.24), but not by cyclic

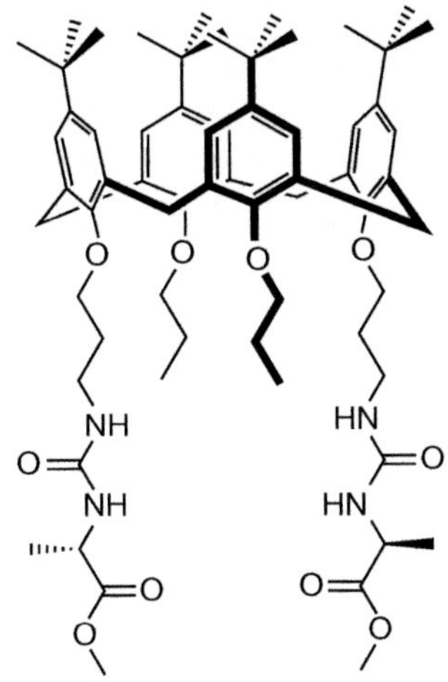

(S, S)-15

Figure 5.11 *Kalchenko and Casnati's chiral bis-urea functionalized calix[4]arene receptor*

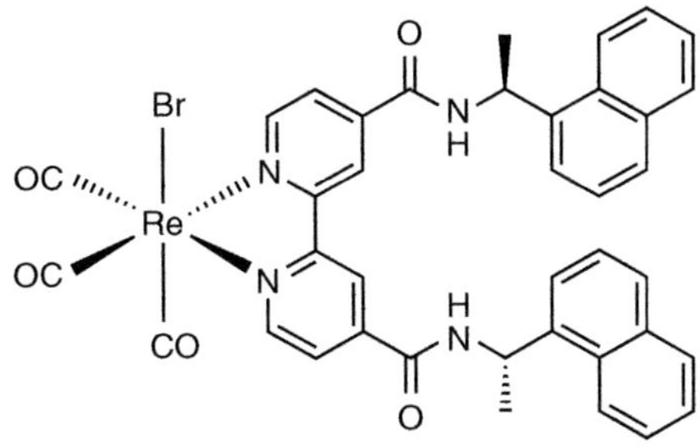

(S, S)-16

Figure 5.12 *Beer's chiral Re complex **(S, S)-16** possessing modest enantioselectivity*

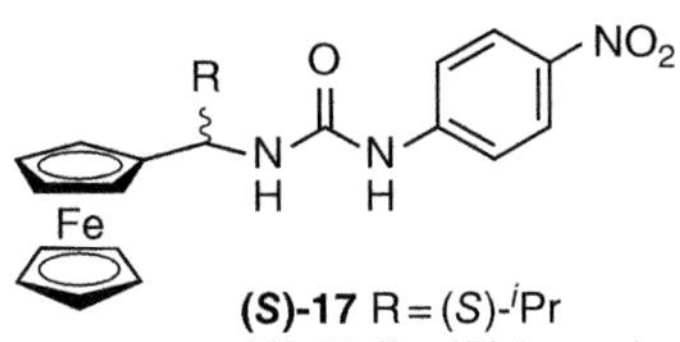

(S)-17 R = (*S*)-^{i}Pr
(*R*)-18 R = (*R*)-benzyl

Figure 5.13 *Moody and Tucker's chiral ferrocene ureas for enantioselective electrochemical anion sensing*

voltammetry in acetonitrile [24]. In a follow-up paper, enantioselectivity was detected electrochemically for enantiomers of *N*-benzenesulfonylproline with receptor **(*R*)-18** [25].

A notable departure from the other receptors discussed in this chapter is a particular class of metal complexes that are conformationally flexible and hence possess no intrinsic chiroptical properties, but upon the binding of a chiral anion provide a chiral optical response [26, 27]. For example, Tsukube's calcium-cyclen complex **19** possesses no circular dichromatic (CD) response in acetonitrile/ethanol solution, due to rapid conversion between *Δ* and *Λ* enantiomers of the complex (Figure 5.14a). Addition of enantiopure, nonchromophoric, Boc-protected amino acids induces a CD signal, due to carboxylate coordination to the calcium cation fixing the orientation of the pendent nitrogen donors. The anion enantiomers induce mirror image CD signals, allowing for **19** to be used to report on the enantiomeric ratio of samples of a given anion [26]. The related terbium-triazacyclononane complex **20**, reported by Parker and coworkers, possesses a displaceable water molecule as a ligand to the lanthanide cation (Figure 5.14b). Lactate can displace this water, with enantio-enriched samples of the anion 'switching on' a circular-polarized luminescence (CPL) signal [27].

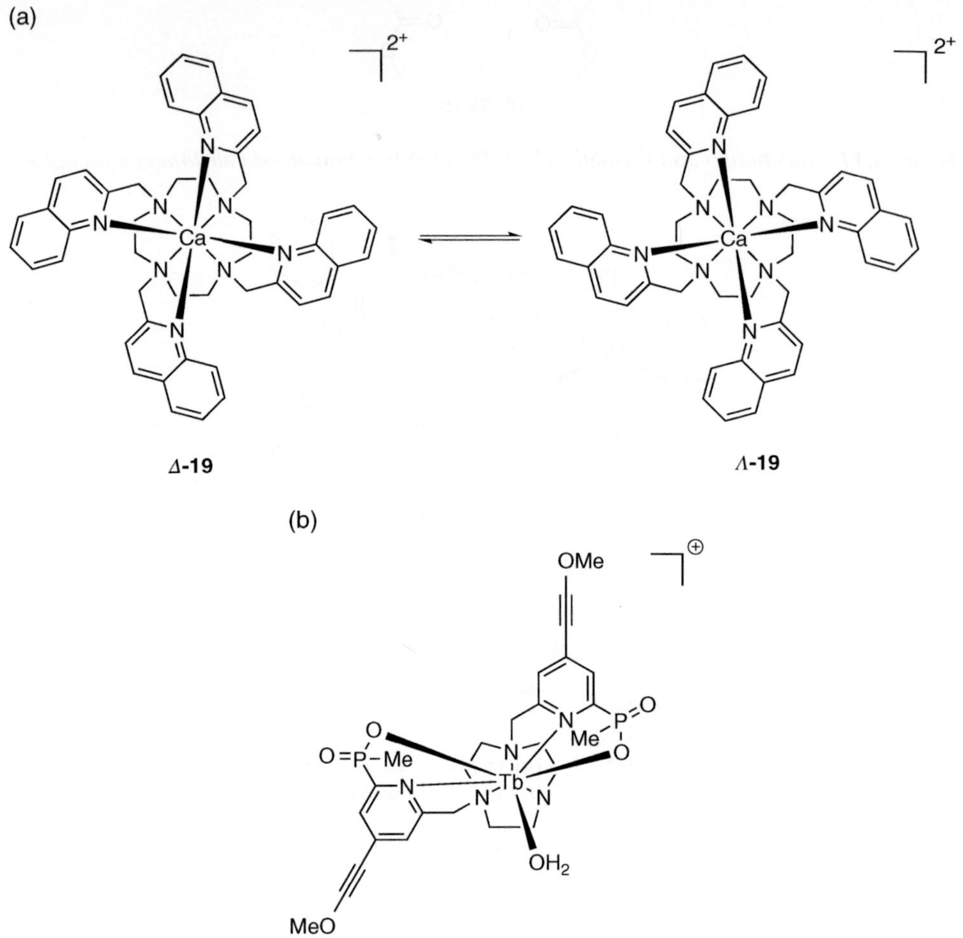

20

Figure 5.14 *Conformationally flexible metal complexes that can provide chiroptical responses upon the binding of a chiral anion: (a) Tsukube's calcium-cyclen complex* ***19*** *and (b) Parker's terbium-triazacyclononane complex* ***20***

5.2.4 Hosts for Chiral Zwitterions and Neutral Molecules

In neutral aqueous solutions, amino acids will exist as zwitterions. Enantioselective binding of such species was demonstrated in a classic study by de Mendoza and coworkers (Figure 5.15) [28]. Receptor **(*S*, *S*)-21** incorporated crown ether and guanidinium functional groups to allow for interaction with the ammonium and carboxylate residues of the amino acid guest. The incorporation of a naphthalene motif allowed for potential π-π stacking interactions with aromatic side chains of appropriate amino acids. Chirality arose from the substitution of the cyclic guanidinium moiety. For **(*S*, *S*)-21**, enantioselective extraction of the L-enantiomer of phenylalanine and tryptophan from racemic solutions was achieved. At the time of its original publication, this was rationalized by invoking the classic three-point model of chiral recognition.

In a later report, the same group looked at a family of such receptors as enantioselective membrane carriers [29]. Of note from this work was a revision of the model of interaction between host and amino acid guest; molecular dynamics calculations suggest that the potential π-π stacking interaction described above is in fact relatively unimportant in the binding of (e.g.) L-trytophan, and that – at least in apolar solvents – hydrophobic interactions are comparatively less important than electrostatic interactions.

You and Lan have described a family of chiral bis-BINOL, bis-imidazolium receptors, capable of reporting by fluorescence the charge-assisted hydrogen bonding of amino acids in aqueous solution (Figure 5.16) [30]. Receptor **(*R*, *R*)-22** exhibited notable enantioselectivity in the binding of tryptophan, K_a (D)/K_a (L) = 6.2. The good level of enantioselectivity was believed to arise from the relatively congested nature of the macrocycle cavity.

Uncharged, neutral molecules may be bound in noncompetitive solvents by neutral hydrogen bond donating hosts. By incorporation of chiral groups into such receptors, it is

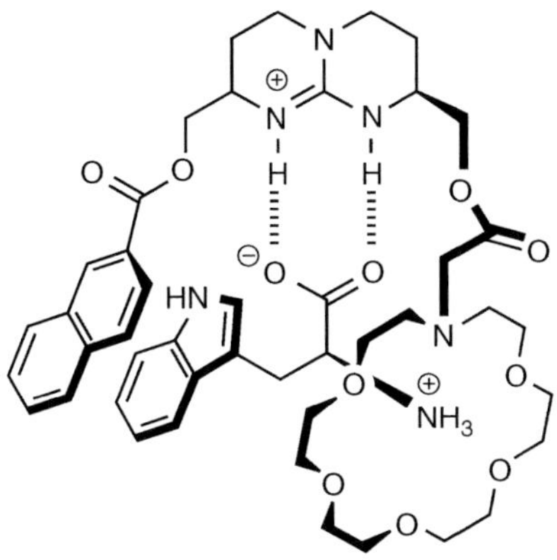

(S, S)-21
binding L-tryptophan

Figure 5.15 *De Mendoza's chiral guanidinium crown ether* ***(S, S)-21****, selective for* L*-tryptophan and* L*-phenylalanine. Reprinted with permission from A. Galán, D. Andreu, A. M. Echavarren et al., A receptor for the enantioselective recognition of phenylalanine and tryptophan under neutral conditions, Journal of the American Chemical Society,* ***114****(4), 1511–1512 (1992). Copyright 1992 American Chemical Society*

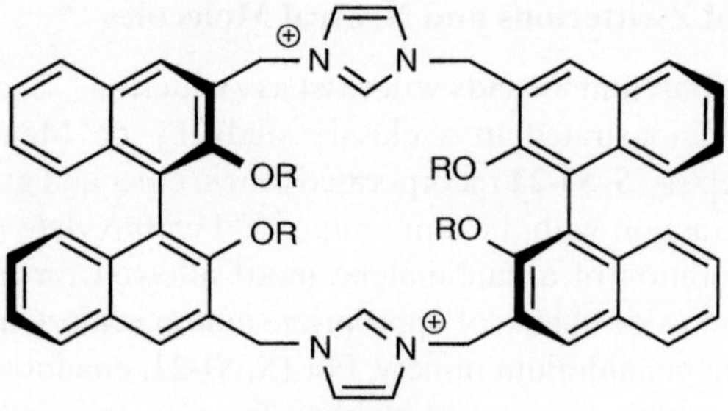

(R, R)-22

Figure 5.16 You and Lan's fluorescent chiral bis-BINOL, bis-imidazolium receptor ***(R, R)-22***

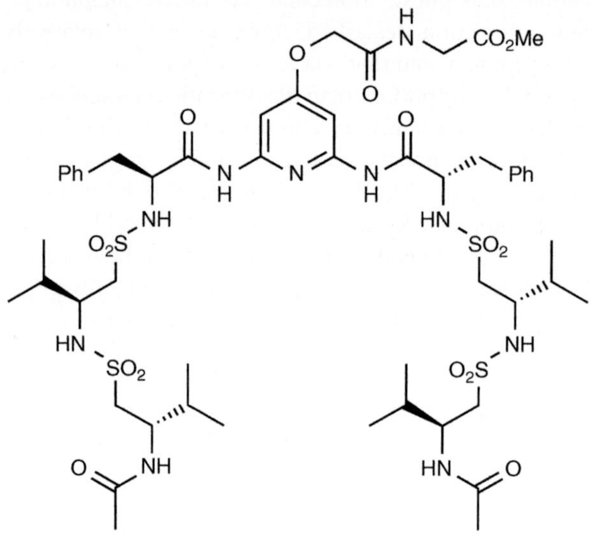

all-(S)-23

Figure 5.17 Gennari and Kilburn's acyclic hydrogen bond donating receptor ***all-(S)-23***

possible to generate enantioselective hosts. In early work on such targets Still bound simple amides [31] and dipeptides [32] using cyclic amide receptors in solvents such as benzene or chloroform. Gennari and Kilburn separately reported on an acyclic receptor **all-(*S*)-23**, that exhibited > 20 : 1 enantioselectivity in the binding of protected dipeptide *N*-Cbz-Ala-AlaOH (K_a (L, L) = 107 M^{-1}, K_a (D, D) = 2404 M^{-1}, $CDCl_3$) (Figure 5.17) [33].

Ema and Sakai have developed a (now commercially available) macrocycle **(*R*)-24**, which can act as a chiral solvating agent at substoichiometric levels (as little as 5 mol % being enough to split the enantiomeric signals of methyl *p*-tolyl-sulfoxide) for a wide range of chiral hydrogen bond-accepting guest molecules (Figure 5.18) [34]. The receptor consists of a well established Hamilton hydrogen bonding motif, incorporated into a macrocycle by use of a chiral binaphthyl unit. ^{1}H NMR titrations revealed the highest

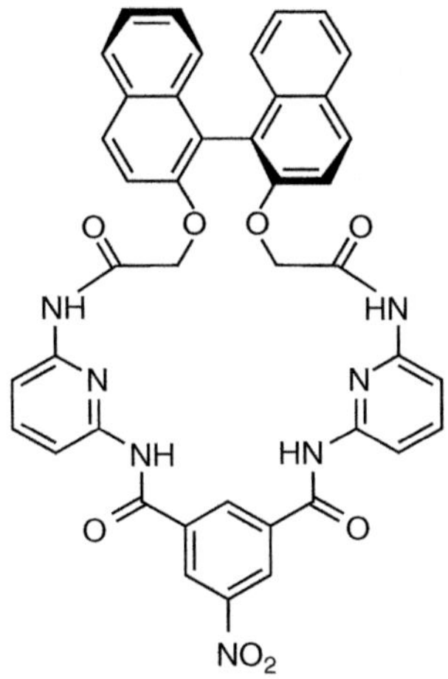

(*R*)-24

Figure 5.18 *Ema and Sakai's chiral solvating agent* ***(R)-24***

enantioselectivity was observed for examples of chiral sulphonamide (K_a (*S*)/K_a (*R*)=4.9) and sulfoxide (K_a (*S*)/K_a (*R*)=4.3) guests. Variants of macrocycle **(*R*)-24** have been incorporated onto silica to allow for the creation of chiral HPLC columns [35].

A large number of enantioselective fluorescent receptors based on BINOL have been investigated by Pu and coworkers to sense chiral carboxylic acids (Figure 5.19) [36]. In early work a BINOL dimer **(*S*, *S*)-25** (Figure 5.19a) exhibited greater fluorescence enhancement upon exposure to the *S* compared to the *R* enantiomer of mandelic acid in benzene (containing 2% dimethoxyethylene) [37]. In another work, the macrocycle **(*S*, *S*)-26** exhibited enhancement of fluorescence only for (*S*)-mandelic acid, and not (*R*)-mandelic acid (Figure 5.19b) [38]. A remarkably simple mono-BINOL receptor **(*S*)-27** was shown to exhibit high enantioselectivity in fluorescence intensity (I_R/I_S ratios of up to 22) to a range of structurally diverse α-hydroxycarboxylic acids (Figure 5.19c) [39].

5.3 Conclusions: Summary and Future Directions

While a number of acyclic and macrocyclic receptors have been prepared that exhibit chiral recognition, this field of research is still undoubtedly ripe for investigation. High levels of enantioselectivity are known, but many systems are still being reported with only modest levels of enantiomeric guest discrimination. We believe that the exploitation of more preorganized receptor architectures, such as chiral interlocked molecules and transition metal templated organic cages, could prove to be highly rewarding areas of research.

We also look forward to the further use of chiral receptors, for example, in 'real-world' applications such as fluorescence and electrochemical sensing. Perhaps most exciting for synthetic chemists are the lessons learnt about chiral host-guest recognition that could have a direct impact on the development of new chiral catalysts for the asymmetric synthesis of biologically and/or medicinally important chiral molecules.

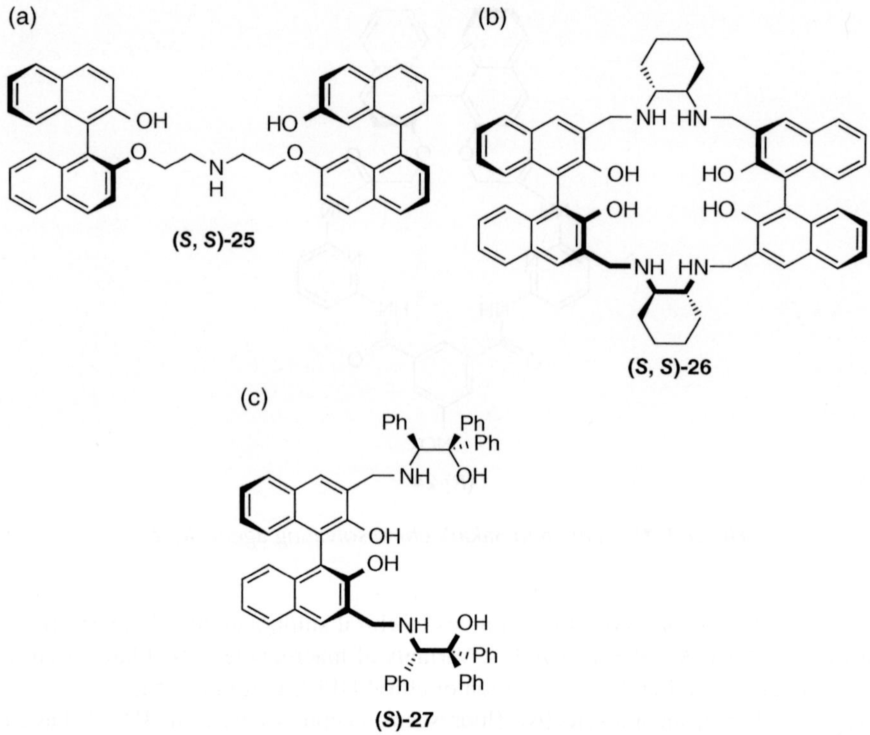

Figure 5.19 *Pu's BINOL receptors for enantioselective fluorescence sensing of chiral carboxylic acids*

References

[1] Y. Okamoto and T. Ikai, Chiral HPLC for efficient resolution of enantiomers, *Chemical Society Reviews*, **37**(12), 2593–2608 (2008).

[2] D. Parker, NMR determination of enantiomeric purity, *Chemical Reviews*, **91**(7), 1441–1457 (1991).

[3] X. Zhang, J. Yin and J. Yoon, Recent advances in development of chiral fluorescent and colorimetric sensors, *Chemical Reviews*, **114**(9), 4918–4959 (2014).

[4] B. Schuur, B. J. V. Verkujil, A. J. Minnaard, *et al.*, Chiral separation by enantioselective liquid-liquid extraction, *Organic and Biomolecular Chemistry*, **9**(1), 36–51 (2011).

[5] A. Bethrod, Chiral recognition mechanisms, *Analytical Chemistry*, **78**(7), 2093–2099 (2006).

[6] E. B. Kyba, K. Koga, L. R. Sousa, *et al.*, Chiral recognition in molecular complexing, *Journal of the American Chemical Society*, **95**(8), 2692–2693 (1973).

[7] S. C. Peacock, L. A. Domeier, F. C. A. Gaeta, *et al.*, Host-Guest Complexation. 13. High Chiral Recognition of Amino Esters by Dilocular Hosts Containing Extended Steric Barriers, *Journal of the American Chemical Society*, **100**(26), 8190–8202 (1978).

[8] G. Dotsevi, Y. Sogah and D. J. Cram, Chromatographic optical resolution through chiral complexation of amino ester salts by a host covalently bound to silica gel, *Journal of the American Chemical Society*, **97**(5), 1259–1261 (1975).

[9] M. Newcomb, J. L. Toner, R. C. Helgeson and D. J. Cram, Host-guest complexation. 20. Chiral recognition in transport as a molecular basis for a catalytic resolving machine, *Journal of the American Chemical Society*, **101**(17), 4941–4947 (1979).

[10] I. Alfonso, B. Dietrich, F. Rebolledo, *et al.*, Optically active hexaazamacrocycles: protonation behavior and chiral-anion recognition, *Helvetica Chimica Acta*, **84**(2), 280–295 (2001).
[11] A. González-Álvarez, I. Alfonso, P. Díaz, *et al.*, A highly enantioselective abiotic receptor for malate dianion in aqueous solution, *Chemical Communications*, **11**, 1227–1229 (2006).
[12] S. Rossi, G. M. Kyne, D. L. Turner, *et al.*, A highly enantioselective receptor for *N*-protected glutamate and anomalous solvent-dependent binding properties, *Angewandte Chemie International Edition*, **41**(22), 4233–4236 (2002).
[13] S. Bartoli, T. Mahmood, A. Malik, *et al.*, Selective binding of L-glutamate derivative in aqueous solvents, *Organic & Biomolecular Chemistry*, **6**(13), 2340–2345 (2008).
[14] L. J. Lawless, A. G. Blackburn, A. J. Ayling, *et al.*, Steroidal guanidines as enantioselective receptors for *N*-acyl α-amino acids. Part 1. 3α-Guanylated carbamates derived from cholic acid, *Journal of the Chemical Society, Perkin Transactions 1*, **11**, 1329–1341 (2001).
[15] B. Baragaña, A. G. Blackburn, P. Breccia, *et al.*, Enantioselective Transport by a Steroidal Guanidinium Receptor, *Chemistry – A European Journal*, **8**(13), 2931–2936 (2002).
[16] J. M. Granda and J. Jurczak, Sweet anion receptors: recognition of chiral carboxylate anions by D-glucuronic-acid-decorated diindolylmethane, *Organic Letters*, **15**(18), 4730–4733 (2013).
[17] J. M. Granda and J. Jurczak, Artificial neural networks for guest chirality classification through supramolecular interactions, *Chemistry – A European Journal*, **20**(39), 12368–12372 (2014).
[18] X. Wu and S. D. Starnes, L-Nipecotic acid-porphyrin derivative: a chiral host with introverted functionality for chiral recognition, *Organic Letters*, **14**(14), 3652–3655 (2012).
[19] V. Nandipati, K. Akinapelli, L. Koya and S. D. Starnes, Recognition of mandelate stereoisomers by chiral porphyrin hosts: prediction of stereopreference in guest binding a priori using a simple binding model? *Tetrahedron Letters*, **55**(5), 985–991 (2014).
[20] J. L. Sessler, A. Andreivsk, V. Král and V. Lynch, Chiral recognition of dicarboxylate anions by sapphyrin-based receptors, *Journal of the American Chemical Society*, **119**(40), 9385–9392 (1997).
[21] H. Miyaji, S.-J. Hong, S.-D. Jeong, *et al.*, A binol-strapped calix[4]pyrrole as a model chirogenic receptor for the enantioselective recognition of carboxylate anions, *Angewandte Chemie International Edition*, **46**(14), 2508–2511 (2007).
[22] A. V. Yakovenko, V. I. Boyko, V. I. Kalchenko, *et al.*, *N*-Linked Peptidocalix[4]arene Bisureas as Enantioselective Receptors for Amino Acid Derivatives, *Journal of Organic Chemistry*, **72**(9), 3223–3231 (2007).
[23] L. H. Uppadine, F. R. Keene and P. D. Beer, Approaches towards the enantioselective recognition of anionic guest species using chiral receptors based on rhenium(I) and ruthenium(II) with amide bipyridine ligands, *Journal of the Chemical Society, Dalton Transactions*, **14**, 2188–2198 (2001).
[24] P. Laurent, H. Miyaji, S. R. Collinson, I. Prokeš, C. J. Moody, J. H. R. Tucker and A. M. Z. Slawin, Asymmetric synthesis of chiral α-ferrocenylalkylamines and their use in the preparation of chiral redox-active receptors, *Organic Letters*, **4**(23), 4037–4040 (2002).
[25] Y. Willener, K. M. Joly, C. J. Moody and J. H. R. Tucker, An exploration of ferrocenyl ureas as enantioselective electrochemical sensors for chiral carboxylate anions, *Journal of Organic Chemistry*, **73**(4), 1225–1233 (2008).
[26] H. Misaki, H. Miyake, S. Shinoda and H. Tsukube, Asymmetric twisting and chirality probing properties of quadruple-stranded helicates: coordination versatility and chirality response of Na^+, Ca^{2+}, and La^{3+} complexes with octadentate cyclen ligand, *Inorganic Chemistry*, **48**(24), 11921–11928 (2009).
[27] R. Carr, R. Puckrin, B. K. McMahon, *et al.*, Induced circularly polarized luminescence arising from anion or protein binding to racemic emissive lanthanide complexes, *Methods and Applications in Fluorescence*, **2**(2), 024007+07 (2014).
[28] A. Galán, D. Andreu, A. M. Echavarren, *et al.*, A receptor for the enantioselective recognition of phenylalanine and tryptophan under neutral conditions, *Journal of the American Chemical Society*, **114**(4), 1511–1512 (1992).
[29] P. Breccia, M. Van Gool, R. Pérez-Fernández, *et al.*, Guanidinium receptors as enantioselective amino acid membrane carriers, *Journal of the American Chemical Society*, **125**(27), 8270–8284 (2003).

[30] L. Yang, S. Qin, X. Su, *et al.*, 1,1′-Binapthyl-based imidazolium chemosensors for highly selective recognition of tryptophan in aqueous solutions, *Organic and Biomolecular Chemistry*, **8**(2) 339–348 (2010).
[31] P. E. J. Sanderson, J. D. Kilburn and W. C. Still, Enantioselective complexation of simple amides by a C_2 host molecule, *Journal of the American Chemical Society*, **111**(21), 8314–8315 (1989).
[32] R. Liu, P. E. J. Sanderson and W. C. Still, Enantioselective complexation of the alanine dipeptide by a C_2 host molecule, *Journal of Organic Chemistry*, **55**(18), 5184–5186 (1990).
[33] E. Botana, S. Ongeri, R. Arienzo, *et al.*, Enantioselective binding of dipeptides using acyclic receptors, *Chemical Communications*, **15**, 1358–1359 (2001).
[34] T. Ema, D. Tanida and T. Sakai, Versatile and practical macrocyclic reagent with multiple hydrogen-bonding sites for chiral discrimination in NMR, *Journal of the American Chemical Society*, **129**(34), 10591–10596 (2007).
[35] T. Ema, K. Hamada, K. Sugita, *et al.*, Synthesis and evaluation of chiral selectors with multiple hydrogen-bonding sites in the macrocyclic cavities, *Journal of Organic Chemistry*, **75**(13), 4492–4500 (2010).
[36] L. Pu, Enantioselective fluorescent sensors: a tale of BINOL, *Accounts of Chemical Research*, **45**(2), 150–163 (2012).
[37] J. Lin, Q.-S. Hu, M.-H. Xu and L. Pu, A practical enantioselective fluorescent sensor for mandelic acid, *Journal of the American Chemical Society*, **124**(10), 2088–2089 (2002).
[38] Z.-B. Li, J. Lin and L. Pu, A cyclohexyl-1,2-diamine-derived bis(binaphthyl) macrocycle: enhanced sensitivity and enantioselectivity in the fluorescent recognition of mandelic acid, *Angewandte Chemie International Edition*, **44**(11), 1690–1693 (2005).
[39] H.-L. Liu, Q. Peng, Y.-D. Wu, *et al.*, Highly enantioselective recognition of structurally diverse α-hydroxycarboxylic acids using a fluorescent sensor, *Angewandte Chemie International Edition*, **49**(3), 602–606 (2010).

6

Chiral Influences in Functional Molecular Materials

David B. Amabilino

6.1 Introduction

Chirality is present in natural systems from atoms to galaxies [1–3] and certain synthetic systems from molecules to materials [4–6]. However, a great majority of materials of all kinds are achiral. Chirality is often seen as a difficulty rather than an opportunity. Yet biological molecular machines and the photosynthetic reaction centres and related electron transfer mechanisms all work through chiral motions and arrangements of the protein's subunits. There is growing evidence that chirality can give unique characteristics to synthetic materials and the devices derived from them. This chapter presents that case with a particular focus on systems containing metal ions.

In organic materials chirality can determine a number of specific characteristics, above all related to optical behaviour, although also related to molecular recognition and conductivity, as reviewed recently [7], where the classic routes of chiral induction to materials were presented. Briefly, they involve either spontaneous symmetry breaking upon aggregation, chiral induction from stereogenic sources (point chirality, planar chirality, or atropoisomerism) in each molecule, or amplification of chirality from minority components in the material, as in the sergeants-and-soldiers effect [8]. These phenomena can also take place in the synthesis of metal complexes [9], although usually the main induction mechanisms are through coordination of metal ions by chiral ligands [10, 11], the formation of racemic mixtures of chiral complexes followed by their spontaneous resolution [12], or the resolution of metal complexes with achiral ligands by chiral counter-ions [13], leading to diastereoisomeric complexes (Figure 6.1).

Chirality in Supramolecular Assemblies: Causes and Consequences, First Edition. Edited by F. Richard Keene.

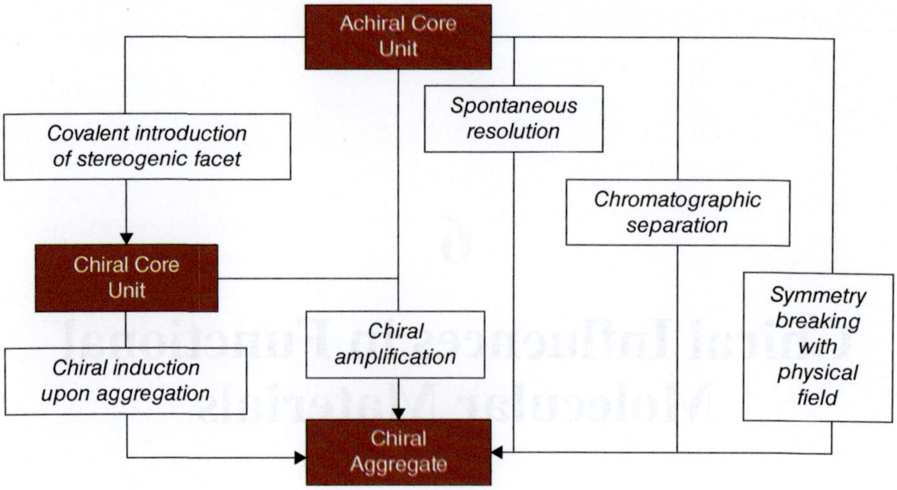

Figure 6.1 *The main routes to chiral metallo-organic materials*

When chirality is introduced through ligands to the metal ion, the induction of a preferred chiral coordination sphere can take place making a complex chiral at the metal [14]. Alternatively, the stereogenic element can induce optical activity in the metal ion by through space interactions, where the metal ion would actually be in an achiral coordination sphere, such as a square planar. Similar scenarios could also be envisaged for diastereoisomeric or spontaneous resolution, although in these cases the chiral in a metal situation is more frequent, as it facilitates the separation of the enantiomeric forms.

Most commonly, chiral molecular materials are studied in their crystalline state: this situation affords precision in terms of the environment of the molecules and their ability to align the samples in what can be anisotropic systems. However, other states of matter are possibly more appealing in terms of their applicability in devices – liquid crystals perhaps being the most outstanding case [15] and soft matter systems in general (notably metallogels [16] and metallo-polymers) [17]. The study of nanomaterials is especially interesting because of the particular properties they can exhibit [18–23].

In most cases the chirality of the materials (as opposed to the molecules) is characterized by circular dichroism (CD) spectroscopy, which is a particularly sensitive probe of supramolecular structure [24, 25]. Usually, this technique is performed in solution, although solid-state CD is a viable option for the study of chiral superstructure [26–28]. Chiroptical spectroscopy is in general a very useful technique for probing coordination compounds [29, 30].

In this chapter, the structural chirality in molecular materials will be discussed, in states of matter that have unique characteristics. Each of these unique states can display behaviour that is a result of both their chirality and the phase the material is in. Then, specific physical properties – such as electrical conduction of chiral functional molecular materials – will be described. The breadth of the field means that the text cannot be comprehensive and the reader is encouraged to explore the fascinating world of chiral functional molecular materials, researching around the cited literature.

6.2 Functional Molecular Materials in Different States

6.2.1 Crystals

The histories of crystals and chirality are, of course, intimately entwined. Pasteur's discovery of the spontaneous resolution of the enantiomers of sodium ammonium tartrate [31] and the diastereomeric salts of various systems [32] is seminal. It laid the foundation for the preparation of chiral crystalline materials by using achiral or racemic compounds, apart from the obvious power of synthetic chemistry to prepare chiral molecules.

As an example, a spontaneous resolution of simple coordination complexes can be achieved, as shown for the ruthenium (II) complex with two 2,2′-bipyridine ligands and one coordinated phenenthroline molecule as its hexafluorophosphate salt [33]. A column of piled phenanthroline units is generated in the crystals of the complex (3.52 Å π-π distance) leading to a homochiral stack that propagates in perpendicular directions. Interestingly, the same metal ion complex with tetrafluoroborate as the counter ion crystallizes as a racemic compound, despite the fact that π stacking interactions are still observed in the crystal, albeit slightly longer than in the hexafluorophosphate version. A racemic compound also crystallizes for tris(2,2′-bipyridine)ruthenium(II) hexafluorophosphate [34]. So, quite minor and certainly subtle packing changes can determine whether a conglomerate or racemic compound is formed.

This spontaneous resolution, which can apparently be nonfunctional, can be extremely useful for deracemization [35] and subsequent absolute asymmetric synthesis: achiral components can be bound together to give chiral complexes and the resulting mixture of enantiomers crystallized under carefully controlled conditions. A very recent example in a string of interesting papers is that of three ruthenium(II) complex conglomerates [36]. The metal ion coordinated bidentate sulfide ligands, and the resulting enantiomerically enriched crystals were oxidized in an enantioselective manner to generate chiral sulfoxide complexes with enantiomeric excesses above 98%. The route is seen as a way of generating sulfoxides in an enantioselective manner more generally. While in this case the resolution was performed by slow crystal growth, it has actually been shown that quite common achiral conglomerate-forming organic molecules can be enriched through Viedma ripening [37], whereby grinding crystals in a supersaturated suspension leads to essentially one enantiomer [38]. This technique may well lead to spontaneous resolution of metal complexes of interest [39, 40].

Naturally, crystals are generally highly ordered entities, with regular arrangements of the component molecules. This situation allows the purification of materials and the precise study of their structure and correlations with their properties – as we shall see over the course of this chapter. However, it is worth mentioning two things before we move on.

Firstly, while good science often requires detail for property-structure correlation, single crystals (or collections of them) are not the state of choice for any device, because they are generally fragile, small and tedious to position. Secondly, most crystals containing chiral compounds have a morphology that is actually achiral; but apart from the morphology chirality can arise from actual twisting of the crystals leading to macroscopically chiral objects [41]. These twisted or curved structures can arise from mineral samples [42] and are prepared increasingly in synthetic inorganic materials [43].

6.2.2 Liquid Crystals

Liquid crystals – materials that are comprised of molecules with orientational order but a degree of positional disorder, that flow like a fluid but are frequently birefringent – are historically important in the study of chirality and also technologically important for flat screens, because of the capability of switching them. Cholesteric and ferroelectric smectic C* phases are especially important in this area. The cholesteric phase is formed by layers of molecules whose long axis is in the plane of the layers with a twist between the layers (either clockwise or anticlockwise). The ferroelectric smectic C* phase comprises molecules at a high angle (say approximately 80°) to the layer plane and whose dipole moments are oriented in one direction, with a twisted piling up of these layers to give polar order. Even though thousands of materials have been discovered and some design features are clear, they are made largely on a trial-and-error basis – although, at least for organic systems, conceptually the noncovalent interactions between the component molecules can be considered a kind of molecular recognition because their conformations are influenced by their environment [44, 45]. Inorganic complexes can be used as rigid components in the promesogenic cores (those that favour the behaviour) of liquid crystals. Often, they are simply a rigid part of the molecule that is locally devoid of chirality, with stereogenic centres located in the aliphatic side chains that provide the fluid character – although the metal ions do endow other characteristics, such as luminescence, paramagnetism, polarization and so on. There are notable exceptions, as we shall show for some specific cases here.

Helical supramolecular arrangements have been seen in the mesophases of a variety of metallomesogens [46]. A particularly interesting example from the supramolecular perspective is that of chiral oxazoline complexes with a variety of metal ions bound to the heterocycle [47]. The six aliphatic side chains surrounding the rigid core are not sufficient to induce stabilization of a mesophase, though mixing the compounds with the π-electron acceptor trinitrofluorenone does give rise to a smectic A liquid crystal phase. Another way to reduce the core-to-core interaction in this kind of molecule is to introduce branching near its centre: when a methyl group is attached stereoselectively next to the ligating nitrogen atom in the oxazoline, the complexes formed with copper(II) and palladium(II) present columnar mesophases in which helical organization of the cores is generated (Figure 6.2) [48]. The steric influence of the methyl groups attached to the stereogenic centres near the centre of the rigid part of the metal complex forces a tilt between what would otherwise be parallel-stacked molecules. X-ray diffraction of oriented mesophases reveals a pattern consistent with a superstructure in which each complex is twisted in its plane by 60° when compared with its neighbours. This interpretation was confirmed through CD measurements on the liquid crystal phase where exciton coupled bands were evident in the spectra, which is a chiroptical indication of a twisted orientation between molecules. In homogeneous solution the complexes reveal single Cotton effects where the twist between chromophores between molecules is not present because the molecule is dissolved in the solvent. While all the measurements were carried out on enantiomers in their mesophases, the racemic material shows very similar structural characteristics, indicating that in the mesophase spontaneous resolution takes place generating homochiral columns.

The kind of helix that is induced by the stereogenic centres can be modified by changing the positions they occupy in the molecule [49]: metallomesogens with β-diketonate cores complexed to oxovanadium(IV), copper(II) and palladium(II) and having ten lactic acid-derived chiral chains at the periphery displayed room-temperature discotic phases with

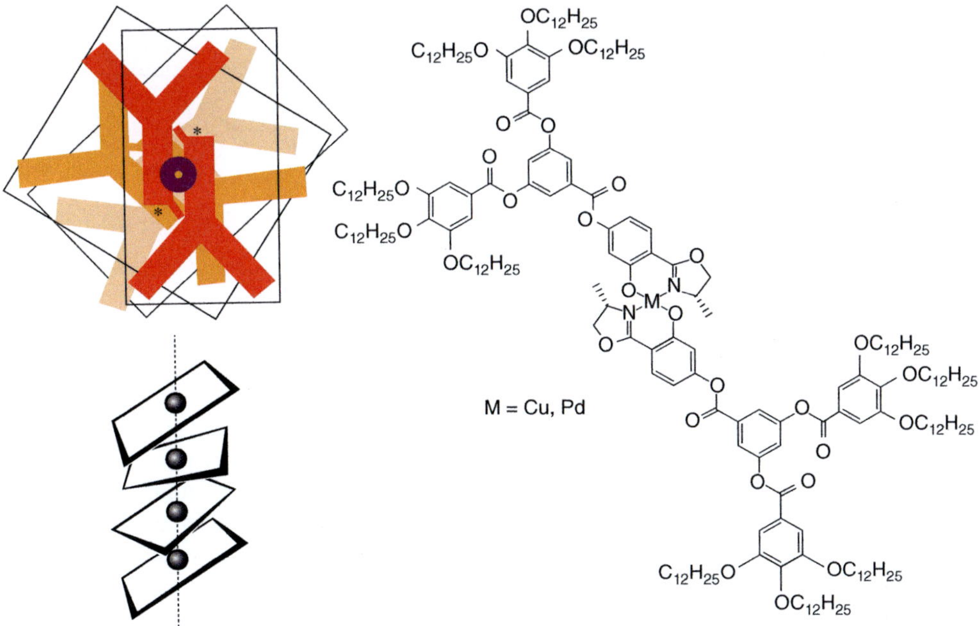

Figure 6.2 *The helical stacks viewed from above (top left) and side-on (bottom left) to the helix axis, formed by the metallomesogen based on a chiral oxazoline core whose chemical formula is shown on the right*

ferroelectric properties [50]. The discotic columnar structure – revealed most characteristically through a Maltese cross pattern seen in polarizing optical microscopy – was shown to have a rectangular columnar structure by X-ray diffraction, which also indicated that the planes of the molecules were tilted with respect to the perpendicular to the column axis as well as to their column neighbours. When this phase is placed between electrodes in an electro-optic cell the organization can be switched upon application of a potential. Although the ferroelectric switching was seen between 80 °C and 150 °C, all the complexes show a response and, provided the viscosity at lower temperatures could be overcome, might even be interesting in devices.

Metal complexes can be used to induce twist chirality on organic mesophases, and the structure of the chiral metal complexes determines the handedness. The enantiomers of the ruthenium(III) complexes of the type shown in Figure 6.3 were used to induce chiral nematic liquid crystal phases with a helical twist between the component mesogens [51]. The complexes differ in their shape – a crucial factor in determining the 'helical twisting power' in the nematic phase. In one complex the C_2 axis is located perpendicular to the direction of the alkyl chains, while in the other it is parallel. The nematic liquid crystal phase undergoes induced helical induction of opposite sense for the two complexes and the helical twisting power is much higher for the compound with the C_2 axis perpendicular to the alkyl chain direction. The origin of the contrasting behaviour of these chiral dopants lies in the orientation that the molecules adopt when hosted in the guest nematic phase. The twist the dopants induce between the layers is transferred in a hierarchical fashion to the bulk liquid crystal material.

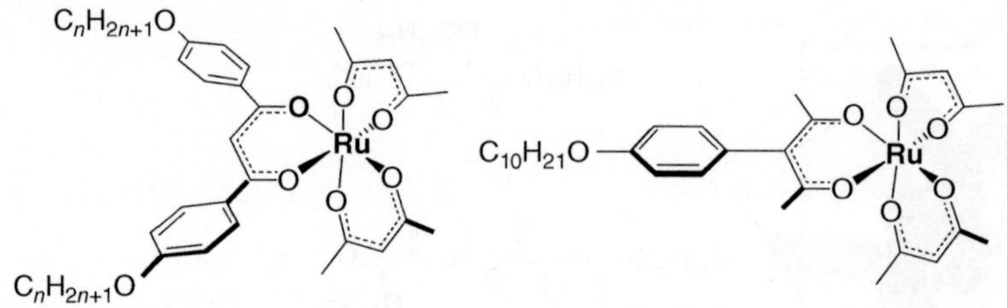

Figure 6.3 Two promesogenic dopants with different dipole directions that induce opposite chirality in nematic liquid crystals

An interesting recent example of a liquid crystal that is chiral on account of doping with an optically active compound is that of an ionic liquid crystal based on an imidazolium salt bearing two rigid components (one of them spaced from the cation by an alkyl chain) whose mesophase is induced to be chiral with a binaphthyl derived promesogenic molecule [52]. The mixture displays a chiral nematic phase whose chirality is determined by that of the dopant, which is effective in its induction in the ionic liquid crystal in as little as 1% molar ratio. The mirror image relationship of the phases doped with different atropoisomeric enantiomers of the binaphthyl derivative was confirmed by their approximately equal and opposite character in cast films (where exact mirror image spectra are very difficult to achieve because of the inhomogeneity of the material). The ionic conductivity of the liquid crystal phase was not appreciably different from that of the undoped salt, with a conductivity of approximately 10^{-7} S cm^{-1} between 76 °C and 94 °C, the temperature range for this phase. Nonetheless, this is a unique case of a chiral ionic conductor that might find interest for the study of alternative chiral charge motion and its possible consequences. For example, a cyclic polyoxothiometalate has been shown to present liquid crystal behaviour [53].

6.2.3 Gels

Gels are a particularly attractive colloidal state – comprising a solid nanofibrous material and a liquid phase that is trapped between the gelator threads – in which to prepare materials [54] because their deposition to give films of the xerogel (the material comprising the gelator once the continuous liquid phase has evaporated) does not affect drastically the initial supramolecular structure in the component material [55] and a number of interesting properties can be displayed by both gels and xerogels [56]. Metallogels are a particularly interesting class of gelators, where the property of the metal ion and the processability of the gel are brought together [57].

The coordination of amino acids and their oligomers to metals can result in the formation of chiral gel materials. Amino acids and peptides are ideal building blocks for coordination to metal ions to give gels. The mixing of glutathione with gold(III) chloride in water results in the formation of a hydrogel that is pH responsive [58]. The coordination of copper(II) by either enantiomer of aspartic acid is also effective [59]. The gels were formed by mixing an aqueous ethanol solution of the amino acid with sodium hydroxide and an aqueous solution of copper nitrate: long fibres formed at the interface which gradually extended through the whole liquid, immobilizing it. Microscopy showed remarkably long and quite uniform fibres with diameter of approximately 200 nm. The optical activity of the sample was proven by solid state CD spectroscopy, where optically active transitions associated with the metal ion are detected clearly in the region of 700 nm, far away from those of the amino acid. This evidence suggests a strong influence of the stereogenic centre on the metal ion coordination sphere. Infrared spectroscopy showed the carboxylate groups to be bound to the copper ion. The size of the fibres can be controlled by the speed of mixing. Silver can also be used to form chiral complexes of a related topology with cysteine [60]. This strategy is believed to be a general one for the preparation of one-dimensional coordination polymers.

The use of copper(II) for the assembly of chiral nanotubes also provides a material that displays asymmetric catalysis for a Diels–Alder reaction [61]. The bolaamphiphilic bis-L-glutamic acid derivative *N,N′*-hexadecanedioyl-di-L-glutamic acid shown in Figure 6.4 forms a hydrogel with copper(II) where nanotubes comprise the colloid. These tubes are formed from lamellae of the bolaamphiphile coordinated with layers of the metal ion. When the copper ions in this chiral nanomaterial were used as a catalyst for the Diels–Alder reaction between an aza-chalcone and cyclopentadiene, enantiomeric excesses of the product of up to 55% were observed, much higher than the value obtained without the nanotube structure. The material is an interesting target for future modification to improve its efficiency.

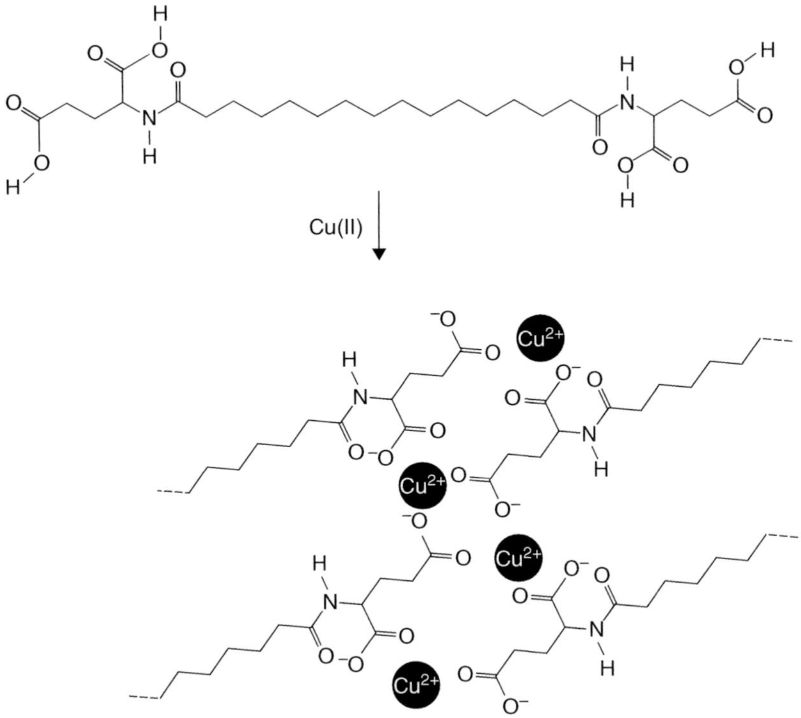

Figure 6.4 *The gelator N,N′-hexadecanedioyl-di-L-glutamic acid that forms nanotubes when complexed with copper(II), which catalyse a Diels–Alder reaction*

A platinum chloride organometallic complex substituted with a steroid group gels a wide range of organic solvents – from *N,N′*-dimethylformamide to tetrahydrofuran and toluene (for $n=1$ in Figure 6.5) – and is able to act as a visual discriminator of chirality of a phosphine that binds to the metal ion [62]. The gelator is formed from a 6-phenyl-2,2′-bipyridine ligand with a carboxy function at the 4-position of the central pyridine ring, to which the lipophilic yet conformationally rigid steroid unit is linked (Figure 6.5), with or without an ethyleneoxy spacer. Mixing even 0.1 equivalents of (*R*)- 2,2′-bis(diphenylphosphino)-1,1′-binaphthyl (binap) with the gel with no spacer in chloroform resulted in a collapse of the gel, while the enantiomer of the binap gave no obvious change. This selectivity was also proven by using CD spectroscopy. Importantly, the compound with a spacer between organometallic unit and steroid was not as sensitive to the chirality of the binap, which coordinates to the platinum(II) ion. It is believed that the steric interaction of the binap in between the fibres of the gelator are responsible for this effect. This kind of responsive gel material forms part of a larger class of these promising metallogel materials [63].

A very interesting recent development is the formation of metallogels with bisdithiolene complexes at their core (surrounded by cholesterol moieties to promote gelation in organic solvents) that show chiroptical effects in the near infrared region [64]. The gelation is essential to the formation of the strong bands because when the complexes are completely

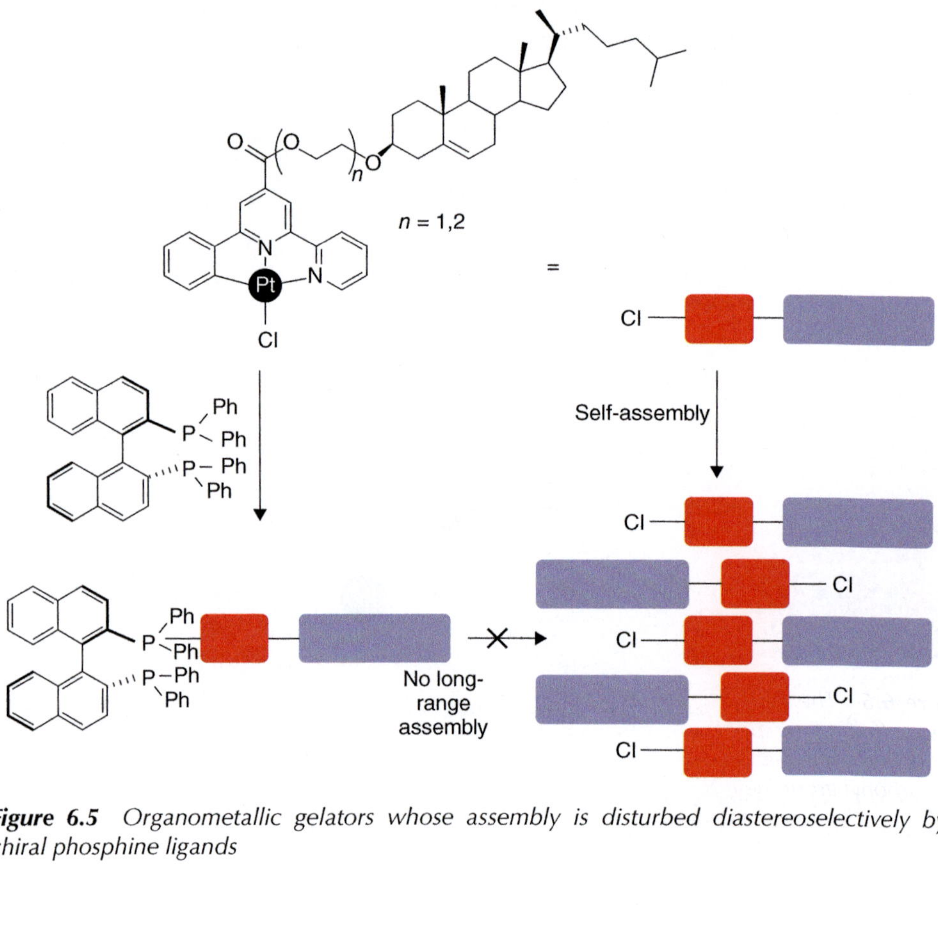

Figure 6.5 *Organometallic gelators whose assembly is disturbed diastereoselectively by chiral phosphine ligands*

dissolved the chiroptical effects in the infrared are quite weak. It is believed that this kind of material may be useful for biological applications because infrared light is not absorbed strongly compared with visible light.

In general, there is a potential role of certain bound metal ions in chiral chromophores on the gelling ability of the complex, even when the organic solubilizing parts are identical across the compounds. This influence was demonstrated clearly for a series of chiral tetraphenylporphyrin derivatives (Figure 6.6) where hydrogen bonds between amides could lead to H-type aggregates [65]. The compounds with palladium(II) or copper(II) both form gels in hexane and methycyclohexane (with a lower critical gel concentration in the latter). Interestingly, the parent free-base porphyrin did not form specific aggregates under the conditions where the metalloporphyrins did so, showing that the metal ion bound at the centre of the aromatic macrocycle plays a determining role in raising the energy of interacting between the porphyrin rings which is made directional by the hydrogen bonds (that on their own are apparently not sufficient to effect aggregation in this system). The zinc(II)

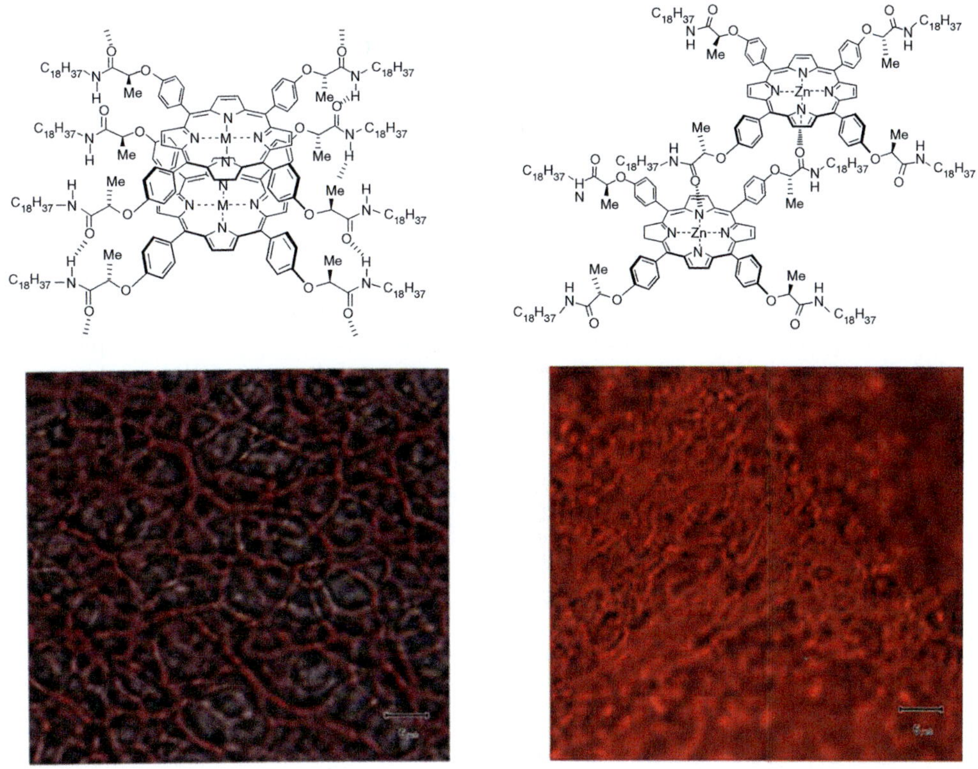

Figure 6.6 *The aggregates formed by the metalloporphyrins Most divalent metals form H-type (fully overlapping stack) aggregates represented at the top left, while the zinc(II) compound forms J-aggregates (staircase like stack) because of a specific interaction between the carbonyl group near the stereogenic centre and the metal ion. Below optical micrographs show the difference in morphology of these two types of aggregates, for the copper(II) compound on the left and zinc(II) on the right*

compound does aggregate strongly in hexane and methylcyclohexane but requires a much higher concentration than the divalent metals cited earlier. The reason for the apparent contradiction in high tendency for aggregation but high critical gel concentration is the formation of a dimer (Figure 6.5) that is formed through the coordination of the metal ion by the carbonyl oxygen atoms of the amide groups – evidenced by NMR and infrared spectroscopy. A similar coordination mode is seen in magnesium porphyrins that mimic chlorosomal bacteriochlorophyll [66]. The formation of this dimer leaves three hydrogen bond-forming moieties available for further aggregation to give the gel. Remarkably, the manganese(III) acetate metalloporphyrin also forms a gel quite effectively, despite the apparent impediment that an axial ligand might present to aggregation. An explanation of this effect is pending.

6.3 Switching

The sensitivity of chiral chromophores to their environment can result in large changes in optical activity for relatively small conformational changes in organic ligand arrangement around metal ions. Three main circular dichroic techniques are appropriate for observing these changes: electronic (absorption), luminescence and vibrational (IR). These changes observed in solution will be discussed here, although surface-based chiral switches and related systems are of enormous interest [67]. The operation of unimolecular chiroptical devices is a huge and potentially very rewarding pursuit with the ultimate aim being nanoscale devices [68].

Perhaps one of the most subtly elegant examples of chiroptical switching is the inversion of helicity of quinoline ligands in a copper complex [69]. The *N,N′*-dialkylmethionine residue – on which the ligand skeleton is based – coordinates copper(II) through the carboxylate ligand (Figure 6.7). It can be appreciated that this five-coordinate complex has an *M* helicity between the quinoline residues. Reduction of this copper(II) complex with ascorbate generates the four-coordinate copper(I) complex, in which the carboxylate group is replaced with the thioether moiety. This rearrangement by rotation around the bond

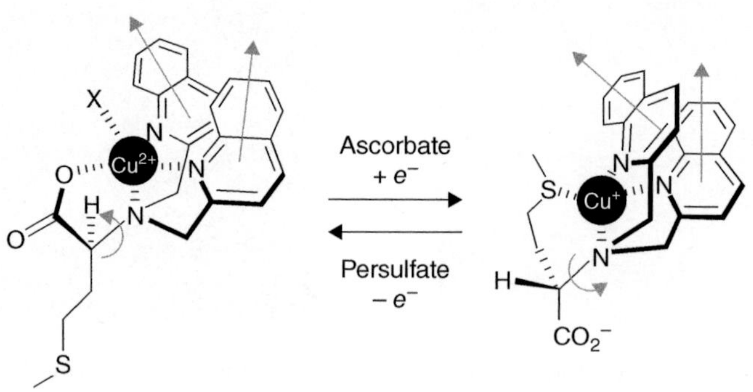

Figure 6.7 A chiroptical switch whose optical activity is inverted upon reducing the copper(II) complex to copper (I), and the process can be reversed upon oxidation. The curly arrows indicate the rotation of the methionine residue, and the solid arrows guide the eye to show the opposite helical chirality of the chromophores in the two states

between the central nitrogen atom of the ligand and the stereogenic centre (indicated with a curly arrow in Figure 6.7) leads to a change in the helical twist of the quinoline residues to *P*, resulting in a practically inverse CD signal. This kind of inversion of twist is reversible, by oxidation of the copper(I) complex with persulfate, and therefore the system is very promising for probing the potential of chiral switching for devices. The foundation of this kind of system was laid in the structural characterization of similar complexes [70, 71] and the generality has also been determined [72, 73].

An interesting aspect of the work discussed in the preceding paragraph is the effect of medium on the complexes, as related copper compounds have shown a dependency of their optical activity on the solvent [74]. Indeed, solvent effects can induce chiral switching in bulk materials, as demonstrated for chiral cyclometallated platinum(II) complexes [75]. The organometallic systems have (−)-4,5-pinene-6′-phenyl-2,2′-bipyridine or its enantiomer as one ligand (of the tridentate variety with the metal ion bonded to one of the phenyl carbon atoms) and phenyl acetylide as the other ligand attached to the platinum centre. Crystals of the compound show slight vapochromism (change of colour upon exposure to chemical vapours) as well as a change in luminescence when exposed to the vapour of dichloromethane, but corresponding changes in both the vibrational and electronic CD spectra are far more notable. Determination of the crystal structures of the compounds showed that a vapour-induced change of polymorph was responsible for the effect. The origin in the change of the ECD was considered to be a result of the different supramolecular interactions in the crystals between the varied π-systems. This kind of complex is also sensitive to mechanical treatment [76]. Grinding the solid converts one polymorph to another, as shown in the CD spectra, which are significantly different for the same reasons as the vapour treated solids. It has been proposed that this type of transition may be useful in sensing.

Simply protonating chiral chromophores can give rise to significant changes in properties (optical, for example) whether they are conformationally flexible or not. The rigid organic edifices derived from helicenes are particularly attractive for optic-electronic properties because of their quite unique twisted but conjugated 'π-way' [77–80]. Their metal complexes are also fascinating [81]. A carbene-osmium complex where the carbon is linked to a helicene showed potential for an acid-base chiroptical switch [82] and an organometallic helicene incorporating platinum also showed a significant effect of this kind [83]. The complex, shown in Figure 6.8, has electronic CD and circularly polarized luminescence that undergo reversible changes upon sequential protonation and treatment with base. In addition, the protonation gives an increase in the phosphorescence quantum yield for the complex and the intensity of the circularly polarized luminescence signal for the protonated form is approximately double that of the neutral complex. While the change is perhaps less dramatic than the ligand on its own, a response of this order is likely to be sufficient should the material prove to be a useful component in devices based on changes of this kind.

The helicene-type molecules are also interesting for single molecule studies, although while purely organic systems have been studied, the potential of coordination complexes has yet to be explored. To probe chiral systems in very small (a few to tens of nanometres) spaces light is also the system of choice, as demonstrated by the observation of the fluorescence and chiroptical signals of single molecules of a bridgedtriarylamine helicene [84–86]. However, the measurements are very complex and prone to difficulties [87].

Among all chiral switches, the sterically hindered alkenes are almost in a class of their own in terms of controlled motion [88]. While other photo-isomerizable compounds flip from *cis* to *trans,* for example, these molecules (because of their inherent chirality) display unidirectional relative movement of the stators between four positions [89]. The rotation rate can be modified using the size and stereoelectronic characteristics of the moieties attached to the alkene. One of the most impressive feats achieved to date is in the modification of a helical backbone of a polymer using the hindered alkene (Figure 6.9) as a 'sergeant' and the macromolecule as a 'soldier', and 'amplifying' that helicity into the phase of a

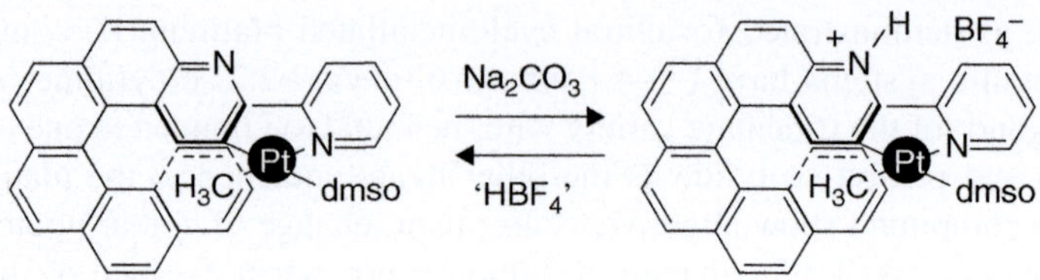

Figure 6.8 *A chiroptical switch based on the protonation of an organometallic helicene incorporating platinum(II)*

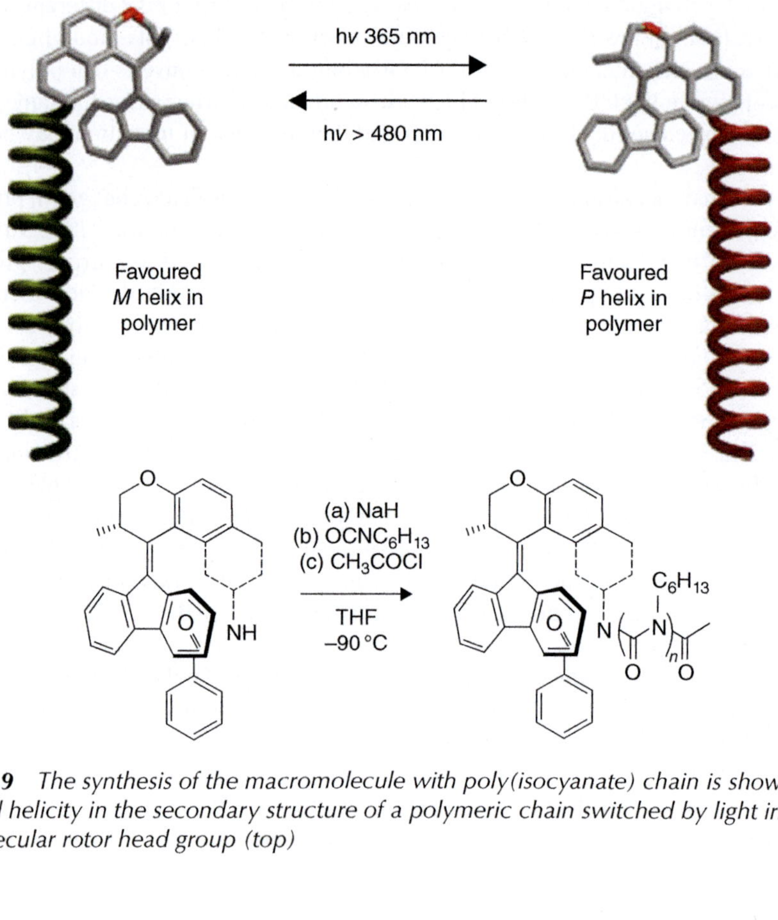

Figure 6.9 *The synthesis of the macromolecule with poly(isocyanate) chain is shown below. Preferred helicity in the secondary structure of a polymeric chain switched by light irradiation of a molecular rotor head group (top)*

liquid crystals [90]. The poly(hexylisocyanate) has conformationally dynamic backbone and changing between the diastereisomers of the hindered alkene causes a change in the handedness of the macromolcule. In turn, the helical polymer induces chirality in a lyotropic liquid crystal.

In a related work with coordination complexes, *cis-trans* isomerization of a diazobenzene appended to a tris(acetylacetonate)ruthenium(III) coordination core was used to modulate the helical twist in a nematic liquid crystal [91]. Here the chirality arises from the metal (lambda or delta) and the switching unit modulates the chiral induction. The induced chiral nematic phase – which comprises layers with their long axis twisted with respect to the normal to their packing axis – had its pitch changed by up to 50% in a reversible manner. The use of azobenzene *cis-trans* isomerization as a switch for the helical sense of liquid crystal phases has also been shown for an organic atropoisomeric compound whereby the colour of the mesophase changed as a result [92]. There, reversible colour changes were achieved, the origin of which is the change in the helical twisting power inducing different helicities, which coincide with certain wavelengths of visible light. There is great potential for coordination complexes to perform similar switching behaviour, especially perhaps the switches that were mentioned above. One case of epimerization in a chromium(III) complex did lead to a 'write and read-out' system, that needed taking to the isotropic state to reset [93].

The helical twist between complex coordination spheres can be 'levered' in a remote fashion by the incorporation of suitable molecular recognition sites [94]. A tetranuclear complex (made up of three zinc(II) Schiff base complexes, where the phenolic oxygen atoms surround a lanthanum(III) ion) incorporating stereogenic centres (at the carbon atoms in the ethylenediamine moiety bound to the zinc(II) in on the molecules' axis of rotation) was decorated with two benzo-18-crown-6 units that form complexes with primary ammonium ions. When shorter diammonium ions (with butyl, pentyl or hexyl spacing units) were bound by the complex, one helicity of the ligand around the lanthanum ion was induced, while longer spacers between the ammonium groups (such as dodecyl) gave a helix inversion at the central metal ion. An organic ligand with a well defined chirality can have its optical activity switched by complex formation with a transition metal ion that locks the orientation of the chromophores in an opposite sense to the free state [95]. This rotation of chromophores is potentially very interesting for the creation of systems that switch with mechanical change as well as chiral-sense switching, as are the hinge-type motion complexes [96, 97].

6.4 Conducting Materials

Electrical conductivity – the transport of electrons through materials – is essential to our everyday activity, for the wires that connect devices and the semiconductor components that are in them. Achiral materials dominate this area but there are fascinating things to be revealed in chiral conductors and perhaps even their applications [98]. Electrical conductivity in many mixed-valence low dimensional molecular materials is highly dependent on the packing in the crystals, [99] whether they are charge transfer complexes (made by combination of donor and acceptor with spontaneous formal charge transfer) or charge transfer salts (made by electrocrystallization). In these ionic materials, electrons flow through stacked molecules affording semiconducting or even metallic behaviour.

Contrastingly, conducting polymers present conduction along the doped (charged) macromolecular chains as well as between them, and have a wide range of potential uses compared with crystalline systems because of their ease of preparation, deposition or processing. In polymers, chirality has been studied extensively with a wealth of interesting phenomena observed in these mainly organic materials [100, 101]. The role of coordination chemistry in molecular conductors is more prominent [102].

The molecular conductors incorporating tetrathiafulvalene derivatives, as the mixed-valence material through which current passes, have often employed inorganic counter-ions in attempts to twist the stack of the organic aromatic component. This role could be purely to influence the packing of chiral organic donors, where only certain anions form crystals of the salts as in the case of the hexafluoroarsenate salts of a chiral donor [103]. There, a 2 : 1 ratio of donor to counter-ion gave crystals for enantiopure and racemic ligands; intriguingly, the conductivity of the enantiopure samples was an order of magnitude higher than that of the racemic material. The reason for the better transport in the case of the enantiopure material is the greater order in the crystal. In the racemic compound the pendant oxazoline ring is disordered so that *R* or *S* enantiomers can be in an equivalent crystallographic position. The enantiopure salts have only slight disorder. The crystals are metallic at room temperature but the racemate less so because the disorder induces localization of the charge carriers.

On the other hand, the counter-ion could incorporate chirality in a metal complex, as achieved for a chiral semiconductor where antimonyltartratewas was used as the counter-on [104]. The donor molecule – bis(ethylenedithio)tetrathiafulvalene, BEDT, Figure 6.10 – was electrocrystallized in the presence of the $K_2[Sb_2(L\text{-tartrate})_2]$ giving the charge transfer salt $BEDT_3Sb_2(L\text{-tart})_2.MeCN$, with BEDT in a mixed-valence state (average charge of two-thirds). The salt crystallizes in the $P2_12_12_1$ space group with alternating layers of donor molecules and chiral counter-ions. A two-fold screw axis relates the BEDT molecules, so no structural chiral nature exists (the molecules have pseudo-centre symmetry).

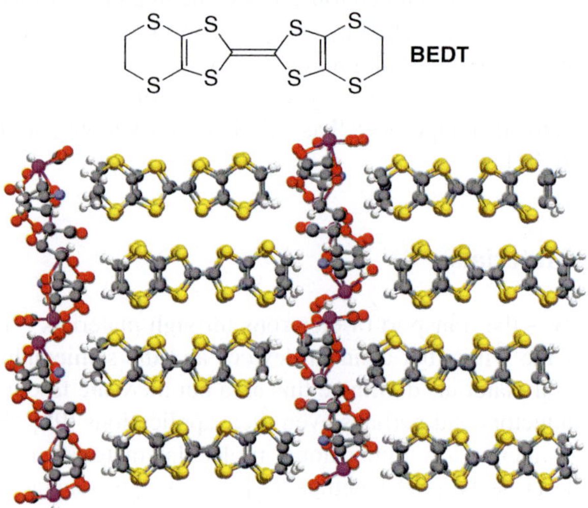

Figure 6.10 *The donor molecule bis(ethylenedithio)tetrathiafulvalene (BEDT) which forms a charge transfer salt ($BEDT_3Sb_2(L\text{-tart})_2.MeCN$) – whose structure is shown – when electrocrystallized in the presence of the $K_2[Sb_2(L\text{-tartrate})_2]$*

In hybrid materials, the lamellar structures formed by oxalate and metal ions (whose structure we shall see in the section on magnetic materials) are attractive because each layer is intrinsically chiral. In some cases, where superconducting or semiconducting behaviour is observed, this chirality alternates from one layer of the oxalate to the next [105]. This situation of ordered alternate left- and right-handed layers of metal oxalate complexes leads to a racemic compound. However, the dynamic nature of the chirality of the oxalates means that chirality can be induced during crystallization. An interesting recent case of the formation of a chiral tetrathiafulvalene derivative salt formation with a mixed oxalate is that of a mixed aluminium-chromium system [106]. A single enantiomer was achieved using the chiral (*R*)-(−)-carvone in the electrolyte along with nitromethane as the solvent, which also incorporates into the crystals as a solvate. No crystals were formed in the absence of the carvone. The oxalates are not configurationally stable, so the chiral additive induces one enantiomer from the racemic mixture of tris(oxalato)aluminate(III) and tris(oxalato)chromate(III) [107]. These salts can be semiconductors as crystals [108]. The chiral additive is not incorporated into the honeycomb formed by the oxalate because the cavities are too small (nitromethane occupies this space) although the carvone surely plays a role during the formation of these chiral materials.

Presently, perhaps the most promising family of molecular materials based on metal complexes for conducting chiral systems, is the dithiolene compounds. In these salts the organic cation can be employed to induce chirality, although the most frequent crystal forms have 2_1 axes, and therefore have parallel stacks of the π-active component. Chiral conducting salts of nickel dithiolene complexes have been electrocrystallized from the radical salt [n-Bu_4N][Ni(II) bis-(4, 5-dimercapto-1, 3-dithiole-2-thione] and enantiopure trimethylammonium salts [109]. Three ammonium ions successfully gave well defined compounds: those of (*R*)-Ph(Me)HC*-NMe_3^+ and (*S*)-(*t*Bu)(Me)HC*-NMe_3^+ gave crystalline salts (space group $P2_12_12_1$) with 1:3 stoichiometry (where the dithiolene unit is in excess and therefore in a mixed-valence state). When (*S*)-(1-naphthyl)MeHC*-NMe_3^+ was used as the counter-ion, a 2:5 stoichiometry salt was formed. The three salts are semiconductors with strong electron localization, which is a result of the dimerization or trimerization of crystallographically independent dithiolene complexes. More needs to be done in this family of promising materials, changing the cations and metals in the complexes. A particularly nice result is the formation of an achiral and neutral paramagnetic gold complex [110]. The control of chiral packing through this kind of strategy seems likely to yield materials with new properties that lead to a deeper understanding of these materials.

A remarkable recent discovery is electrical magnetochiral anisotropy in a crystalline electrical conductor based on enantiomers of a tetrathiafulvalene derivative [111]. The salt has a truly chiral hexagonal space group and is metallic until 40 K. The observation proved the existence of chiral charge transport in the system, the first time seen for a molecular system given the challenges involved in creating helical assembly and conductivity.

6.5 Magnetic Materials

There has been a surge of interest in chiral magnetic materials, largely because of the observation of magnetochiral dichroism [112] – a weak cross effect between electronic (natural optical activity) and magnetic (magnetic optical activity) circular dichroisms – in luminescence from a paramagnetic metal complex in solution. The molecules used were the enantiomers of tris(3-trifluoroacetyl-camphorato)europium(III), and hopes were raised that a material which was intrinsically a magnet would show a larger effect, [113] – the reasoning being that the material that is ordered should have a large internal field and associated anisotropy.

One of the compounds made with the aim of seeing novel synergies between magnetism and chirality was the polymeric complex formed between an enantiopure chiral nitronyl nitroxide radical ((*R*)-3MLNN, Figure 6.11) and manganese(II) hexafluoroacetylacetonate [114]. The coordination polymer crystallizes from solution to give a material whose crystal structure shows that the ligand induces a chiral Λ configuration specifically in the coordination sphere of the manganese ion. Remarkably, the conformational helicities between the imidazolyl ring and the phenyl ring and the twist in the imidazolyl ring itself are similar in

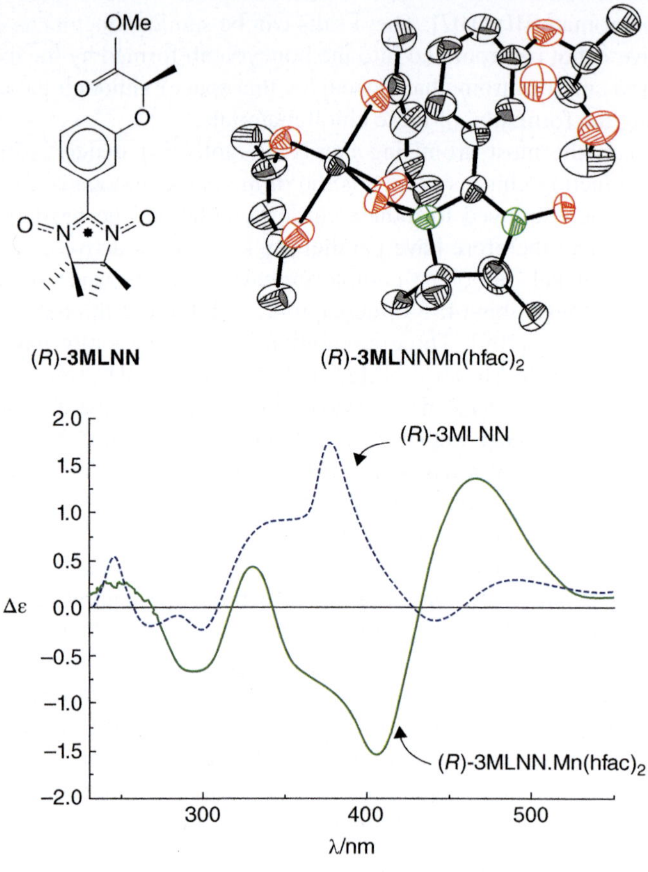

Figure 6.11 *A chiral nitronyl nitroxide (R)-3MLNN that forms a coordination polymer with manganese(II) hexafluoroacetylacetonate (Mn(hfac)$_2$ to give crystals (the unit cell is shown in the centre without hydrogen and fluorine atoms for clarity) of a complex that shows significant optical activity in its electronic natural circular dichroic spectrum. Note the Cotton effects shifted to the red compared with the component ligand, shown in the solid state CD spectra in KBr matrix on the right*

both the isolated free ligand in its crystals, as it is in the complex with the paramagnetic metal ion. While the structure is similar, the optical activity is very different for free ligand and complex, as shown in the CD spectra in Figure 6.11. The solid state spectra reveal a charge transfer taking place between the metal and ligand, and the bisignate nature is an indication of possible helical arrangement [115]. Observation of magnetochiral dichroism would be favourable for the Cotton effect observed at approximately 460 nm in the complex, as this signal is at a wavelength of low absorption [116]. Apart from these features, the compound reveals unique and intriguing magnetic behaviour at low temperatures on account of the antiferromagnetic coupling between the nitroxide and the metal ion, leading to a ferromagnetic chain that reveals collective phenomena [117].

Other nitroxide-containing ligands have been employed in combination with metal ions to prepare molecular magnets, notably bis(nitroxides), which have a triplet ground state [118]. An optically active metamagnet was prepared using this kind of chiral bis(nitroxide), which forms a coordination polymer with manganese(II) hexafluoroacetylacetonate [119–121]. The crystals of the complex show organic ligands and metal ions forming a 2_1 chain with alternating organic and inorganic components. In contrast with previous cases, the chiral complex could even be observed in solution by CD spectroscopy. On the other hand, the chirality of ligands around single molecule magnets has also been demonstrated, both for dodecamanganese clusters [122] and terbium(III) bisphthalocyanines [123], although the consequences for magnetochiral dichroism have not been shown. In the former case, natural optical activity and magnetic circular dichroism were proven, and at low temperatures the intramolecular magnetic ordering was demonstrated, where the magnetic ordering gave a clear increase in the intensity of the magneto-optical activity, and tunnelling events could be detected via the changing intensity of this signal [124]. In the latter, which is a liquid crystalline material, a dependence of the details of the magnetic ordering on the temperature cycling that is used to cool the material was observed; the role that chirality plays there is important for the phase behaviour, but the relationship has not been proven directly for the magnetic behaviour. The strong and highly characteristic magnetic circular dichroism shown by this kind of sandwich complex [125] makes them an interesting target for the study of magneto-chiral dichroism.

Strong magnetochiral dichroism has been observed in a molecular magnet based on a layered chromium-manganese oxalate system [126]. This kind of bimetallic layered compound [127, 128] is particularly interesting because the chirality of the layers can be made enantioselective [129, 130]. Formally, in the two-dimensional sheets of metal oxalate the chirality of neighbouring metal centres is opposite (being identical would result in a three-dimensional structure) but if a bimetallic system is made, each of the two different metal ions could individually be homochiral (Figure 6.12).

A chiral ferromagnet was prepared by chiral induction into the oxalate framework by an optically resolved chiral quaternary ammonium cation [131]. This situation gives rise to one diastereoisomer of the salt (with formula for one of the enantiomers $[N(CH_3)(n\text{-}C_3H_7)_2((S)\text{-}s\text{-}C_4H_9)][(\Lambda)\text{-}Mn(\Delta)\text{-}Cr(ox)_3]$) with a natural CD signal with maximum at 550 nm corresponding to the d–d transitions of the $[Cr(ox)_3]$ unit. Magnetic measurements showed that in the paramagnetic temperature range, the exchange interaction between the chromium(III) and manganese(II) ions is ferromagnetic: The Curie–Weiss temperature in this system is 9.3 K. Both enantiomers show a ferromagnetic phase transition at $T_c = 7$ K. In the ferromagnetic regime, a strong magnetochiral dichroic effect was observed in

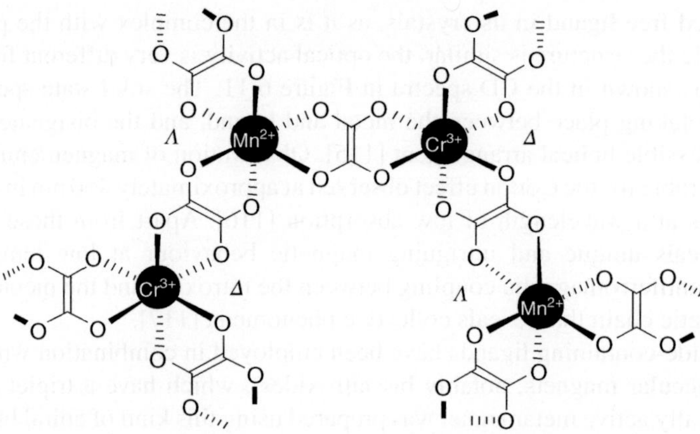

Figure 6.12 *The chemical structure of the globally anionic framework with formula $[Mn^{II}Cr^{III}(ox)_3]^-$ where each metal centre alternates in helical chirality, but where each manganese(II) ion is Λ and each chromium(III) ion is Δ configuration*

absorption with a maximum at approximately 600 nm, which is 'multiplied by a factor of 17 between 11 and 3 K' going from a paramagnetic state (where a weak signal is seen) to the bulk ferromagnetic situation.

Remarkably, this same pair of complexes shows a large increase in the second harmonic generation effect at the paramagnetic-to-ferromagnetic transition [132]. The single crystal nature of the material proved beneficial for a precise analysis of all the parameters leading to this effect. The use of a model that employed hyperpolarizabilities led to the conclusion that the magnetic contributions to the nonlinear susceptibility account for as much as a sixth of the structural factors, and can give rise to big light rotation angles. This result may prove interesting not only for crystalline compounds but perhaps for thin films on magnetic substrates.

Returning to the magneto-chiral dichroic effect (and remembering that bulk magnetic ordering or even paramagnetism are not essential to observe the effect, though they may help) this phenomenon has also been claimed for an artificial light-harvesting antenna system based on *J*-aggregates of zinc chlorins [133]. The effect was seen in the π-π* transition of the aggregates, indicating that the effect should be observable in other self-assembled complexes.

Apart from bulk magnetic ordering, perhaps one of the most important phenomena in paramagnetic materials is that of spin crossover, whereby an ion (most habitually iron) can exist in more than one spin state as a function of the temperature [134]. The main focus of this research is the preparation of bistable systems that can be switched between states at (or near) room temperature, and the main role of chirality is in aiding the structural ordering that leads to the specific spin state in the iron. Several crystalline complexes have been prepared that display spin crossover [135, 136] and even chiral recognition [137, 138]. Here only two cases are presented to illustrate the general interest in the area.

The first-claimed example of a homochiral mixed-valence spin-crossover compound was that of an iron complex with tris{[2-{(imida-zole-4-yl)methylidene}amino]ethyl} amine, which forms a helical twist around the metal ion in both protonated and deprotonated forms [139]. The iron(II) and iron(III) centres are bound to the ligand in either its imidazole or imidazolate forms, which have opposite configuration at the metal, so that overall each single species is homochiral. The hydrogen bonds between the two ligand forms help in the transfer of chirality in the system. While only narrow hysteresis in the change between low and high spin states was observed, it was shown that the complex is optically active, and raised the prospect of temperature-, light-, or pressure-induced changes in the spin state for data recording, because the spin state of the compound – shown in its optical activity – could be 'read' optically far away from its main absorbance band. This situation avoids destructive readout and proves the benefit of chirality for this kind of application.

Enantiopure tetranuclear iron(II) tetrahedral cages were prepared successfully using either enantiomer of the commercially available phenylethylamine in an imine bond forming reaction [140]. The polycrystalline samples are bistable at room temperature with the high spin state existing uniquely at about 350 K and the low spin state below 250 K. Solvent was present in the samples as prepared, and this leads to sharper transitions than the desolvated solid. It was proposed that these compounds could be of interest for electronic switches and devices, although their application is awaited. They do indeed constitute a fascinating nanometre scale complex that might be addressed using state-of-the-art microscopic techniques.

Spin crossover has also been observed in nanoparticles and gels [141]. The particles were formed by a complex of iron with 4-amino-1,2,4-triazole with L-camphorsulfonate as the counter-ion and origin of the induced optical activity. A very sharp change at room temperature in the optical activity was observed using CD spectroscopy, associated with the change in spin state of the material. The sharpness is partially due to the polymeric structure of the complex, where the triazole moieties link metal ions. The same material also formed a gel, which also displayed the spic crossover phenomenon. These materials are promising in the sense that they can be easily processed when compared with crystalline systems.

6.6 Sensors

Very related to switching systems are examples which show a specific response to chemical 'inputs', where that change in signal can be employed to detect – or sense – the analyte. The sensing of the chirality of compounds and its quantification is an important activity because, although chiral chromatography is highly developed and efficient, it is not completely general and there are instances where fluorescent or colorimetric sensors would be a boon [142, 143]. The breadth of activity in this area of research for all kinds of sensors is treated in depth in the cited reviews and some highlights are presented here.

Intrinsically chiral complexes can show diastereoselective binding in which – although the absolute configuration of the ligand is not inverted – the 'chirality at metal' is inverted. The (*S,S,S*) configuration complex shown in Figure 6.13 contains bound europium(III) or terbium(III) ions in the Δ configuration [144]. When these metal complexes were bound to the 'drug II site' of serum albumin the circularly polarized luminescence signal arising from the metal ion (at approximately 550 nm) showed that the configuration of the metal centre had changed from Δ to Λ. Obviously, the absolute configuration of the stereogenic

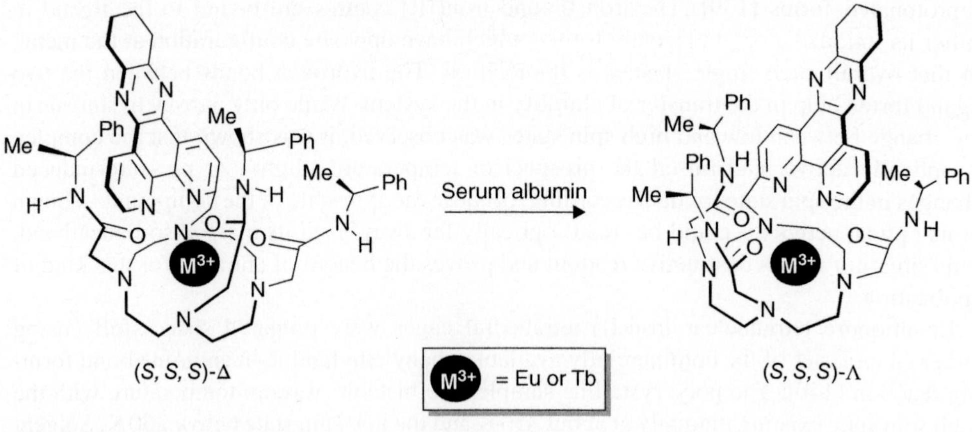

Figure 6.13 *The chemical structures of a chiral lanthanide complex that changes diastereoisomeric conformation – through a rearrangement of the metal coordination sphere that is initially Δ – upon binding to serum albumin*

centres in the ligand is unchanged. Therefore, the binding event to the protein causes a rearrangement in the twist of the ligand leading to an opposite metal ion configuration. However, this situation does not imply that the chirality of the ligand is irrelevant: far from it. The complex with (*S,S,S*) configuration in the ligand shows no change upon binding, and therefore only the (*S,S,S*) variety can be used as a specific sensing molecule. Effectively, the complex is a stereodynamic probe [145].

A europium(III) complex involving coordination of the metal ion by the bidentate ligand 3,3-bis(diphenylphosphoryl)-2,2-bipyridine also displays a reversal of circularly polarized luminescence signal from the metal ion, induced by changing solvent [146]. The origin of the effect is a *cis-trans* type of isomerization by a twisting of the organic fragment, with the origin of the chirality being either tris[3-(trifluoromethylhydroxymethylene)-(+)-camphorate] or tris[3-(heptafluoropropylhydroxymethylene)-(+)-camphorate]. The metal is eight-coordinate in the *cis* complex, including chelate formation involving the phosphoryl oxygen atom coordinating the europium(III) centre. In the *trans* geometry, on the other hand, the coordination of the metal ion is quasi-seven-coordinate because the phosphorus-nitrogen interaction acts like a conformational lock for the ligand. Anion sensing has also been shown for macrocyclic lanthanide complexes, which are solvent dependent in their optical activity [147].

The organization of metallo-organic structures and their coordination environment are strictly related and affect their ability to respond to external agents. A beautiful example is that of metal complexes of the chiral hydroxyquinoline derivative shown in Scheme 6.1 [148]. The tetrahedral complex formed by the oxyquinolate with zinc(II) is a metallogelator that shows a distinct response to the enantiomers of *trans*-1,2-diaminocyclohexane in THF: with the (*R,R*) enantiomer a pale green fluorescence is observed by the naked eye, while the (*S,S*) appears yellow. Interestingly, neither the lithium nor aluminium complexes of the same ligand display any differentiation between the amines to the eye. The lithium complex does show a selective response spectroscopically but the colour is not different for the two enantiomers. The optical activity of the metal complexes of the oxyquinolate is

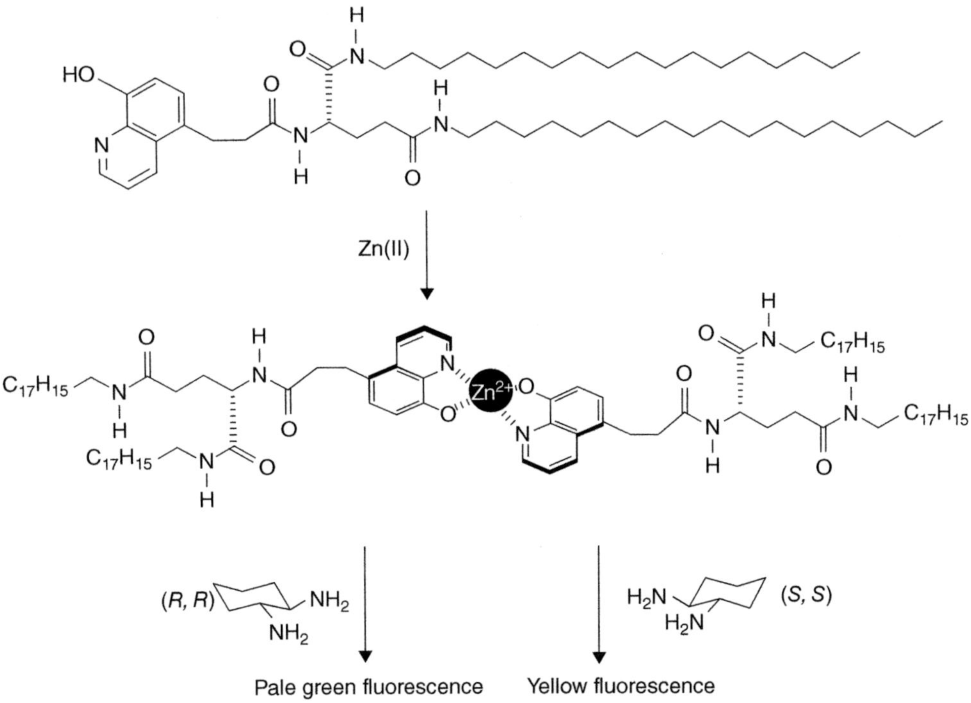

Scheme 6.1 *The synthesis of a zinc(II) complex that gelates organic solvents and displays diastereoselective binding of the enantiomers of* trans-*1,2-diaminocyclohexane*

appreciable for both the aluminium(III) and zinc(II) complexes, indicating a chiral environment of the metal ion. The authors concluded that it was the secondary structure of the complexes that was responsible for the observed selectivity, noting that the aluminium(III) complex is much more sterically encumbered than the zinc(II) complex because of the octahedral coordination of the former and presence of three ligands (in fact, it shows a weak response to the amine when measured spectroscopically).

An achiral metallogel-based sensor, based on luminescence, changes in the presence of certain anions [149] and reaffirms the potential for soft materials to form part of chiroptical sensors with different metal ions and ligands. The use of chiral derivatives employing phenylalanine as a building block is particularly interesting for the creation of chiral sensors based on gelators [150].

A supramolecular polymer comprising a bis-zinc(II) complex with a naphthalenediimide core has been shown to sense the changes in adenosine phosphate (AP) as part of a working enzyme system [151]. The aggregate of the metal complex in aqueous solution shows a distinct chiroptical response for the adenosine triphosphate when compared to the mono- or di-phosphate homologues, and this feature can be employed for sensing in a working environment. The kinetic characteristics of enzymatic hydrolysis of APs were probed *in situ* using circular dichroism of the supramolecular polymer. The phosphate groups from the ribose moiety in the ATP–supramolecular polymer complex were cleaved by calf intestinal alkaline phosphatase, leading to a decrease in the CD intensity before the signal eventually reversed. This observation proved the formation of either ADP or AMP (the complex of these two with the aggregate cannot be discriminated at this point). The aggregate becomes CD-inactive after further hydrolysis, proving the role of the phosphate groups in the chiral induction. It has been pointed out that the use of fluorescence to sense the chirality of the aggregate may improve the sensitivity of the system [152].

6.7 Conclusions and Outlook

Chirality plays an important role in the properties of certain molecular materials containing metal ions, and there is much more that can be explored taking advantage of the different routes available to metallo-organic compounds that are shown in Figure 6.1. This chapter has attempted to highlight the areas where interesting observations have been made, as well as the opportunities that are available for further research where significant advancement in knowledge may be made. The preparation of ordered chiral stacks of molecules that have a genuine helical conformation is a true challenge. Here, the shape of the molecules – already seen to be important in recognition processes [153] – is sure to be important.

Life is very complicated, chemically; while most of the examples presented in this overview are ingenious and often quite hard to achieve they are relatively simple. In coming years, we are sure to witness efforts to incorporate multiple components for the preparation of responsive and ‘intelligent’ materials to emulate biology. Achieving the sequential motion of organic ligands around metals in synthetic systems is a big challenge today [154]. It is an interesting goal from a number of viewpoints; purely the achievement of controlled molecular motion, the creation of multiresponsive systems for data storage or sensing, or for catalysis. The sequential switching of helicity by the introduction of metal ions at each step – first zinc(II), then barium(II) and then lanthanum(III) – is a very interesting development in this sense [155]. This system proved three sequential inversions of helicity. The demonstration of inversion of enantioselectivity of a palladium complex by switching a chiral bisphosphine ligand – which in its own right is a molecular motor – shows that the same chemical components can perform different chiral functions depending on their situation [156]. The redox switching of organocatalysis by urea units in a copper complex has also been demonstrated for the enantioselective addition of diethyl malonate to *trans*-beta-nitrostyrene [157]. The results have interesting potential consequences for the preparation of enantioselective catalysts, where a single catalytic system could realize both enantiomers of a target product independently from different redox states of the same catalyst, in a controlled way. While this method involved the use of the catalyst in a homogeneous medium, building in a recovery system would make this approach highly sustainable because the precious metal could be reused in different reactions aimed at different products. In these complicated systems, the control of ligand lability and the dynamics of the coordination sphere will be necessary and are sure to build on the great body of work on dynamic helical complexes [158].

Supramolecular polymers prepared from a multitopic ligand that formed complexes with europium and/or terbium provide a healable luminescent metallogel [159]. It is interesting to think how these materials might perform as chiroptical materials should a chiral ligand be employed. The formation of dinuclear metal complexes from peptide chains [160] is very relevant in this sense and the idea of ‘chiral information harvesting’ [161] wherein local chirality can affect global structure is a compelling paradigm for soft materials in general [162]. The use of peptides and other natural resources will become increasingly important from another perspective: The incorporation of stereogenic centres into materials will have to be sustainable if they are to prove useful for humanity, and therefore either renewable and recyclable components will have to be used, and systems with cooperative responses will have to be developed [163].

It is important to realize that chirality is a vital aspect of many multifunctional materials, wherein the optical activity does not merely coexist but has synergies with other properties leading to new phenomena, as is the case of magnetochiral dichroism for instance. The future of chiral materials is promising, especially perhaps in the area of electron transport – be it for spintronics or electron transfer for sustainable energy production to highlight two particularly exciting and important areas of research. For example, the assembly of chiral conducting polymer films on ferromagnetic materials leads to a spin-filtering effect and asymmetric magnetoresistance [164]. The employment of metallo-organic structures for these fascinating fields of new research is bound to bring benefits because of the electronic states and anisotropy that metals afford to the relatively limited organic systems.

The development of chiral molecular materials is still in its infancy, and the opportunities for discovering new systems with novel and improved properties, or more sustainable solutions to existing materials, provides us with the motivation to explore this research field vigorously. The advantages of metallo-organic compounds and composites for the preparation of chiral materials – such as their potentially high stability, mechanical properties, high optical activity, anisotropy, paramagnetism, and so on – are good reasons to believe that promise abounds.

Acknowledgements

The author is deeply grateful to the School of Chemistry at The University of Nottingham and the EPSRC for funding his present research into chiral functional molecular materials.

References

[1] G.H. Wagnière, *On Chirality and the Universal Asymmetry*, Verlag Helvetica Chimica Acta, Wiley-VCH, 2007.
[2] D.K. Kondepudi, in *Progress in Biological Chirality*. G. Palyi, C. Zucchi, and L. Caglioti (Eds), pp. 159–171, Elsevier, 2004.
[3] A. González-Campo and D.B. Amabilino, Biomolecules at interfaces – chiral, naturally, *Topics in Current Chemistry*, **333**, 109–156 (2013).
[4] A. Brizard, R. Oda and I. Huc, Chirality effects in self-assembled fibrillar networks, *Topics in Current Chemistry*, **256**, 167–218 (2005).
[5] D.B. Amabilino and J. Veciana, Supramolecular chiral functional materials, *Topics in Current Chemistry*, **265**, 253–302 (2006).
[6] C. Rest, R. Kandanelli and G. Fernandez, Strategies to create hierarchical self-assembled structures via cooperative non-covalent interactions, *Chemical Society Reviews*, **44**, 2543–2572 (2015).
[7] D.B. Amabilino in *Organic Nanomaterials: Synthesis, Characterization, and Device Applications*, Ed. T. Torres, G. Bottari, pp. 59–78, John Wiley & Sons, Inc., 2013.

[8] A.R.A. Palmans and E.W. Meijer, Amplification of chirality in dynamic supramolecular aggregates. *Angewandte Chemie International Edition*, **46**, 8948–8968 (2007).
[9] J. Crassous, Chiral transfer in coordination complexes: towards molecular materials, *Chemical Society Reviews*, **38**, 830–845 (2009).
[10] U. Knof and A. von Zelewsky, Predetermined chirality at metal centers, *Angewandte Chemie International Edition*, **38**, 302–322 (1999).
[11] U. Mamula and A. von Zelewsky, The bright future of stereoselective synthesis of co-ordination compounds, *Journal of the Chemical Society – Dalton Transactions,* 219–231 (2000).
[12] L. Pérez-García and D.B. Amabilino, Spontaneous resolution under supramolecular control, *Chemical Society Reviews*, **31**, 342–356 (2002).
[13] P.D. Knight and P. Scott, Predetermination of chirality at octahedral centres with tetradentate ligands: prospects for enantioselective catalysis, *Coordination Chemistry Reviews*, **242**, 125–143 (2003).
[14] J. Crassous, Transfer of chirality from ligands to metal centres: recent examples, *Chemical Communications*, **48**, 9684–9692 (2012).
[15] J.L. Serrano, *Metallomesogens: Synthesis, Properties, and Applications*, VCH, 1996
[16] A. Tam, Y. Yiu-Yan and W.W. Vivian, Recent advances in metallogels, *Chemical Society Reviews*, **42**, 1540–1567 (2013).
[17] J. Zhang and C.-Y. Su, Metal-organic gels: From discrete metallogelators to coordination polymers, *Coordination Chemistry Reviews*, **257**, 1373–1408 (2013).
[18] D.B. Amabilino (Ed.) *Chirality at the Nanoscale: Nanoparticles, Surfaces, Materials and More*, Wiley-VCH, 2009.
[19] C. Noguez and I.L. Garzon, Optically active metal nanoparticles, *Chemical Society Reviews*, **38**, 757–771 (2009).
[20] A. Ben-Moshe, B. Maoz, A.O. Govorov and G. Markovich, Chirality and chiroptical effects in inorganic nanocrystal systems with plasmon and exciton resonances, *Chemical Society Reviews*, **42**, 7028–7041 (2013).
[21] V.K. Valev, J.J. Baumberg, C. Sibilia and T. Verbiest, Chirality and chiroptical effects in plasmonic nanostructures: fundamentals, recent progress, and outlook, *Advanced Materials*, **25**, 2517–2534 (2013).
[22] Y. Wang, J. Xu, Y. Wang and H. Chen, Emerging chirality in nanoscience, *Chemical Society Reviews*, **42**, 2930–2962 (2013).
[23] H.K. Bisoyi and Q. Li, Light-directing chiral liquid crystal nanostructures: from 1D to 3D, *Accounts of Chemical Research*, **47**, 3184–3195 (2014).
[24] G. Pescitelli, L. Di Bari and N. Berova, Application of electronic circular dichroism in the study of supramolecular systems, *Chemical Society Reviews*, **43**, 5211–5233 (2014).
[25] G. Gottarelli, S. Lena, S. Masiero, S. Pieraccini and G.P. Spada, The use of circular dichroism spectroscopy for studying the chiral molecular self-assembly: an overview, *Chirality*, **20**, 471–485 (2008)
[26] R. Kuroda and T. Honma, CD spectra of solid-state samples, *Chirality*, **12**, 269–277 (2000).
[27] M. Minguet, D.B. Amabilino, K. Wurst and J. Veciana, Circular dichroism studies of crystalline chiral and achiral α-nitronyl nitroxide radicals in KBr matrix, *Journal of the Chemical Society, Perkin Transactions* **2**, 670–676 (2001).
[28] E. Castiglioni, P. Biscarini and S. Abbate, Experimental aspects of solid state circular dichroism, *Chirality*, **21**, E28–E36 (2009).
[29] M. Ziegler and A. von Zelewsky, Charge-transfer excited state properties of chiral transition metal coordination compounds studied by chiroptical spectroscopy, *Coordination Chemistry Reviews*, **177**, 257–300 (1998).
[30] T. Wu, X.Z. You and P. Bour, Applications of chiroptical spectroscopy to coordination compounds, *Coordination Chemistry Reviews*, **284**, 1–18 (2015).
[31] L. Pasteur, Sur les relations qui peuvent exister entre la forme crystalline, la composition chimique et le sens de la polarisation rotatoire, *Annales de chimie et de physique*, **24**, 442–459 (1848).
[32] L. Pasteur, Transformation des acides tartriques en acide racémique. Découverte de l'acide tartrique inactif. Nouvelle méthode de séparation de l'acide racémique en acides tartriques droit et gauche, *Comptes Rendus de l'Académie des Sciences*, **37**, 162–166 (1853).

[33] W. Huang and T. Ogawa, Spontaneous resolution of Δ and Λ enantiomeric pair of [Ru(phen)$(bpy)_2](PF_6)_2$(phen), *Polyhedron*, **25**, 1379–1385 (2006).
[34] J. Breu, H. Domel and A. Stoll, Racemic compound formation *versus* conglomerate formation with $[M(bpy)_3](PF_6)_2$ (M = Ni, Zn, Ru); molecular and crystal structures, *European Journal of Inorganic Chemistry*, 2401–2408 (2000).
[35] D.B. Amabilino and R.M. Kellogg, Spontaneous deracemization. *Israel Journal of Chemistry*, **51**, 1034–1040 (2011).
[36] S. Olsson, P.M. Bjçremark, T. Kokoli *et al.*, Absolute asymmetric synthesis: protected substrate oxidation, *Chemistry – A European Journal*, **21**, 5211–5219 (2015).
[37] D.T. McLaughlin, T. Phuong, T. Nguyen, *et al.*, Viedma ripening of conglomerate crystals of achiral molecules monitored using solid-state circular dichroism, *Crystal Growth and Design*, **14**, 1067–1076 (2014).
[38] C. Viedma, Chiral symmetry breaking during crystallization: complete chiral purity induced by nonlinear autocatalysis and recycling, *Physical Review Letters*, **94**, 065504 (2005).
[39] A. Lennartson, Optical resolution and racemisation of $[Fe(acac)_3]$, *Inorganica Chimica Acta*, **365**, 451–453 (2011).
[40] A. Biswas, C. Estarellas, A. Frontera, *et al.*, Effect of a methyl group on the spontaneous resolution of a square-pyramidal coordination compound: crystal packing and conglomerate formation, *CrystEngComm*, **14**, 5854–5861 (2012).
[41] A.G. Shtukenberg, Y.O. Punin, A. Gujral and B. Kahr, Growth actuated bending and twisting of single crystals, *Angewandte Chemie International Edition*, **53**, 672–699 (2014)
[42] L.J. Spencer, Curvature in Crystals, *Mineralogical Magazine*, **19**, 263–274 (1921).
[43] M. Yanga and N.A. Kotov, Nanoscale helices from inorganic materials, *Journal of Materials Chemistry*, **21**, 6775–6792 (2011).
[44] D.M. Walba, S.C. Slater, W.N. Thurmes, *et al.*, Design and synthesis of a new ferroelectric liquid crystal family – liquid crystals containing a non-racemic 2-alkoxy-1-propoxy unit, *Journal of the American Chemical Society*, **108**, 5210–5221 (1986).
[45] D.M. Walba, H.A. Razavi, A. Horiuchi, *et al.*, Evolution of the Boulder model for the molecular origins of the polarization in ferroelectric liquid crystals, *Ferroelectrics*, **113**, 21–36 (1991).
[46] J.L. Serrano and T. Sierra, Helical supramolecular organizations from metal-organic liquid crystals, *Coordination Chemistry Reviews*, **242**, 73–85 (2003).
[47] M. Lehmann, T. Sierra, J. Barberá, *et al.*, Chiral hexacatenar metallomesogens: supramolecular organization versus steric demand of chiral cores. *Journal of Materials Chemistry*, **12**, 1342–1350 (2002).
[48] J. Barberá, E. Cavero, M. Lehmann *et al.*, Supramolecular helical stacking of metallomesogens derived from enantiopure and racemic polycatenar oxazolines, *Journal of the American Chemical Society*, **125**, 4527–4533 (2003).
[49] J.L. Serrano and T. Sierra, Switchable columnar metallomesogens, *Chemistry – A European Journal*, **6**, 759–766 (2000).
[50] J. Barberá, R. Iglesias, J.L. Serrano, *et al.*, Switchable columnar metallomesogens. New helical self-assembling systems, *Journal of the American Chemical Society*, **120**, 2908–2918 (1998).
[51] J. Yoshida, G. Watanabe, K. Kakizawa, *et al.*, Tris(β-diketonato) Ru(III) Complexes as chiral dopants for nematic liquid crystals: the effect of the molecular structure on the helical twisting power, *Inorganic Chemistry*, **52**, 11042–11050 (2013).
[52] S. Ahn, S. Yamakawa and K. Akagi, Liquid crystallinity-embodied imidazolium-based ionic liquids and their chiral mesophases induced by axially chiral tetra-substituted binaphthyl derivatives, *Journal of Materials Chemistry C*, **3**, 3960–3970 (2015).
[53] N. Watfa, S. Floquet, E. Terazzi, *et al.*, Synthesis, characterization, and tuning of the liquid crystal properties of ionic materials based on the cyclic polyoxothiometalate $[\{Mo_4O_4S_4(H_2O)_3OH_2\}_2(P_8W_{48}O_{184})]^{36-}$, *Soft Matter*, **11**, 1087–1099 (2015).
[54] B. Escuder and J.F. Miravet (Eds), *Functional Molecular Gels*, RSC, 2014.
[55] D.B. Amabilino and J. Puigmartí-Luis, Gels as a soft matter route to conducting nanostructured organic and composite materials, *Soft Matter*, **6**, 1605–1612 (2010).
[56] L. Zhang, X.F. Wang, T.Y. Wang and M.H. Liu, tuning soft nanostructures in self-assembled supramolecular gels: from morphology control to morphology-dependent functions, *Small*, **11**, 1025–1038 (2015).

[57] M.-O.M. Piepenbrock, G.O. Lloyd, N. Clarke and J.W. Steed, Metal- and anion-binding supramolecular gels, *Chemical Reviews*, **110**, 1960–2004 (2010).

[58] I. Odriozola, I. Loinaz, J.A. Pomposo and H.J. Grande, Gold–glutathione supramolecular hydrogels, *Journal of Materials Chemistry*, **17**, 4843–4845 (2007).

[59] I. Imaz, M. Rubio-Martínez, W.J. Saletra, *et al.*, Amino acid based metal-organic nanofibers, *Journal of the American Chemical Society*, **131**, 18222–18223 (2009).

[60] R. Randazz, A. Di Mauro, A. D'Urso, *et al.*, Hierarchical effect behind the supramolecular chirality of silver(I)–cysteine coordination polymers, *Journal of Physical Chemistry B*, **119**, 4898–4904 (2015).

[61] Q. Jin, L. Zhang, H. Cao, *et al.*, Self-assembly of copper(II) ion-mediated nanotube and its supramolecular chiral catalytic behavior, *Langmuir*, **27**, 13847–13853 (2011).

[62] T. Tu, W. Fang, X. Bao, *et al.*, Visual chiral recognition through enantioselective metallogel collapsing: synthesis, characterization, and application of platinum–steroid low-molecular-mass gelators, *Angewandte Chemie International Edition*, **50**, 6601–6605 (2011).

[63] J.H. Jung, J.H. Lee, J.R. Silverman and G. John, Coordination polymer gels with important environmental and biological applications, *Chemical Society Reviews*, **42**, 924–936 (2013).

[64] S. Debnath, J.-F. Bergamini, F. Artzner, *et al.*, Near-infrared chiro-optical effects in metallogels, *Chemical Communications*, **48**, 2283–2285 (2012).

[65] L.N. Feldborg, W.J. Saletra, P. Iavicoli and D.B. Amabilino, Central metal ion determined self-assembly of intrinsically chiral porphyrins, *Journal of Porphyrins and Phthalocyanines*, **15**, 995–1003 (2011).

[66] J. Szmytkowski, J. Conradt, H. Kuhn, *et al.*, Self-assemblies of novel magnesium porphyrins mimicking natural chlorosomal bacteriochlorophylls, *Journal of Physical Chemistry C*, **115**, 8832–8839 (2011).

[67] A. Nuermaimaiti, C. Bombis, M.M. Knudsen, *et al.*, Chiral induction with chiral conformational switches in the limit of low 'sergeants to soldiers' ratio, *ACS Nano*, **8**, 8074–8081 (2014).

[68] J.W. Canary, S. Mortezaei and J. Liang, Transition metal-based chiroptical switches for nanoscale electronics and sensors, *Coordination Chemistry Reviews*, **254**, 2249–2266 (2010).

[69] S. Zahn and J.W. Canary, Electron-induced inversion of helical chirality in copper complexes of *N,N*-dialkylmethionines, *Science*, **288**, 1404–1407 (2000).

[70] S. Zahn, D. Das and J.W. Canary, Redox-induced ligand reorganization and helicity inversion in copper complexes of N,N-dialkylmethionine derivatives, *Inorganic Chemistry*, **45**, 6056–6063 (2006).

[71] D. Das, Z. Dai, A.E. Holmes and J.W. Canary, Exploring the scope of redox-triggered chiroptical switches: Syntheses, X-ray structures and circular dichroism of cobalt and nickel complexes of N,N-bis(arylmethyl)methionine derivatives, *Chirality*, **20**, 585–591 (2008).

[72] H.S. Barcena, B. Liu, M.V. Mirkin and J.W. Canary, An electrochiroptical molecular switch: mechanistic and kinetic studies, *Inorganic Chemistry*, **44**, 7652–7660 (2005).

[73] H.S. Barcena, A.E. Holmes and J.W. Canary, Inversion of helicity in propeller shaped molecules derived from S-methylcysteine and methioninol, *Organic Letters,* 709–711 (2003).

[74] A.C. Chamayou, G. Makhloufi, L.A. Nafie, *et al.*, Solvation-induced helicity inversion of pseudotetrahedral chiral copper(II) complexes, *Inorganic Chemistry*, **54**, 2193–2203 (2015).

[75] X.-P. Zhang, T. Wu, J. Liu, *et al.*, Vapor-induced chiroptical switching in chiral cyclometalated platinum(II) complexes with pinene functionalized C^N^N ligands, *Journal of Materials Chemistry C*, **2**, 184–194 (2014).

[76] X.-P. Zhang, J.-F. Mei, J.-C. Lai, *et al.*. You, Mechano-induced luminescent and chiroptical switching in chiral cyclometalated platinum(II) complexes, *Journal of Materials Chemistry C*, **3**, 2350–2357 (2015).

[77] Y. Shen and C.F. Chen, Helicenes: synthesis and applications, *Chemical Reviews*, **112**, 1463–1535 (2012).

[78] M. Gingras, One hundred years of helicene chemistry. Part 1: non-stereoselective syntheses of carbohelicenes, *Chemical Society Reviews*, **42**, 968–1006 (2013).

[79] M. Gingras, F. Guy and R. Peresutti, One hundred years of helicene chemistry. Part 2: stereoselective syntheses and chiral separations of carbohelicenes, *Chemical Society Reviews*, **42**, 1007–1050 (2013).

[80] M. Gingras, One hundred years of helicene chemistry. Part 3: applications and properties of carbohelicenes, *Chemical Society Reviews*, **42**, 1051–1095 (2013).
[81] N. Saleh, C. Shen and J. Crassous, Helicene-based transition metal complexes: synthesis, properties and applications, *Chemical Science*, **5**, 3680–3694 (2014).
[82] E. Anger, M. Srebro, N. Vanthuyne, *et al.*, Helicene-grafted vinyl- and carbene-osmium complexes: an example of acid–base chiroptical switching, *Chemical Communications*, **50**, 2854–2856 (2014).
[83] N. Saleh, B. Moore II, M. Srebro, *et al.*, Acid/base-triggered switching of circularly polarized luminescence and electronic circular dichroism in organic and organometallic helicenes, *Chemistry – A European Journal*, **21**, 1673–1681 (2015).
[84] R. Hassey, E.J. Swain, N.I. Hammer, *et al.*, Probing the chiroptical response of a single molecule. *Science*, **314**, 1437–1439 (2006).
[85] R. Hassey, K.D. McCarthy, E.S.D. Basak, *et al.*, Single-molecule chiroptical spectroscopy: Fluorescence excitation of individual helicene molecules in polymer-supported thin-films. *Chirality*, **20**, 1039–1046 (2008).
[86] R. Hassey-Paradise, A. Cyphersmith, A.M. Tilley, *et al.*, Dissymmetries in fluorescence excitation and emission from single chiral molecules. *Chirality*, **21**, E265–E276 (2010).
[87] Y.Q. Tang, T.A. Cook and A.E. Cohen, Limits on fluorescence detected circular dichroism of single helicene molecules. *Journal of Physical Chemistry A*, **113**, 6213–6216 (2009).
[88] B.L. Feringa, The art of building small: From molecular switches to molecular motors, *Journal of Organic Chemistry*, **72**, 6635–6652 (2007).
[89] J. Vicario, M. Walko, A. Meetsma and B.L. Feringa, Fine tuning of the rotary motion by structural modification in light-driven unidirectional molecular motors, *Journal of the American Chemical Society*, **128**, 5127–5135 (2006).
[90] D. Pijper, M.G.M. Jongejan, A. Meetsmia and B.L. Feringa, Light-controlled supramolecular helicity of a liquid crystalline phase using a helical polymer functionalized with a single chiroptical molecular switch, *Journal of the American Chemical Society*, **130**, 4541–4552 (2008).
[91] T. Mitsuoka, H. Sato, J. Yoshida, *et al.*, Photomodulation of a chiral nematic liquid crystal by the use of a photoresponsive ruthenium(III) Complex, *Chemistry of Materials*, **18**, 3442–3447 (2006).
[92] J. Ma, Y. Li, T. White, *et al.*, Light-driven nanoscale chiral molecular switch: reversible dynamic full range color phototuning, *Chemical Communications*, **46**, 3463–3465 (2010).
[93] H. Sato, F. Sato and A. Yamagishi, Rewritable optical memory in liquid crystals containing photo-epimerizing Cr(III) complexes, *Chemical Communications*, **49**, 4773–4775 (2013).
[94] S. Akine, S. Hotate and T. Nabeshima, A molecular leverage for helicity control and helix inversion, *Journal of the American Chemical Society*, **133**, 13868–13871 (2011).
[95] G. Haberhauer, A metal-ion-driven supramolecular chirality pendulum, *Angewandte Chemie International Edition*, **48**, 9286–9289 (2010).
[96] G. Haberhauer, Control of planar chirality: The construction of a copper-ion-controlled chiral molecular hinge Record contains structures, *Angewandte Chemie International Edition*, **47**, 3635–3638 (2008).
[97] S. Ernst and G. Haberhauer, A unidirectional open-close mechanism of metal-ion-driven molecular hinges with adjustable amplitude record contains structures, *Chemistry A European Journal*, **15**, 13406–13416.
[98] N. Avarvari and J.D. Wallis, Strategies towards chiral molecular conductors, *Journal of Materials Chemistry*, **19**, 4061–4076 (2009).
[99] R.M. Pinto, E.M.S. Macoas, A.I.S. Neves, *et al.*, Effect of molecular stacking on exciton diffusion in crystalline organic semiconductors, *Journal of the American Chemical Society*, **137**, 7104–7110 (2015).
[100] L. Pu, The study of chiral conjugated polymers, *Acta Polymerica*, **48**, 116–141 (1997).

[101] L.A.P. Kane-Maguire and G.G. Wallace, Chiral conducting polymers, *Chemical Society Reviews*, **39**, 2545–2576 (2010).

[102] E. Coronado and J.R. Galan-Mascaros, Hybrid molecular conductors, *Journal of Materials Chemistry*, **15**, 66–74 (2005).

[103] C. Rethore, N. Avarvari, E. Canadell, *et al.*, Chiral molecular metals: Syntheses, structures, and properties of the AsF_6^- salts of racemic (+/–)-, (R)-, and (S)-tetrathiafulvalene-oxazoline derivatives, *Journal of the American Chemical Society*, **127**, 5748–5749 (2005).

[104] E. Coronado, J.R. Galan-Mascaros, C.J. Gomez-Garcia, *et al.*, A chiral molecular conductor: Synthesis, structure, and physical properties of $[ET]_3$ $[Sb_2(L\text{-tart})_2]$ CH_3CN (ET = Bis(ethylendithio)tetrathiafulvalene; L-tart=(2R,3R)-(+)-tartrate), *Inorganic Chemistry*, **43**, 8072–8077 (2004).

[105] L. Martin, S.S. Turner, P. Day, *et al.*, Crystal chemistry and physical properties of superconducting and semiconducting charge transfer salts of the type $(BEDT\text{-}TTF)_4[A(I)M(III)(C_2O_4)_3]$ PhCN (A(I) = H_3O, NH_4, K; M-III = Cr, Fe, Co, Al; BEDT-TTF = bis (ethylenedithio)tetrathiafulvalene, *Inorganic Chemistry*, **40**, 1363–1371 (2001).

[106] L. Martin, H. Akutsu, P.N. Horton, *et al.*, Chiral radical-cation salts of BEDT-TTF containing a single enantiomer of tris(oxalato)aluminate(III) and -chromate(III), *European Journal of Inorganic Chemistry*, 1865–1870 (2015).

[107] L. Martin, H. Akutsu, P.N. Hortond and M.B. Hursthouse, Chirality in charge-transfer salts of BEDT-TTF of tris(oxalato)chromate(III), *CrystEngComm*, **17**, 2783–2790 (2015).

[108] L. Martin, P. Day, P. Horton, S. Nakatsuji, *et al.*, Chiral conducting salts of BEDT-TTF containing a single enantiomer of tris(oxalato)chromate(III) crystallised from a chiral solvent, *Journal of Materials Chemistry*, **20**, 2738–2742 (2010).

[109] J. Lieffrig, O. Jeannin, P. Auban-Senzier and M. Fourmigue, Chiral conducting salts of nickel dithiolene complexes, *Inorganic Chemistry*, **51**, 7144–7152 (2012).

[110] R. Perochon, C. Poriel, O. Jeannin, L. *et al.*, Chiral, neutral, and paramagnetic gold dithiolene complexes derived from camphorquinone, *European Journal of Inorganic Chemistry*, 5413–5421 (2009).

[111] F. Pop, P. Auban-Senzier, E. Canadell, *et al.*, Electrical magnetochiral anisotropy in a bulk chiral molecular conductor, *Nature Communications*, **5**, 3757 (2014).

[112] G.L.J.A. Rikken and E. Raupach, Observation of magneto-chiral dichroism, *Nature*, **390**, 493–494 (1997).

[113] C. Train, M. Gruselle, and M. Verdaguer, The fruitful introduction of chirality and control of absolute configurations in molecular magnets, *Chemical Society Reviews*, **40**, 3297–3312 (2011).

[114] M. Minguet, D. Luneau, E. L'Hotel, *et al.*, An enantiopure molecular ferromagnet, *Angewandte Chemie International Edition*, **41**, 586–589 (2002).

[115] J. Vidal-Gancedo, M. Minguet, D. Luneau, *et al.*, Stereochemistry and EPR Investigation of a Chiral Molecular Magnet, *Journal of Physics and Chemistry of Solids*, **65**, 723–726 (2004).

[116] M. Minguet, D. Luneau, C. Paulsen, *et al.*, From purely organic to metallo-organic chiral magnetic materials, *Polyhedron*, **22**, 2349–2354 (2003).

[117] E. Lhotel, D.B. Amabilino, C. Sporer, *et al.* Intrinsic avalanches and collective phenomena in a Mn(II)-free radical ferrimagnetic chain, *Physical Review B*, **77**, 064416 (2008).

[118] H. Iwamura, K. Inoue and N. Koga, Tacticity versus dimension of the extended structures in the crystals of heterospin magnets made of transition-metal complexes with the poly(aminoxyl) radical, *New Journal of Chemistry,* 201–210 (1998).

[119] H. Kumagai and K. Inoue, A chiral molecular based metamagnet prepared from manganese ions and a chiral triplet organic radical as a bridging ligand, *Angewandte Chemie International Edition*, **38**, 1601–1603 (1999).

[120] K. Inoue, H. Kumagai and A.S. Markosyan, Synthesis of asymmetric one-dimensional molecular-based magnets, *Synthetic Metals*, **121**, 1772–1773 (2001).

[121] P.S. Ghalsasi, K. Inoue, S.D. Samant and J.V. Yakhmi, A complex of a chiral substituent-based nitroxide triradical having two chiral centres with Mn(hfac)$_2$, *Polyhedron*, **20**, 1495–1498 (2001).
[122] Ph. Gerbier, N. Domingo, J. Gómez-Segura, *et al.*, Chiral, single-molecule nanomagnets: synthesis, magnetic characterization and natural and magnetic circular dichroism, *Journal of Materials Chemistry*, **14**, 2455–2460 (2004).
[123] M. Gonidec, F. Luis, À. Vílchez, *et al.*, A liquid crystalline single-molecule magnet with variable magnetic properties, *Angewandte Chemie International Edition*, **49**, 1623–1626 (2010).
[124] N. Domingo, B.E. Williamson, J. Gómez-Segura, *et al.*, Magnetism of isolated Mn12 single-molecule magnets detected by magnetic circular dichroism: observation of spin tunneling with a magneto-optical technique, *Physical Review B*, **69**, 052405(1–4) (2004).
[125] M. Gonidec, E.S. Davies, J. McMaster, *et al.*, Probing the magnetic properties of three interconvertible redox states of a single-molecule magnet with Magnetic Circular Dichroism, *Journal of the American Chemical Society*, **132**, 1756–1757 (2010).
[126] C. Train, R. Gheorghe, V. Krstic, *et al.*, Strong magneto-chiral dichroism in enantiopure chiral ferromagnets, *Nature Materials*, **7**, 729–734 (2008).
[127] S. Decurtins Multifunctional coordination compounds: design and properties, *Philosophical Transactions of the Royal Society A – Mathematical Physical and Engineering Sciences*, **357**, 3025–3040 (1999).
[128] M. Clemente-Leon, E. Coronado, C. Marti-Gastaldo and F.M. Romero, Multifunctionality in hybrid magnetic materials based on bimetallic oxalate complexes, *Chemical Society Reviews*, **40**, 473–497 (2011).
[129] R. Andres, M. Gruselle, B. Malezieux, *et al.*, Enantioselective synthesis of optically active polymeric homo- and bimetallic oxalate-bridged networks $[M_2(ox)_3]_n$, *Inorganic Chemistry*, **38**, 4637–4646 (1999).
[130] M. Gruselle, C. Train, K. Boubekeur, *et al.*, Enantioselective self-assembly of chiral bimetallic oxalate-based networks. *Coordination Chemistry Reviews*, **250**, 2491–2500 (2006).
[131] R. Gheorghe, L.-M. Chamoreau, J. Kapitan, *et al.* Gruselle, Enantiomerically pure quaternary ammonium salts with a chiral alkyl chain $N(CH_3)(n\text{-}C_3H_7)_2(sec\text{-}C_4H_9)I$: Synthesis and physical studies, *Chirality*, **20**, 1085–1091 (2008).
[132] C. Train, T. Nuida, R. Gheorghe, *et al.*, Large magnetization-induced second harmonic generation in an enantiopure chiral magnet, *Journal of the American Chemical Society*, **131**, 16838–16843 (2009).
[133] Y. Kitagawa, T. Miyatake and K. Ishii, Magneto-chiral dichroism of artificial light-harvesting antenna, *Chemical Communications*, **48**, 5091–5093 (2012).
[134] A. Bousseksou, G. Molnár, L. Salmon and W. Nicolazzi, Molecular spin crossover phenomenon: recent achievements and prospects, *Chemical Society Reviews*, **40**, 3313–3335 (2011).
[135] T. Hashibe, T. Fujinami, D. Furusho, *et al.*, Chiral spin crossover iron(II) complex, fac-Lambda-$[Fe\text{-}II(HLR)_3](ClO_4)_2$ EtOH (HLR=2-methylimidazol-4-yl-methylideneamino-*R*-(+)-1-methylphenyl), *Inorganica Chimica Acta*, **375**, 338–342 (2011).
[136] W. Liu, X. Bao, L.-L. Mao, *et al.* Tong, A chiral spin crossover metal–organic framework, *Chemical Communications*, **50**, 4059–4061 (2014).
[137] C.T. Brewer, G. Brewer, R.J. Butcher, *et al.*, Synthesis and characterization of a spin crossover iron(II)–iron(III) mixed valence supramolecular pseudo-dimer exhibiting chiral recognition, hydrogen bonding, and π–π interactions, *Dalton Transactions*, 295–298 (2007).
[138] Y. Sunatsuki, S. Miyahara, Y. Sasaki, *et al.*, Conglomerate crystallization, chiral recognition and spin-crossover in a host–guest complex consisting of Fe^{III} complexes (host) and $[Cr(ox)_3]^{3-}$(guest), *CrystEngComm*, **14**, 6377–6380 (2012).
[139] Y. Sunatsuki, Y. Ikuta, N. Matsumoto, *et al.*, An unprecedented homochiral mixed-valence spin-crossover compound, *Angewandte Chemie International Edition*, **42**, 1614–1618 (2003).

[140] D.-H. Ren, D. Qiu, C.-Y. Pang, *et al.*, Chiral tetrahedral iron(II) cages: diastereoselective subcomponent self-assembly, structure interconversion and spin-crossover properties, *Chemical Communications*, **51**, 788–791 (2015).

[141] I.A. Gural'skiy, V.A. Reshetnikov, A. Szebesczyk, *et al.*, Chiral spin crossover nanoparticles and gels with switchable circular dichroism, *Journal of Materials Chemistry C*, **3**, 4737–4741 (2015).

[142] X. Zhang, J. Yin and J. Yoon, Recent advances in development of chiral fluorescent and colorimetric sensors, *Chemical Reviews*, **114**, 4918–4959 (2014).

[143] Z. Chen, Q. Wang, X. Wu, *et al.*, Optical chirality sensing using macrocycles, synthetic and supramolecular oligomers/polymers, and nanoparticle based sensors, *Chemical Society Reviews*, **44**, 4249–4263 (2015).

[144] C.P. Montgomery, E.J. New, D. Parker and R.D. Peacock, Enantioselective regulation of a metal complex in reversible binding to serum albumin: dynamic helicity inversion signalled by circularly polarised luminescence, *Chemical Communications*, 4261–4623 (2008).

[145] C. Wolf and K.W. Bentley, Chirality sensing using stereodynamic probes with distinct electronic circular dichroism output, *Chemical Society Reviews*, **42**, 5408–5424 (2013).

[146] J. Yuasa, H. Ueno and T. Kawai, Sign reversal of a large circularly polarized luminescence signal by the twisting motion of a bidentate ligand, *Chemistry, A European Journal*, **20**, 8621–8627 (2014).

[147] A. Gerus, K. Slepokura and J. Lisowski, Anion and solvent induced chirality inversion in macrocyclic lanthanide complexes, *Inorganic Chemistry*, **52**, 12450–12460 (2013).

[148] W. Miao, L. Zhang, X. Wang, *et al.*, Gelation-induced visible supramolecular chiral recognition by fluorescent metal complexes of quinolinol – glutamide, *Langmuir*, **29**, 5435–544 (2013).

[149] Q. Lin, T.-T. Lu, X. Zhu, *et al.*, A novel supramolecular metallogel-based high-resolution anion sensor array, *Chemical Communications*, **51**, 1635–1638 (2015).

[150] A. Tripathi, A. Kumar and P.S. Pandey, Visual chiral recognition of mandelic acid and α-amino acid derivatives by enantioselective gel formation and precipitation, *Tetrahedron Letters*, **53**, 5745–5748 (2012).

[151] M. Kumar, P. Brocorens, C. Tonnele, *et al.*, A dynamic supramolecular polymer with stimuli-responsive handedness for in situ probing of enzymatic ATP hydrolysis, *Nature Communications*, **5**, 5793 (2014).

[152] D.B. Amabilino, Enzyme activity with a twist, *Nature Chemistry*, **7**, 275–277 (2015).

[153] F.R. Keene, The shape of metallosupramolecular assemblies, *Dalton Transactions*, **40**, 2405–2418 (2011).

[154] Z. Dai, J. Lee and W. Zhang, Chiroptical switches: applications in sensing and catalysis, *Molecules*, **17**, 1247–1277 (2012).

[155] S. Akine, S. Sairenji, T. Taniguchi and T. Nabeshima, Stepwise helicity inversions by multisequential metal exchange, *Journal of the American Chemical Society*, **135**, 12948–12951 (2013).

[156] D.P. Zhao, T.M. Neubauer and B.L. Feringa, Dynamic control of chirality in phosphine ligands for enantioselective catalysis, *Nature Communications*, 6, Art. No. 6652 (2015).

[157] S. Mortezaei, N.R. Catarineu and J.W. Canary, A redox-reconfigurable, ambidextrous asymmetric catalyst, *Journal of the American Chemical Society*, **134**, 8054–8057 (2012).

[158] H. Miyake and H. Tsukube, Coordination chemistry strategies for dynamic helicates: time-programmable chirality switching with labile and inert metal helicates, *Chemical Society Reviews*, **41**, 6977–6991 (2012).

[159] M. Martínez-Calvo, O. Kotova, M.E. Möbius, *et al.*, Healable luminescent self-assembly supramolecular metallogels possessing lanthanide (Eu/Tb) dependent rheological and morphological properties, *Journal of the American Chemical Society*, **127**, 1983–1992 (2015).

[160] N. Ousaka, Y. Takeyama and E. Yashima, Dinuclear metal complexes composed of peptide chains: Solvent-induced switching and inversion of the metal-centered chirality, *Chemical Science*, **3**, 466–469 (2012).
[161] N. Ousaka, Y. Takeyama, H. Iida and E. Yashima, Chiral information harvesting in dendritic metallopeptides, *Nature Chemistry*, **3**, 856–861 (2011).
[162] L. Zhang, L. Qin, X. Wang, *et al.*, Supramolecular chirality in self-assembled soft materials: regulation of chiral nanostructures and chiral functions, *Advanced Materials*, **26**, 6959–6964 (2014).
[163] T. Nabeshima, Construction of cooperative and responsive supramolecular systems for molecular functional modulation, *Bulletin of the Chemical Society of Japan*, **83**, 969–991 (2010).
[164] P.C. Mondal, N. Kantor-Uriel, S.P. Mathew, *et al.*, Chiral conductive polymers as spin filters, *Advanced Materials*, **27**, 1924–1927 (2015).

7

Chirality in Network Solids

David R. Turner

7.1 Introduction

Chirality is most often considered as a molecular property, arising either from the presence of an organic stereocentre, from restricted bond rotation leading to conformer trapping or from the arrangement of ligands around a central metal atom. However, chirality also exists in the solid state with 65 of the 230 allowed crystallographic space groups being chiral. Chiral solids can arise in one of two ways; firstly, and most predictably, enantiomerically pure compounds must crystallize in a chiral space group. Secondly, achiral compounds may crystallize in such a way that the arrangement of molecules in the crystal lattice makes the overall arrangement chiral (i.e. spatial or topological chirality). Typically this latter approach gives rise to racemic mixtures of crystalline materials. A discussion on crystallization and chirality can be found elsewhere in this volume.

Discrete species are able to form chiral solids but there is particular recent interest in the formation of infinite chiral networks. These network solids may be purely inorganic, such as zeolites, or may be constructed from a combination of organic and inorganic components – coordination polymers. Although discrete molecules may form extended networks in the solid state through supramolecular interactions, such as hydrogen bonds, these lie outside the scope of this chapter, which will class network solids as those containing metal-ligand or coordination bonds throughout the material and will not examine molecular crystals. Particular emphasis will be placed on 3D networks, especially those containing accessible pore spaces, wherein the entire material is a single network, rather than an ordered agglomeration of networks of lower dimensionality.

There are a plethora of network materials that exist in a chiral space group. A large subset of these comprises achiral components and therefore the materials are synthesized as

---

Chirality in Supramolecular Assemblies: Causes and Consequences, First Edition. Edited by F. Richard Keene.

racemic mixtures. Perhaps the largest subsets of examples of these materials are those with 10,3a or quartz topologies. For the purposes of this chapter, this body of work will not be discussed and attention will be paid only to materials that are able to be formed as bulk homochiral solids. Typically this requires the use of chiral reagents, either directly incorporated into the framework or as templating or induction agents during synthesis. Genuine instances of spontaneous homochiral synthesis of network solids are rare. The term 'spontaneous resolution' relates to the crystallization of a racemic molecular mixture as enantiopure crystals (rather than co-crystallizing in a centrosymmetric setting) although in relation to coordination polymers it confusingly means any *achiral* reagents that give racemic network solids [1]. Spontaneous formation of a homochiral network solid from achiral reagents (in the absence of any chiral species in the reaction) is known as symmetry breaking. There are examples of this occurring during crystallization of molecular species and ionic materials [2, 3], including the notable example of sodium chlorate [4].

For practical application, especially in the fields of enantioselective separations and catalysis, it is necessary to have a bulk homochiral material, and crystal engineering researchers have made significant progress in this area since the mid-1990s. This chapter will discuss the issues surrounding the synthesis of homochiral inorganic network solids, various synthetic routes towards homochiral metal-organic network solids and the practical uses of these chiral materials.

7.2 Chirality in Inorganic Network Solids

Inorganic network solids, such as zeolites, are constructed from purely achiral building blocks yet can be chiral either from the intrinsic topology of the network itself, enforced by a chiral space group, or by the arrangement of heteroatoms. Chiral topologies are not rare and close-packed examples of this include quartz and sodium chlorate (see above) – however, the formation of chiral inorganic porous framework materials – chiral zeolites – is much rarer [5]. Given the dependence on controlling chirality during crystallization, rather than inclusion of chiral subunits, the deliberate synthesis of homochiral inorganic materials is nontrivial with no ability to affect the absolute chirality of the product. This is a somewhat more complicated process than with metal-organic materials as there are less potential directing interactions between ligands and induction agents and the often harsh synthetic conditions may not be suited to induction-based control of chirality [6].

There are a few examples of chiral zeolite-type materials that form as racemic mixtures. One polymorph of zeolite beta, a long-known material that was relatively recently characterized, is intrinsically chiral with a helical pore structure although it intergrows with an achiral polymorph [7]. Also known are a chiral silicogermanate and a gallium germanate [8–10], and the mineral goosecreekite [11], as well as a number of hypothetical frameworks determined using computational methods [12]. An aluminophosphate framework was recently reported, constructed using transition metal heteroatoms as part of the anionic backbone, which displayed a circular dichroism signal, although this is likely to arise from symmetry breaking rather than spontaneous homochiral resolution [13].

Bu and co-workers demonstrated that it is possible to use organic molecules as chiral induction agents for inorganic materials with their preparation of homochiral chiral zincophosphate (CZP, $NaZnPO_4 \cdot H_2O$), which has a zeolite-type topology [14]. To elicit strong complementarity between the induction agent and the framework a nucleotide was used so that the nucleotide phosphate interacts during formation of the phosphate network. Significant enantioenrichment is achieved using this method, although not complete homochiral crystallization, and control experiments showed that the presence of the phosphate group was essential to observe this effect. A homochiral inorganic framework was synthesized in the material [*d*-$Co(en)_3$][$H_3Ga_2P_4O_{16}$] using the racemically pure cationic complex as a template [15]. This latter example is a true template effect, rather than induction, with the templating cation occupying a significant portion of the pore space in the final material. Whilst these are very rare examples of forming homochiral inorganic networks, the result does highlight the potential in this area.

7.3 Synthesis of Chiral Coordination Polymers

Coordination polymers require the presence of metal ions (or clusters thereof) and organic ligands that are able to bridge between these metal-containing nodes [16]. The versatility of these materials has given rise to their popularity as platforms for a variety of applications, including as functional chiral materials [17, 18]. Chirality can be incorporated into these materials by one of four methods:

- Chiral induction or templating. Using achiral components, which would normally lead to a racemic product mixture, it is possible to force the production of a homochiral bulk (or an enhanced yield of one enantiomer) by using a chiral auxiliary species during the synthesis, which may be trapped within the resulting framework (a template) or may play a role in the assembly without being present in the final material (an induction agent).
- Use of small chiral co-ligands. The formation of network solids, particularly those containing open channels, typically requires long and rigid bridging ligands. Small chiral co-ligands, such as amino acids or other chiral carboxylates, can be used alongside the larger bridging ligands to introduce chirality to the network whilst the larger ligands act to space the metals apart.
- Use of chiral spacer ligands. Rather than adopt a two-ligand system, as in the point above, it can be more efficient to embed a chiral component within an organic ligand, thereby ensuring a bulk homochiral product if an enantiopure ligand is used.
- Postsynthetic modification. Achiral networks are typically much easier to synthesize than chiral ones. If there is appropriate functionality within the network, at which a reaction can be conducted once the network has formed, then a chiral group may be grafted into the material to make the overall solid chiral.

These four approaches will be discussed individually below, with some relevant examples to demonstrate each method.

7.3.1 Chiral Induction, Templating and Symmetry Breaking

The use of achiral precursors to synthesize chiral network solids, in which the topological chirality is generated by the arrangement of the components, is an attractive prospect. Achiral reagents are typically easier, and less expensive, to synthesize and the possibility of ligand racemization during the sometimes harsh synthetic conditions used to generate coordination polymers is avoided. The major issue that has to be overcome in this method is the formation of a racemic mixture rather than a bulk homochiral material. In many instances homochirality can be difficult to assign as the presence of a solid-state circular dichroism response indicates an enantiomeric enrichment but not necessarily enantiopurity. On occasion, such systems can be enantiomerically enhanced, or even give rise to homochiral bulk solids, due to symmetry breaking during crystallization based on nucleation events, which can be subject to control kinetically or mechanically [3]. For example, a single nucleation site should produce a bulk homochiral material in much the same way that a seeding event does and mechanical agitation can be used to break up and disperse an early crystal nucleus and therefore bias this handedness in the bulk.

One of the first reports of generating a homochiral coordination polymer from achiral reagents was the synthesis of a 1D coordination polymer containing 5-(9-anthracenyl) pyrimidine, in which the chirality is brought about by the presence of a helical Cd-pyrimidine chain [19]. The authors noted that the crystals were predominantly of a single handedness, brought about by rapid crystallization at a single nucleation site, although the handedness of the sample cannot be controlled in this manner. Homochiral bulk solids could be produced by using a seed crystal during synthesis, effectively providing a nucleation site that determines the handedness of the growing crystals. The presence of seed crystals is perhaps the most reliable method for obtaining an enantiomerically enhanced product, especially given that the handedness can be confirmed prior to seeding and eliminates the 50 : 50 chance associated with a reliance on symmetry breaking. Symmetry breaking has been used to create network solids, with an early report being the homochiral synthesis of a copper-succinate-based 3D network by chemical control of nucleation [20]. Agitation by mechanical stirring has been reported to invoke symmetry breaking in 1D coordination polymers [21, 22] and 3D networks have been reported to show symmetry breaking, although the controlling mechanism is unclear [23, 24]. Perhaps one of the most interesting reports on inducing enantio-enriched network solids is that of Wu *et al.* who used irradiation by circularly polarized light during synthesis to influence the chirality of the solid [25]. The use of auxiliary agents as either templates (which remain trapped in the final material) or induction agents has come to the fore recently to overcome the problem of racemic networks, although the exact mechanism behind these processes often remains unclear [26].

Chiral induction was reported in the mid-2000s separately by the groups of Morris and Bu who used a chiral ionic liquid or spectator chiral solutes, respectively, to induce the production of homochiral network solids [27, 28]. Morris and coworkers used an ionic liquid, containing either L- or D-aspartate as the anionic component, in the ionothermal synthesis of a homochiral material containing 1,3,5-benzenetricarboxylate with the ionic liquid cation (BMIm) residing in the pores. Using an ionic liquid that does not have a chiral component leads to a different, centrosymmetric network, therefore showing the transfer of chirality from the ionic liquid to the network solid. The Bu group used chiral alkaloid spectator species, (−)-cinchonidine or (+)-cinchonine, dissolved in the reaction mixture to obtain bulk homochirality of a thiophene-2,5-dicarboxylate-based network in a process that the authors termed 'chiral catalysis' (Figure 7.1). The handedness of the network obtained depended on the spectator species and, without any chiral alkaloids present, a racemic mixture of coordination polymers was obtained. The same alkaloids have also been used to form other homochiral materials in an analogous manner [29]. Similar effects have been observed for inorganic materials, such as the use of mannitol to homochirally

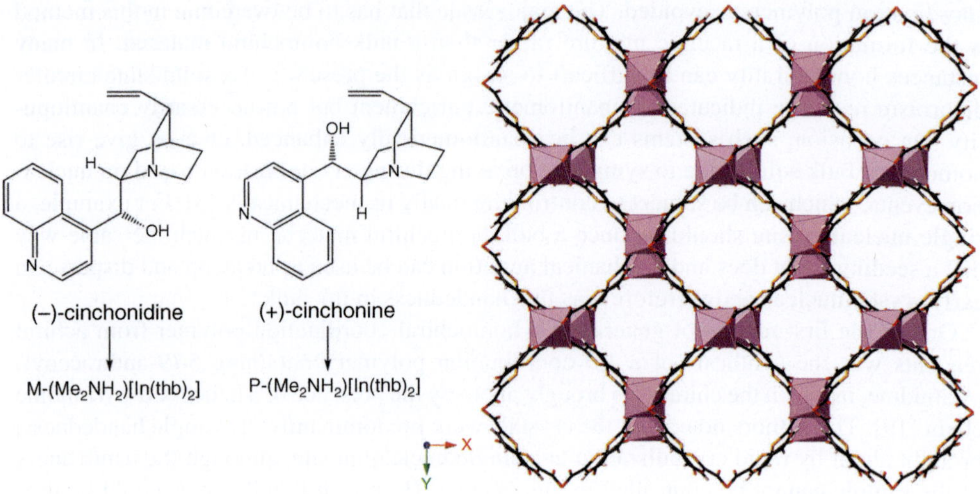

Figure 7.1 *The use of alkaloid spectator species determines the bulk homochirality of the [In(thb)$_2$] network, with the two-fold interpenetrating diamondoid structure of P-(Me$_2$NH$_2$)[In(thb)$_2$] shown*

crystallize sodium chlorate [30]. Enantiopure carvone has been used as an induction agent in the urothermal synthesis of a terephthalate-containing 3D network, which, when synthesized in the absence of the induction agent, gives a racemic product [31]. Carvone was deliberately selected due to the presence of ethyleneurea in the product, acting as a ligand through the carbonyl oxygen atom. Given that carvone also has a potential carbonyl donor, and no other potential coordination sites, it is postulated that it is intimately involved with the framework as it forms.

The use of alkaloids to induce chirality is curious as there is no obvious structural relationship between the chiral molecules and the coordination polymer. A more recent example of chiral induction is the use of camphoric acid to induce chirality in networks containing adamantane-1,3-dicarboxylate (adc) [32]. Use of D-camphoric acid gave an enantioenriched product of [Mn$_3$(HCOO)$_4$(adc)], although not enantiopure, of the opposite handedness to that obtained using L-camphoric acid. Use of DL-camphoric acid gave the anticipated racemic product. Interestingly, reaction in the absence of camphoric acid yielded an entirely different network, clearly highlighting that camphoric acid plays a crucial role in directing the synthesis. The similarity in structure between camphoric acid and adamantane-1,3-dicarboxylic acid (i.e. the spatial arrangement of the carboxylate groups) is postulated to play a significant role during nucleation. Glutamic acid, which also has some structural similarity to adc, showed similar induction behaviour whilst other amino acids did not.

Another recent example of chiral induction was the use of L-*N-tert*-butoxy-carbonyl-2-(imidazole)-1-pyrrolidine or pyrrolidine-2-yl-imidazole (pyi) in forming porous homochiral materials capable of catalysing the asymmetric cyanosilylation of aromatic aldehydes [33]. Whilst these materials show no sign of the chiral agent being incorporated into the crystalline lattice, the pyi molecule was used in another attempted chiral induction and ended up as part of the network solid, highlighting the unpredictability of this method and the necessity of using the 'correct' chiral induction agent for a particular system.

Rather than induction, in which the chiral auxiliary does not feature in the final product, it is also possible to form chiral materials in which the templating species is incorporated although not as part of the coordination framework. A recent example is the formation of an intercalated structure consisting of 2D sheets [34]. In the absence of a chiral agent, sheets form that contain disordered perchlorate anions between the layers. However, introduction of enantiopure sodium pyroglutamate forms a chiral solid that incorporates the chiral anion in place of ClO_4^- and whilst the connectivity of the 2D sheets is analogous to the centrosymmetric system, the spatial orientation of the ligands is different and they are acentric.

7.3.2 Incorporation of Small Chiral Co-Ligands

Perhaps the easiest method of generating chiral network solids is to use small, readily available chiral co-ligands in concert with ligands that have a propensity to generate coordination polymers. Typically these chiral molecules will be able to construct low-dimensionality (1D, 2D) metal-organic assemblies that are then bridged by other organic ligands to form a 3D network solid. Whilst there are many chiral molecules that may be suitable for this application, two general categories of molecule will be discussed below – amino acids and other chiral carboxylates.

The use of chiral co-ligands was demonstrated in 2000 by Rosseinsky using the chiral propane-1,2-diol (1,2-pd) rather than diacids [35]. The material $[Ni_3(btc)_2(py)_6(1,2\text{-pd})_3]$ (where btc = 1,3,5-benzenetricarboxylate, py = pyridine) forms as a twofold interpenetrated 10,3a structure. A similar 10,3a network, which is an intrinsically chiral topology, is obtained using ethyleneglycol (which differs from 1,2-pd only in the absence of a methyl group) although bulk homochirality does not exist, hence the introduction of the chiral diol, which controls the handedness of the helical motif in the network. Whilst this material was not robust to desolvation a second-generation material, replacing pyridine with 3-picoline, was found to have permanent porosity with 47% void space and approximately 1 nm wide pores, although modest enantioselective guest uptake is only observed when the size of the guest is comparable to that of the pores [36].

7.3.2.1 Amino Acid Co-Ligands

Amino acids are Nature's building blocks, and it is unsurprising that chemists have adopted them as simple chiral agents for the synthesis of chiral network solids [37, 38]. Perhaps the most obvious amino acids to use for this purpose are aspartic acid and glutamic acid, both of which possess two carboxylic acid groups and therefore lend themselves well to the construction of polymeric architectures.

Structures have been reported of the compounds $[Ni_{2.5}(OH)(\text{L-Asp})_2]$ and $[Co(\text{L-Glu})(OH_2)]$, which both form microporous, homochiral 3D networks without the need for typical network-inducing co-ligands, highlighting the intrinsic ability of these molecules to form chiral network solids [39, 40]. There is a particularly large body of work concerning coordination polymers containing aspartate in conjunction with 4,4′-bipyridine co-ligands. The first of these reports, by Rosseinsky and coworkers, uses $Ni(\text{L-Asp})\cdot 3H_2O$ as the source of the chiral acid to produce the pillared network of $[Ni_2(\text{L-Asp})_2(4,4'\text{-bpy})]$ (Figure 7.2) [41]. Sheets of $Ni_2(\text{L-Asp})_2$ with carboxylates connecting adjacent metal ions are bridged by the 4,4′-bipyridine ligands to leave 1D channels in the framework. The compound displays selective uptake of chiral diols and ideally demonstrates the approach of

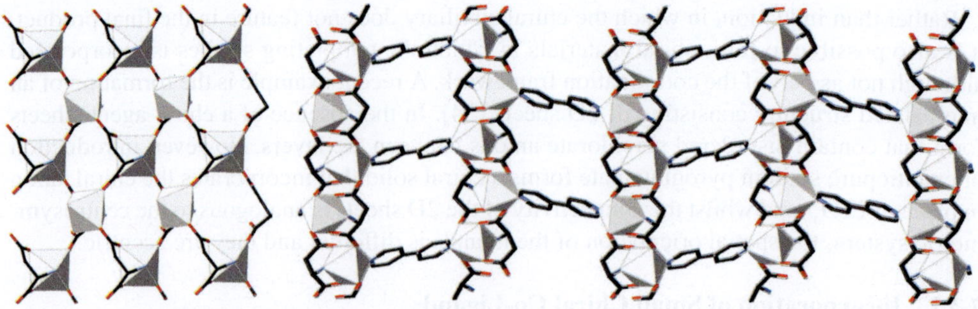

Figure 7.2 The structure of the prototypical [Ni_2(L-Asp)$_2$(4,4′-bpy)] network contains chiral Ni_2(L-Asp)$_2$ sheets (left) pillared by 4,4′-bipyridine (right)

combining small chiral ligands to form substructures and impart chirality on the network with larger achiral ligands to promote porosity to give rise to potential applications in enantioselective storage or separation. The [Ni_2(L-Asp)$_2$(4,4′-bpy)] system can be protonated using HCl to give a material that is able to act as a heterogeneous asymmetric Brønsted acidic catalyst for the methanolysis of *rac*-propylene oxide, although better activity is observed for the structurally related and more porous [Cu_2(L-Asp)$_2$(bpe)] network (bpe=4,4′-bipyridylethene) [42]. A variety of other nitrogen-based co-ligands can be used to create other layered materials containing M_2(L-Asp)$_2$ sheets (M=Ni, Co, Cu) [43–46]. The porosity of these systems has been ably displayed by their incorporation into membranes for selective gas sieving, although this does not exploit their chiral nature [47].

In terms of amino acids as co-ligands in networks, aspartate dominates the literature and the closely related glutamate anion appears considerably less, perhaps due to the extra flexibility that it possesses. Reported network structures of glutamate largely contain simple derivatives, such as N-benzoylglutamate or N-tosylglutamate, which both prevent formation of an N,O-chelating coordination mode that is typically observed for amino acid anions by protecting the nitrogen atom [48, 49]. In the former case the ligand acts as a two-connecting ligand in a 2D network and as a three-connecting ligand in a 3D network {with bis(4-pyridyl)propane and 4,4′-bipyridine, respectively}. The tosylated ligand also forms 2D and 3D networks with 4.4′-bipyridine co-ligands, with a 3D network reported that contains solvent-filled channels [50].

Whilst there is understandably a significant body of work involving aspartate and glutamate as dicarboxylate ligands there are also networks known that contain monocarboxylate amino acids These materials contain coordination polymers that are typically of lower dimensionality due to the limited bridging ability of such ligands although it is worth noting that a 3D achiral network containing racemic D/L-alanine and bis(4-pyridyl)ethene is known [51]. Proline has been shown to form 2D networks in combination with 4,4′-bpy, coordinating as an N,O-chelate to a metal centre and bridging to an adjacent metal through the remaining oxygen atom [52]. A 3D proline-based network has been reported that uses trinuclear copper-proline complexes to act as bridges between Keggin polyoxoanions in a rare example of a homochiral POM-containing framework [53]. Tryptophan has been incorporated into 2D networks using bipyridyl co-ligands with the compound {[Zn_2(L-trp)$_2$(bpe)$_2$(H_2O)$_2$]$\cdot 2H_2O \cdot 2NO_3\}_n$ reported to display a low dielectric constant, that could be modified by exchanging the counter-anion, and an SHG response [54, 55].

7.3.2.2 Chiral Carboxylate Co-Ligands

One of the major drawbacks of using amino acids – with the notable exceptions of aspartate and glutamate – as co-ligands, is that their ability to form extended networks is somewhat limited. In contrast, there are several small chiral carboxylic acids available whose geometry and multiple functional groups better facilitate the formation of polymeric structures such as lactic acid, tartaric acid, malic acid, mandelic acid and camphoric acid (Figure 7.3). Perhaps one of the earliest examples of mixed functionality from a homochiral network solid, the POST-1 system of the Kim group, contains a simple derivative of tartrate and it is unsurprising that similar species have remained a mainstay of the field [56].

Camphoric acid is a relatively rigid molecule and therefore lends itself well to the formation of network solids both by itself and in combination with neutral co-ligands. For example, the network $(Me_4N)[In(D\text{-}cam)_2]$ and its enantiomer can be formed as homochiral bulk materials that not only possess chirality from the ligand but also adopt the inherently chiral quartz topology [57]. Cadmium camphorates can also be isolated as 3D networks in the absence of co-ligands, presumably due to the large ionic radii of the metal allowing coordination by multiple carboxylates [58].

An early example of camphoric acid in combination with a co-ligand is an isostructural series of mixed Ni/Co 3D frameworks of the form $[M_2(cam)_2(1,4\text{-dimb})]$ {where 1,4-dimb = 1,4-di-(1-imidazolyl-methyl)-benzene} [59]. These 3D networks bear similarity to the previously-mentioned $Ni_2(L\text{-}Asp)_2$-based materials, with layers of $M_2(cam)_2$ bridged by the nitrogen-donor ligands to give a pillared structure. The layers adopt a (4,4) topology with paddlewheel nodes that are overall six-connecting. Similar networks can be obtained using 4,4′-bipyridine or bis(4-pyridyl)ethene, with the networks adopting self-penetrating topologies rather than the idealized pcu net [60]. A series of isoreticular α-Po networks, comprising $Zn_2(cam)_2$ sheets bridged by rigid spacer ligands has been reported and although larger spacers are shown to increase the solvent accessible volume of the materials, only the compound with the shortest spacer between layers (1,4-diazabicyclo[2.2.2]octane) is stable to the removal of lattice solvent [61]. Such an approach demonstrates the predictability that can be brought into the synthesis of network solids using well established motifs such as the $M_2(ca)_2$ layer. The use of flexible bis-triazole co-ligands gives rise to more varied structural motifs, rather than the rigid layered networks that results from using linear co-ligands, with the

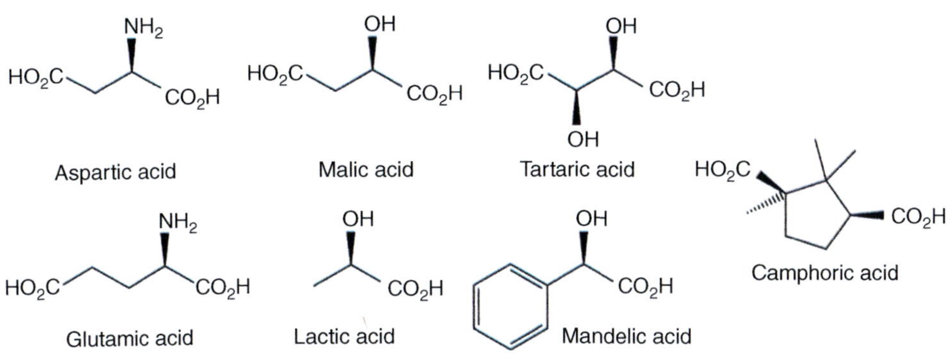

Figure 7.3 *Examples of simple chiral carboxylic acids commonly used in combination with bridging co-ligands for constructing chiral networks*

resulting compounds displaying reasonable secondary harmonic generation (SHG) behaviour [62]. Using charged co-ligands disrupts the formation of $M_2(ca)_2$ sheets, as observed in the case of $[Co_2(D\text{-}ca)(int)_2(H_2O)]_n$ and its enantiomer (where int = isonicotinate) [63]. Using formate as a charged co-ligand gives rise to nonporous 3D networks with a honeycomb structure [64]. Homochiral camphorate-based 3D networks with solvent-accessible channels have also been constructed using s-block and f-block metals [65, 66]. Two 3D camphorate-based materials have been employed in chiral and achiral gas chromatography, with baseline separation of enantiomers for a number of analytes [67–69].

Lactic acid is a simple chiral molecule that is chemically similar to alanine, with a hydroxyl group in place of an amine, although it is slightly more versatile in its coordination chemistry due to this difference and able to act as a dianionic ligand. The efficacy of lactate for forming functional chiral materials is demonstrated by the formation of a 2D europium coordination polymer that possesses ferroelectric properties [70]. Lactate was used as a co-ligand alongside terephthalate to form a 3D network containing Zn-lactate chains that are bridged by the longer terephthalate ligands to give rise to channels within the framework [71]. The pores in the material allowed it to be both catalytically active, albeit with no chiral selectivity, and display enantioselective sorption. Mandelic acid (2-hydroxy-2-phenylacetic acid, Hman) is closely related to lactic acid and has also been reported as a chiral auxiliary ligand. The similarity between the behaviour of lactate and mandelate is exemplified by a report in which 3D networks of $[Zn_2(bpdc)(R\text{-}man)(dmf)]$ (bpdc = 4,4′-biphenyldicarboxylate) were found to have the same topology as $[Zn_2(bdc)(S\text{-}lac)(dmf)]$ with the larger dicarboxylate ligands giving channels suitable for stereoselective sorption and catalysis [72]. Mandelate has also been used alongside 4,4′-bipyridine whereby 1D Co-man chains are bridged into a 3D network [73].

Malic acid (2-hydroxybutanediacarboxylic acid) and tartaric acid (2,3-dihydroxybutanedicarboxylic acid) are both closely related to glutamic acid, with the same alkyl chain length between the two carboxylate groups. The first report of malate in a 3D homochiral network was a pillared structure in combination with 4,4′-bipyridine, as has been observed for other chiral auxiliaries, in which the malate anions are able to bridge three metal ions, using both carboxylate and hydroxyl groups, to form 2D sheets [74]. Whilst this is the common binding mode for malate, it has also been observed without a coordinating hydroxyl group in the mixed-ligand network of $[Mn_2(L\text{-}ma)_{0.4}(suc)_{1.6}(4,4'\text{-}bpy)_2(H_2O)_2]$ (suc = succinate) which forms a similar pillared structure containing the two dicarboxylates in the 2D sheets [75]. The use of longer bipyridyl co-ligands results in similar pillared structures with larger channels that display gas sorption with no framework collapse [76]. Tartrate, which contains an additional hydroxyl group compared to malate, has featured in fewer reports of homochiral coordination polymers, despite a pyridyl derivative of tartrate being used by Kim and coworkers in the seminal POST-1 paper (Figure 7.4) [56].

Tartrate has more diverse coordination chemistry than malate, able to act as a linear bridge through the carboxylates only, or able to chelate to two metals through a carboxylate/hydroxyl ring with bridging to other metal ions through the carboxylate groups. Homochiral networks can be generated in combination with bipyridyl ligands in much the same way as other chiral dicarboxylates [77]. An interesting study that builds on these pillared architectures is the combination of tartrate with rigid dicarboxylate co-ligands in structures of the type $[NaLn(tart)(L)(H_2O)_2]$ (where L = terephthalate or biphenyl-4,4′-dicarboxylate; Ln = lanthanoid), demonstrating that this design philosophy can be extended

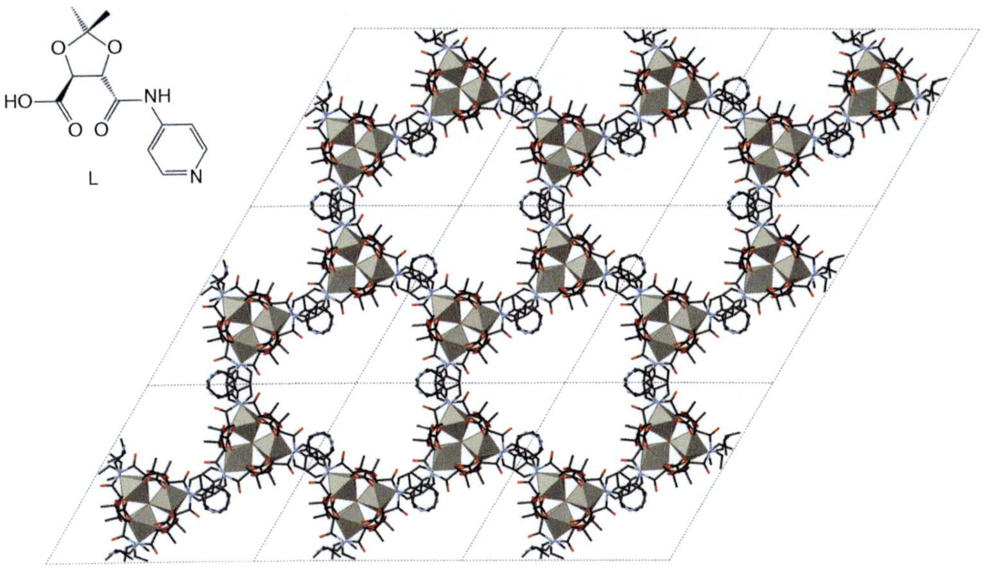

Figure 7.4 The POST-1 network ($[Zn_3(\mu_3\text{-}O)(L\text{-}H)_6]$, the first example of a homochiral metal-organic framework for catalysis and separation, uses a derivative of tartrate (L)

away from transition metals [78]. Longer polyhydroxydicarboxylic acids also have potential for the formation of homochiral networks. This potential is perhaps best highlighted in work by the Robson group using saccharic acid to form $[Zn(C_6H_8O_8)]$ which possesses a chiral 3D network, which contains both hydrophilic and hydrophobic channels, resulting from the arrangement of the noncoordinating hydroxyl groups [79].

7.3.3 Design and Application of Chiral Ligands

Whilst small chiral molecules are useful as co-ligands to induce the formation of homochiral network solids, there has been a substantial body of work exploring the design of bespoke chiral ligands towards the same end. Rationalization for this comes from the ability to embed desired functional groups into the organic backbone for potential applications and a desire to gain better control of the network formation by reducing the number of components present during the self-assembly process. Such an approach necessitates that the potential ligands are synthesized to be enantiopure, which is not always a trivial challenge and there are many different ways in which a chiral functionality can be incorporated. This section outlines some approaches that have been taken towards novel ligand systems in chiral solids and some specific examples.

7.3.3.1 Dipeptides

As seen in section 7.3.2.1, amino acids are a very useful resource as small chiral co-ligands to promote the formation of chiral network solids and there are a large number of synthetic ligands that use amino acids as building blocks. Towards this end there have been several groups that have used dipeptides as ligands using either conventional peptide coupling or as oxalyl retro-peptides.

One of the earliest reports of a 3D chiral coordination polymer containing a dipeptide is a cadmium material using a Gly-Glu peptide in the early 1970s, with the same ligand more recently used with lead [80, 81]. In these instances it is still the glutamate residue that assembles the polymeric structure.

The most significant network solid based on a dipeptide is [Zn(Gly-Ala)$_2$], reported in 2010 by Rosseinsky and coworkers [82]. The coordination framework is a (4,4) sheet containing tetrahedral zinc nodes that are coordinated by two monodentate Ala-carboxylates and two Gly-amines (i.e. by four ligands overall) with the ligands being simple linear bridges. Hydrogen bonding (amine to carbonyl) holds the material together such as to allow full desolvation of the 1D channels with some structural shifting that allows sorption of gas into the materials above a gating pressure. Perhaps the limitation of traditionally coupled peptides is that there remains an N-terminus, which can lead to lower connectivities than if there were two carboxylate termini [83].

Oxalyl retro-peptides contain an N,N′-disubstituted oxamide, coupling two amino acids through the nitrogen atoms and leaving two terminal carboxylic acids that are available for coordination. Some 3D networks have been reported that incorporate the oxalyl retro-peptides of alanine, valine, phenylalanine and leucine, all in combination with the neutral 4,4′-bis-3,5-dimethylpyrazole co-ligand [84]. These structures contain a dinuclear copper unit that is formed from a single retro-peptide in which the amides are deprotonated and coordinating to the metals, giving overall tridentate coordination to each metal from the chiral ligands (carboxylate, amide oxygen and amide nitrogen). These dinuclear units connect to each other through the carboxylate groups to give a (4,4) sheets (i.e. square grids) that are then pillared by the bis-pyrazole ligands to make the 3D pcu network in the cases of Ala, Val and Phe, with the Ala and Val derivatives shown to be stable to solvent removal from the pores. A related material containing the alanine retro-peptide and a more flexible methyl-bridged bis(pyrazole) co-ligand have shown enantioselective uptake of 1-phenylethanol and 1-phenyl-1-propanol [85].

7.3.3.2 Functionalized Amino Acids

There is a limit to the materials that can be obtained from dipeptide ligands, yet there are many more ways in which amino acids can be used to form chiral networks. There has been much research activity into what can be termed ‘simple’ amino acid derivatives, whereby an additional functional group is added to the amino acid to aid in network formation (Figure 7.5).

Some of the earliest examples of this approach are amino acids that are reacted with salicaldehyde. Whilst this class of ligand has been known since the 1950s [86] some of the earliest reports are of the sAla ligand, N-(2-hydroxybenzyl)-L-alanine (abbreviations for this class of ligand will be used as sAla meaning the Schiff base of salicaldehyde with alanine, and so on – see Figure 7.3), forming 1D and 2D coordination polymers [87, 88]. The ligand readily forms a dinuclear M_2(sAla)$_2$ unit with the two metals bridged by two phenolates and each ligand acting as a tridentate chelator through the phenolate, amine and carboxylate groups; the remaining carboxylate is available to coordinate to another metal to propagate a polymeric structure. The coordination of the amine creates an additional chiral centre. The initial report of the hydrated 1D polymer [Cu_2(sAla)$_2$(H_2O)]$_n$ demonstrated that upon dehydration a 3D chiral network of [Cu_2(sAla)$_2$]$_n$, containing small

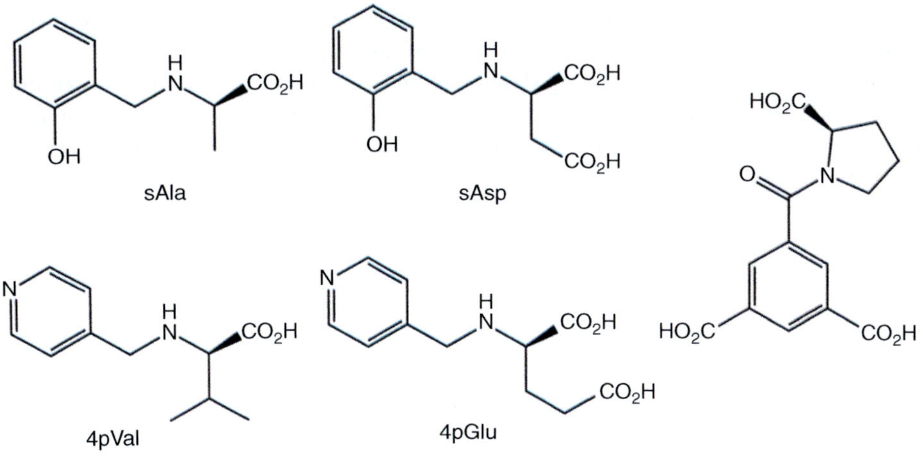

Figure 7.5 *Some examples of 'simple' derivatives of amino acids that have been used in the formation of chiral networks*

channels, could be formed. Recently this material has been used by another group as a surface coating with in a GC column for chiral chromatographic separations [89]. The dimeric chiral $M_2(sAla)_2$ subunit can be bridged by neutral co-ligands to form other network structures, although reports are generally of low dimensionality coordination polymers [90].

A Schiff base ligand that is closely related to sAla, in which the phenolate group is replaced by naphtholate, has been successfully incorporated into a 3D diamondoid network when used in combination with 4,4′-bipyridine, although this has only been isolated as a centrosymmetric structure using the racemic ligand [91]. The aspartate-based ligand sAsp could reasonably be expected to be more fruitful for the formation of 3D networks given the greater number of coordination sites; however, reports of this ligand reveal only 1D polymers, including one containing a Zn_5 cluster that arises from the multiple functional groups on the ligand, particularly the ability of the phenolate to promote cluster formation [92, 93]. The glutamate derivative has been reported in a 1D chiral tube $[Ni(HsGlu)(H_2O)_2]$ that contains a helical arrangement of noncoordinated water molecules [94].

Another major class of ligand, also derived from Schiff base chemistry, is methylpyridyl substituted amino acids. Whilst the salicaldehyde derivatives tend to promote the formation of clusters or low dimensionality networks due to the position of the phenolate group, pyridyl substituents have a greater tendency to form polymeric structures when the nitrogen atom is located in the meta- or para-position. The group of Vittal reported the first of this class of ligand, *N*-(2-pyridylmethyl)-L-alanine (2pAla – abbreviations for this ligand class indicate the amino acid and the substitution position of the methylpyridyl substituent, see Figure 7.5), in a 1D coordination polymer [95]. With a substitution pattern that affords more divergent coordination sites, the 4pSer ligand has been used in the construction of a 2D network, $[Cu_2(4pSer)_2Cl_2]_n$, that packs in the solid state to afford 1D channels [96]. The serine-derived part of this ligand acts as a tridentate chelate to one metal ion with bridging to adjacent metals through the pyridyl group and the carboxylate not involved in bridging (unlike in many networks containing amino acids and their derivatives).

This material has been used as a heterogeneous catalyst for Grignard reactions and displays good to excellent ee% values for a number of substrates.

The 4pVal ligand has been used by Bannerjee and coworkers to form a series of chiral 3D networks with a zeolite-type topology [97]. Members of this series that contain a large water content within the helical channels are observed to display high proton conductivity. The threonine analogue contains a more polar channel environment, displaying the virtue of amino acids as synthons with the ability to retain structure whilst altering physical behaviour by altering the side chain [98]. Use of the 4-methylpyridyl derivatives of aspartate (4pAsp) and glutamate (4pGlu, Figure 7.3) would be expected to have more success in the formation of 3D networks, yet reports of these ligands still show the formation of 1D, 2D and 3D coordination polymers [99]. In combination with cadmium, both 4pAsp and 4pGlu form three-dimensional networks; in the case of 4pAsp the ligand forms a tridentate chelate through the amino acid with the pyridyl moiety then bridging to another metal, whereas 4pGlu forms only a bidentate chelate with both the pyridyl and second carboxylate groups able to coordinate to adjacent metals. Additional coordination sites can be introduced to this class of ligand by replacing the pyridyl group with pyrimidyl, which has been reported in 2D and 3D networks [100, 101].

Whilst the examples above represent the classes of ligands that have the largest body of work in the literature, there are many other ways in which additional functionality can be added to amino acids. Some noteworthy examples include the use of imidazole-functionalized alanine to form a homochiral 3D network with a topology analogous to the naturally occurring and intrinsically chiral β-SnF_2 phase and interpenetrating chiral diamondoid networks [102, 103]. Additional carboxylate groups can be added by reaction of the amino acid amine group, with a terephthalate-derived ligand used to construct a diamondoid network [104], and an S-carboxymethylcysteine ligand used to form a 3D network with SHG behaviour [105]. A proline-derivative of isophthalic acid, which is a reasonably rigid tricarboxylate species, has been used to great effect to form 3D homochiral network solids that are capable of enantioselective catalysis or enantioselective separations (the latter when used alongside a tris(imidazolyl) co-ligand) [106, 107].

7.3.3.3 Poly-Amino Acid Ligands

The previous section demonstrates that simple derivatives of amino acids can be very useful in the creation of homochiral frameworks. The possible disadvantage, at least in terms of forming potentially porous materials for application in catalysis and separation, is that the ligands are rather 'compact' with the coordinating groups having only a small distance between them, leading to a tendency to form close-packed solids. Rather than simple mono-substitution of an amino acid to add additional coordination sites, it is possible to tether two or more amino acids to a central spacer group to yield polycarboxylate ligands with multiple stereocentres (Figure 7.6).

An early example of this approach towards forming infinite chiral metal-organic networks involved a terephthalate-derived bis(amide) using alanine which was incorporated into a 1D chiral 'nanotube' [108]. A similar ligand, based on thiophenedicarboxylic acid, has been used in a 3D homochiral network with europium [109]. Recently, a series of ligands containing two amino acid substituents have been reported that are based on naphthalene- and perylene-diimide cores [110, 111]. Whilst the individual coordination networks in these cases are

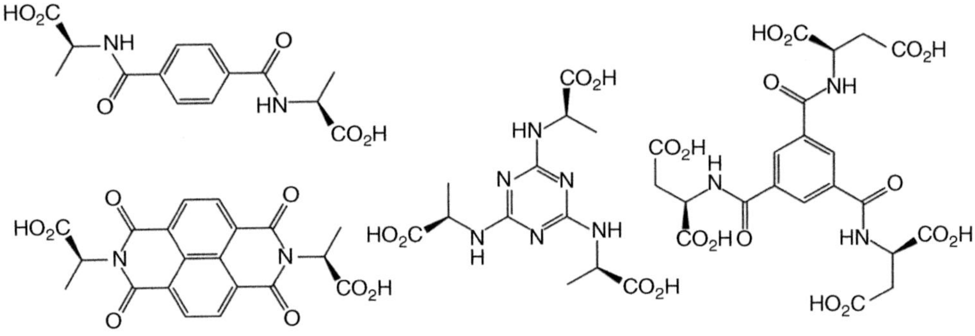

Figure 7.6 *Some examples of poly-amino acid ligands used in the formation of chiral networks*

either 1D or 2D, the large pi-surfaces at the core of the ligands result in interpenetration to form 3D interlocked networks. Two systems, using alanine-substituted naphthalenediimide and phenylalanine-substituted perylenediimide, have been demonstrated to act as stationary phases for the liquid-phase chromatographic resolution of racemic mixtures. These ligands are closely related to amino-acid derived naphthalene monoimides that have been used as co-ligands in the formation of helical 1D coordination polymers or discrete complexes with a large degree of pi-stacking from the naphthalene units, with one such compound showing selective uptake of (S)-ethyl lactate from a racemic vapour mixture [112, 113].

Several chiral frameworks have been reported that incorporate tricarboxylate ligands. Direct attachment of three alanine units to a triazine core yields a tricarboxylate ligand that was used to form a binodal (3,4) connected anionic 3D network [114]. Trimesic acid can also be used as a core group with three alanine units attached by amide linkages and in combination with 4,4′-bipyridine a 3D homochiral network, containing significant pore space, was produced [115]. A very recent example has extended use of the trimesic amide core by appending aspartate groups to give a hexacarboxylate ligand, which is incorporated into a network with small 3D channel systems [116]. Use of a racemic mixture of this ligand gave a different, centrosymmetric 3D structure containing significantly larger pores and suggesting that the chiral networks are intolerant to the presence of the other ligand isomer in the framework lattice.

7.3.3.4 BINOL-Based Networks

Chirality does not originate solely from carbon stereocentres and may be introduced into networks by ligands that contain axially chiral groups. By far the most studied series, particularly by the groups of Lin and Cui, is that which relies on the BINOL (1,1′-bi-2-naphthol) core group, in which axial rotation between the two naphthol units is prohibited.

The first report of homochiral materials based on BINOL was of an isostructural lanthanoid-containing series involving a bisphosphonate ligand with a bis-ethoxy substituted core group [117]. These form as 2D coordination polymers, which pack in a lamellar manner and contain both Brönsted and Lewis acid sites for potential catalytic activity. Catalytic activity was observed, albeit with low enantioselectivity, which demonstrated the potential for this class of material.

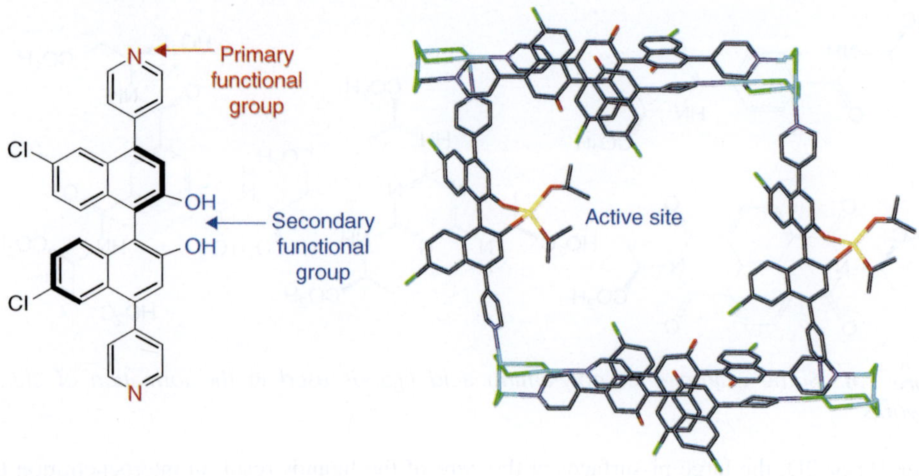

Figure 7.7 *A reactive Ti site can be built into a BINOL-based network to provide eantioselective catalysis [118]. Adapted with permission from J Am Chem Soc,* ***127****, 8940–8941 (2005). Copyright 2005 American Chemical Society*

A 3D network has been reported containing a BINOL-based ligand in which pyridyl substituents are attached so as to form a linear linker with the naphthol OH groups at the centre [118]. This porous material can be postsynthetically reacted with [Ti(O^{i}Pr)$_4$] to give an active catalytic site embedded into the chiral framework with bis(isopropoxide)titanium bound in the binolate pocket (Figure 7.7). The reaction between $ZnEt_2$ and aromatic aldehydes is catalysed with high % ee values that rival the analogous homogeneous catalyst. The TiO^{i}Pr$_2$ catalytic unit has also been incorporated into an isoreticular series of chiral networks using a variety of tetracarboxylate-derived BINOL ligands [119]. Altering the size of the channels within these materials, by varying the size of the ligand, tunes the enantioselectivities of reactions by altering the diffusion rate of the reactants. Whilst the presence of the catalytic site is required for some reactions, other reactions can be conducted with good % ee values simply by virtue of the reactants being confined within a tight chiral channel. A bis(isophthalate)-substituted BINOL was used to construct a 3D homochiral framework in which the photo-induced rearrangement of tropolone ethers proceeds with good enantiomeric excesses [120]. A closely related ligand has been used for asymmetric cyanation of aldehydes displaying high % ee, with the catalytically active unit being a monolithiated binolate [121]. A similar ligand was employed in a 3D network capable of epoxide ring opening with good to high enantioselectivity [122]. Additional functionality can be added to the BINOL group by reaction of the hydroxyl units. For example, it has been shown that a crown ether can be formed at this site to give a potential recognition site [123].

7.3.3.5 Miscellaneous Chiral Ligands

The sections above discuss some the of the major classes of chiral ligands that have been reported although there are of course numerable potential ways in which chirality can be introduced to a network by means of the ligand. A few pertinent examples are given below.

Several prominent groups have used chiral Schiff base ligands to form network solids. The rationale for this strategy comes from the use of similar species in homogeneous chiral catalysis. The first example of this was reported by Hupp and coworkers who embedded a Jacobson-type Mn-salen ligand into a 3D framework containing significant pore space [124]. This material showed comparable % ee values for epoxidation reactions compared to the control homogeneous system and was able to remain active for more cycles, which was postulated to be due to the isolation of catalytic sites from each other, preventing reactive encounters between them. The use of Mn-salen was taken further by Lin and coworkers who synthesized an isoreticular series of homochiral cubic 3D networks, based on the well known MOF-5 system, using salen ligands of differing lengths [125]. As with the previous example, very good % ee values were observed for epoxidation reactions, although the rates for the members of the isoreticular series vary depending on pore size, highlighting that the diffusion of reactants through the pores of a heterogeneous catalyst can be a rate-determining factor.

There are a variety of ways in which chirality can be introduced into a ligand, with perhaps the most conceptually easy route being to embed the chiral centre in the middle of the molecule, allowing synthetic flexibility at the terminus, as seen for BINOL systems and the diaminocyclohexane-derived salen species. Other methods of achieving this goal have been reported with different carbon-centred chiral groups, such as spirodiindane, [126] dioxalane [127] and substituted fluorene [128]. Molecular chirality does not have not have to be centred around a quaternary carbon and a rare example of this in network solids is the use of tris(4-pyridylduryl)borane, which is helically chiral with a propeller-like arrangement minimizing repulsion between the duryl rings [129]. In this case the crystallization did not result in full spontaneous resolution using a racemic ligand, but the crystals were enantio-enriched with one-handedness (approximately 3 : 1). What is interesting about this ligand is that the design is deliberate to induce threefold symmetry in the resulting network solid with the aim of generating strong second-order nonlinear optical behaviour.

4,4′-Bipyridine is a mainstay of coordination polymer synthesis, rigid and predictable in its coordination behaviour, but not chiral; however, enzyme-based synthesis was shown to afford a chirally substituted bipyridine that can be incorporated into 2D frameworks with the same behaviour as its achiral analogue [130]. Decorating a rigid ligand with chiral groups was also shown using 4,4′,4′′-benzene-1,3,5-triyl-tribenzoate appended with oxazolidone substituents immediately adjacent to the carboxylate functionalities in a framework able to act as a Lewis acid catalyst by virtue of free metal coordination sites after framework desolvation [131].

Whilst much focus here has involved ligands using pyridyl or carboxylate coordinating functionalities, reflecting the larger coordination polymer literature, there are other groups that have been incorporated. Bis(triazole) ligands, in which the triazole rings are substituted with chiral substituents, have been used in a 3D silver framework, with the long ligand providing stable pores capable of vapour exchange [132]. Cohen and co-workers adopted a novel approach by using a tris(chelate) metal complex as a source of chirality, using chelating ligands with a nitrile group at the 'back' (i.e. diametrically opposed to the chelate ring) [133]. These three nitrile groups of the metalloligand are able to coordinate to silver to form a heterometallic 3D network with a 10,3a topology (*i.e.* an intrinsically chiral topology also containing either Δ or Λ metal centres). A somewhat more unusual example is the use of quitenine, a ligand derived from an antimalarial agent, which coordinates through carboxylate, quinine and tertiary alkylamine units to form a 3D network [56]. Whilst prediction of this network would be difficult, the resulting material does show enantioselective uptake of 2-butanol.

7.3.4 Post-Synthetic Modification

The main focus of this chapter is the construction of materials that contain a chiral network, either by intrinsic topology or by inclusion of chiral features in the backbone of the framework. However, there has been research exploring the internal functionalization of well known achiral frameworks with pendant chiral groups so that they may be useful for either catalysis or enantioselective separation.

The group of Cohen, one of the pioneers of post-synthetic modification (PSM), used amine-substituted IRMOF-3 (poly-[$Zn_4O(NH_2$-bdc$)_3$], where NH_2-bdc = 2-amino-1,4-benzenedicarboxylate), which they reacted with a chiral anhydride to introduce chiral substituents within the high-symmetry framework [134]. Telfer and co-workers have incorporated proline substituents into an expanded IRMOF framework by first synthesizing a biphenyl-4,4′-dicarboxylate ligand containing a BOC-protected proline substituent [135]. This ligand can be used to form the anticipated α-Po network, containing chiral struts, and later the proline can be thermally deprotected to leave a catalytically active site (Figure 7.8).

The Kim group used the MIL-101 (poly-[$Cr_3(OH)(H_2O)_2O$(bdc$)_3$] where bdc = benzene-1,4-dicarboxylate) framework with a different PSM approach. The MIL-101 system contains coordinated solvent in the framework, which was replaced with a pyridyl-substituted proline ligand through coordination with the pyridyl group [136]. The proline unit was chosen as it is well known for its catalytic activity. The proline-substituted MIL-101 system displayed catalytic activity for the synthesis of chiral aldols with fair to good enantioselectivity. Proline has also been introduced into the MIL-68 (poly-[V(OH)(bdc)]) framework using a solid-phase peptide coupling, rather than by ligand replacement, starting with the amino-substituted network [137].

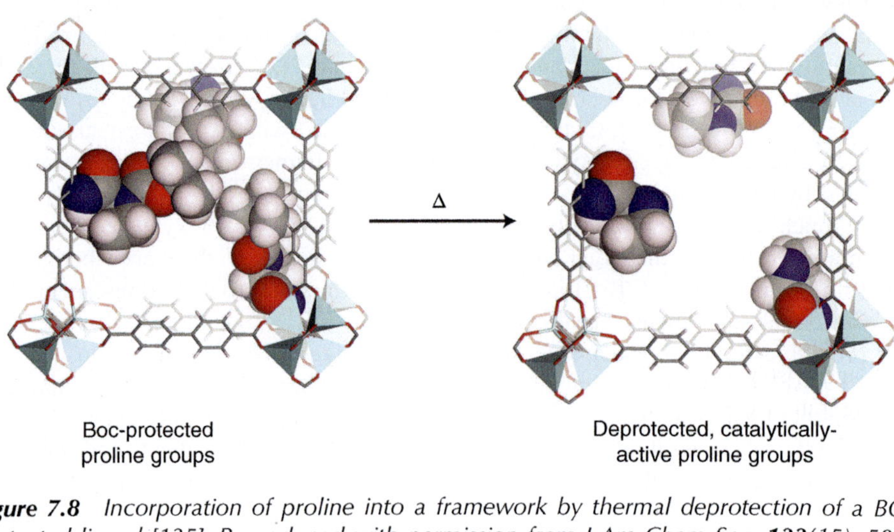

Figure 7.8 Incorporation of proline into a framework by thermal deprotection of a BOC-protected ligand [135]. Reproduced with permission from J Am Chem Soc, **133**(15), 5806–5809 (2011). Copyright 2011 American Chemical Society

7.4 Applications of Chiral Coordination Polymers

The preceding section explored the various methods that have been adopted in the design and synthesis of chiral metal-organic networks. Whilst the applications of these materials have been discussed alongside the myriad of approaches, these uses are collectively summarized below.

7.4.1 Enantioselective Catalysis

One of the dominant applications for which chiral network solids are being investigated is for use as heterogeneous catalysts with several recent reviews and perspectives devoted to this topic [138–141]. Whilst many chiral catalysts are known that operate homogenously, there are many advantages to heterogeneous systems, not least being the ease of separating the catalyst from the reaction mixture. There is a considerable effort expended in the design of catalytically active materials. Two generalized approaches towards this goal can be discerned; embedding a known catalytic group within the framework (either organic or inorganic) or using the size and shape of the pore space to dictate the stereoselectivity of a reaction. Indeed, the combination of these two effects can lead to highly size- and stereo-selective reactions at a chiral catalysis site [119, 125]. The notion of using metal-organic materials for catalysis is in general challenging as the materials must be robust enough to tolerate any reactivity of the substrates and the conditions used.

The use of chiral networks for catalysis was demonstrated early by Kim and coworkers with their seminal paper reporting a modest enantioselectivity in a transesterification reaction attributed to the presence of a noncoordinating pyridyl group that is accessible to species with the channels [56]. Since this initial work there have been reports of numerous types of enantioselective reactions catalysed using chiral network solids including cyanosilylation reactions [33], Grignard reactions [96], ketone hydrogenation [142], sulfoxide oxidation [72, 143], aldol reactions [131, 136], epoxidation [124] and ring opening of epoxides [42].

Although there must be chirality associated with the network material in order for enantioselective catalysis, there are different ways in which this can be achieved. Perhaps the most direct method is to embed chiral catalytic units within the framework, imparting chirality on the network itself at the same time as providing the active sites. Examples of this are the incorporation of a Jacobson's-type salen catalyst and a ruthenium BINAP (2,2′-bis(diphenylphosphino)-1,1′-binaphthyl) site into porous materials which, in both instances, displayed activity comparable to that of the homogeneous analogues [124, 142]. Similar systems can be achieved by placing a reactive chiral site within the as-synthesized network and postsynthetically altering this to be catalytically active, as with the BINOL-containing frameworks that can be reacted with titanium(IV) isopropoxide to give an active titanium binolate [119]. Chiral co-ligands can also be introduced postsynthetically to give to catalytically active sites [136]. An alternative and less predictable method is to add catalytic sites that are not themselves chiral, either organic functionalities or accessible metal sites whose immediate environment is achiral but promote the reaction within the chiral environment provided by the network [33, 56].

7.4.2 Enantioselective Separations

The challenge of chiral catalysis is often touted as the major use of chiral solids given that asymmetric synthesis is a major hurdle in many organic and pharmaceutical syntheses. However, the potential use of such materials for enantioselective separations, both preparative and analytical, should not be understated. Towards this end there has been a body of work exploring chromatographic applications which has recently been reviewed [144, 145]. It is worth noting that there are two distinct ways in which the separation of enantiomers using porous materials is reported. The majority of reports are of sorption into a solid by soaking the porous material in a solution of the racemate, proving that there is an effect and highlighting the viability of materials for separation [146, 147]. There are considerably fewer reports in which the materials have been used as a solid phase within either liquid- or gas-chromatographic separations.

The first report of enantioselective sorption was in the same paper in which Kim first reported enantioselective catalysis, where the selective uptake of Δ-$[Ru(2,2'\text{-bipy})_3]^{2+}$ was demonstrated with 66% ee [56]. Shortly after this, the first instances were reported of the selective sorption of small organic molecules with uptake of 2-butanol in a quitenine-based framework with good selectivity and inclusion of trans-1,2-diaminocyclohexane into a BINOL-containing network showing modest enantio-enrichment of the solution [117, 148]. As with catalytic applications, it is clear that there needs to be complementarity between the size of the guest species and the size of the channels within the network. It is obvious that guests will not penetrate into pores that are of smaller dimensions than themselves. However it is also necessary that the pores are not too large or else the chirality of the channel cannot exert its influence sufficiently to give enantioselective sorption, as demonstrated by Rosseinsky and coworkers [36]. Selective sorption can be of use in catalytic systems, with a demonstration by Kim and colleagues that a catalytic homochiral material that gave poor asymmetric induction was able to yield an enantio-enriched solution of the product by selective sorption of one isomer. Although much emphasis has been placed on the need for solvent-accessible channels for sorption to occur, a recent report suggests that surface activity of metal-organic materials may also play a role. A material based on amino acid diimides showed selective, yet low weight %, sorption of analytes and was subsequently demonstrated to act as a solid-phase support for chiral resolution in small-scale liquid chromatographic separation tests [111].

The use of homochiral metal-organic networks in chromatography has only been reported in the last few years. In 2007, the framework $[Zn_2(bdc)(\text{L-lac})(DMF)]$ (bdc = 1,4-benzenedicarboxylate, lac = lactate) was shown to act as a stationary phase in a preparative chromatographic separation to give baseline separation of sulfoxides [143]. More recently this material has been embedded in a membrane matrix with the resulting composite also shown to display good separation [149]. The first report of the application of a chiral metal-organic framework for gas chromatography was by the Yuan group [89]. For GC applications it is necessary that the chiral material is thermally stable and that the carrier gas will not cause degradation. The open-framework material demonstrated excellent separation for a number of racemic analytes whereas a closely related 1D coordination polymer, containing no accessible pore space, did not. Since this initial report there have been other examples from the same group of thermally stable homochiral materials used in GC separations, primarily based on camphorate networks [67–69]. These results suggest great promise for chiral network solids as chromatographic separation agents, although the field remains in its infancy.

Secondary Harmonic Generation and Electric Properties Nonlinear optical (NLO) materials are those for which the dielectric polarization has a nonlinear response to the electric field of incident light. The most common NLO behaviour is secondary harmonic generation (SHG) in which two photons of the same energy are effectively combined to produce a photon with twice the energy, and therefore half the wavelength, of the incident radiation. The only materials that can display SHG behaviour are those with no inversion symmetry, and hence homochiral materials are ideal candidates and are summarized well in a recent review [150]. One of the prototypical SHG materials is α-quartz which has a noncentrosymmetric diamondoid topology due to nonsymmetric bridges between the 4-connecting silicon nodes.

The prime candidates for metal-organic SHG materials are those that are optically transparent and therefore d^{10} metal ions or main group elements are most commonly employed. There is a significant body of work in the area on noncentrosymmetric materials that are not chiral and instead rely on removing inversion by using nonsymmetric ligands or fortuitous crystallization in a noncentrosymmetric space group. Give that homochiral materials must crystallize in such space groups, the use of the synthetic strategies given in section 7.3 are very applicable towards creating SHG active networks. Examples of network solids that display SHG behaviour are two camphorate-based materials [62] and a cysteine-derived framework [105]. Use of metal-organic systems, rather than purely organic materials, can enhance the hyperpolarizability and hence increase the SHG response, as demonstrated by the incorporation of dinitrotyrosine into a 3D lead-based network [151]. There are many other examples of lower dimensionality coordination polymers or those obtained as racemic materials, although these are outside of the scope of this chapter [150].

Another set of properties that can be present in noncentrosymmetric (not necessarily chiral) crystalline materials are electric properties, particularly ferroelectric and piezoelectric behaviour. Piezoelectric materials accumulate a charge and polarization in response to a mechanical stress being applied whilst ferroelectric materials, such as Rochelle salt (potassium sodium tartrate), are a subset of this have a spontaneous electric polarization in the absence of an applied field and the application of a field can reverse the polarization. Research in these areas is mostly confined to inorganic salts, with limited reports of coordination-based materials, possibly due to the requirement of large single crystals for exploring the electric behaviour. Close-packed coordination polymers such as tartrate, formate and glycinate materials tend to make up the majority of cases, with accentricity arising from the space group in many instances [152]. One of the first reports of a chiral 3D metal-organic network to display piezoelectric and ferroelectric effects, as well as SHG, was made using the herbicide imazethapyr [153]. More recently, a 3D camphorate framework has been reported that displays a stable piezoelectric effect over a wide temperature range [154]. As with SNG behaviour, 3D homochiral materials are not required for these properties yet provide a large scope due to the semipredictable manner in which they can be created and the field is expected to grow.

7.5 Summary and Outlook

Research into the design, synthesis and application of chiral network solids, particularly homochiral porous coordination polymers, is very much an active area of endeavour. There remains a significant challenge in forming bulk homochiral materials, with many of the inventive ways to do so being highlighted in this chapter. Perhaps the area with the most growth potential is that involving the design of bespoke ligands to engineer functional sites within materials. Such activities are largely fuelled by interests in enantioselective catalysis and separations, both of which are of potential industrial application, and the pace at which these areas are growing in the past few years shows no sign of abating.

References

[1] L. Pérez-García and D.B. Amabilino, Spontaneous resolution under supramolecular control, *Chem Soc Rev*, **31**(6), 342–356 (2002).

[2] J.M. McBride and R.L. Carter, Spontaneous resolution by stirred crystallization, *Angew Chem Int Ed Engl*, **30**(3), 293–295 (1991).

[3] D.K. Kondepudi and K. Asakura, Chiral autocatalysis, spontaneous symmetry breaking, and stochastic behavior, *Acc Chem Res*, **34**, 946–954 (2001).

[4] D.K. Kondepudi, R.J. Kaufman and N. Singh, Chiral symmetry breaking in sodium chlorate crystallization, *Science*, **250**, 975–976 (1990).

[5] J. Yu and R. Xu, Chiral zeolitic materials: Structural insights and synthetic challenges, *J Mater Chem*, **18**, 4021–4030 (2008).

[6] M.E. Davis and R.F. Lobo, Zeolite and molecular sieve synthesis, *Chem Mater*, **4**, 756–768 (1992).

[7] J.M. Newsam, M.M.J. Treacy, W.T. Koetsier and C.B.d. Gruyter, Structural characterisation of zeolite beta, *Proc R Soc Lond A*, **420**, 375–405 (1988).

[8] L. Tang, L. Shi, C. Bonneau, *et al.*, A zeolite family with chiral and achiral structures built from the same building layer, *Nature Mater*, **7**, 381–385 (2008).

[9] J. Sun, C. Bonneau, Á. Cantín, *et al.*, The itq-37 mesoporous chiral zeolite, *Nature*, **458**, 1154–1158 (2009).

[10] T.E. Gier, X. Bu, P. Feng and G.D. Stucky, Synthesis and organization of zeolite-like materials with three-dimensional helical pores, *Nature*, **395**, 154–157 (1998).

[11] R.C. Rouse and D.R. Peacor, Crystal structure of the zeolite mineral goosecreekite, $CaAl_2Si_6O_{16}{\cdot}5H_2O$, *Am Mineral*, **71**, 1494–1501 (1986).

[12] Y. Li, J. Yu, Z. Wang, *et al.*, Design of chiral zeolite frameworks with specified channels through constrained assembly of atoms, *Chem Mater*, **17**, 4399–4405 (2005).

[13] X. Song, Y. Li, L. Gan, Z. Wang, *et al.*, Heteroatom-stabilized chiral framework of aluminophosphate molecular sieves, *Angew Chem Int Ed*, **48**, 314–317 (2009).

[14] J. Zhang, S. Chen and X. Bu, Nucleotide-catalyzed conversion of racemic zeolite-type zincophosphate into enantioenriched crystals, *Angew Chem Int Ed Engl*, **48**(33), 6049–6051 (2009).

[15] S.M. Stalder and A.P. Wilkinson, Synthesis and characterization of a chiral 3D-framework material: D-$Co(en)_3[H_3Ga_2P_4O_{16}]$, *Chem Mater*, **9**, 2168–2173 (1997).

[16] S.R. Batten, S.M. Neville and D.R. Turner. *Coordination Polymers. Design, Analysis and Application*. Cambridge: RSC Publishing; 2009.

[17] C. Wang, D. Liu and W. Lin, Metal-organic frameworks as a tunable platform for designing functional molecular materials, *J Am Chem Soc*, **135**(36), 13222–13234 (2013).

[18] W. Lin, Homochiral porous metal-organic frameworks: Why and how? *J Solid State Chem*, **178**(8), 2486–2490 (2005).

[19] T. Ezuhara, K. Endo and Y. Aoyama, Helical coordination polymers from achiral components in crystals. Homochiral crystallization, homochiral helix winding in the solid state, and chirality control by seeding, *J Am Chem Soc*, **121**, 3279–3283 (1999).

[20] S.T. Wu, Y.R. Wu, Q.Q. Kang, *et al.*, Chiral symmetry breaking by chemically manipulating statistical fluctuation in crystallization, *Angew Chem Int Ed Engl*, **46**(44), 8475–8479 (2007).

[21] S. Ding, Y. Gao, Y. Ji, *et al.*, Homochiral crystallization of single-stranded helical coordination polymers: Generated by the structure of auxiliary ligands or spontaneous symmetry breaking, *CrystEngComm*, **15**(28), 5598–5601 (2013).

[22] A. Bhattacharyya, B.N. Ghosh, S. Herrero, *et al.*, Formation of a novel ferromagnetic end-to-end cyanate bridged homochiral helical copper(II) Schiff base complex via spontaneous symmetry breaking, *Dalton Trans*, **44**(2), 493–497 (2015).
[23] S.C. Chen, J. Zhang, R.M. Yu, *et al.*, Spontaneous asymmetrical crystallization of a three-dimensional diamondoid framework material from achiral precursors, *Chem Commun*, **46**(9), 1449–1451 (2010).
[24] F.Y. Yi, J. Zhang, H.X. Zhang and Z.M. Sun, Chiral transformations of achiral porous metal-organic frameworks via a stepwise approach, *Chem Commun*, **48**(84), 10419–10421 (2012).
[25] S.T. Wu, Z.W. Cai, Q.Y. Ye, *et al.*, Enantioselective synthesis of a chiral coordination polymer with circularly polarized visible laser, *Angew Chem Int Ed Engl*, **53**(47), 12860–12864 (2014).
[26] R.E. Morris and X. Bu, Induction of chiral porous solids containing only achiral building blocks, *Nature Chem*, **2**(5), 353–361 (2010).
[27] Z. Lin, A.M.Z. Slawin and R.E. Morris, Chiral induction in the ionothermal synthesis of a 3-D coordination polymer, *J Am Chem Soc*, **129**, 4880–4881 (2007).
[28] J. Zhang, S. Chen, T. Wu, *et al.*, Homochiral crystallization of microporous framework materials from achiral precursors by chiral catalysis, *J Am Chem Soc*, **130**, 12882–12883 (2008).
[29] X. Jing, C. He, D. Dong, *et al.*, Homochiral crystallization of metal-organic silver frameworks: Asymmetric [3+2] cycloaddition of an azomethine ylide, *Angew Chem Int Ed Engl*, **51**(40), 10127–10131 (2012).
[30] R.M. Pagni and R.N. Compton, Asymmetric synthesis of optically active sodium chlorate and bromate crystals, *Cryst Growth Des*, **2**(4), 249–253 (2002).
[31] Y. Kang, S. Chen, F. Wang, *et al.*, Induction in urothermal synthesis of chiral porous materials from achiral precursors, *Chem Commun*, **47**(17), 4950–4952 (2011).
[32] J. Zhang, S. Chen, R.A. Nieto, *et al.*, A tale of three carboxylates: Cooperative asymmetric crystallization of a three-dimensional microporous framework from achiral precursors, *Angew Chem Int Ed Engl*, **49**(7), 1267–1270 (2010).
[33] D. Dang, P. Wu, C. He, *et al.*, Homochiral metal-organic frameworks for heterogeneous asymmetric catalysis, *J Am Chem Soc*, **132**, 14321–14323 (2010).
[34] Y. Wen, T. Sheng, S. Hu, *et al.*, Intercalation of chiral molecules into layered metal-organic frameworks: A strategy to synthesize homochiral mofs, *Chem Commun*, **49**(90), 10644–10646 (2013).
[35] C.J. Kepert, T.J. Prior and M.J. Rosseinsky, A versatile family of interconvertible microporous chiral molecular frameworks: The first example of ligand control of network chirality, *J Am Chem Soc*, **122**, 5158–5168 (2000).
[36] D. Bradshaw, T.J. Prior, E.J. Cussen, *et al.*, Permanent microporosity and enantioselective sorption in a chiral open framework, *J Am Chem Soc*, **126**, 6106–6114 (2004).
[37] I. Imaz, M. Rubio-Martinez, J. An, *et al.*, Metal-biomolecule frameworks (MBIOFs), *Chem Commun*, **47**(26), 7287–7302 (2011).
[38] X.-L. Yang and C.-D. Wu, Recent advances on porous homochiral coordination polymers containing amino acid synthons, *CrystEngComm*, **16**(23), 4907–4918 (2014).
[39] E.V. Anokhina, Y.B. Go, Y. Lee, *et al.*, Chiral three-dimensional microporous nickel aspartate with extended Ni-O-Ni bonding, *J Am Chem Soc*, **128**, 9957–9962 (2006).
[40] Y. Zhang, M.K. Saha and I. Bernal, [cobalt(II)L-glutamate$(H_2O)\cdot H_2O]_\infty$: A new 3D chiral metal–organic interlocking network with channels, *CrystEngComm*, **5**(5), 34–37 (2003).
[41] R. Vaidhyanathan, D. Bradshaw, J.N. Rebilly, *et al.*, A family of nanoporous materials based on an amino acid backbone, *Angew Chem Int Ed Engl*, **45**(39), 6495–6499 (2006).
[42] M.J. Ingleson, J.P. Barrio, J. Bacsa, *et al.*, Generation of a solid Brønsted acid site in a chiral framework, *Chem Commun*, (11), 1287–1289 (2008).
[43] J.P. Barrio, J.N. Rebilly, B. Carter, *et al.*, Control of porosity geometry in amino acid derived nanoporous materials, *Chem Eur J*, **14**(15), 4521–4532 (2008).
[44] M.P. Yutkin, M.S. Zavakhin, D.G. Samsonenko, *et al.*, Porous homo and heterochiral cobalt(II) aspartates with high thermal stability of the metal organic framework, *Russ Chem Bull, Int Ed*, **59**(4), 733–740 (2010).
[45] A.C. Kathalikkattil, K.K. Bisht, N. Aliaga-Alcalde and E. Suresh, Synthesis, magnetic properties, and structural investigation of mixed-ligand Cu(II) helical coordination polymers with an

amino acid backbone and N-donor propping: 1-D helical, 2-D hexagonal net (hcb), and 3-D ins topologies, *Cryst Growth Des*, **11**(5), 1631–1641 (2011).

[46] P. Zhu, W. Gu, F.-Y. Cheng, *et al.*, Design of two 3d homochiral Co(II) metal–organic open frameworks by layered-pillar strategy: Structure and properties, *CrystEngComm*, **10**(8), 963–967 (2008).

[47] Z. Kang, M. Xue, L. Fan, *et al.*, Highly selective sieving of small gas molecules by using an ultra-microporous metal–organic framework membrane, *Energy Environ Sci*, **7**(12), 4053–4060 (2014).

[48] Q. Feng, M.-J. Yan, H.-H. Song and S.-K. Shi, Self-assembly of two chiral zinc coordination polymers based on N-benzoyl-L-glutamic acid: Influence of different N-donor ancillary ligands on structural arrangement and photoluminescent properties, *Inorg Chim Acta*, **415**, 75–80 (2014).

[49] R. He, H.-H. Song, Z. Wei, *et al.*, Coordination polymers with the chiral ligand N-p-tolylsulfonyl-L-glutamic acid: Influence of metal ions and different bipyridine ligands on structural chirality, *J Solid State Chem*, **183**(9), 2021–2026 (2010).

[50] Q. Feng, M. Yan, L. Wu, *et al.*, An unprecedented chiral porous 3D metal–organic framework self-assembled with 1D helical and double chain, *Inorg Chem Commun*, **43**, 1–4 (2014).

[51] J. Guo, X. Li, C.N. Moorefield and G.R. Newkome, Novel twofold interpenetrating channel framework based on metal tetramer subunits, *Cryst Growth Des*, **11**(7), 2695–2697 (2011).

[52] M.J. Ingleson, J. Bacsa and M.J. Rosseinsky, Homochiral H-bonded proline based metal organic frameworks, *Chem Commun*, (29), 3036–3038 (2007).

[53] H.Y. An, E.B. Wang, D.R. Xiao, *et al.*, Chiral 3D architectures with helical channels constructed from polyoxometalate clusters and copper-amino acid complexes, *Angew Chem Int Ed Engl*, **45**(6), 904–908 (2006).

[54] B. Wisser, Y. Lu and C. Janiak, Chiral coordination polymers with amino acids, *Z Anorg Allg Chem*, **633**(8), 1189–1192 (2007).

[55] S. Mendiratta, M. Usman, T.-T. Luo, *et al.*, Anion-controlled dielectric behavior of homochiral tryptophan-based metal–organic frameworks, *Cryst Growth Des*, **14**(4), 1572–1579 (2014).

[56] J.S. Seo, D. Whang, H. Lee, *et al.*, A homochiral metal-organic porous material for enantioselective separation and catalysis, *Nature*, **404**, 982–986 (2000).

[57] J. Zhang, S. Chen, A. Zingiryan and X. Bu, Integrated molecular chirality, absolute helicity, and intrinsic chiral topology in three-dimensional open-framework materials, *J Am Chem Soc*, **130**, 17246–17247 (2008).

[58] J. Zhang and X. Bu, Temperature dependent charge distribution in three-dimensional homochiral cadmium camphorates, *Chem Commun*, (4), 444–446 (2008).

[59] M.-H. Zeng, B. Wang, X.-Y. Wang, *et al.*, Chiral magnetic metal-organic frameworks of dimetal subunits: Magnetism tuning by mixed-metal compositions of the solid solutions, *Inorg Chem*, **45**, 7069–7076 (2006).

[60] J. Zhang, Y.-G. Yao and X. Bu, Comparative study of homochiral and racemic chiral metal-organic frameworks built from camphoric acid, *Chem Mater*, **19**, 5083–5089 (2007).

[61] D.N. Dybtsev, M.P. Yutkin, E.V. Peresypkina, *et al.*, Isoreticular homochiral porous metal-organic structures with tunable pore sizes, *Inorg Chem*, **46**, 6843–6845 (2007).

[62] X.-Q. Liang, D.-P. Li, X.-H. Zhou, *et al.*, Metal – organic coordination polymers generated from chiral camphoric acid and flexible ligands with different spacer lengths: Syntheses, structures, and properties, *Cryst Growth Des*, **9**(11), 4872–4883 (2009).

[63] J. Zhang and X. Bu, Absolute helicity induction in three-dimensional homochiral frameworks, *Chem Commun*, (2), 206–208 (2009).

[64] J. Zhang, S. Chen, H. Valle, *et al.*, Manganese and magnesium homochiral materials: Decoration of honeycomb channels with homochiral chains, *J Am Chem Soc*, **129**(46), 14168–14169 (2007).

[65] J.A. Rood, W.C. Boggess, B.C. Noll and K.W. Henderson, Assembly of a homochiral, body-centered cubic network composed of vertex-shared Mg_{12} cages: Use of electrospray ionization mass spectrometry to monitor metal carboxylate nucleation, *J Am Chem Soc*, **129**, 13675–13682 (2007).

[66] M.L. Sun, J. Zhang, Q.P. Lin, *et al.*, Multifunctional homochiral lanthanide camphorates with mixed achiral terephthalate ligands, *Inorg Chem*, **49**(20), 9257–9264 (2010).

[67] S.-M. Xie, X.-H. Zhang, Z.-J. Zhang, *et al.*, A 3-D open-framework material with intrinsic chiral topology used as a stationary phase in gas chromatography, *Anal Bioanal Chem*, **405**(10), 3407–3412 (2013).
[68] S.-M. Xie, X.-H. Zhang, Z.-J. Zhang and L.-M. Yuan, Porous chiral metal-organic framework $InH(D\text{-}C_{10}H_{14}O_4)_2$ with anionic-type diamond network for high-resolution gas chromatographic enantioseparations, *Anal Lett*, **46**(5), 753–763 (2013).
[69] S. Xie, B. Wang, X. Zhang, *et al.*, Chiral 3D open-framework material Ni(D-cam)$(H_2O)_2$ used as GC stationary phase, *Chirality*, **26**(1), 27–32 (2014).
[70] Z.R. Qu, Q. Ye, H. Zhao, *et al.*, Homochiral laminar europium metal-organic framework with unprecedented giant dielectric anisotropy, *Chem Eur J*, **14**(11), 3452–3456 (2008).
[71] D.N. Dybtsev, A.L. Nuzhdin, H. Chun, *et al.*, A homochiral metal-organic material with permanent porosity, enantioselective sorption properties, and catalytic activity, *Angew Chem Int Ed Engl*, **45**(6), 916–920 (2006).
[72] D.N. Dybtsev, M.P. Yutkin, D.G. Samsonenko, *et al.*, Modular, homochiral, porous coordination polymers: Rational design, enantioselective guest exchange sorption and ab initio calculations of host-guest interactions, *Chemistry*, **16**(34), 10348–10356 (2010).
[73] M.-H. Zeng, S. Gao and X.-M. Chen, A novel 3D coordination polymer containing pentacoordinate cobalt(II) dimers with antiferromagnetic ordering, *Inorg Chem Commun*, **7**(7), 864–867 (2004).
[74] L.-M. Duan, F.-T. Xie, X.-Y. Chen, *et al.*, Syntheses, structures, and magnetic properties of three novel metal-malate-bipyridine coordination polymers with layered and pillared topology, *Cryst Growth Des*, **6**(5), 1101–1106 (2006).
[75] A. Zingiryan, J. Zhang and X. Bu, Cooperative self-assembly of chiral l-malate and achiral succinate in the formation of a three-dimensional homochiral framework, *Inorg Chem*, **47**, 8607–8609 (2008).
[76] C.M. Nagaraja, R. Haldar, T.K. Maji and C.N.R. Rao, Chiral porous metal–organic frameworks of Co(II) and Ni(II): Synthesis, structure, magnetic properties, and CO_2 uptake, *Cryst Growth Des*, **12**(2), 975–981 (2012).
[77] J.A. Rood, B.C. Noll and K.W. Henderson, Homochiral frameworks derived from magnesium, zinc and copper salts of L-tartaric acid, *J Solid State Chem*, **183**(1), 270–276 (2010).
[78] Z. Amghouz, S. Garcia-Granda, J.R. Garcia, *et al.*, Series of metal organic frameworks assembled from Ln(III), Na(I), and chiral flexible-achiral rigid dicarboxylates exhibiting tunable UV-vis-IR light emission, *Inorg Chem*, **51**(3), 1703–1716 (2012).
[79] B.F. Abrahams, M. Moylan, S.D. Orchard and R. Robson, Zinc saccharate: A robust, 3D coordination network with two types of isolated, parallel channels, one hydrophilic and the other hydrophobic, *Angew Chem Int Ed*, **42**, 1848–1851 (2003).
[80] R.J. Flook, H.C. Freeman, C.J. Moore and M.L. Scudder, Model compounds for metal-protein interaction: Crystal structures of seven cadmium(II) complexes of amino-acids and peptides, *J Chem Soc, Chem Commun*, 573–574 (1973).
[81] R. Ferrari, S. Bernés, C. Barbarín, *et al.*, Interaction between glyglu and Ca^{2+}, Pb^{2+}, Cd^{2+} and Zn^{2+} in solid state and aqueous solution. Crystal structures of poly[aqua-1,2-κ-O-di[lead(glygluH)]bis(perchlorate)] and poly[bisglycylglutamic-cadmium(II) tetrahydrate], *Inorg Chim Acta*, **339**, 193–201 (2002).
[82] J. Rabone, Y.-F. Yue, S.Y. Chong, *et al.*, An adaptable peptide-based porous material, *Science*, **329**, 1053–1057 (2010).
[83] H.-Y. Lee, J.W. Kampf, K.S. Park and E.N.G. Marsh, Covalent metal-peptide framework compounds that extend in one and two dimensions, *Cryst Growth Des*, **8**(1), 296–303 (2008).
[84] L. Lin, R. Yu, W. Yang, *et al.*, A series of chiral metal–organic frameworks based on oxalyl retro-peptides: Synthesis, characterization, dichroism spectra, and gas adsorption, *Cryst Growth Des*, **12**(6), 3304–3311 (2012).
[85] L. Lin, R. Yu, X.-Y. Wu, *et al.*, Enantioselective inclusion of alcohols by solvent-controlled assembled flexible metal–organic frameworks, *Inorg Chem*, **53**(10), 4794–4796 (2014).
[86] A. Nakahara, The preparation and properties of *N*-salicylideneglycinato-aqua-copper(II), sodium *N*-salicylideneglycylglycinato-cuprate(II) and related compounds, *Bull Chem Soc Jpn*, **32**(11), 1195–1199 (1959).

[87] J.D. Ranford, J.J. Vittal and D. Wu, Topochemical conversion of a hydrogen-bonded three-dimensional network into a covalently bonded framework, *Angew Chem Int Ed*, **37**(8), 1114–1116 (1998).

[88] J.D. Ranford, J.J. Vittal, D. Wu and X. Yang, Thermal conversion of a helical coil into a three-dimensional chiral framework, *Angew Chem Int Ed*, **38**(23), 3498–3501 (1999).

[89] S.M. Xie, Z.J. Zhang, Z.Y. Wang and L.M. Yuan, Chiral metal-organic frameworks for high-resolution gas chromatographic separations, *J Am Chem Soc*, **133**(31), 11892–11895 (2011).

[90] B.-Y. Lou, D.-Q. Yuan, S.-Y. Gao, *et al.*, A chiral supramolecular architecture $[Cu_2(4,4'$-bipyridine$)_2(sala)_2]_n \cdot 4.5H_2O$ (sala=n-(2-hydroxybenzyl)-L-alanine anion), *J Mol Struc*, **707**(1–3), 231–234 (2004).

[91] Z.Z. Li, L. Du, J. Zhou, *et al.*, Rational design of two bpy-bridged 3D and 2D Co(II) open frameworks with similar amino-acid-based schiff bases, *Dalton Trans*, **41**(47), 14397–14403 (2012).

[92] L. Jia, N. Tang and J.J. Vittal, Zinc(II) and copper(II) 1D coordination polymeric complexes of a reduced schiff base ligand, *Inorg Chim Acta*, **362**(8), 2525–2528 (2009).

[93] B. Sreenivasulu and J.J. Vittal, Hydrogen-bonded copper(II) and nickel(II) complexes and coordination polymeric structures containing reduced schiff base ligands, *Inorg Chim Acta*, **362**(8), 2735–2743 (2009).

[94] B. Sreenivasulu and J.J. Vittal, Helix inside a helix: Encapsulation of hydrogen-bonded water molecules in a staircase coordination polymer, *Angew Chem Int Ed Engl*, **43**(43), 5769–5772 (2004).

[95] X. Wang and J.J. Vittal, Nature of the reactants and influence of water on the supramolecular assembly, *Inorg Chem*, **42**, 5135–5142 (2003).

[96] M. Wang, M.H. Xie, C.D. Wu and Y.G. Wang, From one to three: A serine derivate manipulated homochiral metal-organic framework, *Chem Commun*, (17), 2396–2398 (2009).

[97] S.C. Sahoo, T. Kundu and R. Banerjee, Helical water chain mediated proton conductivity in homochiral metal-organic frameworks with unprecedented zeolitic unh-topology, *J Am Chem Soc*, **133**(44), 17950–17958 (2011).

[98] T. Kundu, S.C. Sahoo and R. Banerjee, Relating pore hydrophilicity with vapour adsorption capacity in a series of amino acid based metal organic frameworks, *CrystEngComm*, **15**(45), 9634–9640 (2013).

[99] X.-L. Yang, M.-H. Xie, C. Zou, *et al.*, A series of metal–organic coordination polymers containing multiple chiral centers, *CrystEngComm*, **13**(5), 1570–1579 (2011).

[100] P. Arranz-Mascarós, M.L. Godino-Salido, R. Lo´pez-Garzón, *et al.*, Coordination modes of N-2-(4-amino-1-methyl-5-nitroso-6-oxo-1,6-dihydropyrimidinyl) potassium L-alaninate toward Ag(I), Cd(II) and Pd(II). Crystal structure of $\{Ag((\eta^4\text{-}\mu_3\text{-}L)] \cdot 3H_2O\}$, first example of a silver 3D-polymer with a pyrimidine-derivative ligand, *Polyhedron*, **18**, 793–797 (1999).

[101] P. Arranz-Mascarós, M.D. Gutiérrez-Valero, R. López-Garzón, M.D. *et al.*, Study of protonation and Zn(II), Cd(II), Cu(II) and Mn(II) complexation with a glutamic acid n-pyrimidine derivative: Crystal structure of a neutral Cd(II) complex of the bianionic ligand, *Polyhedron*, **27**(2), 623–632 (2008).

[102] J. Qin, C. Qin, X.L. Wang, *et al.*, Metal-organic replica of chiral natural material beta-SnF(2): An enantiopure binodal (3,5)-connected net based on triangular and trigonal bipyramidal coordination, *Chem Commun*, **46**(4), 604–606 (2010).

[103] G. Zhang, S.-Y. Yao, D.-W. Guo and Y.-Q. Tian, Noncentrosymmetric and homochiral metal-organic frameworks of (s)-2-(1H-imidazole1-yl) propionic acid, *Cryst Growth Des*, **10**, 2355–2359 (2010).

[104] H.Y. Lee, J. Park, M.S. Lah and J.-I. Hong, One-dimensional double helical structure and 4-fold type [2+2] interpenetration of diamondoid networks with helical fashion, *Cryst Growth Des*, **8**(2), 587–591 (2008).

[105] W. Xu, W. Liu, F.-Y. Yao and Y.-Q. Zheng, Synthesis, crystal structure and properties of the novel chiral 3D coordination polymer with s-carboxymethyl-L-cysteine, *Inorg Chim Acta*, **365**, 297–301 (2011).

[106] Z.X. Xu, Y.X. Tan, H.R. Fu, *et al.*, Homochiral metal-organic frameworks with enantiopure proline units for the catalytic synthesis of beta-lactams, *Inorg Chem*, **53**(22), 12199–12204 (2014).

[107] Z.-X. Xu, Y.-X. Tan, H.-R. Fu, *et al.*, Integration of rigid and flexible organic parts for the construction of a homochiral metal–organic framework with high porosity, *Chem Commun*, **51**, 2565–2568 (2015).

[108] H.-T. Zhang, Y.-Z. Li, T.-W. Wang, *et al.*, A znii-based chiral crystalline nanotube, *Eur J Inorg Chem*, **2006**(17), 3532–3536 (2006).

[109] J. Liu, M. Wang, Z.-W. Zhang, G.-M. Xia, *et al.*, Hydrothermal synthesis, crystal structure and properties of a Eu(III) coordination polymer assembled by (-)-2,2′-[thiophene-2,5-diylbis(carbonylimino)] dipropanoic acid and 4,4′-dipyridine, *Chin J Inorg Chem*, **28**(1), 50–54 (2012).

[110] S.A. Boer, C.S. Hawes and D.R. Turner, Engineering entanglement: Controlling the formation of polycatenanes and polyrotaxanes using pi interactions, *Chem Commun*, **50**(9), 1125–1127 (2014).

[111] S.A. Boer, Y. Nolvachai, C. Kulsing, *et al.*, Liquid-phase enantioselective chromatographic resolution using interpenetrated, homochiral framework materials, *Chem Eur J*, **20**(36), 11308–11312 (2014).

[112] D.L. Reger, J. Horger, M.D. Smith and G.J. Long, Homochiral, helical metal-organic framework structures organized by strong, non-covalent pi-pi stacking interactions, *Chem Commun*, **41**, 6219–6221 (2009).

[113] D.L. Reger, J.J. Horger, A. Debreczeni and M.D. Smith, Syntheses and characterization of copper(II) carboxylate dimers formed from enantiopure ligands containing a strong pi···pi stacking synthon: Enantioselective single-crystal to single-crystal gas/solid-mediated transformations, *Inorg Chem*, **50**(20), 10225–10240 (2011).

[114] Q. Zhu, T. Sheng, R. Fu, C. Tan, *et al.*, Two luminescent enantiomorphic 3D metal-organic frameworks with 3D homochiral double helices, *Chem Commun*, **46**(47), 9001–9003 (2010).

[115] Z. Chen, X. Liu, C. Zhang, *et al.*, Structure, adsorption and magnetic properties of chiral metal-organic frameworks bearing linear trinuclear secondary building blocks, *Dalton Trans*, **40**(9), 1911–1918 (2011).

[116] I. Burneo, K.C. Stylianou, I. Imaz and D. Maspoch, The influence of the enantiomeric ratio of an organic ligand on the structure and chirality of metal-organic frameworks, *Chem Commun*, **50**(89), 13829–13832 (2014).

[117] O.R. Evans, H.L. Ngo and W. Lin, Chiral porous solids based on lamellar lanthanide phosphonates, *J Am Chem Soc*, **123**, 10395–10396 (2001).

[118] C.-D. Wu, A. Hu, L. Zhang and W. Lin, A homochiral porous metal-organic framework for highly enantioselective heterogeneous asymmetric catalysis, *J Am Chem Soc*, **127**, 8940–8941 (2005).

[119] L. Ma, J.M. Falkowski, C. Abney and W. Lin, A series of isoreticular chiral metal-organic frameworks as a tunable platform for asymmetric catalysis, *Nature Chem*, **2**(10), 838–846 (2010).

[120] Y. Peng, T. Gong and Y. Cui, A homochiral porous metal-organic framework for enantioselective adsorption of mandelates and photocyclizaton of tropolone ethers, *Chem Commun*, **49**(74), 8253–8255 (2013).

[121] K. Mo, Y. Yang and Y. Cui, A homochiral metal-organic framework as an effective asymmetric catalyst for cyanohydrin synthesis, *J Am Chem Soc*, **136**(5), 1746–1749 (2014).

[122] S. Regati, Y. He, M. Thimmaiah, *et al.*, Enantioselective ring-opening of meso-epoxides by aromatic amines catalyzed by a homochiral metal-organic framework, *Chem Commun*, **49**(84), 9836–9838 (2013).

[123] C. Valente, E. Choi, M.E. Belowich, *et al.*, Metal-organic frameworks with designed chiral recognition sites, *Chem Commun*, **46**(27), 4911–4913 (2010).
[124] S.H. Cho, B. Ma, S.T. Nguyen, *et al.*, A metal-organic framework material that functions as an enantioselective catalyst for olefin epoxidation, *Chem Commun*, **24**, 2563–2565 (2006).
[125] F. Song, C. Wang, J.M. Falkowski, *et al.*, Isoreticular chiral metal-organic frameworks for asymmetric alkene epoxidation: Tuning catalytic activity by controlling framework catenation and varying open channel sizes, *J Am Chem Soc*, **132**, 15390–15398 (2010).
[126] K. Gedrich, I. Senkovska, I.A. Baburin, *et al.*, New chiral and flexible metal-organic framework with a bifunctional spiro linker and Zn_4O-nodes, *Inorg Chem*, **49**(10), 4440–4446 (2010).
[127] S.M. Shin, D. Moon, K.S. Jeong, *et al.*, Homochiral 3D metal-organic frameworks from chiral 1D rods: 6-way helical packing, *Chem Commun*, **47**(33), 9402–9404 (2011).
[128] N.G. Pschirer, D.M. Ciurtin, M.D. Smith, *et al.*, Noninterpenetrating square-grid coordination polymers with dimensions of 25 x 25 $Å^2$ prepared by using N,N-type ligands: The first chiral square-grid coordination polymer, *Angew Chem Int Ed*, **41**(4), 583–585 (2002).
[129] Y. Liu, X. Xu, F. Zheng and Y. Cui, Chiral octupolar metal-organoboron nlo frameworks with (14,3) topology, *Angew Chem Int Ed Engl*, **47**(24), 4538–4541 (2008).
[130] L. Sbircea, N.D. Sharma, W. Clegg, *et al.*, Chemoenzymatic synthesis of chiral 4,4′-bipyridyls and their metal-organic frameworks, *Chem Commun*, **43**, 5538–5540 (2008).
[131] K. Gedrich, M. Heitbaum, A. Notzon, *et al.*, A family of chiral metal-organic frameworks, *Chem Eur J*, **17**(7), 2099–2106 (2011).
[132] S.S. Iremonger, P.D. Southon and C.J. Kepert, A nanoporous chiral metal–organic framework material that exhibits reversible guest adsorption, *Dalton Trans*, 6103–6105 (2008).
[133] D.L. Murphy, M.R. Malachowski, C.F. Campana and S.M. Cohen, A chiral, heterometallic metal-organic framework derived from a tris(chelate) coordination complex, *Chem Commun*, **44**, 5506–5508 (2005).
[134] S.J. Garibay, Z. Wang, K.K. Tanabe and S.M. Cohen, Postsynthetic modification: A versatile approach toward multifunctional metal-organic frameworks, *Inorg Chem*, **48**(15), 7341–7349 (2009).
[135] D.J. Lun, G.I. Waterhouse and S.G. Telfer, A general thermolabile protecting group strategy for organocatalytic metal-organic frameworks, *J Am Chem Soc*, **133**(15), 5806–5809 (2011).
[136] M. Banerjee, S. Das, M. Yoon, *et al.*, Postsynthetic modification switches an achiral framework to catalytically active homochiral metal-organic porous materials, *J Am Chem Soc*, **131**, 7524–7525 (2009).
[137] J. Canivet, S. Aguado, G. Bergeret and D. Farrusseng, Amino acid functionalized metal-organic frameworks by a soft coupling-deprotection sequence, *Chem Commun*, **47**(42), 11650–11652 (2011).
[138] Y. Liu, W. Xuan and Y. Cui, Engineering homochiral metal-organic frameworks for heterogeneous asymmetric catalysis and enantioselective separation, *Adv Mater*, **22**(37), 4112–4135 (2010).
[139] G. Nickerl, A. Henschel, R. Grünker, *et al.*, Chiral metal-organic frameworks and their application in asymmetric catalysis and stereoselective separation, *Chem Ing Tech*, **83**(1–2), 90–103 (2011).
[140] M. Yoon, R. Srirambalaji and K. Kim, Homochiral metal-organic frameworks for asymmetric heterogeneous catalysis, *Chem Rev*, **112**(2), 1196–1231 (2012).
[141] L. Ma, C. Abney and W. Lin, Enantioselective catalysis with homochiral metal-organic frameworks, *Chem Soc Rev*, **38**(5), 1248–1256 (2009).
[142] A. Hu, H.L. Ngo and W. Lin, Chiral porous hybrid solids for practical heterogeneous asymmetric hydrogenation of aromatic ketones, *J Am Chem Soc*, **125**, 11490–11491 (2003).
[143] A.L. Nuzhdin, D.N. Dybtsev, K.P. Bryliakov, *et al.*, Enantioselective chromatographic resolution and one-pot synthesis of enantiomerically pure sulfoxides over a homochiral Zn-organic framework, *J Am Chem Soc*, **129**, 12958–12959 (2007).

[144] P. Peluso, V. Mamane and S. Cossu, Homochiral metal-organic frameworks and their application in chromatography enantioseparations, *Journal of Chromatography A*, **1363**, 11–26 (2014).
[145] X. Li, C. Chang, X. Wang, *et al.*, Applications of homochiral metal-organic frameworks in enantioselective adsorption and chromatography separation, *Electrophoresis*, **35**(19), 2733–2743 (2014).
[146] B. Van de Voorde, B. Bueken, J. Denayer and D. De Vos, Adsorptive separation on metal-organic frameworks in the liquid phase, *Chem Soc Rev*, **43**(16), 5766–5788 (2014).
[147] K.A. Cychosz, R. Ahmad and A.J. Matzger, Liquid phase separations by crystalline microporous coordination polymers, *Chem Sci*, **1**, 293–302 (2010).
[148] R.-G. Xiong, X.-Z. You, B.F. Abrahams, *et al.*, Enantioseparation of racemic organic molecules by a zeolite analogue, *Angew Chem Int Ed*, **40**(23), 4422–4425 (2001).
[149] W. Wang, X. Dong, J. Nan, *et al.*, A homochiral metal–organic framework membrane for enantioselective separation, *Chem Commun*, **48**, 7022–7024 (2012).
[150] C. Wang, T. Zhang and W. Lin, Rational synthesis of noncentrosymmetric metal-organic frameworks for second-order nonlinear optics, *Chem Rev*, **112**(2), 1084–1104 (2012).
[151] Q. Ye, Y.-H. Li, Q. Wu, *et al.*, The first metal (Nd^{3+}, Mn^{2+}, and Pb^{2+}) coordination compounds of 3,5-dinitrotyrosine and their nonlinear optical properties, *Chem Eur J*, **11**, 988–994 (2005).
[152] W. Zhang and R.G. Xiong, Ferroelectric metal-organic frameworks, *Chem Rev*, **112**(2), 1163–1195 (2012).
[153] D.-W. Fu, W. Zhang and R.-G. Xiong, The first metal–organic framework (MOF) of imazethapyr and its SHG, piezoelectric and ferroelectric properties, *Dalton Trans*, 3946–3948 (2008).
[154] P. Yang, X. He, M.-X. Li, *et al.*, The first homochiral coordination polymer with temperature-independent piezoelectric and dielectric properties, *J Mater Chem*, **22**(6), 2398–2400 (2012).

8
Chiral Metallosupramolecular Polyhedra

Jack K. Clegg and John C. McMurtrie

8.1 Introduction

Metallosupramolecular polyhedra have captured the imagination and attention of chemists since they were first discovered in the late 1980s [1]. There are arguably three main reasons for their proliferation. Firstly, they are inherently beautiful. By combining the various topological coordination environments of metals (tetrahedral, square planar, octahedral etc.) with ligands that have appropriate donor atoms, disposed with specific predesigned directional binding modes, a wide array of high symmetry polymetallic polyhedra have been produced. Secondly, as metallosupramolecular chemistry developed, it quickly became apparent that designed modifications could be used to vary the size, shape and symmetry of polyhedral architectures. As such, this chemistry is a principal exponent of supramolecular design from which fundamental approaches to the production – not just of metallosupramolecular polyhedra, but also of the wide array of other materials envisaged by supramolecular chemists, including MOFs, nanoballs and clusters to name a few. Lastly, metallosupramolecular polyhedra, by their very nature, usually contain space within them. This space is often accessible by guests and the design of such polyhedra can indeed be employed to create spaces of specific sizes and interior surface functionality so that they show selectivity for specific guests. This of course creates a huge array of possibilities as selective encapsulation can in turn be exploited for applications such as separation, catalysis and sensing.

With this in mind it is only a small jump to consider the implications of chirality in supramolecular polyhedra. The presence of multiple bridging ligands and metal centres

Chirality in Supramolecular Assemblies: Causes and Consequences, First Edition. Edited by F. Richard Keene.

results in the possibility of multiple chiral centres (including metal-based and organic-based stereocentres) in the polyhedra that they produce. This has led to a number of challenges for the synthetic chemist. Can polyhedra be designed to display specific chirality? Can this chirality be induced at the metal centres, or at the ligands, or both? Can it be induced by inclusion of a chiral guest? Will creation of a chiral cavity within polyhedra make them selective for chiral guests and thus make them useful for enantiomeric separations, or perhaps applications as chiral catalysts? In later sections we will elaborate on these factors, describing the key steps in the evolution of the chemistry of chiral metallosupramolecular polyhedra and the current (and future) applications for these remarkable complexes. However, before that, it is worthwhile considering the basic design principles that underpin the construction of metallosupramolecular polyhedra.

8.2 Basic Design Principles

In geometrical terms a polyhedron is a three-dimensional object with flat polygonal faces that are joined by straight edges and vertices. A polyhedron can be regular (all the faces consist of the same polygonal shape), uniform (all the faces are regular polygons), or many other combinations. In a purely mathematical treatment, polyhedra fit into one of the following categories: (i) Platonic solids, of which there are five; (ii) Archimedean solids, of which there are 13; and (iii) prisms and antiprisms of which there are an infinite number. Platonic solids have cubic symmetry in which all faces are identical polygons, and Archimedean solids are made up of two or more regular polygons. Prisms and antiprisms are made from two regular polygons (say n-gons) that are parallel and bisected by a rotation axis of the same order as the symmetry of the polygons (C_n). In a prism the polygons are congruent, while in an antiprism the polygons are rotated by an angle of 180/n. The simplest regular polyhedron is a trigonal pyramid or tetrahedron in which four triangular faces are linked by six edges and four vertices forming a closed three-dimensional solid.

In simple terms, therefore, a metallosupramolecular polyhedron is produced when a specific combination of bridging ligands and metals results in the formation of a polymetallic complex where the disposition of apex positions define a polyhedron. The apices could be occupied by either organic or metallic components. The scope then for the construction of metallosupramolecular polyhedra is immense. The particular geometry of a metallosupramolecular polyhedron results from the specific combination of the angles produced at the coordination centres, as well as the directional characteristics of the donor groups of the ligands. Significant deviation of either of these features would render the production of this polyhedron impossible, although perhaps another may be produced instead. Due to the vagaries of molecular motion and, in the case of solid-state structures, crystal packing forces, it is common for metallosupramolecular polyhedra to deviate from these strict definitions. For the purposes of this review – and as is common in the literature – the molecular species will be described by the polyhedron to which it most closely approximates.

Many approaches have been employed for the systematic design of metallosupramolecular polyhedral [2–15]. For example, Fujita *et al.* have investigated the preparation of a series of regular and semiregular polyhedra based on simple square-planar palladium(II) metal ions and ditopic di-pyridyl linkers in a M_nL_{2n} ratio [16]. Based on simple geometric restraints, it was predicted that values of n that would produce stable polyhedra are limited

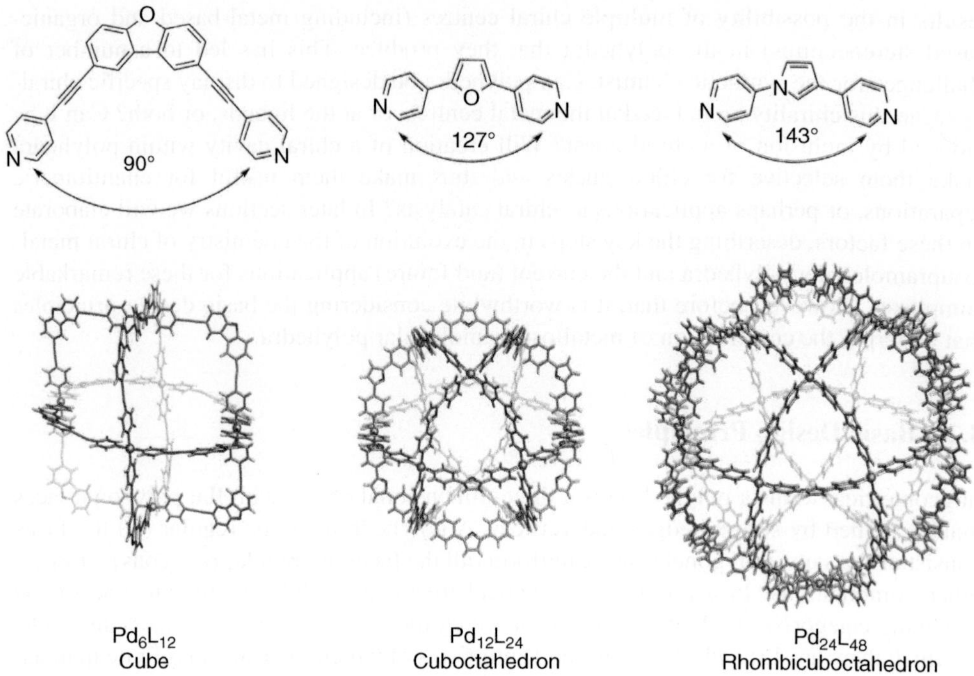

Figure 8.1 *Fujita* et al. *have successfully prepared a series of M_nL_{2n} metallosupramolecular polyhedra based on square planar metal palladium(II) ions and ditopic pyridyl ligands through the subtle variation in the ligands' bite angles*

to 6, 12, 24, 30 or 60. It was hypothesized that changing the bite angle of the connecting ligands would lead to the selective formation of each of the different polyhedra. While the investigations are ongoing, so far an M_6L_{12} cube, $M_{12}L_{24}$ cuboctahedra, and $M_{24}L_{48}$ rhombicuboctahedra have all been successfully prepared (Figure 8.1), illustrating that a systematic approach involving explicit geometric design can be employed to produce desired polyhedral architectures [17, 18].

In a related approach that has also been widely exploited, the primary design consideration is the overall point symmetry of the desired polyhedron. Molecular components (i.e. metals and ligands) are selected such that they will confer the required symmetry elements on the desired polyhedron. For example, a tetrahedron can by constructed by incorporating metals and ligands that will give rise to production of the necessary combination of C_2 and C_3 axes of tetrahedral point groups.

With few exceptions, Platonic, Archimedean, prismatic and antiprismatic solids are intrinsically achiral due to the presence of mirror planes within their symmetry operators. Chiral polyhedra can be produced from achiral components provided that their aggregation is associated with the formation of chiral centres. Most commonly this occurs at the metal centres. For example, the formation of a metallosupramolecular tetrahedron can be imagined from the self-assembly of four metals and six di-bidentate bridging ligands. Since the ligands coordinate to the metals in a bidentate chelating fashion and the metals have an octahedral coordination sphere it follows that each metal centre is *tris*-chelated with 'local'

C_3 symmetry and therefore either Δ or Λ stereochemistry – even though the starting materials are achiral the final assembly contains stereogenic centres, which have formed spontaneously during the ligation process. Perhaps less obviously, chirality can also occur where the connecting ligands are twisted due to the steric constraints of coordination at each end. The resulting induced axial chirality in the ligands imparts overall chirality on the metallosupramolecular polyhedron.

Chiral polyhedra can also be produced by using chiral ligands. Where those ligands have been resolved into one enantiomer then the overall chirality of the metallosupramolecular architecture can be controlled. Another option for producing chiral polyhedra is by employing chiral auxiliary ligands (e.g. 2,2′-bis(diphenylphosphino)-1,1′-binaphthyl, also known as BINAP, and related species). These ligands are not involved in the direct connection of metal centres but instead occupy terminal or capping positions in the coordination spheres of the metals. Nevertheless their inclusion imparts intrinsic chirality on the overall polyhedral architecture. Finally, inclusion of chiral counterions and/or encapsulation of chiral species may induce chirality in the supramolecular polyhedron through host-guest type interactions, provided of course that the coordination environment is labile enough to result in dynamic stereocentric conversion. Importantly, as many metallosupramolecular polyhedra encapsulate a defined space, a chiral polyhedron would define a chiral space, which might be used for myriad applications including selective guest binding or catalysis [19].

A detailed discussion of chiral metallosupramolecular polyhedra follows with particular focus on chiral polyhedra produced from achiral components, chiral polyhedra incorporating chiral molecular components, separation of resolved polyhedra and chiral induction and applications such as chiral encapsulation and catalysis.

8.3 Chiral Polyhedra from Achiral Components

There are many potential ways to design metallosupramolecular polyhedra, with an overarching approach being the consideration of the inherent geometric and chemical information present in each of the metal and organic components. A significant proportion of the systems developed to date have employed bridging di-bidentate ligands and *tris*-chelated octahedral metal-ions in their formation. The direct consequence of this design approach is the introduction of metal-based stereocentres. Metallosupramolecular polyhedra containing *tris*-chelated octahedral metal centres, therefore, will be inherently chiral objects, as long as the symmetry relationships between linked metal centres are derived from only proper rotation axes; the chirality of the assembly and its symmetry are inherently linked. For example, the point groups O_h and O both posses octahedral (cubic) symmetry; however, the presence of an improper rotation axis in the O_h point group (and therefore a mirror plane) results in an achiral object, despite the presence of a number of stereocentres, whereas an object with O-symmetry is chiral as it possesses only a combination of C_2, C_3 and C_4 rotation axes. Similarly, of the tetrahedral point groups, only T is chiral as objects with this symmetry lack centres of inversion or mirror planes, while the point groups T_d and T_h are achiral (although objects with these point groups may still contain stereocentres).

As implied by the above discussion, it is also possible that not all metallosupramolecular polyhedra are homochiral – that is, not all stereocentres within an assembly need be the same hand, or even if they are, two enantiomers of homochiral assemblies are possible.

Indeed, this yields the possibility of multiple diastereomers of a given assembly, some of which – despite the presence of multiple stereocentres – are achiral (i.e. *meso*). The tendency of a particular architecture that contains stereocentres to adopt a chiral configuration (if self-assembled from achiral starting materials) is dependent on a number of factors including the number, nature and geometry of the self-assembled components, statistical probability, as well as the degree of stereochemical coupling between chiral centres. Indeed, it is well understood that a strong degree of stereochemical communication between a metal stereocentre and a coordinated organic component (ligand) can induce the formally achiral organic molecules to adopt conformations that display atropisomerism and are chiral despite the lack of a stereogenic centre. In many cases, therefore, in homochiral metallosupramolecular polyhedra both the metallic and organic components adopt chiral configurations.

The majority of chiral metallosupramolecular polyhedra form from achiral components and most of these complexes form such that their metal centres are homochiral; that is, with each of the metal centres within each discrete species adopting the same (either Δ or Λ) configuration. As many metallosupramolecular polyhedra contain central cavities it follows that chiral analogues encapsulate a volume of chiral space. However, a far smaller number do exist (particularly in solution) as a mixture of diastereomers, each of which contains a different proportion of both right-handed and left-handed metal stereocentres. A consequence of the spontaneous generation of chirality in these systems is, of course, that a racemic mixture of the chiral polyhedra is produced regardless of whether or not each assembly is homochiral. Furthermore, while each individual molecule may be chiral, the bulk of a particular material formed in this way is not optically active. Predictably, most homochiral polyhedra therefore crystallize in centrosymmetric space groups although a small number of materials spontaneously resolve upon crystallization.

There are a wide variety of geometric architectural types of relevance, including tetrahedra, cuboids, cubes and prisms – with tetrahedra representing the largest (and conceptually simplest) of these. The principles of symmetry and chirality that apply to metallosupramolecular tetrahedra can be extended directly to apply to more complex assemblies. Although we draw on selected examples of more complex polyhedra, the tetrahedra in this chapter feature more predominately than other polyhedra simply because there are more examples. The synthesis, design features and properties (including host-guest recognition) of metallosupramolecular polyhedra have also been extensively reviewed and are therefore not discussed in detail here [2–15].

8.3.1 Tetrahedra

The simplest metallosupramolecular polyhedral architecture, in both mathematical description (possessing the smallest number of linked vertices possible) and in the chemical approach (having the smallest number of metal centres bridged by organic molecules) is a tetrahedron (or trigonal pyramid). In assemblies of this topology, four metal centres are linked by organic ligands. Two approaches have commonly been employed for the synthesis of these species where *facially* coordinated octahedral metal centres are situated on the vertices of the tetrahedron. In the first, di-bidentate chelating ligands with idealized C_2-symmetry are combined with octahedral transition metals to form M_4L_6 tetrahedra in which the ligands are situated on the edges of the polyhedral

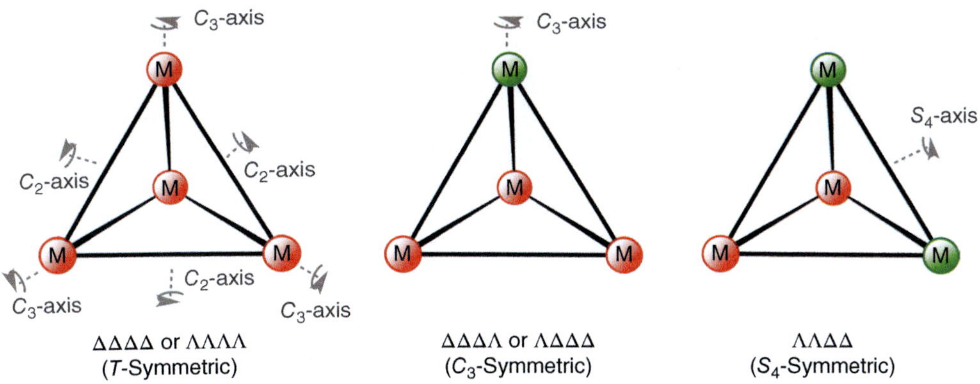

Figure 8.2 *Schematic representations of three possible diastereomers of chiral metallo-supramolecular tetrahedra. Green and red balls indicate metal centres of opposite chirality. These diastereomers occur in both M_4L_6 edge-bridged and M_4L_4 face-capped tetrahedra*

assembly. In the second approach, tri-bidentate ligands with idealized C_3-symmetry are employed resulting in a more closed "face-capped" M_4L_4 assembly in which the ligands are situated on the faces of the prismatic assembly; this approach also allows the formation of an M_4L_4 tetrahedron from a tri-tridentate ligand and nine-coordinate lanthanoid ions [20]. Despite the use of achiral starting materials in both M_4L_6 and M_4L_4 tetrahedra, the combination of the metals and ligands results in the metal centres becoming chiral, each with three chelating ligands. With four metal stereocentres within each polyhedron there are four potential diastereomers (Figure 8.2). Each tetrahedron can be either homochiral, where all of the metal stereocentres within each cluster are of the same hand (i.e. ΔΔΔΔ or ΛΛΛΛ) and will therefore display idealised *T*-symmetry, or may have a mixture of chiral centres within each assembly resulting in either C_3- (ΔΔΔΛ or ΛΛΛΔ) or S_4- (ΛΛΔΔ) symmetry. The latter case is a *meso* form and achiral despite the presence of four stereocentres.

As mentioned above, in many cases the ligands within each polyhedral structure will also adopt various chiral or achiral conformations due to steric requirements and stereochemical coupling with the metal centres to which they are coordinated. In the lower symmetry diastereomers (C_3 and S_4) it is expected that there will be more than one ligand conformation present within each assembly; this has been observed a number of times in solution through NMR experiments and in the solid state.

8.3.1.1 *T-Symmetric Tetrahedra*

By far the greatest number of both M_4L_4 and M_4L_6 metallosupramolecular polyhedra reported to date have homochiral metal centres resulting in chiral molecules with *T*-point symmetry, despite their preparation from achiral starting materials. Indeed, Saalfrank and coworkers [1, 21, 22] reported the first such example in 1988 and other notable examples have been produced by the groups of Raymond [23, 24], Ward [25, 26], Nitschke [27], Lindoy [28–30] and others [31–34]. Due to their high symmetry, these products also display relatively straightforward NMR spectra in solution. For example, the reaction of

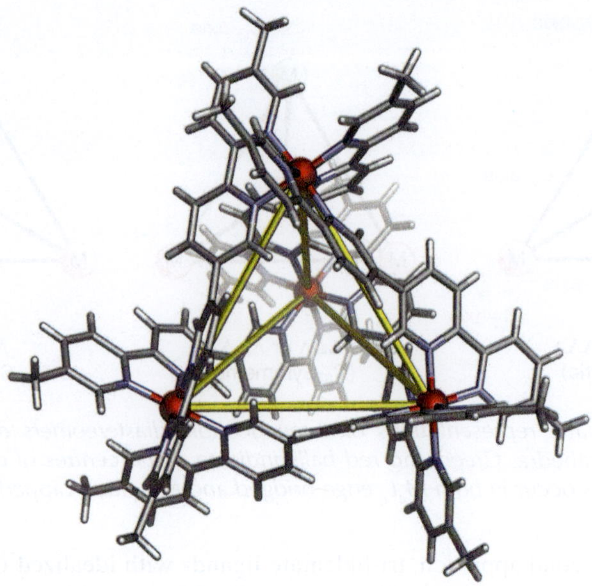

Figure 8.3 *Crystal structure of the [Fe_4**1**$_6$]$^{8+}$ (where n = 0) homochiral T-symmetric tetrahedron viewed down a C_3-axis. Each ligand adopts the same twisted chiral conformation. The ΔΔΔΔ enantiomer is shown*

ligand **1** (n = 0) with iron(II) in acetonitrile produced a homochiral [Fe_4**1**$_6$]$^{8+}$ tetrahedron that displays only seven signals in its ^{1}H NMR spectrum consistent with *T*-point symmetry such that only 1/12 of the molecule is symmetry independent and all the ligands must adopt identical C_2-symmetric conformations [29]. The single crystal structure of this material (Figure 8.3) confirms that each of the metal centres within each assembly have the same chirality, although both enantiomeric forms are present in the crystal lattice. The high symmetry of the space group in which the molecule crystallized ($P\bar{4}3n$) results in only 1/12 of the molecule being present in the asymmetric unit. This assembly does not require the presence of a template to form and is capable of encapsulating a variety of anions including BF_4^-, PF_6^- and $FeCl_4^-$ [29, 30].

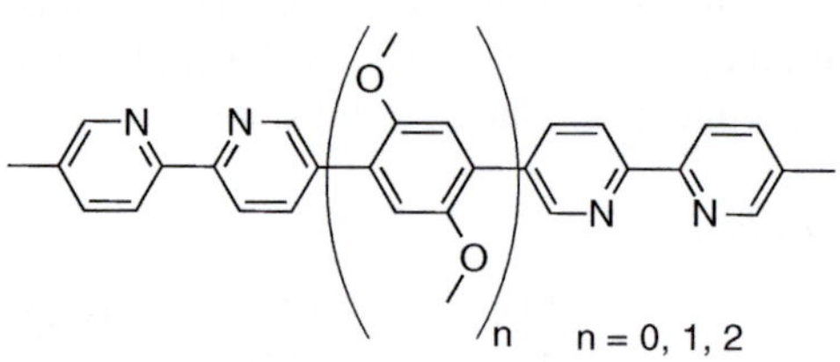

1

8.3.1.2 C_3-Symmetric Tetrahedra

Changing the chirality of one metal within a tetrahedral assembly to either ΔΔΔΛ or ΛΛΛΔ configurations produces an assembly that is still chiral but has lower symmetry compared to a T-symmetric cage – that is, rather than 1/12 of the molecule being symmetry independent, only 1/3 of the molecule is, requiring at least two different ligand arrangements to be present. While a number of C_3-symmetric M_4L_6 tetrahedra have been observed in solution as part of a mixture of diastereomers [35–38], there is only one example, $[Fe_4\mathbf{2}_6]^{8+}$, that has been characterized crystallographically [39].

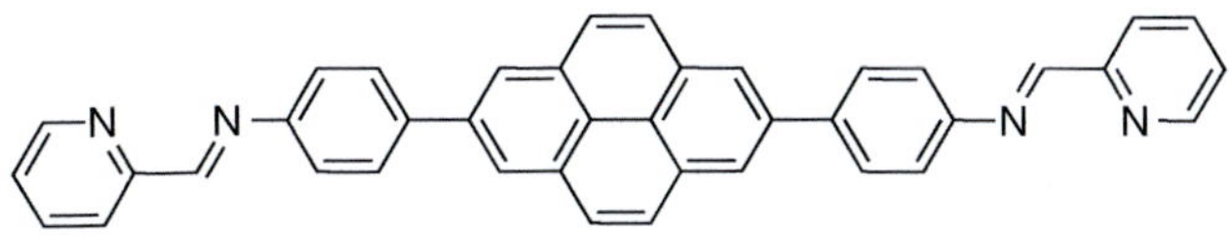

2

The reaction of six equivalents of **2**, which was formed *in situ* from the condensation of 2-pyridylcarboxaldehyde and a pyrene spaced diamine, with four equivalents of iron(II) in acetonitrile produced $[Fe_4\mathbf{2}_6]^{8+}$ in high yield. The ^1H and ^{13}C NMR spectra of the reaction mixture confirmed the formation of the complex and suggested that each of the potential T, C_3 and S_4 diastereomers were present in solution in a ratio of 12% : 45% : 43%, which is close to the expected statistical distribution (12.5% : 50% : 37.5%). However, crystallographic analysis of the material (Figure 8.4) showed that in the solid state the material exists exclusively as the C_3 symmetric diastereomer with equal quantities of the two ΔΔΔΛ and ΛΛΛΔ enantiomeric forms present [39].

There are two ligand conformers present in equal ratios. Each metal centre of opposite chirality is bridged by ligands, which adopt a close to flat achiral *syn*-conformer, while

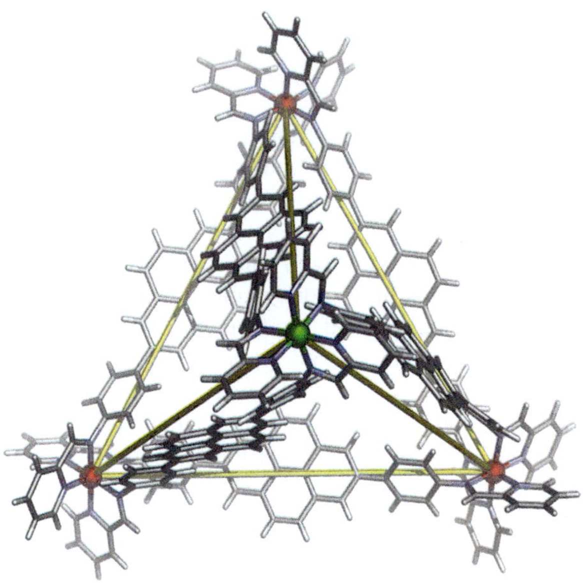

Figure 8.4 *The crystal structure of $[Fe_4\mathbf{2}_6]^{8+}$ heterochiral C_3-symmetric tetrahedron viewed down the C_3-axis. Two different ligand conformations are present in equal number, an achiral* syn-form *and a chiral* anti-form. *The ΔΔΔΛ enantiomer is shown (Λ metal centre is green, Δ metal centres are red)*

the three ligands that bridge homochiral metal centres are present in a twisted *anti*-arrangement. The length of the ligand results in metal-metal separations of close to 21 Å and a significant internal cavity volume results. In the solid state, this large void binds benzene solvent molecules; however, no evidence of guest binding was observed in solution, potentially due to the large apertures of the cavity not sufficiently constraining guest molecules.

8.3.1.3 S_4-Symmetric Tetrahedra

The third potential diastereomer of a metallosupramolecular tetrahedron containing four *facially* coordinated octahedral stereocentres is the achiral (ΛΛΔΔ) S_4 species; 1/4 of such a molecule is symmetry independent. When four equivalents of the tri-bidentate ligand **3** (which has idealized C_3 symmetry) was reacted in the presence of base with four equivalents of indium(III), a neutral [In$_4$**3**$_4$] tetrahedron is produced [40, 41] in which the triply-deprotonated ligands lie on faces of the tetrahedron defined by the four octahedral metal centres.

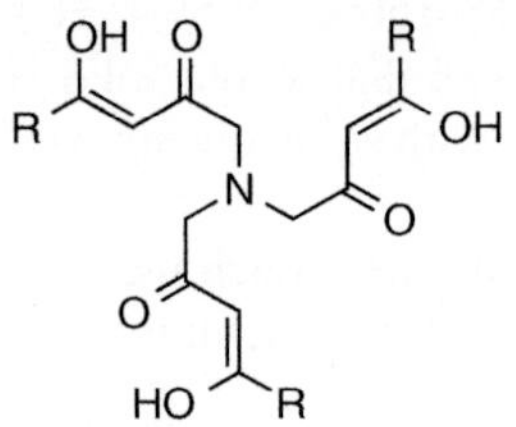

3

C_3-symmetric

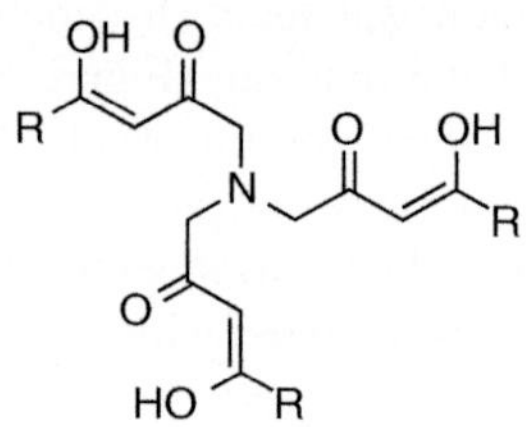

3

C_1-symmetric

The ^{1}H NMR of the [In$_4$**3**$_4$] material is complex, with three signals present for each proton within the ligand. Such a result is consistent with S_4 symmetry where, in the polyhedral complex, the ligand adopts a low symmetry conformation with a C_1 point group. Indeed, the crystal structure (Figure 8.5) confirms both the overall S_4-symmetry of the tetrahedral [In$_4$**3**$_4$] assembly with two Λ and two Δ stereocentres present in each complex and also that the ligands lie in nonsymmetric conformations, consistent with the solution measurements. This ligand conformation is expected given the steric demands of bridging three different octahedral metal centres, two of which are of one hand and the third the opposite. The ligands, were also observed to adopt chiral conformations, with two of C_1-symmetric having *P* stereochemistry and two displaying *M*. The two *P* ligands each bridge two Δ and one Λ metal centres, while the two *M* ligands each bridge two Λ and one Δ.

In all examples of S_4-symmetric edge-bridged M_4L_6 polyhedral cages characterized in the solid state two different conformations of the di-bidentate ligands have been observed in a 2:4 ratio, consistent with the restrictions of symmetry independence imposed by the point group. As can be seen from Figure 8.2 in a M_4L_6 cage with S_4-symmetry, four ligands must bridge metal centres of opposite stereochemistry (one Λ and one Δ) and two ligands bridge homochiral centres, that is one bridges two Λ metal ions and one bridges two Δ

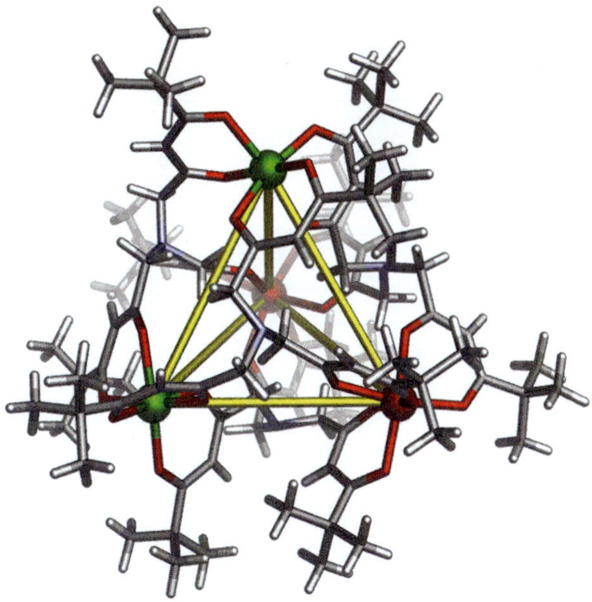

Figure 8.5 *The crystal structure of the face-capped [In_4**3**$_4$] S_4-symmetric tetrahedron. Each ligand adopts a chiral C_1-symmetric conformer so the overall stereochemistry can be defined as P,P,M,M-(ΛΛΔΔ). Λ metal centres shown as green, Δ metal centres as red*

metal ions. For example, the charge neutral metallosupramolecular tetrahedra formed from the reaction of either four Fe(III) or four Ga(III) centres with the doubly deprotonated form of **4** (with either n = 1 or 2) crystallized around $\bar{4}$ special positions such that only ¼ of each molecule was symmetry independent [28, 42]. The X-ray structures (Figure 8.6) reveal the presence of both Λ and Δ metal centres and ligands adopting two different conformations in the expected 2 : 4 ratio.

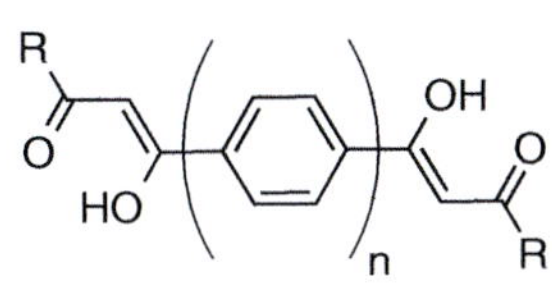

4 n = 1, 2

In these examples, the two ligands that bridge the homochiral metal centres adopt helical *anti*-conformations, while the four other ligands, bridging the metals of opposite chirality adopt *syn*-conformations, which are close to planar and hence achiral; similar chiral and achiral ligand conformations are present in every example in the literature (including those formed with ligands **1** (n = 2) [43], **5** [44] and **6** [45]) and reflect the steric requirements imposed by bridging chiral octahedral metal ions with rigid ligands (discussed in more detail below).

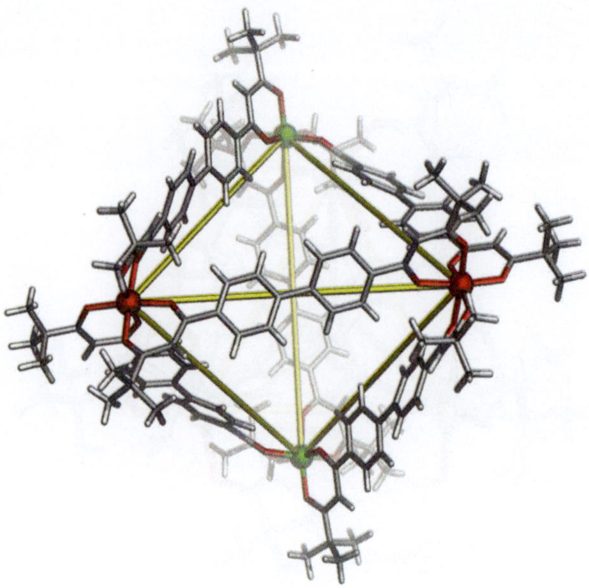

Figure 8.6 *Crystal structure of the edge-bridged [Fe_4**4**$_6$] (where n = 1) S_4-symmetric tetrahedron viewed down an S_4-axis. Two ligand conformers are present. Four ligands adopt an achiral* syn-*form and two adopt a chiral* anti-*form. Δ metal centres shown as red, Λ metal centres as green*

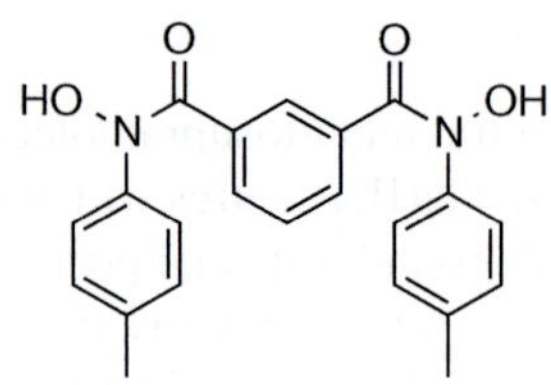

5

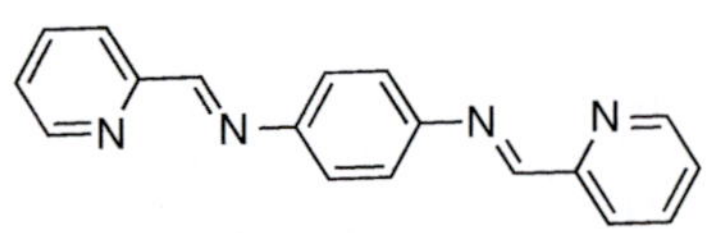

6

Each [M_4**4**$_6$] polyhedron encapsulates a significant void, with the largest approaching 844 Å^3 and binds neutral tetrahydrofuran guests within their central cavities. Intriguingly, while each of [M_4**4**$_6$] [42], [Fe_4**1**$_6$]$^{8+}$ (n = 2) [43] and [M_4**5**$_6$] [44] crystallize with S_4 symmetry, in solution studies at room temperature they exclusively display ^{1}H NMR spectra consistent with T symmetry. [Fe_4**6**$_6$]$^{8+}$ [45], on the other hand, displays S_4 symmetry both in solution and the solid state. Expanding ligand **6**, by replacing the central phenyl ring with a biphenyl spacer (**7**) produced an [Fe_4**7**$_6$]$^{8+}$ cage that was present as a mixture diastereomers in solution with only 49% S_4, but that crystallised exclusively as this isomer with triflimide counter-ions [37]. Finally, a [Fe_4L_6]$^{8+}$ tetrahedron formed from a 2,6-disubsituted naphthalene ligand (**8**) crystallizes such that it has two Λ and Δ metal centres, but due to steric hindrance the polyhedron has no-symmetry (and correspondingly a C_1 point group) in both solution and the solid state [46]. The complex formed with the anthracenyl spaced ligand **9**, $[Fe_4\mathbf{9}_6]^{8+}$, displays S_4 symmetry both in solution and the solid state [46].

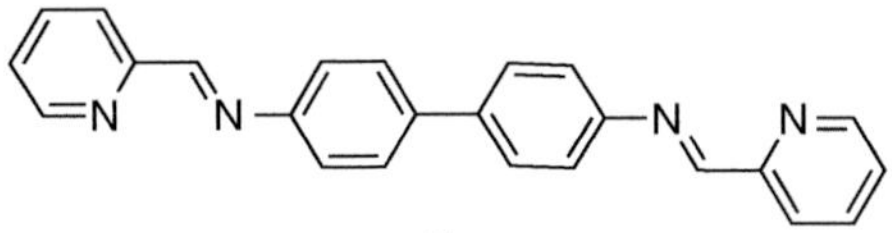

7

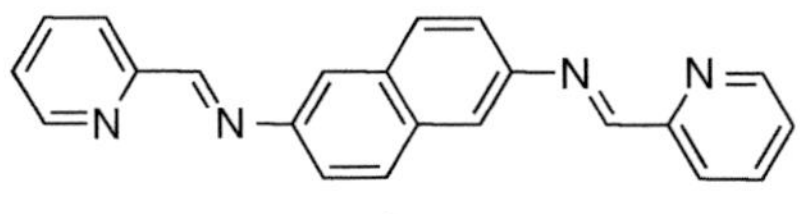

8

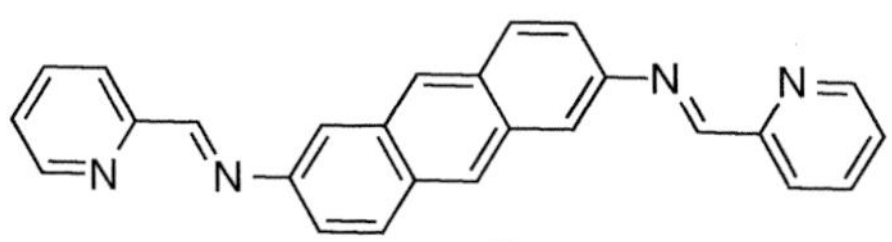

9

8.3.2 Higher Order Polyhedra

The general principles discussed above for the formation of chiral metallosupramolecular tetrahedra from achiral components also apply to higher order polyhedra; noting, of course, that the number of metal-based stereocentres per assembly increases. The geometric restraints inherent in larger polyhedra additionally constrain the possibilities of the shapes that can be formed from the combination of *facially* coordinated *tris*-bidentate octahedral metals with a single ligand (although Fujita *et al.* have elegantly demonstrated that subtle differences in ligand angles can have dramatic effects on the order of achiral polyhedra produced [16]). Accordingly, to form higher order chiral polyhedra a number of research groups have employed either a combination of metals of more than one geometry (often termed the *metallo-ligand* approach [47–50]), a combination of more than one ligand (each with different geometric properties), *meridionally* coordinated metal centres and a combination of *meridionally* and *facially* coordinated metal centres. Following these approaches a diverse range of regular and nonregular chiral metallosupramolecular polyhedra, including M_6L_9 trigonal prisms [51], M_8L_6 [52, 53] and M_8L_{12} cubes [54–63], $M_{10}L_{15}$ pentagonal prisms [62, 64–66], $M_{12}L_{12}$ icosahedra [67], $M_{12}L_{18}$ hexagonal prisms [68], mixed ligand $M_{12}L^A{}_{12}L^B{}_4$ cubeoctahedra [69], as well as $M_{12}L_{18}$ and $M_{16}L_{24}$ truncated tetrahedra [7, 70–72] and various stellated systems [73, 74]. In a number of instances, these architectures have been observed in solution as mixtures of polyhedra under equilibrium.

The larger number of stereocentres also results in the potential for higher order polyhedra to form larger numbers of diastereomers. For example, while there are four potential diastereomers of a tetrahedron, a cube has 14 different possible arrangements including six achiral (*meso*) diastereomers. As with tetrahedra, the formation of one diastereomer over another is often unpredictable. For example, the reaction of **10**, which has coordination vectors of 120°, with iron(II) was expected to form an $[Fe_8\mathbf{10}_{12}]^{16+}$ cube with the *facially* coordinated octahedral metal centres located on the vertices and the ligands defining the edges [63]. If the metal centres were homochiral, and the ligands adopted a twisted conformation (consistent with observations in tetrahedra as they bridge metals of the same hand),

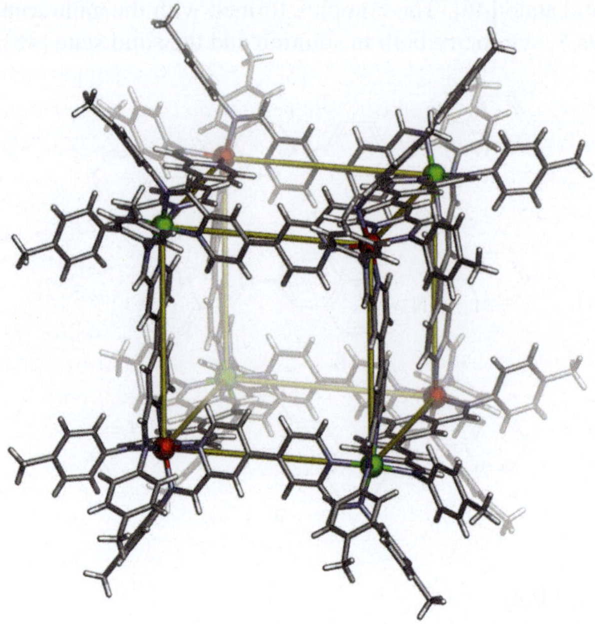

Figure 8.7 *The crystal structure of the O-symmetric* $\Lambda\Lambda\Lambda\Lambda\Delta\Delta\Delta\Delta$-$[Fe_8\mathbf{10}_{12}]^{16+}$ *cube. Each ligand adopts a flat achiral conformation.* Δ *metal centres shown as red,* Λ *metal centres as green*

then the assembly would have idealized O (chiral cubic) point symmetry. The ^{1}H NMR spectrum was consistent with the formation of a high symmetry polyhedron with only seven signals present (one for each symmetry independent proton on the ligand). The synchrotron crystal structure (Figure 8.7) revealed that while the molecule had idealized high symmetry consistent with the solution studies, that the assembly was indeed achiral with four Λ and four Δ metal centres and T_d point symmetry. The 12 ligands, therefore, each bridge metal centres of different stereoisomers and adopt flat achiral conformations. This observation was particularly unexpected as this ligand conformation brings the pyridyl protons into eclipsed positions where transannular strain will occur. To partially alleviate this strain the ligands bow away from completely flat arrangements. Other cubes that are either homochiral (O point symmetry) or achiral with S_6 point symmetry have also been prepared [52–63].

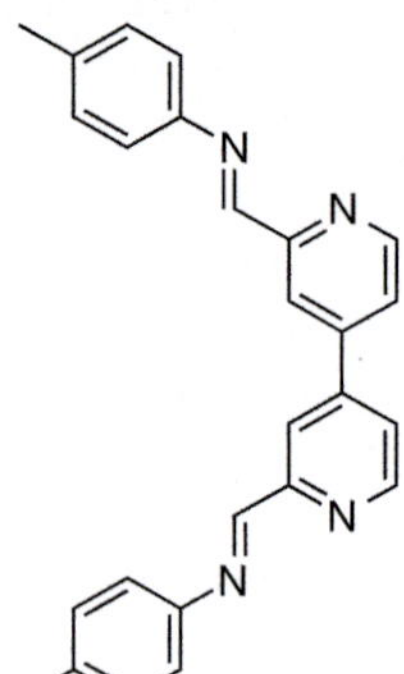

10

8.4 Stereochemical Communication

The transmission and recognition of stereochemical information between chiral centres mediated by either intermolecular interactions (i.e. where stereocentres are not mechanically linked through covalent or coordinate bonds and are influenced through supramolecular interactions) or intramolecular interactions (i.e. through covalent or coordinate bonds) has significant implications for diverse areas of science including catalysis, nonlinear optics and biological processes. In both cases a significant determining factor in the successful transfer of information is the separation of the physical stereocentres through space. In the presence of a mechanical (organic) bridge, the size, shape and properties of the linker also play significant roles as these features strongly influence the relative energies between diastereomers and the barriers of enantiomerization of individual stereocentres [75, 76], (particularly where the stereocentres are *tris*(bidentate) octahedral metals).

As mentioned above, the vast majority of metallosupramolecular polyhedra formed from achiral components adopt homochiral configurations, despite these arrangements being statistically disfavoured. For example, from a purely statistical approach four stereocentres of either hand randomly arranged into tetrahedra produces a $T:C_3:S_4$ diastereomer ratio of 12.5 : 50 : 37.5, but in the vast majority of cases tetrahedra prepared to date exist exclusively as homochiral T-symmetric diastereomers in solution and in the solid state. The extent to which the proportions of diastereomers present in a given solution differ from the expected purely statistical distribution is related to the efficiency of transfer of stereochemical information between the metal stereocentres. That is to say we can define *the relative strength of the stereochemical coupling* present within a chiral metallosupramolecular polyhedron to be the degree to which the ratio of diastereomers present deviates from the statistical ideal. For the sake of convenience we can also define *the sign* (in the sense of the direction of a vector) of this deviation. We therefore define *positive stereochemical coupling* to be the deviation of the proportion of diastereomers present towards chiral diastereomers from the statistical ideal and *negative stereochemical coupling* to be the deviation of the proportion of diastereomers present towards the achiral (*meso*) forms.

It can therefore be expected that a particular bridging ligand could be designed to favour either strong or weak positive or negative stereochemical coupling between the metal centres and hence favour the formation of either homochiral or achiral metallosupramolecular polyhedra. For example, rigid ligands that possess substituents that favour the formation of atropisomeric conformations would facilitate positive stereochemical coupling. Less rigid, longer ligands, or those with large areas devoid of the potential for atropisomerism would favour negative stereochemical coupling. It could also be expected that the presence of a suitable guest molecule within the cavity of a polyhedron might provide an additional intramolecular pathway for stereochemical communication.

When tetrahedral $[Fe_4\mathbf{7}_6]^{8+}$ was prepared with different anions, each of which was bound inside the central cavity with different strengths (and some were too large to be encapsulated), each assembly was observed to adopt different ratios of diastereomers in solution [37]. When the anion was too large to be encapsulated the cage adopted a $T:S_4:C_3$ diastereomer ratio of 32 : 49 : 19, a deviation from expected statistical distribution towards the homochiral T diastereomer. The addition of small achiral anionic guests (halides, NO_3^-, BF_4^-, ClO_4^-, PF_6^-, etc.) changed the ratios of diastereomers present with the strongest bound guests drastically increasing the strength of positive stereochemical communication

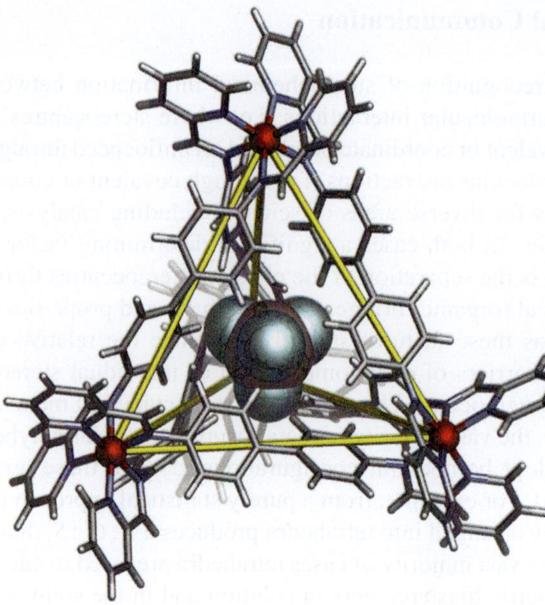

***Figure 8.8** Crystal structure of the $[Fe_4 7_6]^{8+}$ homochiral T-symmetric tetrahedron encapsulating a tetrafluoroborate anion viewed down a C_3-axis. Upon addition of suitable guest molecules this complex adapts its diastereomeric composition through an increase in stereochemical communication. The ΔΔΔΔ enantiomer is shown*

present such that only the homochiral *T* diastereomer remained (Figure 8.8). These cages were found to act as a system, adapting to the addition of guest molecules, tailoring the binding pocket to the particular guest.

Similarly, the addition of guests to, or protonation of the ligand in, $[In_4\mathbf{3}_4]$ significantly changed the stereochemical coupling present. As mentioned above, this complex was observed to form exclusively as the S_4 diastereomer in solution demonstrating strong negative stereochemical communication between stereocentres. The simple expedients of either protonating the central tertiary nitrogen of the ligand or the addition of a caesium ion guest resulted in a dramatic reversal of the sign of the stereochemical communication. The $[In_4(H\mathbf{3})_4]^{4+}$ and $[Cs \subset In_4\mathbf{3}_4]^+$ do not adopt the achiral diastereomer, but instead are observed exclusively as the homochiral *T* species [40]. Less dramatic, but still significant, are changes to the diastereomer ratios upon the addition of neutral guest molecules that have also been observed in a polyhedral cage formed from a pyrene-spaced ligand (an isomer of **2**), which showed a decrease in the proportion of the S_4 diastereomer present with an amplification of both the *T*- and C_3-symmetric complexes [39].

8.4.1 Stereocontrol through Ligand Modification

Nitschke *et al.* have investigated the control of stereochemical coupling between metal centres through the modification of a ligand's steric properties [77, 78]. Four terphenylene-spaced ligands (**11–14**), each substituted with different numbers and positions of methyl groups, and their corresponding $[Fe_4L_6]^{8+}$ tetrahedral cage complexes were prepared [36]. The position and degree of methylation was expected to influence the degree to which each of the organic components would adopt either atropisomeric or achiral conformers and hence the efficiency with which they would act to transfer chiral information between the metal stereocentres.

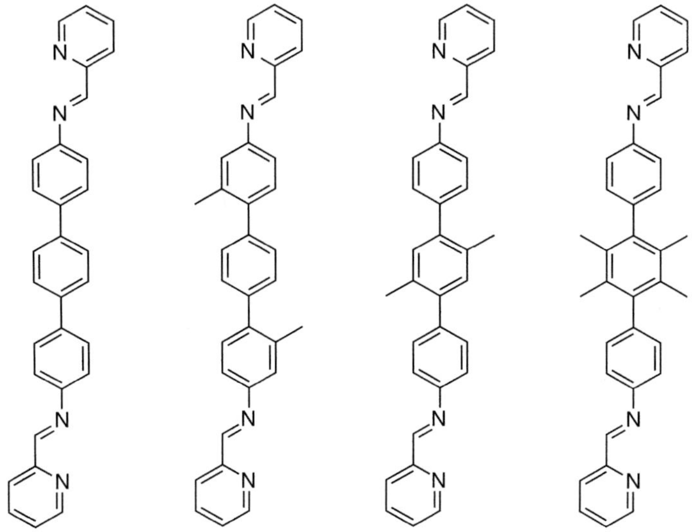

11 **12** **13** **14**

The different diastereomer ratios obtained for each of the polyhedral cages formed are given in Figure 8.9. While each of the ligands was found to produce complexes that showed stereochemical coupling, the magnitude of the communication and its sign varied with the ligand substitution pattern. When **11** was employed, a relatively small increase in the proportion of T diastereomer (compared to the statistical distribution) was observed, while the introduction of two methyl groups on the terminal phenylene rings of the ligand (**12**) further increased the proportion of the homochiral T-symmetric species present, suggesting a positive enhancement of stereochemical communication. Increasing the temperature of the solution of $[Fe_4\mathbf{11}_6]^{8+}$ resulted in a shift in the proportions towards the statistical distribution.

Twofold methylation on the central ring, however, resulted in a reduction in efficiency of the transfer of stereochemical information as a mixture approximating the statistically predicted ratios was produced (with the C_3 diastereomer dominating). Fourfold methylation resulted in the complete removal of the homochiral cages from the mixture with 88% of the cages observed to adopt the achiral S_4 form suggesting strong negative stereogenic coupling. The influence of the substitution patterns on the stereochemical communication was attributed to the relative energetic preferences of atropisomeric ligand conformers available due to steric effects; that is helical conformers (those expected to bridged homochiral metal centres) are favoured in **11–12** while flat *meso* conformers (those expected to bridge heterochiral metal centres) are favoured in **13–14**.

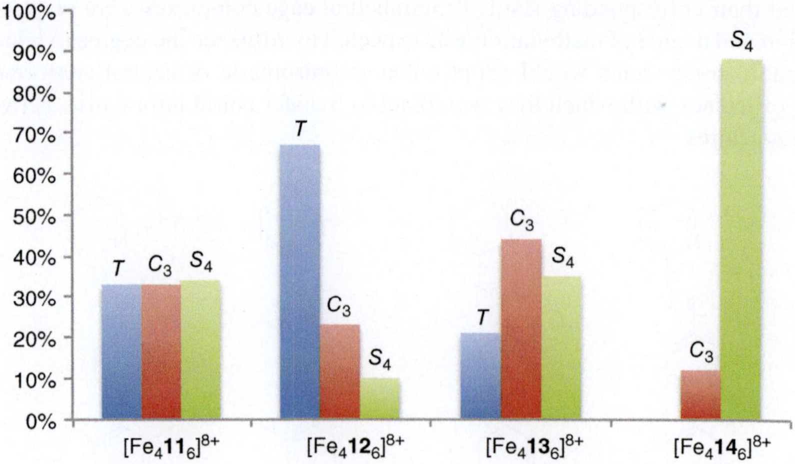

Figure 8.9 *Diastereomer ratios in solution of tetrahedra formed from ligands **11–14**. Changing the properties of the ligand affects the stereochemical communication between metal centres in turn favouring different diastereomers*

8.4.2 Mechanisms of Interconversion between Diastereomers

As the origin of chirality in chiral metallosupramolecular polyhedra assembled from achiral components is the arrangement of bidentate ligands around an octahedral metal centre, it is not surprising that mechanisms for the interconversion between various diastereomers, the racemization of a spontaneously resolved material or the enantiomerization of a chiral polyhedron are directly related to the mechanisms of racemisation of mononuclear octahedral complexes. Thus, the handedness of each individual metal stereocentre within a polyhedron can switch via two potential processes: either a dissociative mechanism (which could be complete or partial) or a nondissociative Bailar twist.

For the complete enantiomerization of a homochiral (T-symmetric) four-metal tetrahedron (i.e. $\Delta\Delta\Delta\Delta \leftrightharpoons \Lambda\Lambda\Lambda\Lambda$) to occur, four individual stereocentres must enantiomerize. This process could be either sequential, in which the enantiomerization proceeds through a series of diastereomeric intermediates (i.e. $\Delta\Delta\Delta\Delta$ (T) $\leftrightharpoons$ $\Delta\Delta\Delta\Lambda$ (C_3) $\leftrightharpoons$ $\Delta\Delta\Lambda\Lambda$ (S_4) $\leftrightharpoons$ $\Delta\Lambda\Lambda\Lambda$ (C_3) $\leftrightharpoons$ $\Lambda\Lambda\Lambda\Lambda$ (T)) or synchronous (cooperative) whereby the handedness of all four metal ions changes simultaneously proceeding through a single transition state. The steric demands of the ligands, inherent lability of the metal ions and magnitude of the relative activation energies will determine which mechanism will dominate in a particular system; however, it seems likely that the higher the stereochemical coupling between stereocentres, the more likely that a cooperative process will occur, particularly where the interconversion is nondissociative.

Raymond *et al.* [35] found that their [Ga_4**5**$_6$] tetrahedron crystallized as its S_4 diastereomer while studies at 300 K of the same material in solution determined that the T diastereomer was the unique product. Upon cooling to 220 K, however, each of the C_3, S_4 and T diastereomers could be observed in approximately the statistically expected distribution, suggesting minimal stereochemical coupling between metal centres and that each diastereomer is of approximately equal energy. Further variable temperature studies allowed the authors to calculate that the activation barrier for the interconversion between the two hands of each metal centre (and hence each diastereomer) was 58 kJ mol^{-1} and that the enantiomerization of the entire assembly proceeded in a stepwise fashion via a series of sequential nondissociative Bailar twists.

Saalfrank *et al.* were able to examine the mechanism of diastereomer interconversion in their [Mg$_4$**15**$_6$]$^{4-}$ and [Mg$_4$**16**$_6$]$^{4-}$ tetrahedral cage systems (in which the ligands **15** and **16** are both doubly deprotonated) using similar variable temperature studies [79]. Due to the steric demands of ligand **16** no diastereomer interconversion was observed between 223–343 K in [Mg$_4$**16**$_6$]$^{4-}$; however, upon cooling, the ethyl protons in [Mg$_4$**15**$_6$]$^{4-}$ become diastereotopic indicating that enantiomerization between the T-symmetric ΔΔΔΔ ⇋ ΛΛΛΛ forms was occurring.

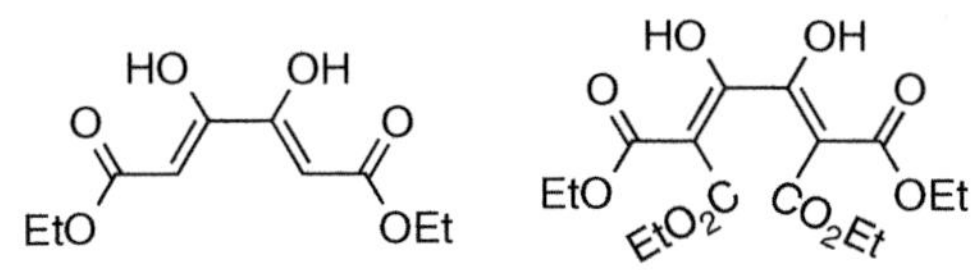

15 **16**

In contrast to the [Ga$_4$**5**$_6$]$^{12-}$ tetrahedron, this process was found to proceed through four simultaneous Bailar twists via a cooperative mechanism with a transition state energy of 60 kJ mol^{-1}. It was possible to assign this mechanism as synchronous rather than sequential as no other diastereomeric intermediates were observed. Accompanying the simultaneous enantiomerization of the metal ions are six simultaneous atropenantiomerization steps – one for each ligand, which occur via a twisting mechanism. The bulky substituents on **16** prevent a similar process occurring accounting for the different behaviour of [Mg$_4$**16**$_6$]$^{4-}$. In a later study the same group observed a related mesomerization in their S_4-symmetric achiral [In$_4$**3**$_4$] tetrahedron [41]. In this case, the ΔΔΛΛ ⇋ ΛΛΔΔ process was again found to feature four simultaneous Bailar twists and four helical ligand isomerizations with a cooperative transition state energy of 74 kJ mol^{-1}.

Barriers of diastereomer interconversion were found to be even higher for the metallosupramolecular polyhedra formed from **11–14** despite the presence of weaker stereochemical coupling between metal centres [36]. Activation barriers of at least 85 kJ mol^{-1} for the stepwise interconversion between diastereomers were determined although it was not possible to distinguish between a dissociative or nondissociative mechanism. Diastereomer interconversion in the related [Fe$_4$**7**$_6$]$^{8+}$ tetrahedron, however, was found to take place via a dissociative de-ligation mechanism with activation energies of between 78–107 kJ mol^{-1} depending on the counterion and the diastereomeric identity of the starting materials and products [37]. The similarity of the materials suggests that de-ligation is a likely mechanism for the interconversions of diastereomers in the iron(II) complexes of **11–14** and other related iron(II) complexes of pyridyl-imine ligands as well.

8.5 Resolution of Racemic Metallo-Supramolecular Polyhedra

It has long been established that simple chiral metal complexes can be separated (resolved) into their enantiomerically pure forms if the metal ion is sufficiently kinetically inert. This has been achieved in a number of ways but most commonly by the selective crystallization of one hand through the aid of an enantiomerically-pure counter ion or other precipitating agent, or the use of a chiral stationary phase and/or eluent in chromatography. These techniques, particularly the use of chiral chromatographic methods, have also been exploited for the separation and analysis of more complex multimetallic complexes, including metallosupramolecular helicates, for the investigation of chiral recognition, stereochemical communication and electron transfer processes [80–87]. Despite the plethora of racemic mixtures of chiral metallosupramolecular tetrahedra just discussed, there are very few examples of these materials that have been successfully resolved and a majority of those rely on spontaneous resolution of the materials into chiral crystals.

It was noted above that the addition of either protons or a caesium guest to the achiral S_4-symmetric complex $[In_4\mathbf{3}_4]$ resulted in the transformation of the host molecule to a chiral T-symmetric diastereomer. While in solution these complexes are present as a racemic mixture of the ΛΛΛΛ and ΔΔΔΔ enantiomers, upon crystallization the enantiomers spontaneously separate, resolving within each single crystal so that crystals containing either all Λ metal centres or all Δ metal centres result [40]. The bulk material remains racemic in that there are equal quantities of the two enantiomeric forms of the crystal.

A similar effect was observed in the $[Fe_4\mathbf{1}_6]^{8+}$ tetrahedron of Lindoy *et al.* While this cage was exclusively observed as the T diastereomer, crystallization of the material with an $[FeCl_4]^-$ guest resulted in the formation of chiral crystals [30]. The face-capped tetrahedra $[Fe_4\mathbf{17}_4]^{8+}$ and $[Fe_4\mathbf{18}_4]^{8+}$ were also found to crystallize in chiral space groups although the bulk remained racemic [67, 88].

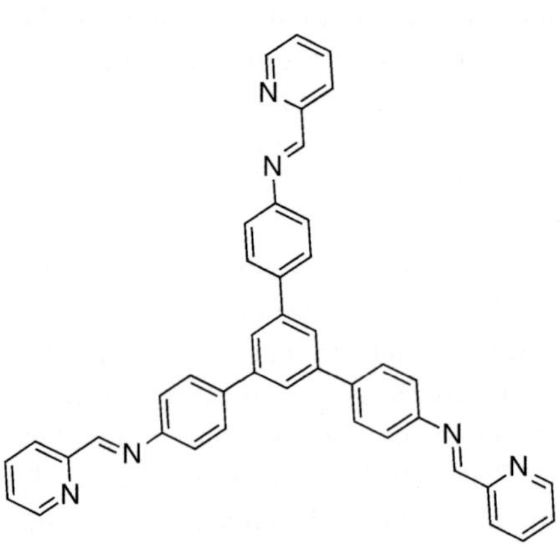

17

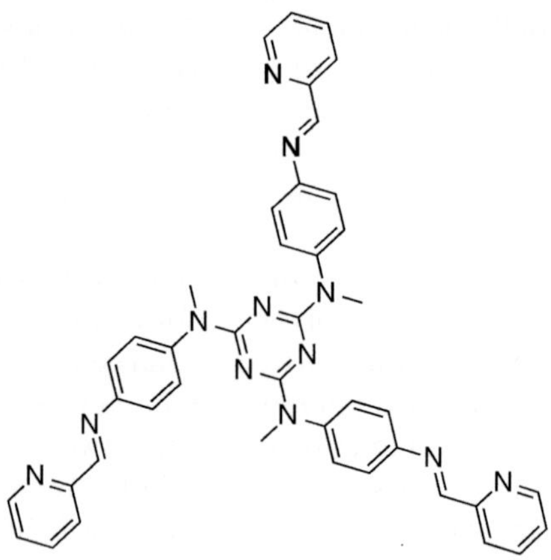

18

The reaction of the anthracene-spaced ligand **19** with a number of transition metals resulted in the formation of $[M_6\mathbf{19}_9]^{12+}$ assemblies that have trigonal prismatic geometries (Figure 8.10) in which two $M_3\mathbf{19}_3$ triangular circular helicates are bridged by three additional ligands [51].

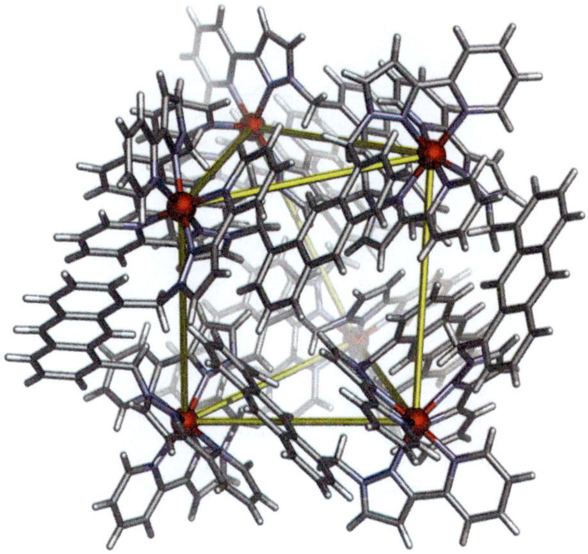

Figure 8.10 *Crystal structure of the [Zn$_6$**19**$_9$]$^{12+}$ homochiral trigonal prism. The metal centres are meridionally coordinated and the complex spontaneously resolves in to chiral crystals on crystallization. The ΔΔΔΔ enantiomer is shown*

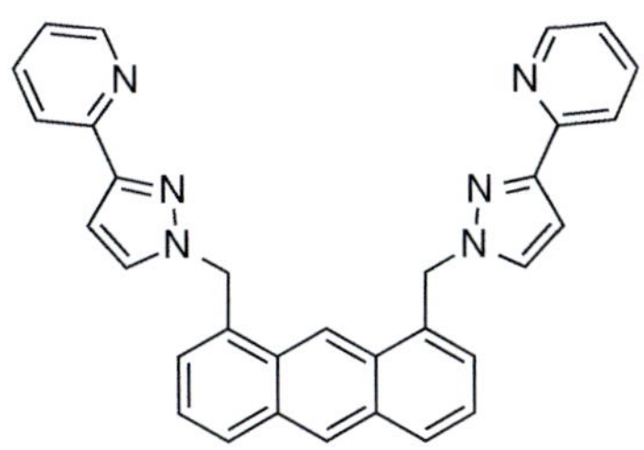

19

Each of the six octahedral metal ions in the structures is bridged by three different ligands that are coordinated in a *meridional* fashion and the assembly encapsulates a tetrafluoroborate anion. All of the metal centres have the same stereochemistry resulting in a chiral structure, which again resolves upon crystallization into enantiomerically pure crystals. While not producing chiral crystals, the same approach of linking circular helicates with additional ligands and *meridional*-coordinated metal stereocentres has been used to form higher order polyhedral, including rectangular prisms, pentagonal prisms and hexagonal prisms [57, 62, 64–66, 68].

Raymond *et al.* employed kinetically inert metal centres to prepare a series of anionic chiral [M$_4$**20a**$_6$]$^{12-}$ (where **20a** is quadruply deprotonated) *T*-symmetric metallosupramolecular polyhedra (Figure 8.11) that encapsulates a wide variety of guests within its hydrophobic central cavity and catalyses a variety of transformations [89]. Interestingly, the closely related ligand **20b** forms exclusively [M$_2$**20b**$_3$] triple helicates rather than tetrahedra [90]. Upon treatment of a racemic mixture of the [M$_4$**20a**$_6$] cages with the chiral cation

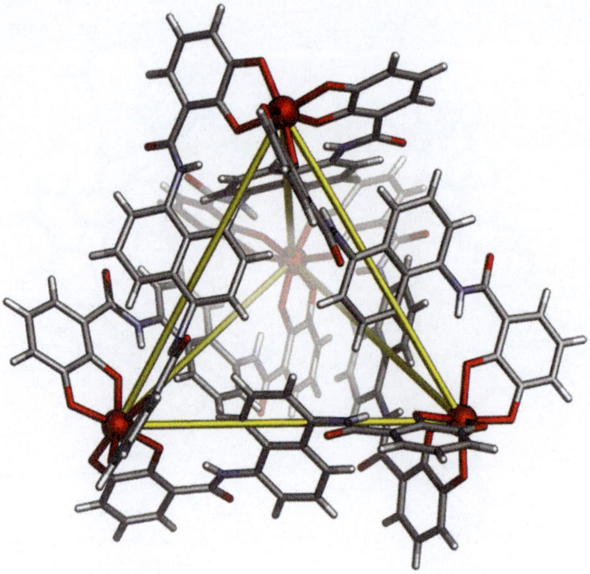

Figure 8.11 *Crystal structure of the* $[Fe_4\mathbf{20a}_6]^{12-}$ *homochiral tetrahedron. The two enantiomeric forms of the compound could be resolved by the addition of a chiral cation. The ΔΔΔΔ enantiomer is shown*

S-nicotinium, the ΔΔΔΔ enantiomer of $[M_4\mathbf{20a}_6]^{12-}$ selectively precipitated leaving the antipodean cage in solution and allowing for the separation of the two enantiomerically pure cages [91–93]. After cation exchange the cages were found to remain enatiomerically pure after 8 months in solution and displayed some enantioselective guest binding and catalytic properties [92, 94].

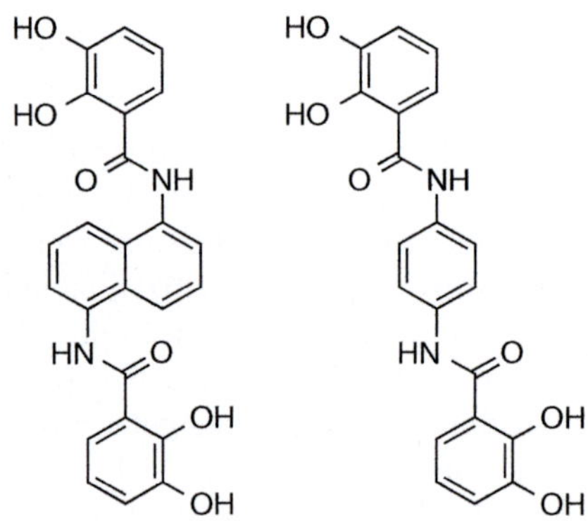

20a **20b**

There is clearly scope for further investigation into the resolution of these materials – for example, using chromatographic methods that have proved viable on closely related ruthenium-containing helicates such as $[Ru_2\mathbf{1}_3]^{4+}$ [86].

8.6 Chiral Polyhedra from Chiral Molecular Components

As has been discussed above, many coordination polyhedra comprise metal centres with axial local chirality. Specific chirality (Δ or Λ) at the metal centres can be induced through stereochemical communication resulting from the use of resolved chiral ligands with the overall result being formation of resolved homochiral polyhedra. But this is not the only source of chirality. Chirality is possible in coordination polyhedra in which the metal centres themselves are achiral but where the connecting ligands are chiral, or where the other ligands in the coordination sphere (sometimes referred to as capping ligands) are chiral. There are examples of both types but these are not particularly common due to the normal difficulties of enantiomeric separation required prior to assembly of the polyhedra. Nevertheless, this is an important approach to the formation of chiral polyhedra and the benefits of this as a strategy for producing such targets result primarily from the fact that the chirality of the polyhedra is predetermined by the chirality of the components and as such they are produced in enantiomerically pure forms. Even where the metal centres are labile, the chirality of the polyhedra is almost always preserved.

Adamantoid polyhedra typically form by combination of a tritopic linker and a ditopic linker (in a ratio of 6 : 4). The first chiral polyhedra reported were adamantoid architectures prepared by the combination of a bis(platinum) complexes **22** as the ditopic linker with an enantiomerically pure chiral tritopic organic linker **21** (Figure 8.12) [95]. Minor variations in the structure of the achiral bis(platinum) complex (**22a** and **22b**) resulted in two such examples. The formation of these polyhedra is described by the authors as instantaneous and studies of optical rotation confirmed that they are indeed optically active. No crystal structures were obtained but MM2 forcefield simulations were carried out and these visualizations showed that the symmetry of the adamantoid structure is reduced from *T*d to D_2 due to the presence of the four enantiomerically equivalent tritopic linkers.

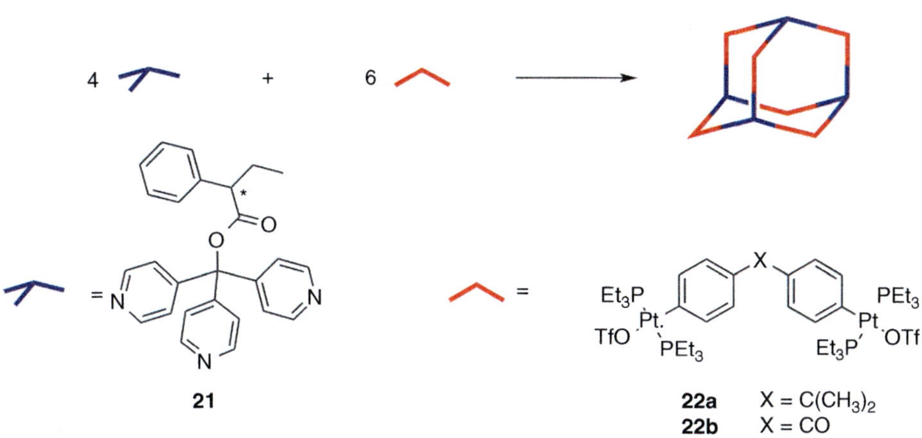

Figure 8.12 *Combination of four chiral tritopic nodes, **21** (represented schematically in blue), with ditopic linkers **22a** and **22b** resulted of the first chiral polyhedra, in this case an adamantoid architecture*

In an early and elegant piece of work on this topic, Ward *et al.* describe the first diastereoselective synthesis of a chiral M_4L_6 tetrahedral cage complex. The ligand, **23**, contains a pinene group fused to a pyridyl group, which is the chiral structural motif of the "CHIRAGEN" ligands. Reaction with Co(II) or Zn(II) salts resulted in the formation of discrete tetrahedral cages with NMR evidence indicating formation of only one diastereomer [96]. In the crystal structure of $[Zn_4\mathbf{23}_6](BF_4)_8$, the complex is pseudo-$T$-symmetric (Figure 8.13) with the symmetry reduced by crystal packing effects. The ligands are twisted and minimization of steric interference between the terminal pinene groups causes the twist on each ligand to occur with the same handedness. The result is a marked amplification of the optical activity associated with the ligands. The α_D (589 nm) for the complex was –300°, while that for the pure ligand was –74°. This equates to a 30-fold increase in the optical rotation per mole of complex (which contains six ligands) which is thus five times greater per mole ligand. The chiral induction that occurs at the different metal centres and which is also driven by the interactions between chiral ligands also leads to formation of a single stereoisomer. As explained above, this phenomenon is attributed to stereochemical communication and is a feature of many of the systems and studies that have evolved in this area.

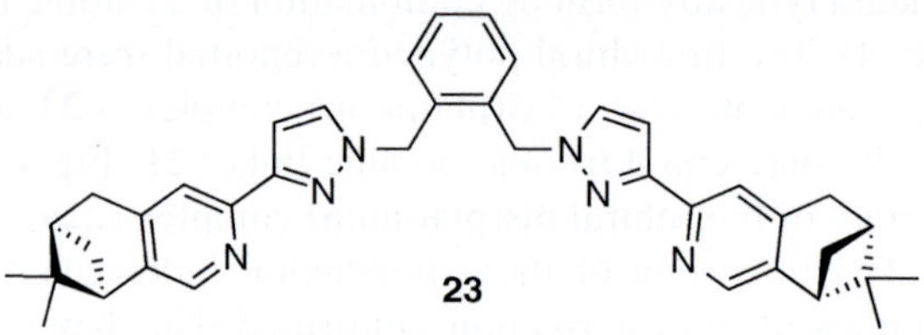

23

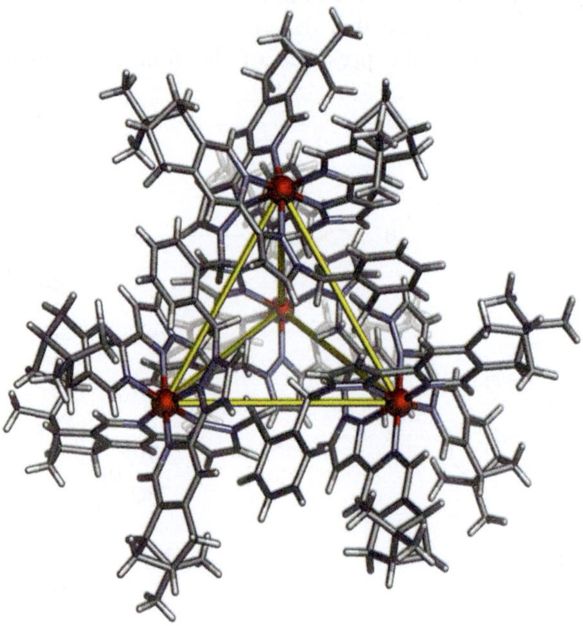

Figure 8.13 *The crystal structure of the diastereotopically pure* $[Zn_4\mathbf{23}_6](BF_4)_8$ *tetrahedron. Metal centres are all* Δ

In 2008, the group of Albrecht described the first enantiomerically pure M_4L_4 tetrahedron[97]. The container molecule is constructed from the combination of an enantiomerically pure tri-catecholate triangular ligand (**24**), which deprotonates at the phenol groups and combines with Ti(IV) metal centres to give the chiral $[Ti_4\mathbf{24}_4]^{8-}$ metallosupramolecular ion. The chiral component of the ligand (a phenylethylamide group) is in close proximity to the coordination site and it is believed that this, in turn, forces chiral induction at the metal centre so that not only are the ligands themselves a source of chirality but that their structure imparts particular axial chirality at the *tris*(bidentate) octahedral coordination centres. Although a crystal structure for the chiral tetrahedron was not determined (crystals not being obtained) the authors noted that the *S*-phenylethylamide substituent was previously shown to produce Λ stereochemistry, which comes about through a concerted arrangement of hydrogen bonds between the amide N–H (donor) and the neighbouring catecholate oxygen (acceptor). This, combined with circular dichroism studies, strongly suggests that the enantiomerically pure *tris*(catecholate) induces a Λ-configuration at all four of the Ti(IV) coordination centres.

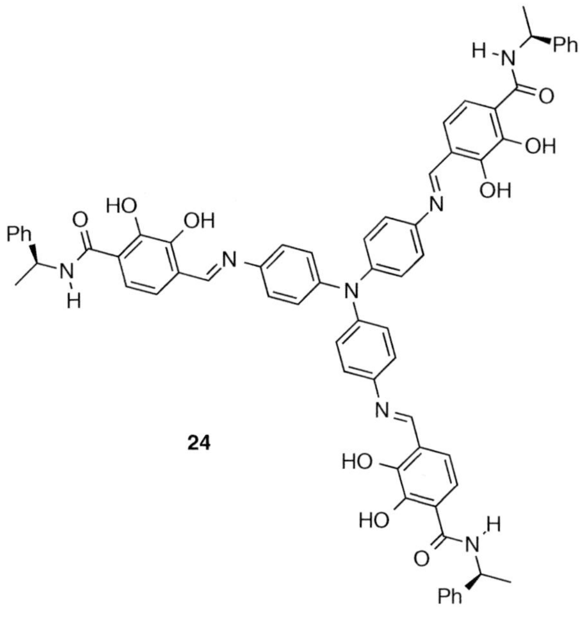

24

The group of Fujita has taken a somewhat different approach to the chiral ligand-mediated generation of chirality in metallosupramolecular capsules [98]. In this case, the inclusion of chiral amino acids and peptides as functional groups on the bridging ligand, **25**, was used to create chiral nano-sized $Pd_{12}\mathbf{25}_{24}$ 'molecular spheres' (cubeoctahedra) of the type shown in Figure 8.1. By design, the chiral peptides were placed on the ligand so that they would project in to the interior of the sphere and thus generate a chiral enclosed space. Combination of the nine different bridging ligands led to the creation of combinatorial libraries of spheres containing different types and ratios of the different chiral ligands. The cavity in the spheres was reasoned to resemble chiral space in enzyme pockets which could be used for asymmetric molecular recognition and as a vessel for asymmetric synthesis.

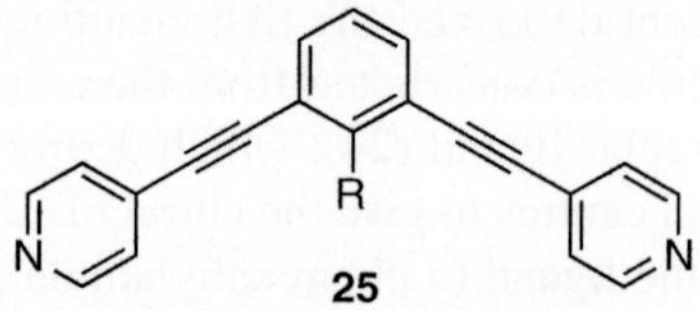
25

R = chiral amino acid/peptide

In follow-up work, the group of Fujita reported the synthesis of similar $M_{12}L_{24}$ spheres ($M=Pd^{2+}$), but where the chiral peptide substituents were placed on the bridging ligand, with the scaffold **26**, they projected from the surface of the sphere [99]. Of specific interest was the minTBP-1 hexapeptide aptamer (Arg-Lys-Leu-Pro-Asp-Ala) which is known for its ability to bind to Ti surfaces. The mode of binding is through electrostatic interaction between the cationic Arg and Lys residues and oxidised Ti surfaces under aqueous conditions. It was reasoned that incorporation of this aptamer on the surfaces of nano-spheres would provide a means of binding those spheres to inorganic surfaces and that the high density of those aptamers on the surface would provide multiple binding points for attachment. A single-crystal plate of TiO_2 was immersed in a solution of the spheres. Analysis of the surface by AFM showed that the spheres were bound uniformly on surface. Surprisingly the spheres were bound irreversibly and could not be removed by repeated (aqueous) washing. Typically, the adhesion of a single minTBP-1 aptamer is weak. It appears that the high density of aptamers on the surface of the spheres and thus the adhesion through multiple binding points significantly enhances the binding strength. Interestingly, spheres produced with a ligand using a different aptamer, with anionic components, show no particular propensity to bind to the Ti surface, highlighting the specificity of the minTBP-1 aptamer. The irreversible binding of the spheres opens new opportunities for surface chemistry and the authors identify possible applications whereby functional nanoparticles or molecules could be enclathrated within the sphere and then bound to surfaces where they may act as reaction centres and/or receptors.

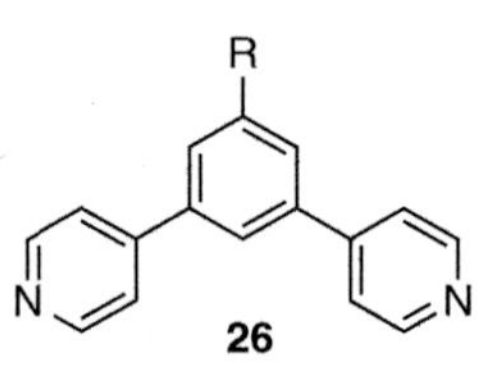
26

R = chiral peptide

Ligands with axial chirality have been shown by the group of Cui to induce diastereoselective synthesis of M_4L_6 tetrahedral cages [100]. The di-acetylacetenato ligand with biphenylene backbone 5,5′,6,6′-tetramethyl-3,3′-diketone-2,2′-bis(methoxymethoxy)-biphenyl, **27**, forms metallosupramolecular tetrahedra with Fe(III) and Ga(III) with high enantiopurity at the metal centres (Figure 8.14). The axial chirality of the ligand occurs through restricted rotation around the bond connecting the phenyl rings caused by the pendant methyl and ether groups. The authors synthesized both *R* and *S* forms of the ligand from commercially available enantiopure 5,5′,6,6′-tetramethyl-3,3′-di-*tert*-butyl-1,1′-biphenyl-2,2′-diol. On reaction of the *S* hand of the ligand (in a 3 : 2 ratio) with the metal the tetrahedron forms with Λ stereochemistry at all four metal centres. Conversely, when the *R*

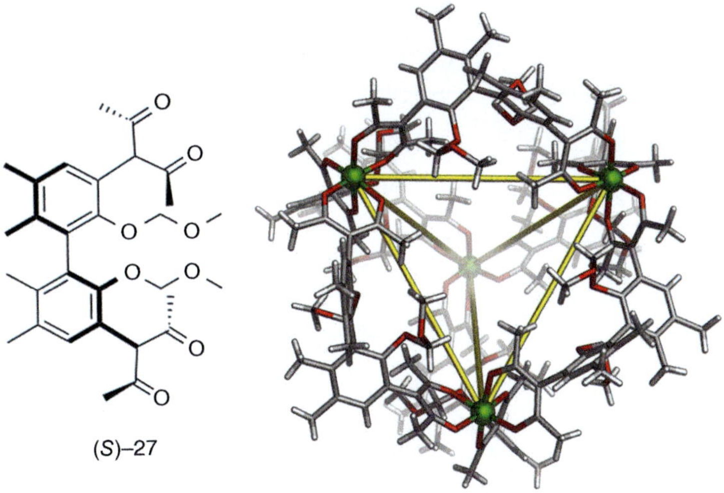

(*S*)–27

Figure 8.14 *The ligand* (S)-***27*** *(left) exclusively forms a tetrahedral metallosupramolecular complex ΛΛΛΛ-*[M_4((S)-*27*)$_6$] *enantiomer (M = Fe or Ga) while the* R *form of the ligand produces the ΔΔΔΔ enantiomer. The crystal structure of ΛΛΛΛ-*[Fe_4((S)-***27***)$_6$] *(right)*

hand of the ligand is used the homochiral Δ enantiomer is produced. The enantiopurity of the tetrahedra was confirmed through a combination of NMR and circular dichroism spectroscopy. Access to the cavity by guests is facilitated by relatively large apertures in the tetrahedral faces. Cocrystallization of the evacuated tetrahedra with racemic 2-butanol in chloroform was performed. The desorbed 2-butanol from [Fe_4((*S*)-**27**)$_6$] was found to be of the *S* hand and present in 98.8% *ee* (based on GC with a chiral support). Similar enantioselective behaviour was also observed for the uptake of 3-methyl-2-butanol. In this case, cocrystallization with [Fe_4((*R*)-**27**)$_6$] occurred with 99.5% *ee* of the *R* enantiomer.

The elegant transfer of chiral information from the ligand to the metal has been illustrated through so-called 'sergeants and soldiers' experiments by Nitschke *et al.* [101]. The di-pyridylimine bridging ligand, (*S*)-**28**, formed an M_4L_6 tetrahedral cage on reaction with iron(II). Importantly, the species was produced in a one-pot reaction by combination of 12 equivalents of resolved (*S*)-**29** and six equivalents of **30** in the presence of iron(II) as shown in Figure 8.15. The resulting [Fe_4((*S*)-**28**)$_6$]$^{8+}$ complex had homochiral Δ-*tris*(bidentate) coordination environments.

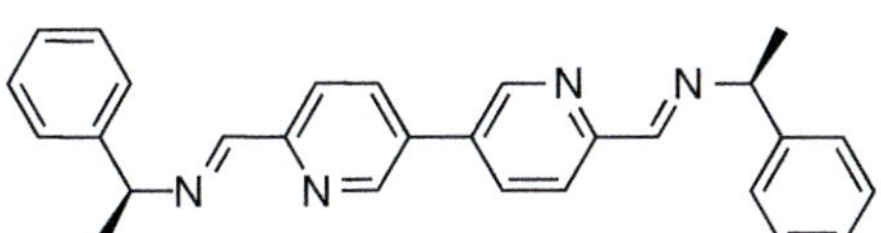

(*S*)–**28**

By contrast, the related ligand **31**, which is achiral produces a racemic mixture of homochiral Δ- and Λ-[Fe_4**31**$_6$]$^{8+}$ tetrahedra [102]. The reversible nature of imine bond formation provides the potential for *in situ* replacement of the imine substituent by introduction of alternate amines. The Nitschke group exploited this to carry out 'sergeants and soldiers' experiments whereby the chiral amine (*S*)-**29** was introduced to a racemic

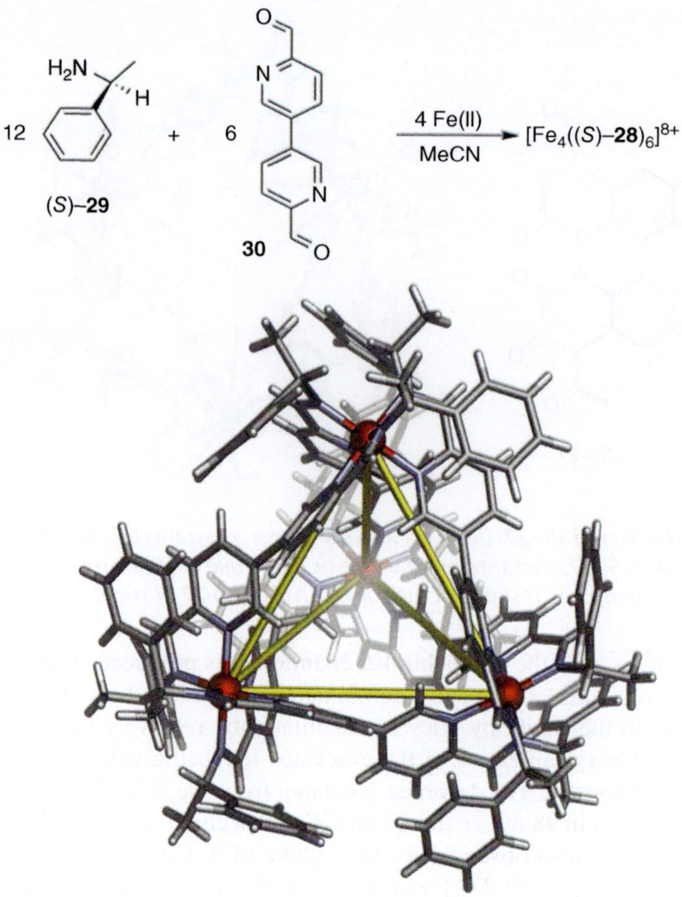

Figure 8.15 *One-pot synthesis of [Fe$_4$((S)-**28**)$_6$]$^{8+}$ (top) and the crystal structure of the triflate salt of ΔΔΔΔ-[Fe$_4$((S)-**28**)$_6$] (bottom)*

mixture of [Fe$_4$**31**$_6$]$^{8+}$. It was found that the more nucleophilic aliphatic amine was able to replace the aromatic imine component of **31** to create (*S*)-**28**. Interestingly, the CD intensities increased nonlinearly with respect to increasing amounts of (*S*)-**29** with maximum CD intensity achieved when 50% of (*S*)-**28** had been incorporated implying that two (*S*)-**28** components (out of six) on each tetrahedron is sufficient to induce Δ-handedness at all four centres. Changing the steric bulk of the terminal chiral group (from phenyl to ethyl) and increasing the length of the bridging ligand were both found to reduce stereochemical coupling effects [38].

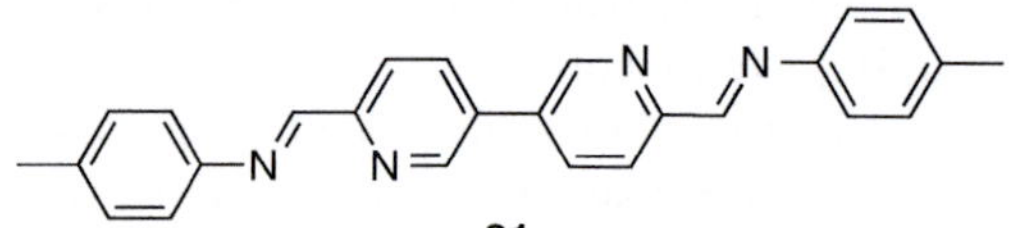

31

Chiral polyhedra can, in some instances, retain stereochemistry even after ligand substitution through a chiral memory effect. The first report of this was made by the group of Raymond [90]. The chiral metallosupramolecular tetrahedron ΔΔΔΔ-$[(Et_4N^+)\subset Ga_4\mathbf{20a}_6]^{11-}$ retained homochirality after solution-state ligand exchange with **20b**. The progress of ligand exchange was followed by NMR and CD spectroscopies over 24 hours at 75 °C. The NMR experiments clearly indicated ligand exchange while the CD spectroscopy confirmed no significant racemization of the metal coordination environments.

More recently, the group of Nitschke reported the formation of a chiral Fe_4L_4 tetrahedral capsule by combination of a tritopic trialdehyde with a chiral amine to produce a tri-bidentate ligand (**32**) [103]. The *S* form of the chiral imine produces the ΔΔΔΔ-$[Fe_4((S)\text{-}\mathbf{32})_4]^{8+}$ tetrahedral cation (Figure 8.16).

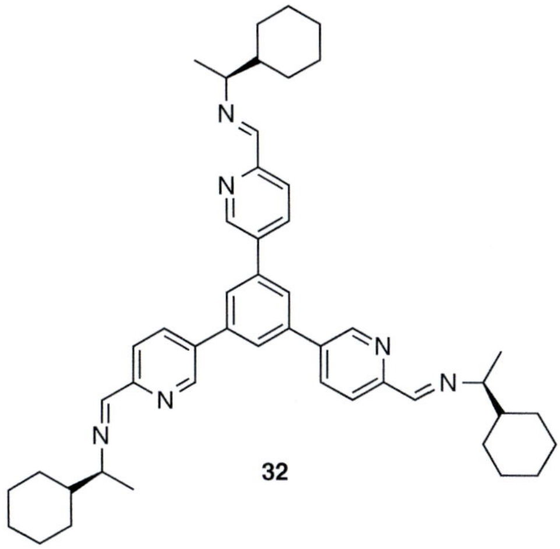

32

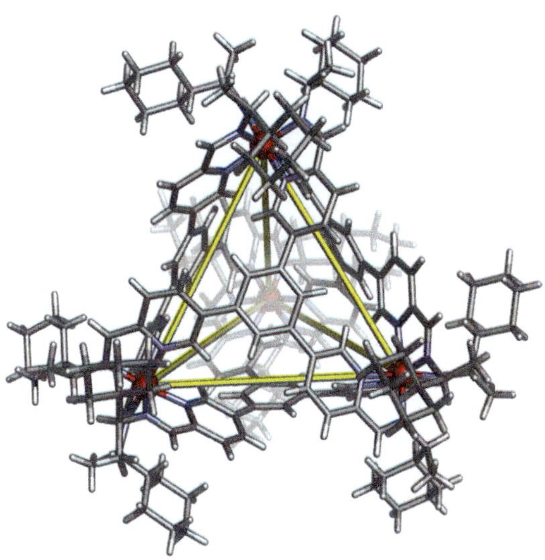

Figure 8.16 *The crystal structure of the diastereoptopically pure ΔΔΔΔ-$[Fe_4((S)\text{-}\mathbf{32})_4]^{8+}$ face-capped tetrahedral complex*

Introduction of the achiral tripodal amine, **33**, to a solution of $[Fe_4((S)\text{-}\mathbf{32})_4]^{8+}$ results in threefold displacement of the chiral amine groups at each metal centre and thus links terminal groups of the ligand to create a covalently connected tetrahedral capsule while maintaining the ΔΔΔΔ chirality at each metal centre through a remarkable chiral memory effect. This is despite the fact that the ligand is now achiral. After four days at 90 °C the enantiomeric excess is 99%. The fact that each terminal group is covalently linked effectively limits the racemization pathway to a twisting mechanism as opposed to a ligand dissociation/association pathway. Mechanistic studies indicated that the chiral memory is a consequence of stereochemical communication between the metal centres resulting from the threefold symmetry of the ligand and its relatively rigid structure combined with the fact that the chiral/achiral substitution occurs through a stepwise mechanism. This is further confirmed by the fact that removal of the metal centres results in loss of optical activity. The importance of the threefold connectivity in the chiral memory effect (in this instance at least) is that similar experiments were conducted using a ditopic ligand to produce an M_4L_6 tetrahedron. In this case the chiral memory effect was not observed, despite what the authors describe as the presence of strong stereochemical communication between the metal centres.

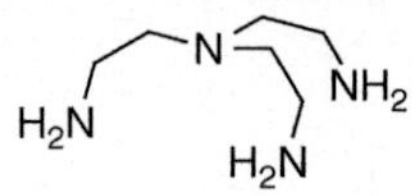

33

The chiral nature of M_4L_6 tetrahedra with *tris*(bidentate) metal coordination environment extends to the internal cavity. This is demonstrated by the enantio-differentiation of achiral tetrafluoroborate anions encapsulated inside chiral tetrahedra [104]. The ligands **34** and **35** both combine with Co(II) to produce $[Co_4\mathbf{34}_6]^{8+}$ and $[Co_4\mathbf{35}_6]^{8+}$ cations. In both cases the tetrahedra form with *T* symmetry and therefore exist as a racemic mixture of the homochiral ΔΔΔΔ and ΛΛΛΛ isomers. The central cavity tightly binds one BF_4^- counterion.

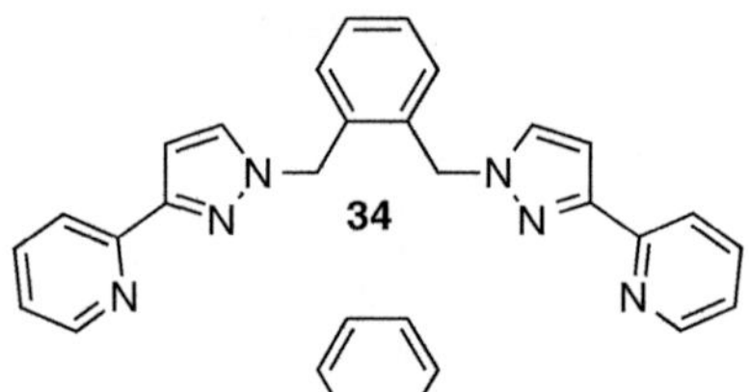

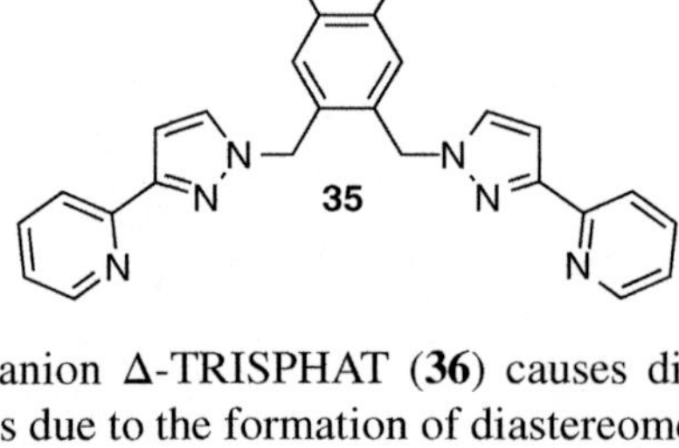

Addition of the chiral anion Δ-TRISPHAT (**36**) causes differentiation of the proton NMR signals of the ligands due to the formation of diastereomeric ion-pairs with different binding affinities. The effect is dependent on the concentration of Δ-TRISPHAT and differences as large as 0.4 ppm in the chemical shifts for the diastereomeric ion pairs were observed. Remarkably though, the effect of the chiral differentiation is also very evident on the internally bound achiral BF_4^- ion. The ^{19}F NMR signal, which occurs at approximately

–245 ppm, moves towards higher frequencies on addition of Δ-TRISPHAT and with the shift associated with the more tightly-bound diastereomeric ion-pair being significantly more perturbed than the other. A difference of 2 ppm between the signals associated with the bound anions was observed. This is particularly interesting given that the cavity is well isolated from the external surface of the capsule which is in contact with the TRISPHAT.

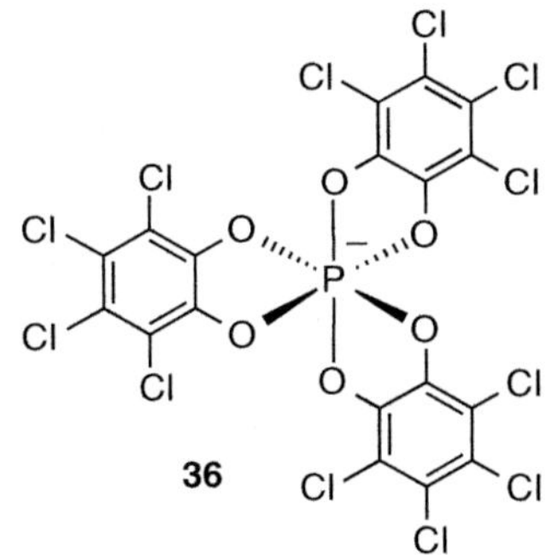

36

Nitschke *et al.* prepared both (*R*,*R*)-**37** and (*S*,*S*)-**37** and reacted them with iron(II) to produce enantio- and diastereo-pure tetrahedra $[Fe_4((R,R)\text{-}\mathbf{37})_6]^{8+}$ and $[Fe_4((S,S)\text{-}\mathbf{37})_6]^{8+}$ respectively [105]. The resolved glycerol groups were found to direct the chirality of the octahedral metal centres to exclusively produce either ΛΛΛΛ-$[Fe_4((R,R)\text{-}\mathbf{37})_6]^{8+}$ or ΔΔΔΔ-$[Fe_4((S,S)\text{-}\mathbf{37})_6]^{8+}$. Although no crystal structure was obtained, the chiral relationship between the metal centres and the chiral ligands was inferred through observation of the sign of the Cotton effect in the circular dichroism spectra. A racemic mixture of both cages resulted in no optical activity. As well as imparting resolved chirality onto the cage assembly the glycerol groups also imparted water solubility and assisted in enclosing the 418 Å^3 internal cavity. The tetrahedral capsules were demonstrated to bind a wide variety of hydrophobic guest molecules. While the assembly is enantio-pure, no enantio-selective guest binding was observed for the chiral guest limonene. Despite ΔΔΔΔ-$[Fe_4((S,S)\text{-}\mathbf{37})_6]^{8+}$ encapsulating both (*R*)-limonene and (*S*)-limonene with similar binding constants, differences were observed in the respective NMR spectra allowing for this chiral metallosupramolecular polyhedron to be used as encapsulative chiral shift reagents.

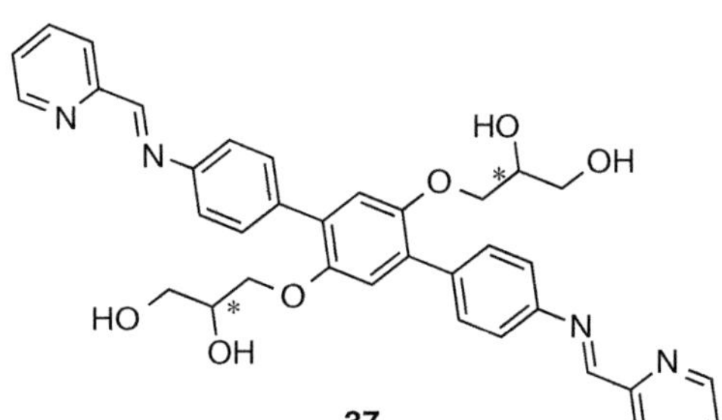

37

In 2013, Raymond *et al.* described the synthesis of a diastereo- and enantio-pure M_4L_6 tetrahedron [106]. The di-catecholate ligand, **38**, terminates with chiral *t*-butylmethylamide groups. The authors prepared both the *R* and *S* forms of the ligand and subsequently

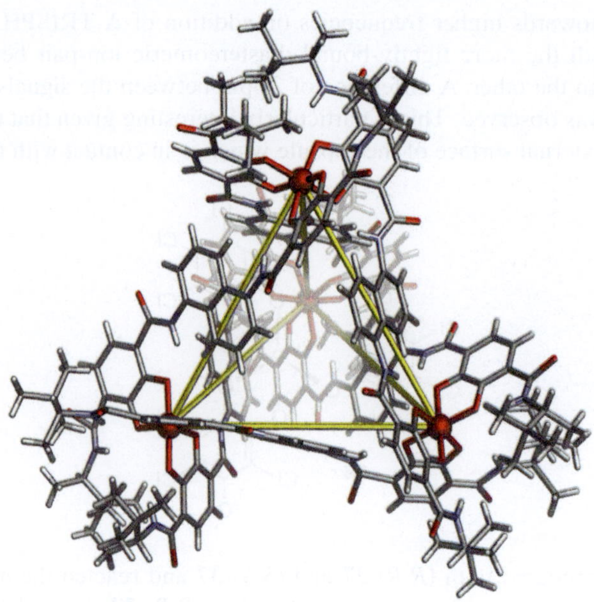

Figure 8.17 *The crystal structure of ΔΔΔΔ-[Ga((R)-**38**)$_6$]$^{12-}$ tetrahedral complex*

prepared metallosupramolecular tetrahedra of the form [Ga_4((*R*)-**38**)$_6$]$^{12-}$ and [Ga_4((*S*)-**38**)$_6$]$^{12-}$ by reaction of the ligand with Ga(III). Circular dichroism spectroscopy confirmed that the *R* enantiomer of the ligand produces the ΔΔΔΔ form of the tetrahedron whereas the *S* form of the ligand produces the ΛΛΛΛ tetrahedron. These results were further confirmed by the crystal structure of the ΔΔΔΔ tetrahedron produced from the *R* enantiomer of the ligand (Figure 8.17).

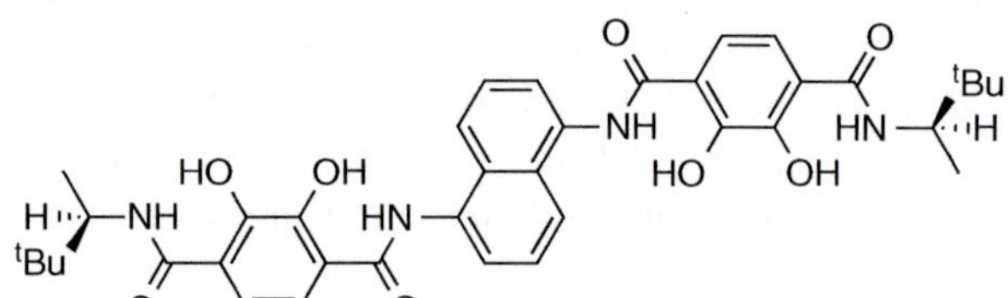

The host-guest chemistry of these species was tested with chiral ammonium salts and it was found that their interaction resulted in substantial enantiomeric enrichment of the salt. Similar tetrahedra had been produced by Raymond *et al.* prior to this using ligands without the chiral centre [93]. These were resolved by column chromatography but proved unstable to racemisation in solution. The complexes containing the chiral ligands resulted in permanent helical chirality at the metal centres and this makes them suitable as enantioselective catalysts. Indeed, the compounds do exhibit enantioselective and chemoselective catalytic properties [106, 107]. In the carbonyl-ene cyclization of **39** to produce the enantiomers **40** and **41** (Figure 8.18) the cyclization proceeds with 92% yield with chemoselectivity leading to *trans*:*cis* ratio of 8 : 1 and 61% enantiomeric excess of **40**.

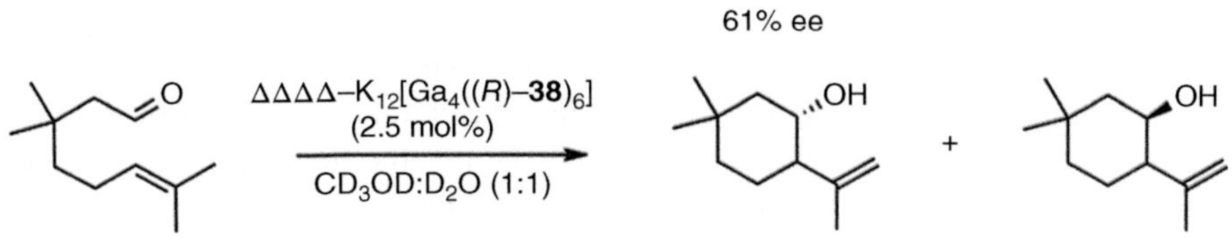

39 40 41

Figure 8.18 *The cyclization of **39** in the presence of a catalytic quantity of the Δ-homochiral tetrahedron ΔΔΔΔ-$[Ga_4((R)\text{-}\mathbf{38})_6]^{12-}$ produces the enantiomers **40** and **41** but with **40** in 61% enantiomeric excess*

Chiral auxiliary or capping ligands (ligands not involved in bridging the metal centres) also lead to the generation of chiral polyhedra [108]. The first example came from the group of Stang and involved the use of *R*-(+)-BINAP, **42**, to fill the coordination sphere of square planar Pd(II) and Pt(II) centres, connected in each case in to an M_4L_4 tetrahedral capsule by the bridging ligand **43**.

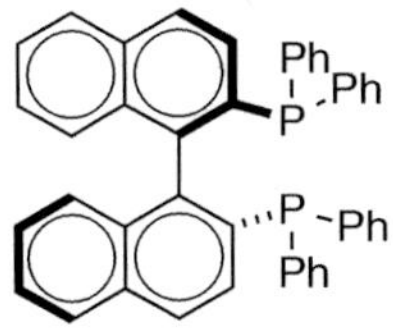

42

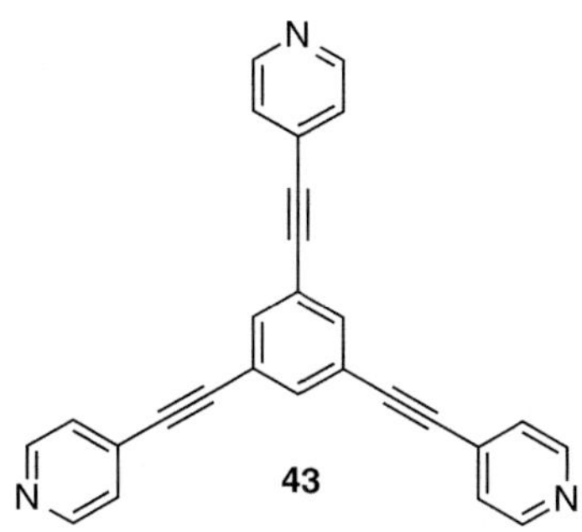

43

Even with the chiral auxiliary ligands located on the outside of the polyhedral capsule, there are implications for the chirality of the interior space. The M_6L_4 octahedral capsule produced by the reaction of **44** with Pd(II) was exploited as a reaction vessel for [2+2] olefin cross photoaddition reactions [109]. In this capsule, each square planar Pd(II) centre coordinates in a *cis* arrangement to two of the bridging ligands. When the two remaining coordination sites are occupied by the achiral diamine **45**, the overall architecture is achiral. The normally inert compound fluoroanthene underwent efficient cross [2+2] photoaddition with *N*-cyclohexylmaleimide. This reaction proceeds inside the capsule.

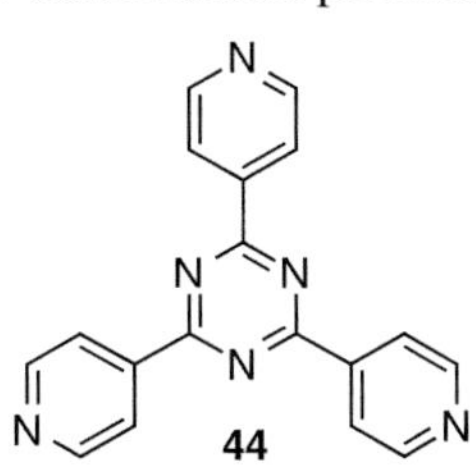

44

This unusual reaction led the group to examine the potential for asymmetric synthesis through chiral induction. The same capsule was used but with the achiral auxiliary ligands replaced with chiral (resolved) (1*R*,2*R*)-*N*,*N*′-diethyl,-1,2-diaminocyclohexane (**46**). Under these conditions the cross [2+2] photoaddition reaction occurs but now proceeds with 40% enantiomeric excess of one enantiomer of the product. This is a remarkable result given that the reaction takes place inside the capsule while the chiral components of the capsule (the auxiliary ligands) project on the outside. It appears that the steric bulk of the auxiliary ligands plays an important role. While the photoaddition reaction occurs when the capsule employs the auxiliary ligands **47** and **48**, which have progressively less steric bulk around the nitrogen atoms, the enantiomeric excess of product is reduced significantly to 20% and 5%, respectively.

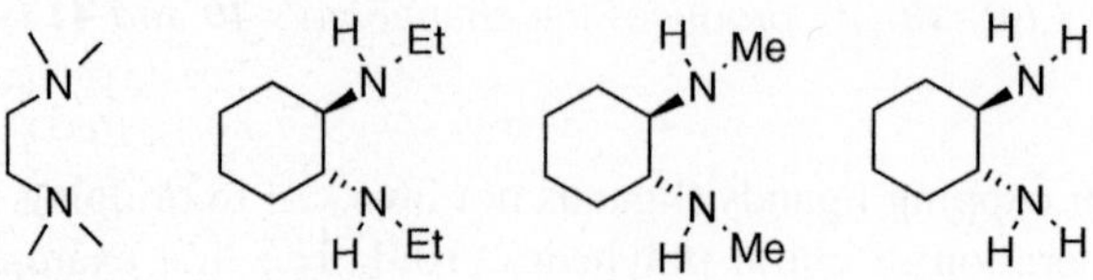

45 **46** **47** **48**

This research has since been extended to asymmetric [2+4] cycloadditions[110]. Given the unique ability of metallosupramolecular capsules to act as reaction vessels it can be expected that there will be many more such examples in the years to come.

8.7 Conclusions and Outlook

There is enormous scope for the creation of chiral metallosupramolecular polyhedra. This is in large part because the polyhedra, by their very nature, are constructed of multiple components (metals, bridging ligands, auxiliary ligands, counterions, guests) and each of these, as we have described in the preceding pages, is a potential source of chirality. Where the polyhedra contain resolved components, the bulk sample is optically active. Chiral polyhedra can also form from achiral components and indeed most known examples are of this type. In such cases, optical activity is achieved by resolution and in many cases the optical activity is retained even after substitution or exchange of the resolving agent. Polyhedra of all types are fascinating for their symmetry and beauty and there is no doubt that this alone has been a significant motivator for many researchers interested in the creation of supramolecular analogues. The deliberate generation of chirality in such systems adds a further challenge but amplifies their beauty. Perhaps even more importantly, efforts to create such species help us to develop and understand the fundamental principles of supramolecular design allowing the creation of more structurally sophisticated examples. While many of the endearing features of supramolecular polyhedra could be viewed as esoteric it is clear that these remarkable materials have great commercial potential. Recent advances in the field are beginning to focus on the applications of these coordination polyhedra for asymmetric synthesis and catalysis. It is likely that their use for these purposes will expand significantly over the coming years, particularly in light of the need for efficient synthesis of chirally resolved pharmaceuticals and biologically active agents. Whatever the future holds for these materials, one thing is certain – their design and construction has been at the forefront of supramolecular chemistry – where it will stay while ever there are chemists fascinated by their beauty and the challenges of their creation.

References

[1] R. W. Saalfrank, A. Stark, K. Peters, H. G. von Schnering, Adamantoid Chelate Complexes.1. The First Adamantoid Alkaline-Earth Metal Chelate Complex – Synthesis, Structure, and Reactivity, *Angewandte Chemie, International Edition*, **27**(6), 851–853 (1988).

[2] L. F. Lindoy, I. Atkinson (2000) *Self-Assembly in Supramolecular Systems*, Royal Society of Chemistry, Cambridge.

[3] R. W. Saalfrank, H. Maid, A. Scheurer, Supramolecular Coordination Chemistry: The Synergistic Effect of Serendipity and Rational Design, *Angewandte Chemie, International Edition*, **47**(46), 8794–8824 (2008).

[4] D. L. Caulder, K. N. Raymond, Supermolecules by Design, *Accounts of Chemical Research*, **32**(11), 975–982 (1999).

[5] T. R. Cook, Y.-R. Zheng, P. J. Stang, Metal-Organic Frameworks and Self-Assembled Supramolecular Coordination Complexes: Comparing and Contrasting the Design, Synthesis, and Functionality of Metal-Organic Materials, *Chemical Reviews*, **113**(1), 734–777 (2013).

[6] R. Chakrabarty, P. S. Mukherjee, P. J. Stang, Supramolecular Coordination: Self-Assembly of Finite Two- and Three-Dimensional Ensembles, *Chemical Reviews*, **111**(11), 6810–6918 (2011).

[7] M. D. Ward, Polynuclear Coordination Cages, *Chemical Communications* (30), 4487–4499 (2009).

[8] C. Richardson, J. K. Clegg, L. F. Lindoy. Bioinspired Self-Assembly I: Self-Assembled Structures. In: Swiegers GF, editor. *Bioinspiration and Biomimicry in Chemistry*. Wiley-VCH, Weinheim, pp. 17–46 (2012)

[9] S. J. Dalgarno, N. P. Power, J. L. Atwood, Metallo-Supramolecular Capsules, *Coordination Chemistry Reviews*, **252**(8–9), 825–841 (2008).

[10] D. J. Tranchemontagne, Z. Ni, M. O'Keeffe, O. M. Yaghi, Reticular Chemistry of Metal-Organic Polyhedra, *Angewandte Chemie, International Edition*, **47**(28), 5136–5147 (2008).

[11] M. Yoshizawa, J. K. Klosterman, M. Fujita, Functional Molecular Flasks: New Properties and Reactions within Discrete, Self-Assembled Hosts, *Angewandte Chemie, International Edition*, **48**(19), 3418–3438 (2009).

[12]. N. J. Young, B. P. Hay, Structural Design Principles for Self-Assembled Coordination Polygons and Polyhedra, *Chemical Communications*, **49**(14), 1354–1379 (2013).

[13] M. M. J. Smulders, I. A. Riddell, C. Browne, J. R. Nitschke, Building on Architectural Principles for Three-Dimensional Metallosupramolecular Construction, *Chemical Society Reviews*, **42**(4), 1728–1754 (2013).

[14] M. Fujita, Metal Directed Self-Assembly of Two and Three Dimensional Synthetic Receptors, *Chemical Society Reviews*, **27**, 417 (1998).

[15] M. Han, D. M. Engelhard, G. H. Clever, Self-Assembled Coordination Cages Based on Banana-Shaped Ligands, *Chemical Society Reviews*, **43**(6), 1848–1860 (2014).

[16] K. Harris, D. Fujita, M. Fujita, Giant Hollow M_nL_{2n} Spherical Complexes: Structure, Functionalisation and Applications, *Chemical Communications*, **49**(60), 6703–6712 (2013).

[17] B. Olenyuk, J. A. Whiteford, A. Fechtenkötter, P. J. Stang, Self-Assemblty of Nanoscale Cuboctahedra by Coordination Chemistry, *Nature*, **398**, 796–799 (1999).

[18] M. Fujita, D. Oguro, M. Miyazawa, *et al.*, Self-Assembly of Ten Molecules into Nanometre-Sized Organic Host Frameworks, *Nature*, **378**(6556), 469–471 (1995).

[19] T. D. Hamilton, L. R. MacGillivray, Enclosed Chiral Environments from Self-Assembled Metal-Organic Polyhedra, *Crystal Growth and Design*, **4**(3), 419–430 (2004).

[20] B. El Aroussi, L. Guenee, P. Pal, J. Hamacek, Lanthanide-Mediated Supramolecular Cages and Host-Guest Interactions, *Inorganic Chemistry*, **50**(17), 8588–8597 (2011).

[21] R. W. Saalfrank, B. Hörner, D. Stalke, J. Salbeck, Adamantanoid Chelate Complexes. 3. The First Neutral Adamantanoid Iron(III) Chelate Complex: Spontaneous Formation, Structure, and Electrochemistry, *Angewandte Chemie, International Edition*, **32**(8), 1179–1182 (1993).

[22] R. W. Saalfrank, H. Glaser, B. Demleitner, *et al.*, Self-Assembly of Tetrahedral and Trigonal Antiprismatic Clusters $[Fe_4(L4)_4]$ and $[Fe_6(L5)_6]$ on the Basis of Trigonal Tris-Bidentate Chelators, *Chemistry – A European Journal*, **8**(2), 493–497 (2002).

[23] D. L. Caulder, R. E. Powers, T. N. Parac, K. N. Raymond, The Self-Assembly of a Predesigned Tetrahedral M_4L_6 Supramolecular Cluster, *Angewandte Chemie, International Edition*, **37**(13–14), 1840–1843 (1998).
[24] C. Brückner, R. E. Powers, K. N. Raymond, Symmetry-Driven Rational Design of a Tetrahedral Supramolecular Ti_4L_4 Cluster, *Angewandte Chemie, International Edition*, **37**(13–14), 1837–1839 (1998).
[25] A. J. Amoroso, J. C. Jeffery, P. L. Jones, *et al.*, Self-Assembly of a Ferromagnetically Coupled Manganese(II) Tetramer, *Angewandte Chemie, International Edition*, **34**, 1443–1446 (1995).
[26] J. S. Fleming, K. L. V. Mann, C.-A. Carraz, *et al.*, Anion-Templated Assembly of a Supramolecular Cage Complex, *Angewandte Chemie, International Edition*, **37**(9), 1279–1281 (1998).
[27] T. K. Ronson, S. Zarra, S. P. Black, J. R. Nitschke, Metal-Organic Container Molecules through Subcomponent Self-Assembly, *Chemical Communications*, **49**(25), 2476–2490 (2013).
[28] J. K. Clegg, L. F. Lindoy, B. Moubaraki, *et al.*, Triangles and Tetrahedra: Metal Directed Self-Assembly of Metallo-Supramolecular Structures Incorporating Bis-β-Diketonato Ligands, *Dalton Transactions*, 2417–2423 (2004).
[29] C. R. K. Glasson, G. V. Meehan, J. K. Clegg, *et al.*, A New Fe^{II} Quaterpyridyl M_4L_6 Tetrahedron Exhibiting Selective Anion Binding, *Chemical Communications*, 1190–1192 (2008).
[30] C. R. K. Glasson, J. K. Clegg, J. C. McMurtrie, *et al.*, Unprecedented Encapsulation of a $[Fe^{III}Cl_4]^-$ Anion in a Cationic $[Fe^{II}L_6]^{8+}$ Tetrahedral Cage Derived from 5, 5‴-Dimethyl-2, 2′:5′, 5″:2″, 2‴-Quaterpyridine, *Chemical Science*, **2**(3), 540–543 (2011).
[31] S. Yi, V. Brega, B. Captain, A. E. Kaifer, Sulfate-Templated Self-Assembly of New M_4L_6 Tetrahedral Metal Organic Cages, *Chemical Communications*, **48**(83), 10295–10297 (2012).
[32] R. Custelcean, J. Bosano, Peter V. Bonnesen, *et al.*, Computer-Aided Design of a Sulfate-Encapsulating Receptor, *Angewandte Chemie, International Edition*, **48**(22), 4025–4029 (2009).
[33] M. Albrecht, I. Janser, S. Meyer, *et al.*, A Metallosupramolecular Tetrahedron with a Huge Internal Cavity, *Chemical Communications*, 2854–2855 (2003).
[34] A. Ferguson, M. A. Squire, D. Siretanu, *et al.*, A Face-Capped $[Fe_4L_4]^{8+}$ Spin Crossover Tetrahedral Cage, *Chemical Communications*, **49**(16), 1597–1599 (2013).
[35] T. Beissel, R. E. Powers, T. N. Parac, K. N. Raymond, Dynamic Isomerization of a Supramolecular Tetrahedral M_4L_6 Cluster, *Journal of the American Chemical Society*, **121**(17), 4200–4206 (1999).
[36] W. Meng, J. K. Clegg, J. D. Thoburn, J. R. Nitschke, Controlling the Transmission of Stereochemical Information through Space in Terphenyl-Edged Fe_4L_6 Cages, *Journal of the American Chemical Society*, **133**(34), 13652–13660 (2011).
[37] J. K. Clegg, J. Cremers, A. J. Hogben, *et al.*, A System of Self-Assembled Anion-Binding $Fe_4L_6^{8+}$ Cages That Adapts to External Stimuli *Chemical Science*, **4**(1), 68–76 (2013).
[38] N. Ousaka, S. Grunder, A. M. Castilla, *et al.*, Efficient Long-Range Stereochemical Communication and Cooperative Effects in Self-Assembled Fe_4L_6 Cages, *Journal of the American Chemical Society*, **134**(37), 15528–15537 (2012).
[39] T. K. Ronson, A. B. League, L. Gagliardi, *et al.*, Pyrene-Edged $Fe^{II}_4L_6$ Cages Adaptively Reconfigure During Guest Binding, *Journal of the American Chemical Society*, **136**(44), 15615–15624 (2014).
[40] R. W. Saalfrank, H. Maid, A. Scheurer, *et al.*, Template and Ph-Mediated Synthesis of Tetrahedral Indium Complexes [Cs. $\{In_4(L)_4\}]^+$ and $[In_4(HL)_4]^{4+}$: Breaking the Symmetry of N-Centered C_3 $(L)^{3-}$ to Give Neutral $[In_4(L)_4]$, *Angewandte Chemie, International Edition*, **47**, 8941–8945 (2008).
[41] R. W. Saalfrank, H. Maid, A. Scheurer, *et al.*, Mesomerization of S_4-Symmetric Tetrahedral Chelate Complex $[In_4(L^3)_4]$: First-Time Monitored by Temperature-Dependent ^{1}h Nmr Spectroscopy, *European Journal of Inorganic Chemistry*, **19**, 2903–2906 (2010).
[42] J. K. Clegg, F. Li, K. A. Jolliffe, *et al.*, An Expanded Neutral M_4L_6 Cage That Encapsulates Four Tetrahydrofuran Molecules, *Chemical Communications*, **47**(21), 6042–6044 (2011).
[43] C. R. K. Glasson, G. V. Meehan, C. A. Motti, *et al.*, New Nickel(II) and Iron(II) Helicates and Tetrahedra Derived from Expanded Quaterpyridines, *Dalton Transactions*, **40**(40), 10481–10490 (2011).

[44] T. Beissel, R. E. Powers, K. N. Raymond, Symmetry-Based Metal Complex Cluster Formation, *Angewandte Chemie, International Edition*, **35**(10), 1084 (1996).

[45] C. Giri, F. Topic, P. Mal, K. Rissanen, Self-Assembly of a M_4L_6 Complex with Unexpected S_4 Symmetry, *Dalton Transactions*, **43**(48), 17889–17892 (2014).

[46] W. Meng, T. K. Ronson, J. R. Nitschke, Symmetry Breaking in Self-Assembled M_4L_6 Cage Complexes, *Proceedings of the National Academy of Sciences*, **110**(26), 10531–10535 (2013).

[47] M. M. J. Smulders, A. Jiménez, J. R. Nitschke, Integrative Self-Sorting Synthesis of a $Fe_8Pt_6L_{24}$ Cubic Cage, *Angewandte Chemie, International Edition*, **51**(27), 6681–6685 (2012).

[48] M. B. Duriska, S. M. Neville, J. Lu, *et al.*, Systematic Metal Variation and Solvent and Hydrogen-Gas Storage in Supramolecular Nanoballs, *Angewandte Chemie, International Edition*, **48**, 8919–8922 (2009).

[49] H.-B. Wu, Q.-M. Wang, Construction of Heterometallic Cages with Tripodal Metalloligands, *Angewandte Chemie, International Edition*, **48**(40), 7343–7345 (2009).

[50] F. Reichel, J. K. Clegg, K. Gloe, *et al.*, Self-Assembly of an Imidazolate-Bridged FeIII/CuII Heterometallic Cage, *Inorganic Chemistry*, **53**(2), 688–690 (2014).

[51] A. M. Najar, C. Avci, M. D. Ward, Self-Assembly of Trigonal Prismatic $M_6(L)_9$ Coordination Cages, *Inorganic Chemistry Communications*, **15**, 126–129 (2012).

[52] W. Meng, B. Breiner, K. Rissanen, *et al.*, A Self-Assembled M_8L_6 Cubic Cage that Selectively Encapsulates Large Aromatic Guests, *Angewandte Chemie, International Edition*, **50**(15), 3479–3483 (2011).

[53] W. J. Ramsay, T. K. Ronson, J. K. Clegg, J. R. Nitschke, Bidirectional Regulation of Halide Binding in a Heterometallic Supramolecular Cube, *Angewandte Chemie, International Edition*, **52**(50), 13439–13443 (2013).

[54] S. Turega, M. Whitehead, B. R. Hall, *et al.*, Shape-, Size-, and Functional Group-Selective Binding of Small Organic Guests in a Paramagnetic Coordination Cage, *Inorganic Chemistry*, **52**(2), 1122–1132 (2013).

[55] A. Stephenson, M. D. Ward, An Octanuclear Coordination Cage with a 'Cuneane' Core-a Topological Isomer of a Cubic Cage, *Dalton Transactions*, **40**(31), 7824–7826 (2011).

[56] M. H. Alkordi, J. L. Belof, E. Rivera, *et al.*, Insight into the Construction of Metal-Organic Polyhedra: Metal-Organic Cubes as a Case Study, *Chemical Science*, **2**(9), 1695–1705 (2011).

[57] A. Stephenson, M. D. Ward, Molecular Squares, Cubes and Chains from Self-Assembly of Bis-Bidentate Bridging Ligands with Transition Metal Dications, *Dalton Transactions*, **40**(40), 10360–10369 (2011).

[58] S. Roche, C. Haslam, S. L. Heath, J. A. Thomas, Self-Assembly of a Supramolecular Cube, *Chemical Communications*, **16**, 1681–1682 (1998).

[59] B. F. Abrahams, S. J. Egan, R. Robson, A Very Large Metallosupramolecular Capsule with Cube-Like 43 Topology Assembled from Twelve Cu(II) Centers and Eight Tri-Bidentate Tri-Anionic Ligands Derived from 2,4,6-Triphenylazo-1,3,5-Trihydroxybenzene, *Journal of the American Chemical Society*, **121**(14), 3535–3536 (1999).

[60] I. S. Tidmarsh, T. B. Faust, H. Adams, *et al.*, Octanuclear Cubic Coordination Cages, *Journal of the American Chemical Society*, **130**(45), 15167 (2008).

[61] J.-P. Lang, Q.-F. Xu, Z.-N. Chen, B. F. Abrahams, Assembly of a Supramolecular Cube from a Preformed Incomplete Cubane-Like Compound, *Journal of the American Chemical Society*, **125**(42), 12682–12683 (2003).

[62] I. A. Riddell, Y. R. Hristova, J. K. Clegg, *et al.*, Five Discrete Multinuclear Metal-Organic Assemblies from One Ligand: Deciphering the Effects of Different Templates, *Journal of the American Chemical Society*, **135**(7), 2723–2733 (2013).

[63] C. Browne, S. Brenet, J. K. Clegg, J. R. Nitschke, Solvent-Dependent Host–Guest Chemistry of an Fe_8L_{12} Cubic Capsule, *Angewandte Chemie, International Edition*, **52**(7), 1944–1948 (2013).

[64] I. A. Riddell, M. M. J. Smulders, J. K. Clegg, *et al.*, Anion-Induced Reconstitution of a Self-Assembling System to Express a Chloride-Binding $Co_{10}L_{15}$ Pentagonal Prism, *Nature Chemistry*, **4**(9), 751–756 (2012).

[65] I. A. Riddell, T. K. Ronson, J. K. Clegg, *et al.*, Cation- and Anion-Exchanges Induce Multiple Distinct Rearrangements within Metallosupramolecular Architectures, *Journal of the American Chemical Society*, **136**(26), 9491–9498 (2014).

[66] S. Zarra, J. K. Clegg, J. R. Nitschke, Selective Assembly and Disassembly of a Water-Soluble $Fe_{10}L_{15}$ Prism, *Angewandte Chemie, International Edition*, **52**(18), 4837–4840 (2013).
[67] R. A. Bilbeisi, T. K. Ronson, J. R. Nitschke, A Self-Assembled $[Fe^{II}L_{12}L_{12}]$ Capsule with an Icosahedral Framework, *Angewandte Chemie, International Edition*, **52**(34), 9027–9030 (2013).
[68] W. Meng, T. K. Ronson, J. K. Clegg, J. R. Nitschke, Transformations within a Network of Cadmium Architectures, *Angewandte Chemie, International Edition*, **52**(3), 1017–1021 (2013).
[69] N. K. Al-Rasbi, I. S. Tidmarsh, S. P. Argent, *et al.*, Mixed-Ligand Molecular Paneling: Dodecanuclear Cuboctahedral Coordination Cages Based on a Combination of Edge-Bridging and Face-Capping Ligands, *Journal of the American Chemical Society*, **130**(35), 11641–11649 (2008).
[70] A. Stephenson, S. P. Argent, T. Riis-Johannessen, *et al.*, Structures and Dynamic Behavior of Large Polyhedral Coordination Cages: An Unusual Cage-to-Cage Interconversion, *Journal of the American Chemical Society*, **133**(4), 858–870 (2011).
[71] S. P. Argent, H. Adams, T. Riis-Johannessen, *et al.*, High-Nuclearity Homoleptic and Heteroleptic Coordination Cages Based on Tetra-Capped Truncated Tetrahedral and Cuboctahedral Metal Frameworks, *Journal of the American Chemical Society*, **128**(1), 72–73 (2006).
[72] Y.-R. Zheng, W.-J. Lan, M. Wang, *et al.*, Designed Post-Self-Assembly Structural and Functional Modifications of a Truncated Tetrahedron, *Journal of the American Chemical Society*, **133**(42), 17045–17055 (2011).
[73] T. K. Ronson, J. Fisher, L. P. Harding, *et al.*, Stellated Polyhedral Assembly of a Topologically Complicated Pd_4L_4 'Solomon Cube', *Nature Chemistry*, **1**(3), 212–216 (2009).
[74] Q.-F. Sun, S. Sato, M. Fujita, An M18l24 Stellated Cuboctahedron through Post-Stellation of an $M_{12}L_{24}$ Core, *Nature Chemistry*, **4**(4), 330–333 (2012).
[75] J. Clayden, A. Lund, L. Vallverdu, M. Helliwell, Ultra-Remote Stereocontrol by Conformational Communication of Information along a Carbon Chain, *Nature*, **431**(7011), 966–971 (2004).
[76] M. Albrecht, I. Janser, H. Houjou, R. Froehlich, Long-Range Stereocontrol in the Self-Assembly of Two-Nanometer-Dimensioned Triple-Stranded Dinuclear Helicates, *Chemistry – A European Journal*, **10**, 2839–2850 (2004).
[77] A. M. Castilla, W. J. Ramsay, J. R. Nitschke, Stereochemistry in Subcomponent Self-Assembly, *Accounts of Chemical Research*, **47**(7), 2063–2073 (2014).
[78] A. M. Castilla, W. J. Ramsay, J. R. Nitschke, Stereochemical Communication within Tetrahedral Capsules, *Chemistry Letters*, **43**(3), 256–263 (2014).
[79] R. W. Saalfrank, B. Demleitner, H. Glaser, *et al.*, Chelate Complexes (Metalla Coronates), Part 21. Enantiomerisation of Tetrahedral Homochiral Clusters: Synchronized Four Bailar Twists and Six Atropenantiomerisation Processes Monitored by Temperature-Dependent Dynamic ^1H NMR Spectroscopy, *Chemistry – A European Journal*, **8**(12), 2679–2683 (2002).
[80] R. M. Yeh, K. N. Raymond, Supramolecular Asymmetric Induction in Dinuclear Triple-Stranded Helicates, *Inorganic Chemistry*, **45**, 1130–1139 (2006).
[81] M. Albrecht, Helicates Making Head-Head-Tails of It, *Nature Chemistry*, **6**(9), 761–762 (2014).
[82] A. D. Faulkner, R. A. Kaner, M. A. AbdallahQasem, *et al.*, Asymmetric Triplex Metallohelices with High and Selective Activity against Cancer Cells, *Nature Chemistry*, **6**(9), 797–803 (2014).
[83] M. Albrecht, 'Let's Twist Again' – Double-Stranded, Triple-Stranded, and Circular Helicates, *Chemical Reviews*, **101**(11), 3457 (2001).
[84] F. R. Keene, J. A. Smith, J. G. Collins, Metal Complexes as Structure-Selective Binding Agents for Nucleic Acids, *Coordination Chemistry Reviews*, **253**, 2021–2035 (2009).
[85] D. M. D'Alessandro, F. R. Keene, Stereochemical Effects on Intervalence Charge Transfer, *Pure and Applied Chemistry*, **80**, 1–16 (2008).
[86] C. R. K. Glasson, G. V. Meehan, J. K. Clegg, *et al.*, Microwave Synthesis of a Rare $[Ru_2L_3]^{4+}$ Triple Helicate and its Interaction with DNA, *Chemistry – A European Journal*, **14**, 10535–10538 (2008).
[87] G. Rapenne, J.-P. Sauvage, B. T. Patterson, F. R. Keene, Resolution, X-Ray Structure and Absolute Configuration of a Double-Stranded Helical Diiron(Ii) Bis(Terpyridine) Complex, *Chemical Communications*, 1853–1854 (1999).

[88] J. L. Bolliger, T. K. Ronson, M. Ogawa, J. R. Nitschke, Solvent Effects Upon Guest Binding and Dynamics of a $Fe^{II}_4L_4$ Cage, *Journal of the American Chemical Society*, **136**(41), 14545–14553 (2014).

[89] M. D. Pluth, R. G. Bergman, K. N. Raymond, Proton-Mediated Chemistry and Catalysis in a Self-Assembled Supramolecular Host, *Accounts of Chemical Research*, **42**, 1650–1659 (2009).

[90] M. Ziegler, A. V. Davis, D. W. Johnson, K. N. Raymond, Supramolecular Chirality: A Reporter of Structural Memory, *Angewandte Chemie, International Edition*, **42**(6), 665–668 (2003).

[91] A. J. Terpin, M. Ziegler, D. W. Johnson, K. N. Raymond, Resolution and Kinetic Stability of a Chiral Supramolecular Assembly Made of Labile Components, *Angewandte Chemie, International Edition*, **40**(1), 157160 (2001).

[92] D. Fiedler, D. H. Leung, R. G. Bergman, K. N. Raymond, Enantioselective Guest Binding and Dynamic Resolution of Cationic Ruthenium Complexes by a Chiral Metal-Ligand Assembly, *Journal of the American Chemical Society*, **126**(12), 3674–3675 (2004).

[93] A. V. Davis, D. Fiedler, M. Ziegler, *et al.*, Resolution of Chiral, Tetrahedral M4l6 Metal-Ligand Hosts, *Journal of the American Chemical Society*, **129**, 15354–15363 (2007).

[94] C. J. Brown, R. G. Bergman, K. N. Raymond, Enantioselective Catalysis of the Aza-Cope Rearrangement by a Chiral Supramolecular Assembly, *Journal of the American Chemical Society*, **131**, 17530–17531 (2009).

[95] M. Schweiger, S. R. Seidel, M. Schmitz, P. J. Stang, *Organic Letters*, **2**, 1255–1257 (2000).

[96] S. P. Argent, T. Riis-Johannessen, J. C. Jeffery, *et al.*, Diastereoselective Formation and Optical Activity of an M_4L_6 Cage Complex, *Chemical Communications* (37), 4647–4649 (2005).

[97] M. Albrecht, S. Burk, P. Weis, Chiral Confined Space: Induction of Stereochemistry in a M_4L_4 Metallosupramolecular Container, *Synthesis*, **18**, 2963–2967 (2008).

[98] K. Suzuki, M. Kawano, S. Sato, M. Fujita, Endohedral Peptide Lining of a Self-Assembled Molecular Sphere to Generate Chirality-Confined Hollows, *Journal of the American Chemical Society*, **129**(35), 10652–10653 (2007).

[99] M. Ikemi, T. Kikuchi, S. Matsumura, *et al.*, Peptide-Coated, Self-Assembled $M_{12}L_{24}$ Coordination Spheres and Their Immobilization onto an Inorganic Surface, *Chemical Science*, **1**(1), 68–71 (2010).

[100] T. Liu, Y. Liu, W. Xuan, Y. Cui, Chiral Nanoscale Metal–Organic Tetrahedral Cages: Diastereoselective Self-Assembly and Enantioselective Separation, *Angewandte Chemie, International Edition*, **49**(24), 4121–4124 (2010).

[101] N. Ousaka, J. K. Clegg, J. R. Nitschke, Nonlinear Enhancement of Chiroptical Response through Subcomponent Substitution in M_4L_6 Cages, *Angewandte Chemie, International Edition*, **51**(6), 1464–1468 (2012).

[102] Y. R. Hristova, M. M. J. Smulders, J. K. Clegg, *et al.*, Selective Anion Binding by a 'Chameleon' Capsule with a Dynamically Reconfigurable Exterior, *Chemical Science*, **2**(4), 638–641 (2011).

[103] A. M. Castilla, N. Ousaka, R. A. Bilbeisi, *et al.*, High-Fidelity Stereochemical Memory in a $Fe^{II}_4L_4$ Tetrahedral Capsule, *Journal of the American Chemical Society*, **135**(47), 17999–18006 (2013).

[104] R. Frantz, C. S. Grange, N. K. Al-Rasbi, *et al.*, Enantiodifferentiation of Chiral Cationic Cages Using Trapped Achiral BF_4^- Anions as Chirotopic Probes, *Chemical Communications* (14), 1459–1461 (2007).

[105] J. L. Bolliger, A. M. Belenguer, J. R. Nitschke, Enantiopure Water-Soluble $[Fe_4L_6]$ Cages: Host–Guest Chemistry and Catalytic Activity, *Angewandte Chemie, International Edition*, **52**(31), 7958–7962 (2013).

[106] C. Zhao, Q.-F. Sun, W. M. Hart-Cooper, *et al.*, Chiral Amide Directed Assembly of a Diastereo- and Enantiopure Supramolecular Host and its Application to Enantioselective Catalysis of Neutral Substrates, *Journal of the American Chemical Society*, **135**(50), 18802–18805 (2013).

[107] C. Zhao, F. D. Toste, K. N. Raymond, R. G. Bergman, Nucleophilic Substitution Catalyzed by a Supramolecular Cavity Proceeds with Retention of Absolute Stereochemistry, *Journal of the American Chemical Society*, **136**(41), 14409–14412 (2014).

[108] P. J. Stang, B. Olenyuk, D. C. Muddiman, R. D. Smith, Transition-Metal-Mediated Rational Design and Self-Assembly of Chiral, Nanoscale Supramolecular Polyhedra with Unique T Symmetry, *Organometallics*, **16**, 3094–3096 (1997).

[109] Y. Nishioka, T. Yamaguchi, M. Kawano, M. Fujita, Asymmetric [2+2] Olefin Cross Photoaddition in a Self-Assembled Host with Remote Chiral Auxiliaries, *Journal of the American Chemical Society*, **130**(26), 8160 (2008).

[110] T. Murase, S. Peschard, S. Horiuchi, *et al.*, Remote Chiral Transfer into [2+2] and [2+4] Cycloadditions within Self-Assembled Molecular Flasks, *Supramolecular Chemistry*, **23**(3 and 4), 199–208 (2011).

9

Chirality at the Solution/ Solid-State Interface

Iris Destoop and Steven De Feyter

Over the last few years, the number of investigations dealing with the emergence of chirality and chirality amplification at surfaces has increased significantly. These studies provide insight into the complexities of the origin of life but are also very useful regarding applications such as heterogeneous catalysis and chiral separation. In this chapter, the focus lies on chirality expression and amplification at the solution/solid-state interface, with an emphasis on the developments since 2011. The chapter will be concluded with some studies that take steps towards chiral separation.

9.1 Self-Assembly at the Solution/Solid-State Interface

Molecule-based self-assembly is a process that is characterized by the spontaneous ordering of molecules in (well defined) structures or patterns [1–3]. In solution, many molecular systems self-assemble into a variety of structures. Such spontaneous ordering also happens at the interface between liquids and solid substrates. The adsorption of molecules on such surfaces goes along with a two-dimensional (2D) crystallization process, where 2D nuclei might develop into well ordered domains. In this chapter, we focus on molecular systems that interact with the substrate via physisorption and not chemisorption. While chemisorbed systems such as alkyl thiols on gold [4, 5] are very popular, their importance in forming chiral structures on surfaces is limited.

Among the characterization tools to visualize molecular patterns with (sub)molecular resolution, scanning probe methods such as scanning tunneling microscopy (STM) and

Chirality in Supramolecular Assemblies: Causes and Consequences, First Edition. Edited by F. Richard Keene.

atomic force microscopy (AFM) are definitely the more popular ones [6–9]. A common characteristic of these techniques is that a sharp probe scans the surface, often even in liquid, with a resolution that approaches the atomic level. In contrast to scattering techniques, these scanning probe techniques are local imaging techniques, as they scan areas ranging from a few thousand square micrometers to several square nanometers. Among the substrates that are selected for the surface-assisted self-assembly process, those that are (electro)chemically inert and are characterized by atomically flat terraces are the most popular ones. If STM is used as imaging tool, these substrates should be conductive as well, but not the liquid.

At the liquid/solid interface, the 2D assemblies are created by simply depositing the solution of the compound of interest and do not require complicated or expensive infrastructure. This is applicable to a large spectrum of compounds as no thermal treatment is required to deposit molecules on a surface. The presence of the solvent provides a dynamic environment for adsorption and desorption of molecules. Conditions for reaching thermodynamic equilibrium are often favorable and self-repair of defects in these self-assembled layers is possible. In addition to the properties of the substrate, the dynamics of the system are inherently linked to the solubility, viscosity, polarity and so forth of the chosen solvent.

However, not all combinations of molecules, liquids, and substrates will turn out to be successful in achieving 2D self-assembly. Both technical restrictions, as mentioned above, and conceptual constraints are at play. A good balance between molecule-molecule and molecule-substrate interactions is important. For instance, too strong molecule-substrate interactions will slow down the in-plane translational and rotational dynamics and out-of-plane adsorption/desorption dynamics and decrease the reversibility of the self-assembly process, a requisite to reach thermodynamic equilibrium. Very weak molecule-substrate interactions, in combination with weak molecule-molecule interactions. also decrease the probability of surface-confined molecular self-assembly. The enthalpic and/or entropic gain of self-assembly on surfaces must be favorable, under conditions where the kinetics of molecular adsorption and desorption are in balance.

9.2 Chirality Expression at the Solution/Solid-State Interface

Chirality can manifest in two distinct ways at a surface: by the ordering and packing of the molecules themselves within the layer or by the orientation of the monolayer with respect to the symmetry of the underlying substrate. Another important factor is the inherent chirality of the molecule itself, as this will decide whether local or global chirality will be attained. If the chirality is uniquely maintained across the surface global chirality is generated, while coexistence of mirror domains will lead to locally chiral but globally achiral surfaces. For this reason, the chirality expression of chiral (enantiopure) molecules, racemates and achiral molecules will be treated separately.

9.2.1 Enantiopure Molecules at the Solution/Solid-State Interface

Many reports deal with the self-assembly of enantiopure molecules at the solution/solid-state interface and emphasize that globally chiral surfaces are generated upon surface-confinement [10, 11]. Mirror image structures are consequently not present and can only be formed by deposition of the opposite enantiomer. However, in order for an enantiomerically pure molecule to transfer its intrinsic chirality, the stereogenic center should influence the molecular adsorption effectively. Otherwise, no diastereomeric adsorbed state can be formed with the surface. For example, a diacetylene-containing isophthalic acid derivative carrying an (*S*)-stereogenic center does not express global organizational chirality as the chiral 2-methylbutoxy group is bent away from the surface [12]. Analysis of the orientation of the lamella with respect to the underlying substrate symmetry reveals a symmetric distribution around 0°, which is typical behavior for achiral molecules.

9.2.2 Racemates at the Solution / Solid-State Interface

The self-assembly of racemates is a lot more complicated compared to enantiopure molecules. In total they can crystallize in three different ways: as a conglomerate, a racemic compound or a pseudoracemate [13]. In case of a conglomerate both enantiomers crystallize separately. This process is also called *spontaneous resolution* as it allows pure or nearly pure enantiomers to be obtained by simply sorting them. A racemic compound means that, within the crystal, both enantiomers are present in equal amounts in a well defined arrangement. And lastly, a pseudoracemate is obtained when both enantiomers are randomly built in into the crystal lattice. Most commonly racemates self-assemble into 2D conglomerates and pseudoracemates are hardly encountered.

Interestingly, there are a few examples of systems where conglomerate and racemic structures exist side by side. A first example is the self-assembly of an asymmetrically functionalized 5-amino[6]helicene (AH). The enantiopure *P*-AH is known to form a three-dot $p3$-(P_3) pattern at the 1,2,4-trichlorobenzene/Au(111) interface [14] (Figure 9.1). This pattern is analogous to the self-assembly pattern of the unsubstituted *P*-[7]helicene on Cu(111) and suggests a face-on orientation of the amino helicene [15]. The exact position of the amino group cannot be discerned as the helicenes appear as an unstructured blob that is believed to represent the uppermost part of the molecule. However, density functional theory (DFT) calculations provide evidence that the configuration with the amino group close to the gold surface is significantly more stable. The *p3*-pattern is in fact a chiral packing, as the unit cell vectors and the sides of the trimers are not parallel. The mirror image pattern is not present and is only obtained by self-assembly of the opposite enantiomer *M*-AH. In addition to the $p3$-(P_3) and $p3$-(M_3) pattern, the racemate *MP*-AH self-assembles into a new polymorph with *p6* symmetry (see Figure 9.1b). This $p6$-(P_6M_6)-pattern is built up from racemic dimers (*M*-AH/*P*-AH). Even though this pattern has a racemic composition, two enantiomorphous domains are formed. As the *p3*- and *p6*-pattern co-exist, AH thus represents an intermediate case between unsubstituted [7]helicene and bis-functionalized 6,13-dicyano[7]helicene, which form exclusively an enantiomorphous racemic mixture and a 2D conglomerate, respectively [16, 17]. Most likely, the co-existence of both patterns for this system reflects the slow surface diffusion and the high barrier for desorption leaving a kinetically trapped phase on the gold surface.

The second example deals with oligo-(phenylenevinylene) (OPV) derivatives that form large domains consisting of tetramers at the interface between 1-phenyloctane and highly oriented pyrolytic graphite (HOPG), a high-quality form of graphite [18]. The windmill-like structures contain a shift between the OPV molecules in each tetramer creating two chiral structures, clockwise (CW) and counterclockwise (CCW) tetramers. The absolute

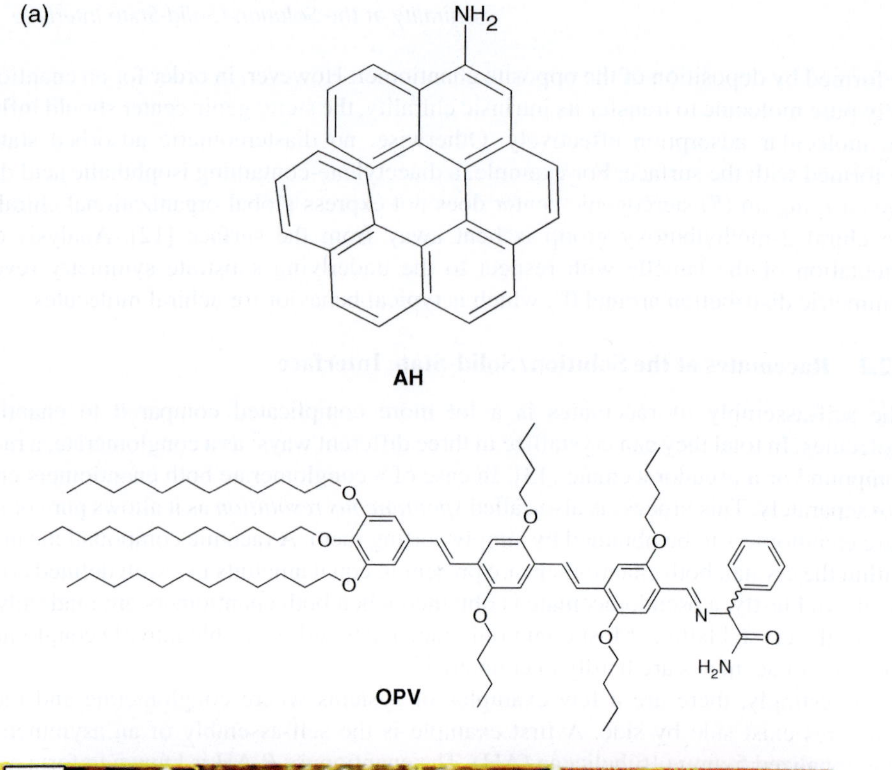

Figure 9.1 *(a) Chemical structures of AH and chiral OPV. (b) STM image showing the self-assembly of racemic MP-AH at the interface between 1,2,4-trichlorobenzene and Au(111). Co-existence of conglomerate and racemic structures. Also some defects are present. (Adapted from Ref. [14] with permission of The Royal Society of Chemistry). (c–d) Scanning tunneling microscopy images of (rac)-OPV at the 1-phenyloctane/HOPG interface. Yellow and red rectangles in (c) are covering the tetramer lamellae, which are formed by (S)-OPV or (R)-OPV3I, respectively. In (d) (R)-OPV is indicated in red and (S)-OPV in yellow. [(R)-OPV] = [(S)-OPV] = 0.5 mM. (Adapted from Ref. [18] with permission of The Royal Society of Chemistry.)*

configuration of the stereogenic center determines the monolayer chirality: (*R*)-OPV and (*S*)-OPV self-assemble into domains with CCW and CW chirality, respectively. Deposition of the racemate leads to monolayers composed of intercalated tetramer lamellae as depicted in Figure 9.1c. A single lamella contains all CCW-rotated (4*R*)-tetramers or either all CW-rotated (4*S*)-tetramers pointing to conglomerate formation. On the other hand, at the level of the lamellae, this construction can be considered as a racemic compound.

9.2.3 Achiral Molecules at the Solution / Solid-State Interface

For the discussion on how chirality is manifested at the solution/solid-state interface for achiral molecules, a distinction will be made between prochiral and nonchiral molecules. If a molecule is considered to be prochiral, it means that this object can become chiral in a single desymmetrization step by addition of substituents to one or both of their heterotopic faces [19]. This signifies the disappearance of one or more symmetry elements, which preclude chirality. When 2D self-assembly is regarded, molecules can only be prochiral if the surface can act as a symmetry breaker. As planar molecules frequently do not possess a symmetry axis, they can give rise to the formation of localized 2D chiral motifs upon surface confinement. As a consequence, prochiral molecules behave in a similar way to racemates. This means that they will most likely form conglomerates when homochiral associations are more favorable, and racemic compounds when heterochiral associations are favored. However, the assumption that only molecule-molecule interactions are responsible for guiding a system towards conglomerate or racemate formation is not always correct.

As an example, in the case of prochiral alkylated anthracene derivatives the outcome of the self-assembly is controlled by the molecule-substrate interactions [20]. The number of carbon atoms in the side chains determines the morphology of the self-assembled monolayers on graphite. Derivatives with an alkyl side chain containing an odd number of carbons form a 2D racemate as neighboring rows contain anthracenes adsorbed via opposite enantiotopic faces, while those with an even number always adsorb via the same enantiotopic face and thus form a 2D conglomerate. The well known odd-even effects in alkane self-assembly imply that alkyl chain-substrate interactions play a crucial role in the chirality expression of these anthracene derivatives [21]. So generally, the self-assembly of prochiral molecules will produce either a conglomerate, racemate or pseudoracemate. Nonetheless, some special cases of 2D crystallization have been reported.

Prochiral bisisophthalic acid (BISA) derivatives, flat molecules with a C_{2h}-symmetric crankshaft geometry, can self-assemble both as a conglomerate and as a 2D racemate, as described by Zhou *et al.* [22]. The deposition of BISA in heptanoic acid onto the freshly cleaved HOPG surface yields large ordered domains stabilized by hydrogen bonding between the acid groups of neighboring molecules. Close inspection reveals that two different motifs coexist on the surface (Figure 9.2c). The parallel motif is formed, as the name already hints, when the molecules are packed parallel with respect to each other. In order to form a long-range periodic pattern, the molecules within a domain must all have the same configuration. Consequently, the localized 2D chiral motifs phase separate and therefore mirror image domains are formed. On the other hand, the chevron pattern is built up by alternating both 2D configurations and is thus racemic in nature. The ratio of both patterns is approximately 7:3 in favor of the racemic chevron motif, which reflects the slightly higher coverage density and its inherent twofold statistical advantage.

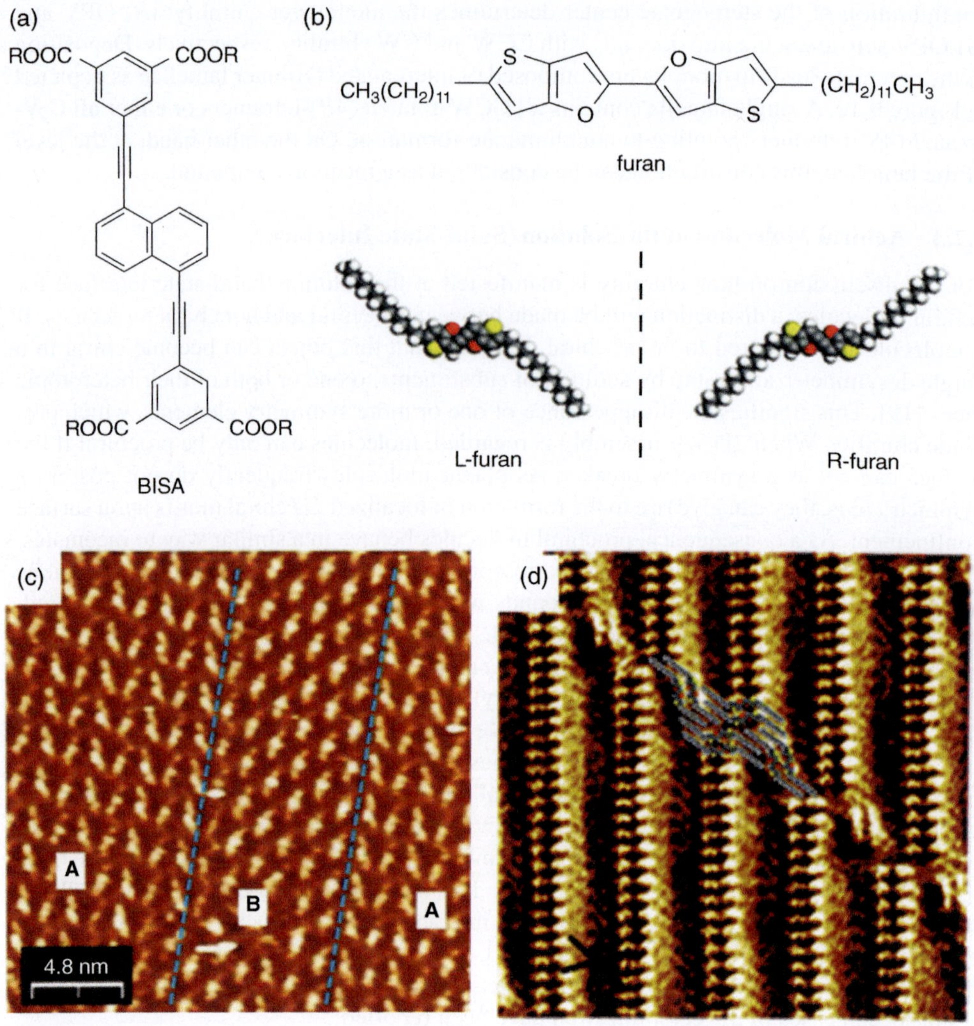

Figure 9.2 Molecular structure of (a) BISA with R = H and (b) furan and its 2D enantiomers. (c) STM image of the adsorption of BISA at the heptanoic acid/HOPG interface showing smoothly intersecting domains composed of parallel motif (domain B) and chevron motif (domain A). (Adapted with permission from Ref. [22]. Copyright 2013, American Chemical Society). (d) STM image of the 2D assembly of furan at the 1-phenyloctane/HOPG interface showing a conglomerate with periodic enantiomer inclusion together with an overlaid computed model. (Adapted from Ref. [23] with permission of Royal Society of Chemistry.)

Another example of exceptional 2D crystallization behavior was discovered by Matzger and co-workers. They investigated the self-assembly behavior of an achiral conjugated oligomer 2,2′-didodecyl-5,5′-bithieno[3,2-*b*]furan at the 1-phenyloctane/HOPG interface over a range of concentrations from 0.10 mM to 20.0 mM [23] (Figure 9.2). Upon surface confinement, this molecule becomes chiral as the presence of the surface cancels out the C_{2h}-symmetry. As a result two 2D enantiomers named L-furan and R-furan can be distinguished. For the lowest probed concentration regime (0.10 mM), the majority of the surface is covered with a well ordered 2D pattern stabilized by hydrogen bonding and van der Waals interactions. Preferentially enantiomers with the same orientation (L or R) are built in to obtain maximal molecule-molecule interactions, and hence conglomerate formation is obtained. Interestingly, a small amount of the surface is covered with a pattern that encloses periodic insertions of two molecules causing an alignment shift between the adjoining columns (Figure 9.2d). This behavior resembles one-dimensional (1D) cocrystals characterized by the periodic insertion along one axis of one component between assemblies formed by the other component [24]. The high-resolution STM images indicate that the inserted molecules express opposite handedness compared to the surrounding molecules. Therefore, aggregates of R-furan are able to interrupt the 2D assembly of L-furan by coadsorption and vice versa. Consequently, the crystals display random mixing of the two enantiomers along one axis and periodic order along the other axis. The frequency of the random mixing in 1D co-crystals is typically tuned by modifying the molar ratio of the two components. This, however, cannot be done in this case as it is a one-component system. As the random insertion is primarily induced by kinetic effects, supersaturated solutions (10–20 mM) of the furan are deposited on graphite. An increase in the frequency of random insertions is observed. These insertions are randomly positioned along the nonperiodic axis and can vary in dimensions ranging from one to four molecule aggregates. To maintain close packing, the insertions are periodically imbedded along the periodic axis. Quite similar behavior is reported for an enantiopure 2,6-dihydroxy naphthalene derivative [25], where the lamellae formed are shifted up and down at random positions.

Achiral molecules, which are not prochiral, can also form locally chiral domains. Usually, the chirality arises from molecule-substrate interactions that lead to symmetry breaking with respect to the underlying substrate. In some cases, asymmetric intermolecular interactions can be formed between the achiral units, resulting in the formation of a prochiral hierarchical building unit and thus two 2D enantiomers [26]. Therefore, two mirror image patterns can be obtained and as these configurations are iso-energetic, they will appear in equal abundances on the surface.

9.2.4 Other Factors Influencing 2D Chirality

As we are dealing with self-assembly at the solution/solid-state interface, it is important to realize that factors like solvent coadsorption and molar ratio in solution can have a profound influence on the self-assembly process and thus on the chirality expression. The group of Li-Jun Wan showed how tuning the molar ratio of three components (see Figure 9.3A) in the liquid phase can influence the self-assembly process resulting in a series of flower-like chiral hierarchical structures with a tunable periodicity in a size range of 7 to 14 nm [27]. A triangular shaped building unit (TBU), consisting of three 2,3,7,8,12,13-hexahexyloxy-truxenone (TrO23), one 1,3,5-tris(10-carboxydecyloxy)benzene (TCDB) and three copper phthalocyanine (CuPc), can be considered as the secondary building block. The TBUs arrange in an edge-to-edge fashion by sharing common CuPc lines as a separator to form a rhombic unit cell. The relative alignment of the TBUs results in the formation of 2D enantiomers, as depicted in Figure 9.3B. Each flower structure consists of six lobes, which are composed of several TBU rows. In total, five different flower-like structures are observed

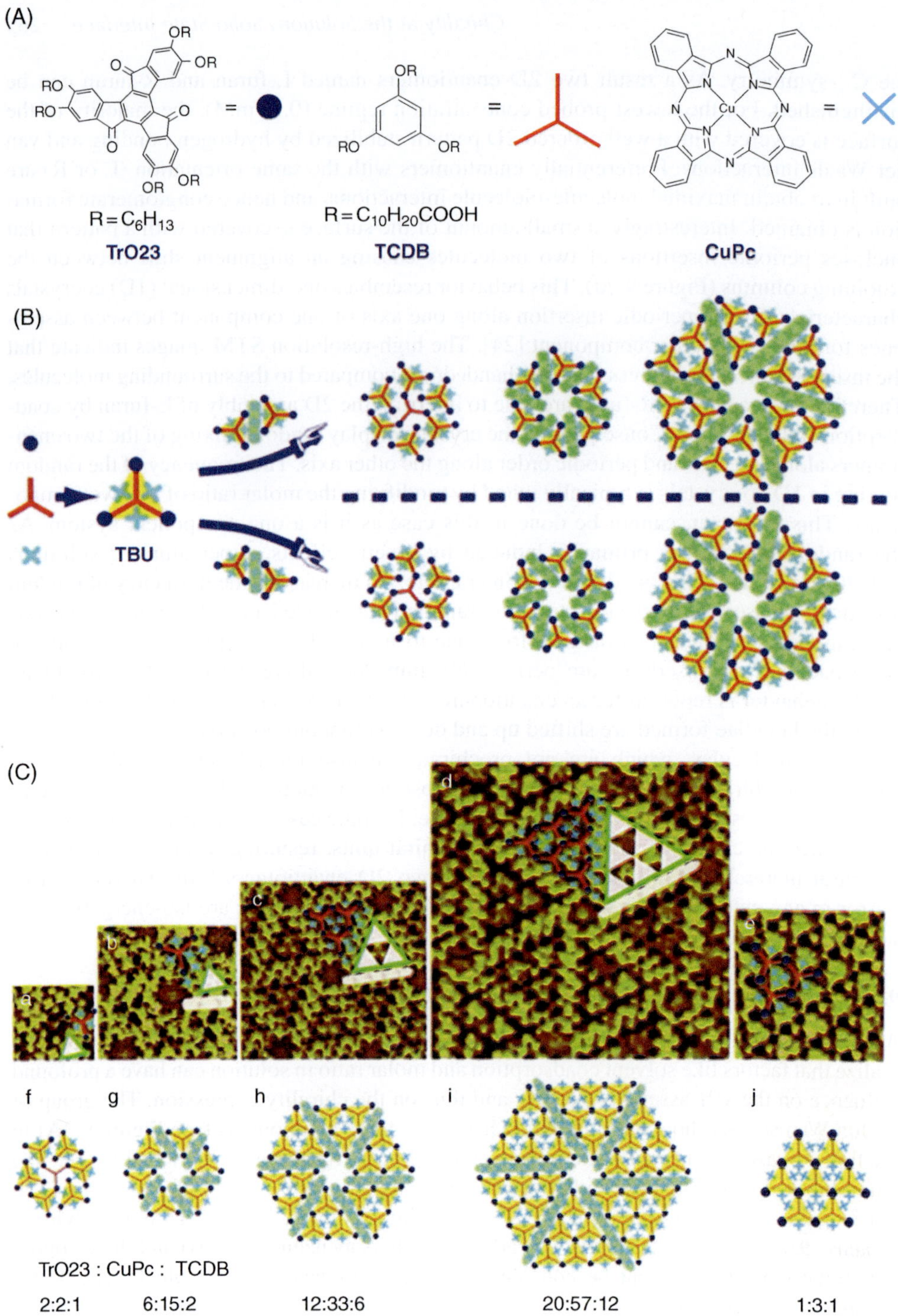

Figure 9.3 *(A) Molecular structures of TrO23, TCDB and CuPc. (B) Illustration of the hierarchical self-assembly formation process. The secondary building block TBU forms surface-confined enantiomers that can self-assemble into several flower-like structures. (C) The structural evolution of hierarchical structures of different generations. (a)–(e) STM images of single flowers of Flowers 0, 1, 2, 3, and ∞. (f)–(j) schematic models for Flowers 0, 1, 2, 3, and ∞. Image sizes: (a) 11 × 11 nm^2; (b) 19 × 19 nm^2; (c) 25 × 25 nm^2; (d) 40 × 40 nm^2; (e) 20 × 20 nm^2. The relative molecular ratios between TrO23, CuPc and TCDB in each self-assembled structure are shown below each STM image. (Adapted with permission from Ref. [27]. Copyright 2011, American Chemical Society.)*

at the 1-phenyloctane/HOPG interface (Figure 9.3C). The number of TBU rows in a lobe defines the generation of the flower structure. So, flower n ($n>0$) is created by lobes with n rows of TBUs separated by $2n+1$ CuPc molecules in a separator line. A special case is flower ∞, which lacks the CuPc separator lines. Organizational chirality is expressed for these flower structures as the lobes spiral around the center voids. The TBUs in each unit cell are offset, which breaks the mirror symmetry of the TBUs. To show that the hierarchical assemblies can be tuned by the relative amount of the components, a set of experiments are performed where the concentration of CuPc is kept constant (CuPcs are believed to be preferentially adsorbed due to their low solubility), while the concentration of the two other components is varied. Typically, flowers 1 and 2 are favored when the concentration of TCDB and TrO23 are both at medium level. The observation of concentration-dependent multilevel hierarchical assemblies indicates that the solution/solid-state interface plays an important role in the self-assembly process.

Another interesting example shows how the interplay between steric and van der Waals molecule-substrate and intermolecular interactions can influence the emergence of chirality. Sleczkowski *et al.* investigated the self-assembly of 2,3,6,7,10,11-hexapentyloxytriphenylene (H5T) at the n-tetradecane/Au(111) interface by STM and DFT [28]. The molecular structure of H5T is portrayed in Figure 9.4: The molecule presents threefold symmetry and

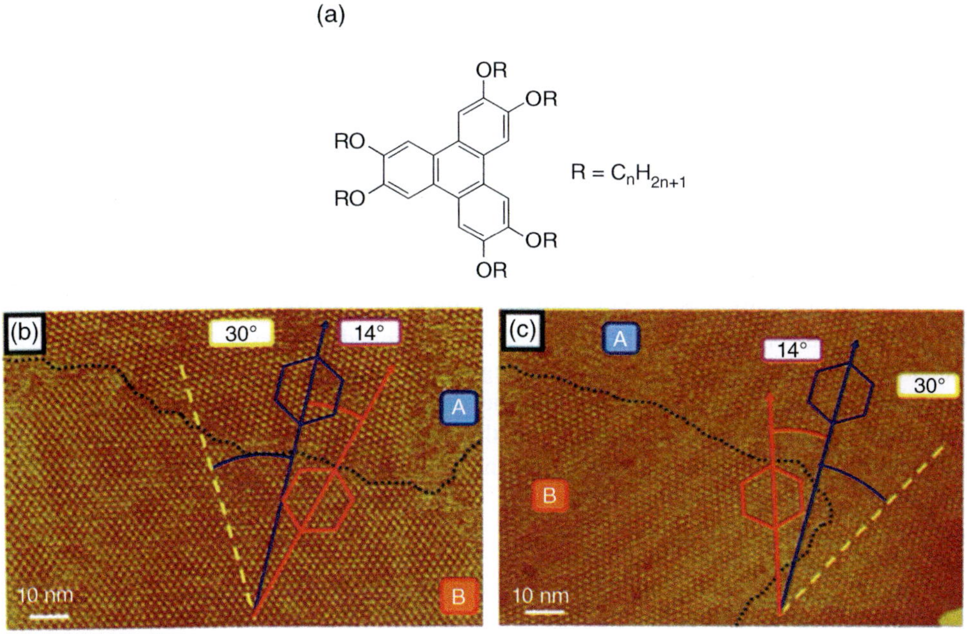

Figure 9.4 *(a) Molecular structure of H5T. STM images of twofold orientation of H5T molecules self-assembled at the n-tetradecane/Au(111) interface. (7×7) domains A and (√52×√52)R13.9° domains B are depicted by blue and red graphics, respectively. Yellow dashed lines highlight the main direction of Au(111) reconstruction, i.e., the ⟨112⟩ direction. Black dotted lines highlight the domain boundaries. (b–c) STM pictures are taken on the same Au(111) monocrystal and evidence the two possible orientations of B domains, rotated by ±13.9° from the A ones. (Adapted with permission from Ref. [28]. Copyright 2014, American Chemical Society.)*

is composed of an aromatic core symmetrically substituted with six alkoxy chains. Deposition of H5T directly after flame annealing of the gold substrate, leads to the formation of a (7×7) hexagonal network for which the molecular rows are aligned along the <110>direction. Surprisingly, a second hexagonal network with a ($\sqrt{52}\times\sqrt{52}$)R13.9° mesh coexists on the surface when H5T is deposited on a cold gold substrate. Both networks are believed to be commensurate with the substrate, as moiré patterns are absent in the STM images. The unit cell vectors of the latter network are rotated by 13.9° with respect to the thermodynamically stable honeycomb pattern. Figure 9.4b–c shows that both CW and CCW rotated domains of the ($\sqrt{52}\times\sqrt{52}$)R13.9° mesh are present on the surface as a result of the symmetry breaking with respect to the substrate. Unfortunately, it is not possible to visualize H5T with submolecular resolution and hence the exact orientation of the aromatic core and the alkyl chains with respect to the substrate cannot be determined experimentally. In order to obtain more insight into the geometry of the system, DFT calculations were performed. It seems that maximization of the molecular density leads to a molecule disorientation around 14° with respect to the substrate symmetry axes in order to avoid steric repulsion between neighboring alkyl chains. As a result, two enantiomorphous patterns are formed. In addition, the molecular network of the ($\sqrt{52}\times\sqrt{52}$)*R*13.9° mesh also breaks the symmetry with respect to the underlying substrate. This implies that theoretically there should be four diastereomeric patterns, but due to unfavorable aromatic core-substrate interactions only two diastereomers should exist. The subtle interplay between aromatic core-substrate and alkyl chain-substrate interactions is responsible for favoring either the ($\sqrt{52}\times\sqrt{52}$)*R*13.9° or (7×7) mesh.

9.3 Chiral Induction/Amplification at the Solution/Solid-State Interface

For industrial purposes such as heterogeneous enantioselective catalysis, where the chirality of the modified substrate plays a crucial role, globally chiral surfaces are a necessity. This means that enantiopure organic molecules must be used for self-assembly as only in this case, global organizational chirality will be obtained. However, the preparation of enantiopure molecules is not the most straightforward task. Therefore, different routes were explored to achieve global organizational chirality from the self-assembly of achiral molecules and racemates. An overview of these chirality induction methods and some examples will be given in the following section.

9.3.1 Sergeants and Soldiers

By merging a small amount of chiral dopant with a supramolecular network of achiral molecules, the achiral molecules can be forced to assemble into domains of a particular handedness. The ability of chiral molecules (sergeants) to control the self-assembly of a large number of achiral units (soldiers) has been termed the *sergeants-and-soldiers principle*. The sergeants-and-soldiers principle was pioneered by Green *et al.* for solution-based systems and has since been expanded to supramolecular self-assembled systems [29, 30]. The sergeant molecule is always a chiral analog of the soldier, which guarantees coassembly in the 2D network. Typically, only a small percentage of these sergeants are necessary to generate a homochiral surface. Tahara *et al.* demonstrated how perfect chiral induction can be achieved for surface-confined porous molecular networks by doping with 9 mol% of a chiral sergeant [31]. Alkoxylated dehydrobenzo[12]annulene (DBA) derivatives, as shown in Figure 9.5a, are used to construct the porous hexagonal network. This porous pattern is stabilized by van der Waals interactions between the interdigitated alkoxy chains of adjacent DBA molecules. The chirality is governed by the relative alignment of the four interdigitated alkoxy chains per DBA pair (+ or − type interdigitation) as indicated in Figure 9.5b. A virtual "clockwise" (CW) or "counterclockwise" (CCW) nanowell is obtained by combining six − type or + type interdigitation patterns, respectively. The rotation of the nanowell, CW or CCW, is determined by the orientation of the alkoxy chains making up the rim of the nanowell. The chirality expression of the DBA nanowells is domain specific and, in the case of achiral derivatives, CW and CCW domains are expressed equally. Chirality is also expressed at the level of orientation with respect to the underlying graphite substrate: CW domains have their unit cell vector rotated clockwise with respect to the main axis of graphite, while for CCW domains a counterclockwise rotation has to be exerted. The molecular structure of the three different sergeant molecules is depicted in Figure 9.5a. The chiral sergeant molecules differ from the soldiers as they bear a methyl group on every second carbon of the alkoxy chains. The three sergeants diverge from each other at the level of absolute chirality of the stereogenic center and at the level of alkoxy chain lengths. The chirality of the self-assembled monolayers of cDBA-OC12-(*S*) and cDBA-OC13-(*R*) are investigated by STM and lead to the exclusive formation of CW and CCW nanowells, respectively. The preferential formation of a honeycomb structure with a well defined handedness, is caused by an energy difference between the two interdigitation patterns. As all the stereogenic centers have the same absolute chirality, the methyl groups will be pointing alternatively up and down. This implies that, in the case of cDBA-OC12-(*S*) for instance, for the − type interdigitation pattern the two methyl groups of the two middle alkoxy chains point away from the surface, while the methyl groups of the outer chains point towards the surface. For the + type interdigitation pattern, the direction of the methyl groups is reversed causing a local distortion of the alkoxy chains. As a result, optimal van der Waals interactions between the chains of adjacent DBAs are hindered. To investigate the ability of cDBA-OC12-(*S*) to act as a chiral inducer, cDBA-OC12-(*S*) is mixed at different molar ratios with achiral DBA-OC12. About 30 mol% of the sergeant is needed in order to obtain homochirality, which is not efficient at all. It seems that the low surface affinity of the chiral sergeant and the small energy difference between CW and CCW nanowells are the basis for this low efficiency. To improve the induction efficiency, a third more sophisticated sergeant molecule, cDBA-OC12-(*S*)-OC13-(*R*), was designed. This molecule bears alternatively (*S*)-OC12 and (*R*)-OC13 alkoxy side chains. The alternating chirality of the stereogenic centers ensures that the methyl groups all point in the same direction with respect to the plane of the molecule and consequently allow the control of the adsorption face as the adsorption geometry with all methyl groups pointing towards the graphite substrate must be unfavorable. Mixing cDBA-OC12-(*S*)-OC13-(*R*) with DBA-OC12 at different molar ratios reveals a rise in the CW nanowell fraction when increasing the molar ratio of the sergeant. When 9 mol% of the sergeant is present, homochirality is already achieved.

Rather than adding a chiral sergeant, mixing in a giant chiral molecule that resembles the supramolecular structure formed by the achiral units has been investigated. This strategy is termed the *giant sergeants-and-soldiers* approach. As a proof of principle, an achiral

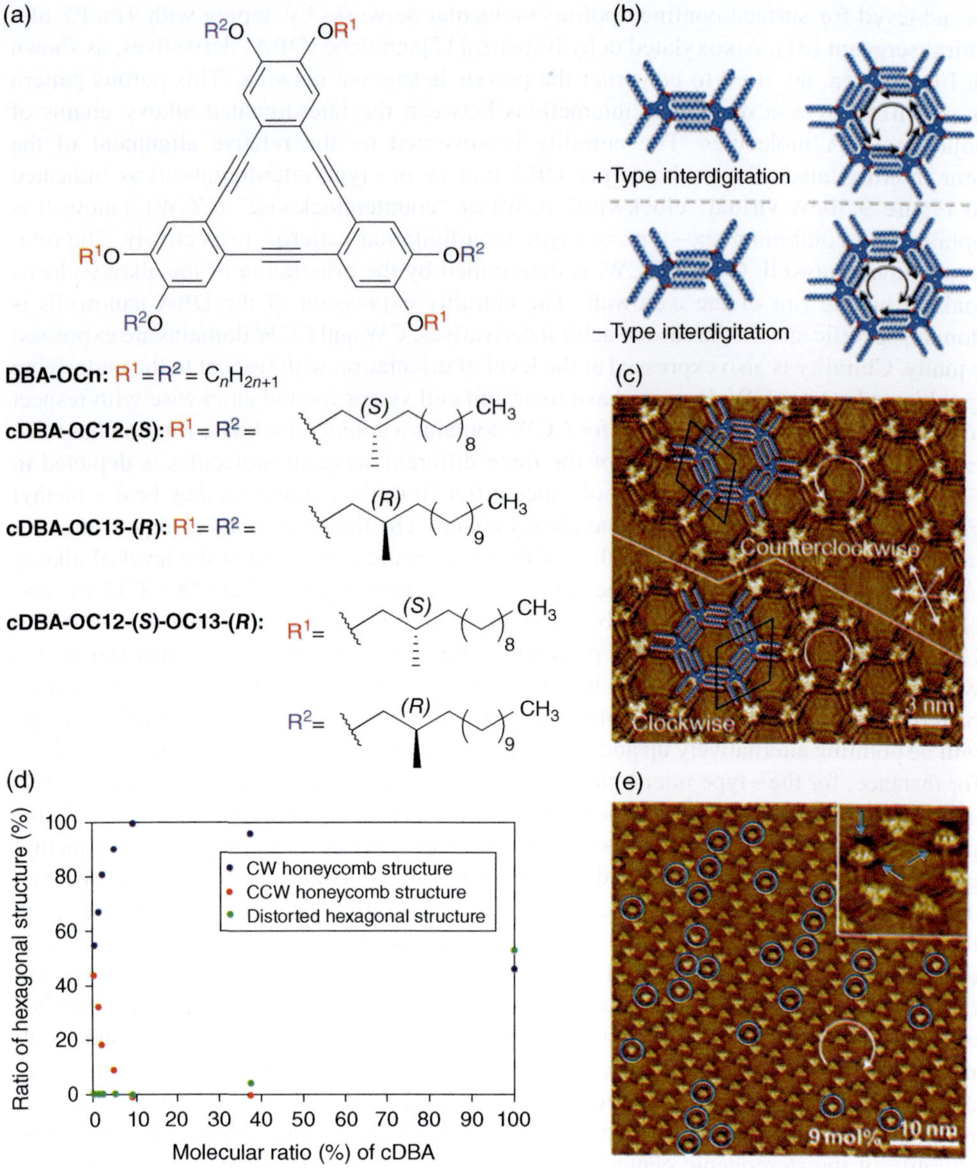

Figure 9.5 *(a) Molecular structures of achiral and chiral DBAs. (b) Molecular models of + and − type interdigitation patterns, and CW and CCW nanowells. (c) STM image of the honeycomb structure of DBA-OC12 at the 1-phenyloctane/graphite interface (I_{set} = 0.32 nA, V_{bias} = −0.17 V, Concentration = 3.0 × 10^{-6} M). The white line indicates a domain boundary. Unit cells are indicated in black and the main symmetry axes of graphite in white. (d) Composition-dependent induction of 2D chirality of the honeycomb structure, showing the population of the CW nanowell (blue), the CCW nanowell (red) and the distorted nanowell (green), with respect to mol% of cDBA-OC12-(S)-OC13-(R) in a solution containing DBA-OC12. (e) STM image of a monolayer of a mixture of cDBA-OC12-(S)-OC13-(R) and DBA-OC12 at 1:10 mixing ratios (9 mol% of the chiral DBA, I_{set} = 0.30 nA, V_{bias} = −0.22 V). Blue circles indicate DBAs with black contours. (Adapted with permission from Ref. [31]. Copyright 2011, Macmillan Publishers Ltd.)*

oligo-(*p*-phenylene vinylene) (OPV4T) derivative end-capped with diaminotriazine groups is mixed with a large chiral molecule consisting of six covalently linked OPV units [32] (Figure 9.6). The achiral OPV4T is known to form hydrogen-bonded cyclic hexamers or rosettes driven by hydrogen bonding between the triazine end groups. Based on the relative orientation of the OPV units in the rosettes, two different kinds of rosettes are identified on the surface : CW and CCW rosettes. The chirality of these rosettes is beside the monolayer symmetry also expressed at the level of symmetry breaking with the underlying graphite surface. As is expected for achiral molecules, both enantiomorphous motifs are present in equal amounts on the surface. The giant sergeant molecule containing 24 stereogenic centers, on the other hand, forms exclusively CW domains if the stereogenic centers are *S* and CCW if they are *R*. The unit cell of the giant sergeant is slightly smaller compared to the achiral OPV4T, which is in favor of their coassembly. Simple drop-casting of a premixed solution of both components on the basal plane of graphite does not appear to be a good approach, as the amount of the chiral sergeant on the surface is ten times less than excepted based on the solution composition. Therefore, a different assembly strategy was applied – namely to deposit chiral OPVs prior to the deposition of achiral OPV4Ts. So first, the sergeant is deposited on the surface, then the excess molecules are removed by rinsing with ethanol, leaving small chiral islands on the surface that can act as chiral seeds, and lastly the achiral OPV4Ts are deposited. For the (*S*)-derivative a clear bias (75 : 25) for CCW rosettes is observed. The opposite bias (32 : 68) is observed when the (*R*)-enantiomer is used. A detailed analysis of the STM data shows that co-adsorption of the giant sergeants only appears in the induced domains and at the domain boundaries.

9.3.2 Chiral Auxiliaries

In the case of the sergeants-and-soldiers principle, the sergeant molecule is always the chiral analog of the achiral building block and therefore it can be built in easily into the monolayer. However, it is also possible to use chiral molecules as "sergeants" that have a very different molecular structure from the achiral molecules. This type of molecule is termed a *chiral auxiliary* and usually it transfers its chirality to the monolayer via the formation of noncovalent interactions with the building blocks. As the molecular structure of the auxiliary is so different, there is no guarantee that it will be built in into the monolayer. Consequently, these auxiliaries can coadsorb and become part of the chiral matrix, which is characterized by different unit cell parameters or they can act from the solution phase. To illustrate these two cases, some distinct examples of chirality induction for OPV derivatives will be discussed here.

Inspired by experiments carried out in solution, where a chiral tartaric acid (TA) derivative-induced homochirality in stacks of an achiral oligo(*p*-phenylene vinylene) derivative equipped with a ureido-triazine end group (AOPV3), De Cat *et al.* showed that this approach is also valid for 2D surface-confined supramolecular layers [33]. The self-assembly of AOPV3s at the 1,2,4-trichlorobenzene / graphite interface led to dimer formation by hydrogen bonding [34]. The dimers, in their turn, stacked into lamella that are stabilized via two-by-two interdigitation of the dodecyl chains. Due to symmetry breaking of the underlying graphite, two mirror image dimers were obtained: CW dimers and CCW dimers with an angle of $-9 \pm 3°$ and $+11 \pm 3°$ (respectively) with respect to the main symmetry axis of the graphite (Figure 9.7b). In addition, the chirality could also be recognized by examining the angle between the dimer direction and the normal of the lamella direction. Careful

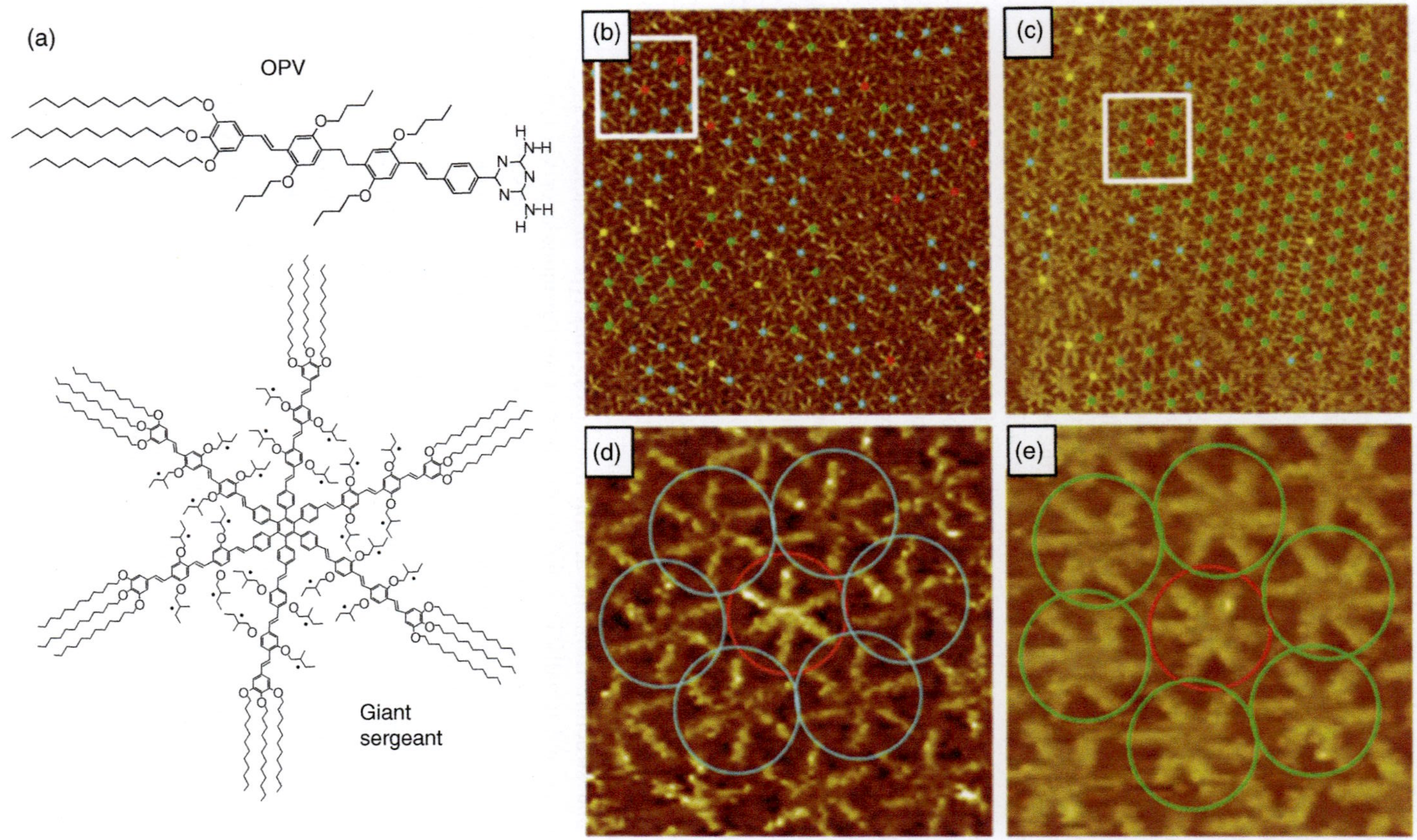

Figure 9.6 *(a) Chemical structures of the achiral oligo(p-phenylene vinylene) OPV4T and the chiral OPV-substituted hexaarylbenzene or giant sergeant molecule. Scanning tunneling microscopy images of a monolayer formed by OPV4T and the giant sergeant molecule, after rinsing the surface containing a monolayer of the chiral sergeant with ethanol, and subsequent exposure of the seeded surface to OPV4T at the 1-octanol/graphite interface. (b) Domain formed by (S)-enantiomer of the chiral sergeant and OPV4T. (c) Domain formed by (R)-enantiomer of the chiral sergeant and OPV4T. (d) Enlarged image of the area highlighted with a white square in (b). (d) enlarged image of the area highlighted with a white square in (c). Chiral sergeant and supramolecular rosettes of OPV4T are color coded: CCW (blue), CW (green), cyclic hexamers with no identifiable orientation (yellow) and chiral sergeant (red). (Adapted with permission from Ref. [32].)*

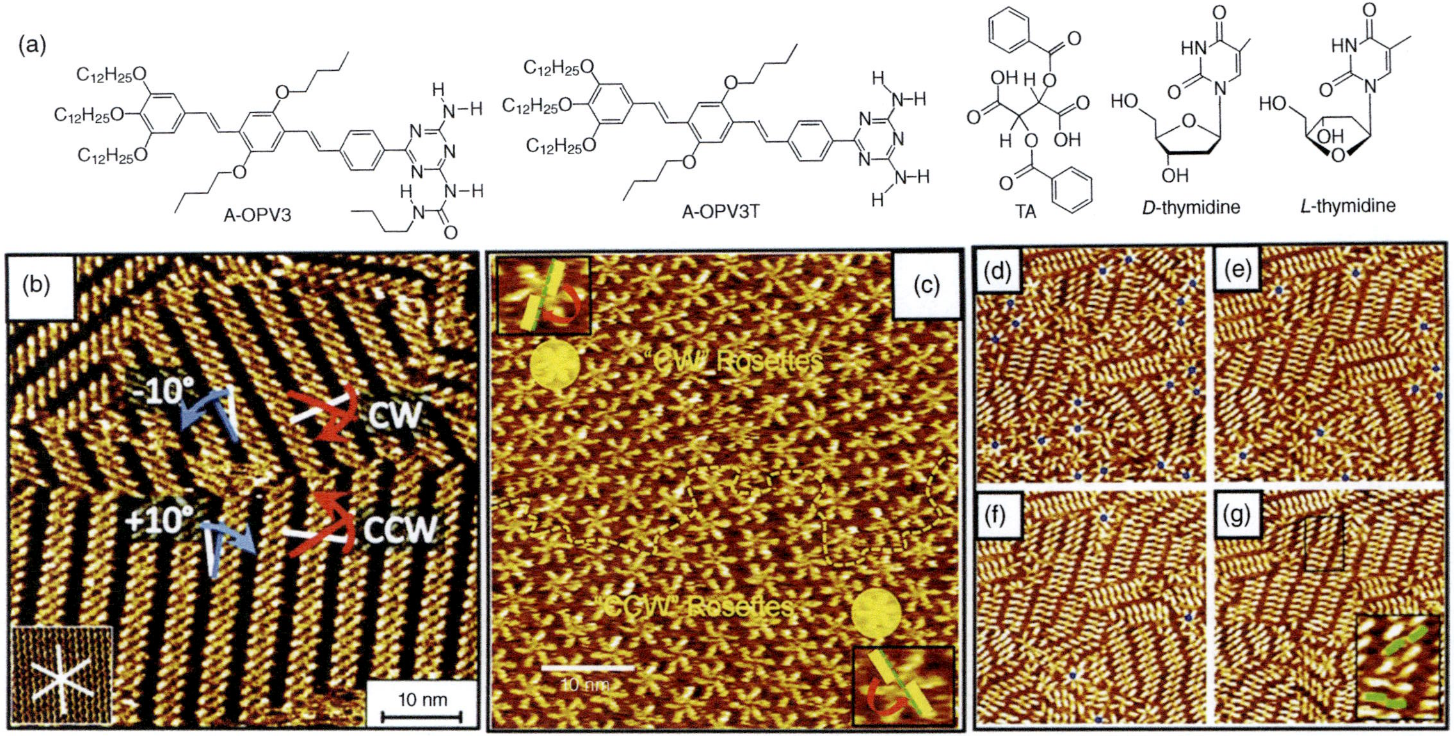

Figure 9.7 *(a) Molecular structure of AOPV3, AOPV3T, TA and L- and D-thymidine. (b) Scanning tunneling microscopy image of a monolayer of pure AOPV3 at the 1,2,4-trichlorobenzene/HOPG interface (I_{set} = 0.109 nA, V_{bias} = −0.86 V). The blue arrows indicate the angle between a symmetry axis of graphite and the lamella direction. The red arrows show the angle between the normal of the lamella direction and the direction of the dimers. Inset: graphite and set of main symmetry axes of graphite. (Adapted with permission from Ref. [34]. Copyright 2012, American Chemical Society.) (c) Scanning tunneling microscopy images of AOPV3T at the 1-octanol/HOPG interface (I_{set} = 0.28 nA, V_{bias} = 0.30 V). The insets highlight the structure of CW and CCW rosettes. The yellow dashed line separates a CW and a CCW rosette domain. Yellow rods cover OPV units to show where the OPV is located. (d–g) Sequence of STM images of AOPV3T in the presence of D-thymidine with a time interval of 11 min. The first image was recorded 30 min after drop casting. Blue disks indicate rosettes. (Adapted with permission from Ref. [35]. Copyright 2011, American Chemical Society.)*

inspection of the monolayer revealed that four hydrogen bonding sites per dimer remain free. So, the chiral auxiliary should preferentially be able to coordinate with all four hydrogen bonding sites and a TA derivative fits this requirement perfectly. The self-assembly of a premixed solution of AOPV3 and D-TA, led to the formation of lamellar structures that are similar to those observed for the pure AOPV3. Analysis of the lamellar structures revealed a clear bias (85 ± 8%) for CW dimers in the case of D-TA. Similarly, a premixed solution of AOPV3 and L-TA led to the preferential formation of CCW dimers (81 ± 9%). As the unit cell parameters remain unaltered upon addition of TA, it can be concluded that TA is not coadsorbed. As preorganization in solution can be excluded from fluorescence measurements, the TA derivatives must interact from the solution phase. Accordingly, the graphite surface must play a crucial role in the interactions between these two molecules. In addition, De Cat *et al.* showed that the chiral auxiliary can be removed by adding a base that can form a strong acid-base complex with TA, resulting in a globally chiral monolayer exclusively formed from achiral molecules.

To illustrate coadsorption of the chiral auxiliary, the self-assembly of a slightly modified achiral OPV derivative (AOPV3T) with the chiral nucleoside thymidine is considered [35]. Here, the diaminotriazine end-capped AOPV3T formed hydrogen-bonded cyclic hexamers or rosettes driven by hydrogen bonding at the 1-octanol/HOPG interface. Based on the relative orientation of the OPV units in the rosettes, CW and CCW rosettes can be identified on the surface. The chirality is also expressed at the level of symmetry breaking with the underlying graphite surface. The CW and CCW rosette monolayers should be rotated by −9° and +10°, respectively, in order to have their unit cell vector coincide with the main symmetry axis of the graphite. Upon mixing AOPV3T with enantiopure thymidine, a dynamic transformation from rosettes to dimers was observed (Figure 9.7d–g). In case of D-thymidine the dimers were all oriented CCW with respect to the normal of the row axis. For L-thymidine, dimers of opposite chirality (CW) were obtained. Time-dependent STM measurements revealed that initially dimers, rosettes and defects coexist on the surface and that the surface coverage by rosettes and defects gradually decreased in favor of the dimers. This means that the dimers are thermodynamically more stable. The pattern transformation from rosettes to dimers indicates that the thymidine interacts with the AOPV3T via noncovalent interactions by coadsorption in or on top of the monolayer.

9.3.3 Chiral Solvents

The potential role of solvents in amplification of chirality and emergence of homochirality at surfaces has received considerable attention recently. Like chiral auxiliaries, they can interact with achiral molecules via noncovalent interactions either by coadsorption or by complexation from the solution phase. Obviously, the presence of the solvent makes this strategy of chirality amplification uniquely applicable at the solution/solid-state interface. Here we will illuminate how chiral solvents can induce homochirality in self-assembled monolayers built from achiral units and how they can overrule the molecular chirality of enantiopure compounds.

The first demonstration of emergence of homochirality at an interface dates from 2008 [36]. Katsonis *et al.* selected a hydrogen-bonding OPV derivative (AOPV4T) that self-assembled from achiral solvents in the form of cyclic hexamers, which could be oriented either CW or CCW, and showed no bias towards either CW or CCW rosettes.

1-Phenyl-1-octanol was chosen as the chiral solvent as it can be considered as the hybrid of two very commonly used STM solvents: 1-phenyloctane and 1-octanol. Analysis of around 2000 rosettes revealed a clear bias towards CCW rotating rosettes in (*R*)-1-phenyl-1-octanol and CW rotating rosettes in (*S*)-1-phenyl-1-octanol. A homochiral surface, however, was never obtained. Katsonis *et al.* assigned this as the result of the slow kinetics of the ordering process. Their time-dependent STM studies revealed that nonordered rosettes gradually turn into rosettes of a particular handedness. However, STM experiments with enantiopure 1-phenyl-1-octyl acetate did not reveal a chirality bias, demonstrating that hydrogen bonding interactions are key in inducing surface homochirality. Circular dichroism measurements proved that no preorganization took place in solution and, furthermore, co-adsorption of the solvent is out of the question as the unit cell parameters were not altered upon coassembly. Probably, multiple factors come into play like solvent-rosette interactions, chiral desolvatation processes, and steric restrictions in the monolayer. It is most probable that the chiral solvent favors the formation of a transient complex of a particular handedness. To establish whether chiral solvents can be used as a general strategy to induce 2D chirality, several other chiral solvents have been examined. Deposition of AOPV4T on graphite from enantiopure 2-octanol and β-citronellol also revealed preferential formation of rosettes of a particular handedness [32].

The effects of chiral solvents as auxiliaries have also been tested for low-density porous phases [37, 38]. As these monolayers leave empty spaces on the surface, solvent coadsorption and dynamics become extremely important. The chirality amplification of two low-density porous hexagonal networks built up from different building blocks will be discussed here. Both deal with chirality induction by enantiopure 2-octanol, but solvent coadsorption is only visible for one hexagonal network.

The first honeycomb network is built up from achiral DBAs by interdigitation of the alkoxy side chains (as previously discussed in the sergeants-and-soldiers section – see Figure 9.5) [31]. Clockwise and CCW nanowells are, as expected, equally expressed on the surface. An alkoxy chain length-dependent chirality induction is observed for the DBAs when drop-casted from enantiomerically pure 2-octanol: CCW nanowells are induced in (*S*)-2-octanol and CW honeycombs in (*R*)-2-octanol [37]. For this system, homochiral surfaces are only obtained for the derivatives with short alkoxy chains (e.g. DBA-OC6).

The second hexagonal porous network is formed by 5-(benzyloxy)-isophthalic acid (BIC) derivatives [38]. The molecular structure of BIC is shown in Figure 9.8a. In this case, three BIC molecules merge to form a trimer stabilized via hydrogen bonding between the acid groups and the trimers then come together by alkyl chain interdigitation in order to form the hexagonal pattern. Two enantiomorphous hexagons can be formed as shown in Figure 9.8b. For the BIC self-assembly in enantiopure 2-octanol, globally chiral porous networks are always obtained: exclusively CW domains are obtained in (*R*)-2-octanol and CCW domains in (*S*)-2-octanol. To get insights into the origin of chiral bias, high resolution STM images are scrutinized for both systems. For the DBA network, solvent co-adsorption is not detected in the STM images. It seems that the solvent dynamics are too fast to follow by STM. The STM images of the BIC monolayers, on the other hand, do reveal co-adsorption. From the high resolution images (Figure 9.8c–d) it becomes clear that four solvent molecules are entrapped in the small gap between the interdigitated alkyl chains. On the basis of these results, models of the CW and CCW patterns are proposed. Three hydrogen bonds are formed here between the neighboring BICs in the trimer. The leftover hydrogen

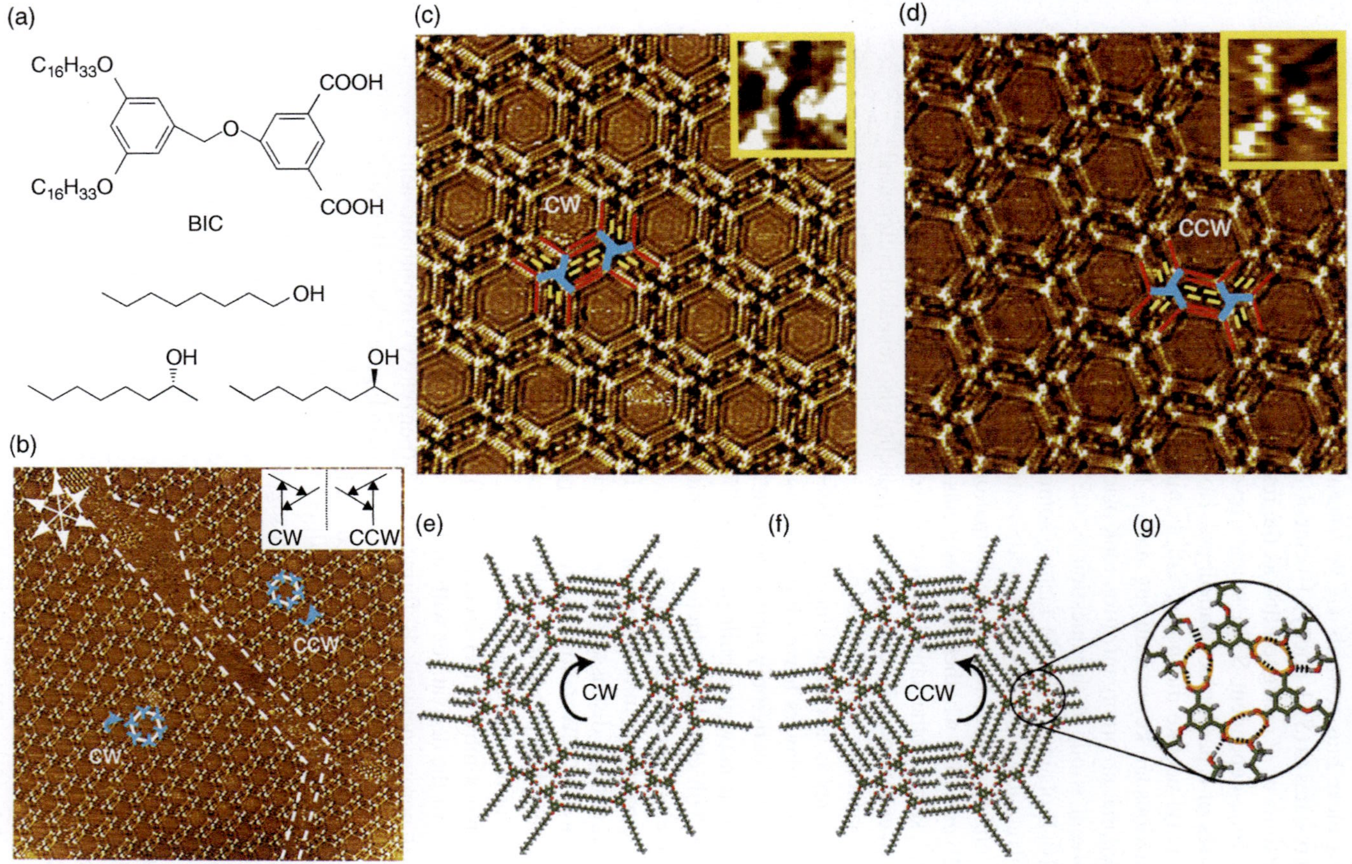

Figure 9.8 *(a) Chemical structure of BIC, 1-octanol and (R)- and (S)-2-octanol. (b) Large-scale STM image ($I_{set=}$ 0.205 nA, V_{bias} = 0.900 V) of BIC monolayer at the 1-octanol/HOPG interface. The dotted lines indicate domain boundaries. The white hexagons indicate the hexagonal units in BIC assembly. The blue windmills and arrows indicate the BIC trimers and their rotational directions, respectively. The white arrows indicate the substrate lattice of HOPG. Inset, CW and CCW rotation of a BIC trimer. (c) High-resolution STM image (I_{set} = 0.095 nA, V_{bias} = 0.929 V) of BIC monolayer at (R)-2-octanol/HOPG interface. (d) High-resolution STM image (I_{set} = 0.100 nA; V_{bias} = 0.900 V) of BIC monolayer at (S)-2-octanol/HOPG interface. The blue, red and yellow lines outline the bright aromatic cores, the alkoxy chains in BIC and the solvent alkyl chains, respectively. Structural models of (e) CW network at (R)-2-octanol/HOPG interface and of (f) CCW network at (S)-2-octanol/HOPG interface. (g) Illustration of the intermolecular interactions within the molecular trimer. The black dotted lines represent the hydrogen bonds. The orange circle indicates the ten-membered ring. (Adapted with permission from Ref. [38]. Copyright 2013, Macmillan Publishers Ltd.)*

bonding sites interact with the hydroxyl group of the co-adsorbed solvent molecule. In this way, a ten-membered ring stabilized by hydrogen bonding is formed. Closer inspection reveals that the ten-membered ring arranges in a chiral manner due to the directionality of the hydrogen bonding and hence two mirror image configurations can be formed. As the solvent molecules like to have their methyl group pointing away from the surface due to steric effects, the enantiopure solvent molecules prefer to coadsorb in a particular oriented BIC trimer and consequently a homochiral surface is attained.

For the BIC derivatives, remote chiral communication has also been investigated [39]. To probe the influence of the distance between the 2D chirality-determining stereogenic center and the point of chirality transfer (i.e. the interaction site between chiral solvent and BIC), BICs were self-assembled from a series of enantiopure methyl substituted 1-heptanol derivatives. The BIC derivative used in this set of experiments has shorter alkyl chains. Consequently, only two solvent molecules, instead of four, coadsorb in the space enclosed by neighboring BIC trimers. Also, deposition from enantiopure 2-octanol, which can be considered as 1-methyl-1-heptanol, led to the formation of a homochiral porous network. Counterclockwise and CW nanowells were exclusively obtained in (*S*)- and (*R*)-2-octanol, respectively. Upon moving the stereogenic center along the heptyl backbone, an odd-even alternation rule was followed for the handedness. So for (*S*)-2-methyl-1-heptanol only CW domains were formed, while for (*S*)-5-methyl-1-heptanol only CCW domains were present on the surface. In addition, the transmission efficiency was tested by diluting the chiral solvent with achiral 1-heptanol. Homochirality was obtained starting from 20 vol% of the chiral solvent and in the lower concentration region a nonlinear dependency between chirality and concentration appeared. Interestingly, the dilution curve of (*S*)-2-octanol was higher than that of (*S*)-5-methyl-1-heptanol, implying that (*S*)-2-octanol is a more efficient chirality transmitter. To verify this statement, (*R*)-2-octanol was mixed with (*S*)-5-methyl-1-heptanol in a 1 : 1 ratio. Of the BIC networks 79% displayed CW chirality and the remaining 21% were CCW, clearly indicating that the stereochemical information is more efficiently transmitted when the stereogenic center is positioned at the 1-position.

The previous examples clearly illustrate how chiral solvents can influence the self-assembly of achiral molecules. A similar observation is made for achiral amide-substituted porphyrin derivatives [40]. Their self-assembly leads to the formation of domains consisting of regular rows stabilized by van der Waals interactions. In achiral solvents, two mirror domains are observed in equal abundances. When drop-casted from enantiopure 2-octanol, mainly CW domains (92%) in (*S*)-2-octanol and CCW domains (80%) in (*R*)-2-octanol are formed. Once again, the chiral solvent is able to create chiral monolayers from achiral building blocks and it poses the question of whether chiral solvents can also overrule the intrinsic chirality of chiral derivatives. Therefore a so-called *sergeants-and-corporals* experiment is performed. A porphyrin derivative bearing one chiral center is used as sergeant (Figure 9.9), and (*S*)- and (*R*)-2-octanol are used as corporals. The chiral derivative with a chiral (*R*)-chain, is known to form exclusively CW rotating domains in achiral solvents. In (*R*)-2-octanol, the CCW chirality is induced and covers about 20% of the entire surface. A similar observation is made for the derivative with a chiral (*S*)-chain in enantiopure (*S*)-2-octanol: about 20% of the domains on the surface have a CW rotation. From this, it seems that the corporals are capable of inverting the chirality of the sergeant to some extent. However, chirality inversion is not observed for porphyrin derivatives with more than one stereogenic center. As the unit cells remain unaltered, the solvent cannot be part

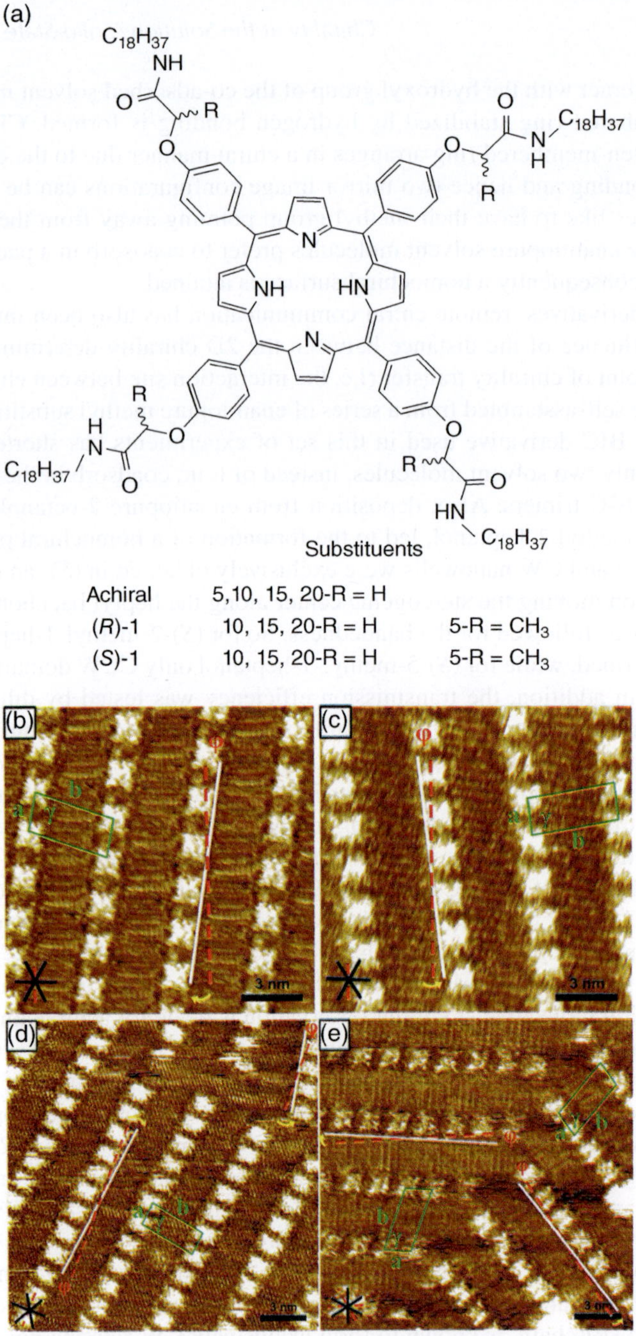

__Figure 9.9__ (a) Molecular structures of the achiral and chiral porphyrin derivatives. STM images of the achiral porphyrin physisorbed at the (b) (S)-2-octanol/HOPG interface and (c) (R)-2-octanol/HOPG interface (I_{set} = 0.6 nA, V_{bias} = −1.0 V). STM images of chiral porphyrin (d) (R)-1 physisorbed at the (R)-2-octanol/HOPG interface and (e) (S)-1 physisorbed at the (S)-2-octanol/HOPG interface (I_{set} = 0.280 nA, V_{bias} = −230 V). Both CW and CCW domains are present. The black solid lines indicate the direction of the main symmetry axes of the underlying graphite. The dashed red lines indicate the reference axis of graphite (−1 1 0 0). The white lines indicate the lamellar direction. Unit cells are indicated in green. φ is the angle between the reference axis and the unit cell vector a. The yellow arrows display the rotation direction. (Adapted from Ref. [40] with permission of The Royal Society of Chemistry.)

of the monolayer matrix. Molecular dynamics have revealed that the solvent is responsible for stabilizing the self-assembled monolayers via hydrogen-bonding with the amide group of the porphyrin side chains [41]. So most probably the hydrogen bonding between the chiral solvent and the porphyrin amide group alters the orientation of the area around the amide group, allowing the formation of the opposite handedness. It seems reasonable to assume that there will be an energetic barrier for the change in orientation and that this barrier would increase upon enhancing the number of stereogenic centers. Hence, for porphyrins with one stereogenic center this barrier can be overcome but derivatives with more than one stereogenic center may find it difficult to prevail over this energetic barrier.

9.3.4 Majority Rules

Chirality amplification is not only possible for achiral compounds – racemates can also undergo amplification. By introducing a small enantiomeric excess ($ee = (R-S)/(R+S)$), a bias towards one organizational handedness can be created. The majority enantiomer will enforce the chiral molecules to follow its self-assembly behavior, and hence this amplification strategy is termed *majority rules*. So far, majority rules were only observed once at the solution / solid-state interface. The group of Li-Jun Wan demonstrated S-shaped nonlinear behavior of the BIC network self-assembly on the enantiomeric excess of the chiral coadsorber ((*S*)- and (*R*)-2-octanol) [38]. Depending on the nature of the majority enantiomer, the mirror symmetry was broken: CCW and CW chirality are formed in (*S*)- and (*R*)-2-octanol, respectively. When an excess of 5.2% of the chiral coadsorber was present, a homochiral surface was already achieved. The driving force is believed to be entropy driven at the point of nucleation. As there is always an entropy increasing driving force in the liquid phase to reach equal amounts of (*S*)- and (*R*)-2-octanol, the majority solvent will tend to adsorb on the surface. As a result, the majority component can dictate the outcome of the self-assembly process and create homochiral networks of its preferred chirality.

9.3.5 Magnetic Fields

The last method that will be described to induce chiral organization of molecules does not involve a chiral input. Patrick *et al.* showed that mirror symmetry can be broken in a layer of achiral molecules by controlling their arrangement with the use of an external magnetic field [42, 43]. They studied the self-assembly of 4-cyano-4′-octylbiphenyl (8CB) on graphite. This compound is a liquid crystal (LC) that forms a smectic A phase at room temperature. The self-assembly of 8CB on graphite in the absence of a magnetic field resulted in the formation of commensurate domains. As a result of the threefold symmetry of the graphite and the symmetry breaking that occurs upon adsorption, the domains will distribute along six energetically degenerate orientations. To break the degeneracy, the graphite sample with a drop of 8CB on top was placed in a 1.2 T magnetic field parallel to the surface. The sample was subsequently heated to 100 °C for 2 min and then allowed to gradually cool back down to room temperature. Ultimately, a molecular film with uniaxial order was obtained as the orientational order in the bulk LC supernate created by the presence of the magnetic field becomes imprinted on the monolayer. So, by simply changing the orientation of the field the absolute chirality and enantiomeric purity of the film can be engineered. The only drawback of this imprinting technique is obviously that it only applies to liquid crystals. Still, recent research showed that these LC monolayers can be displaced by other organic molecules such as n-tetracontane preserving the original alignment orientation [44, 45].

9.4 Towards Applications

First and foremost, 2D molecular self-assembly on solid surfaces aims at understanding molecular recognition. Among these studies, chirality at surfaces has gained widespread interest because of its fundamental understanding and potential applications. Chirality plays an important role in crystallization and 2D studies of chirality expression and amplification promise an understanding of enantioselective heterogeneous catalysis and chiral separations. The separation of racemates into enantiomers is of great importance in pharmaceutical and material technologies. The interaction of molecules with interfaces provides interesting opportunities to resolve chiral molecules. For instance, modified surfaces can be used as a stationary phase in enantioselective chromatography [46]. Moreover, spontaneous resolution on surfaces is observed for the majority of the racemates, while in solution only about 5–10% of the racemates crystallize as a conglomerate. Nonetheless, isolation and collection of the enantiomers remains a very difficult task, as the domain sizes often have nanometer scale dimensions. From a practical point of view, enantioselective adsorption or crystallization is more relevant for separation purposes. Recently, some STM studies emerged that accentuate the potential of self-assembled monolayers as platforms for chiral resolution [47–49]. These studies, which will immediately be discussed in detail, illustrate chiral separation by either preferential surface-assisted diastereomer adsorption or by enantioselective adsorption.

9.4.1 Chiral Resolution at the Solution/Solid-State Interface

One of the most frequently performed techniques to separate enantiomers in solution is diastereomeric resolution. This process relies on differences in solubility and crystallization propensity of the formed diastereomers. Typically, the diastereomers are formed by adding a complementary chiral compound that can form a complex or a salt with the racemate. A drawback is that the identification process of the resolving agent relies on empirical screening, which can be quite time consuming. A similar Pasteurian resolution has been performed for 1,2-diaminocyclohexane (DACH) at the solution/solid-state interface by Xu *et al.* based on one diastereomeric complex having a higher affinity for adsorption [47]. They selected an alkylated resorcinol derivative with a lactate group (Figure 9.10a) as a potential resolving agent because the lactate moiety should be able to form hydrogen bonds with the amino groups. The resorcinol derivative itself displayed spontaneous resolution and the observed lamella structure was stabilized by interdigitation of the alkyl chains and C-H···π interactions between the alkyl chains and the graphite surface. Prior to the actual Pasteurian resolution experiment, all possible resorcinol/DACH combinations were characterized. In case of like enantiomers (solution ratio of 2:1 resorcinol/DACH, Figure 9.10c–f), the alkyl chains in the self-assembled layer were no longer perpendicular to the row axis and they did not display interdigitation with molecules of adjacent rows. As a consequence the lamella were wider. The difference in width, however, could not be solely assigned to the different arrangement of the alkyl chains. In the STM images fuzzy features appeared in between the lactate groups of adjacent resorcinol rows, which were attributed to coadsorption of DACH. Most probably they appeared streaky as the STM tip knocks the molecular segments that were standing above the plane of the resorcinol alkyl chains. The chirality is expressed at the level of monolayer organization (orientation of the alkyl chains

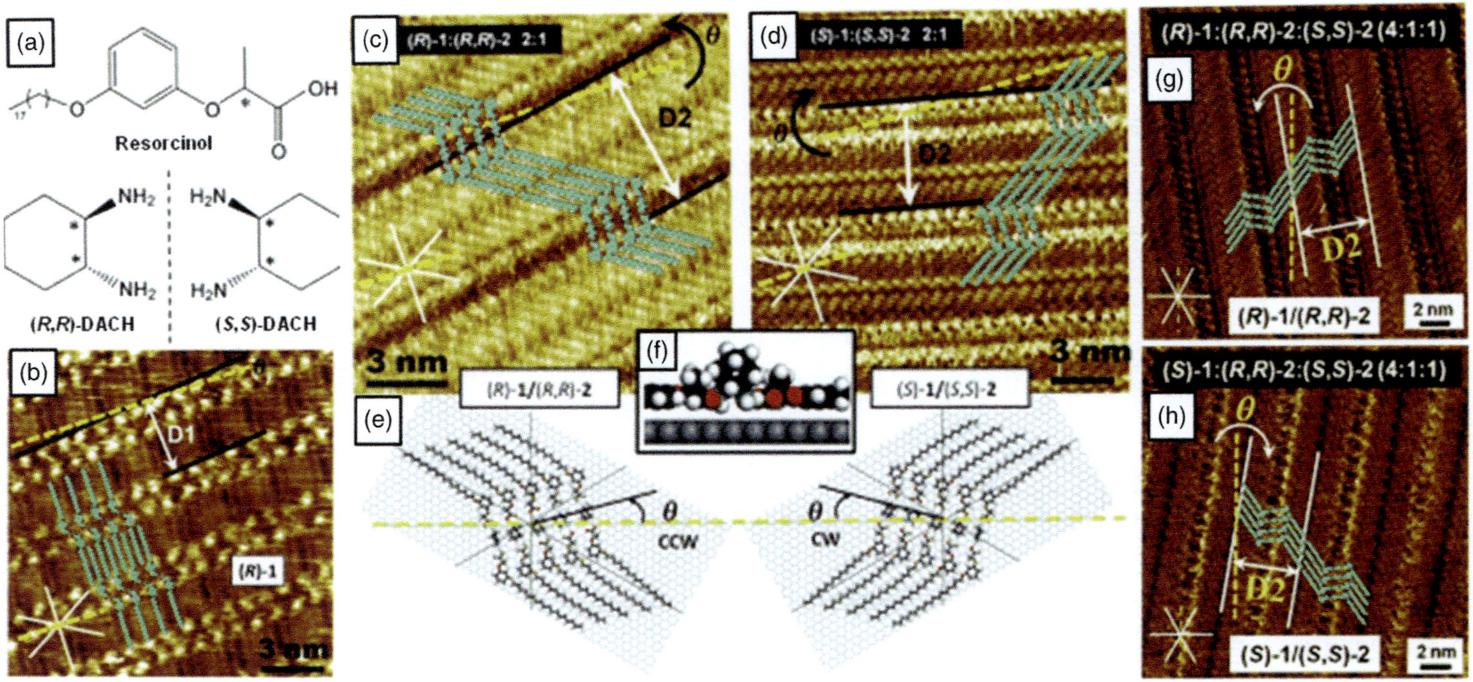

Figure 9.10 *(a) Chemical structures of the chiral resorcinol derivative and 1,2-diaminocyclohexane. * indicates the location of the stereogenic center. Scanning tunneling microscopy images of monolayers at the 1-phenyloctane/HOPG interface formed by (b) (R)-resorcinol (I_{set}=0.70 nA, V_{bias}=−0.50 V) and (c–d) upon premixing resorcinol and DACH at a 2:1 ratio. (c) (R)-resorcinol:(R,R)-DACH (2:1) I_{set}=0.25 nA, V_{bias}=-1.00 V), (d) (S)-resorcinol:(S,S)-DACH (2:1) (I_{set}=0.20 nA, V_{bias}=−1.00 V). The solution composition is shown in a black text box, whereas the domain composition is given in a white text box. White solid lines indicate the main symmetry axes of graphite. Yellow dashed lines indicate the selected graphite reference axis. Black solid lines indicate the direction of molecular rows. θ is the angle between the row axis and the HOPG reference axis. Double-headed arrows show the width D of the rows. Models of resorcinol and diamine molecules are superimposed on the STM image. (e) Molecular model of the z-shaped (R)-resorcinol/(R,R)-DACH self-assembled structure on graphite obtained by molecular mechanics simulations. The model at the right is the s-shaped (S)-resorcinol/(S,S)-DACH structure which is obtained by reflection. The yellow dashed line is the graphite reference axis. (f) Snapshot of representative structure (side view) of the core of the (R)-resorcinol/(R,R)-DACH assembly (N blue, O red) (g–h) STM images of monolayers formed upon premixing one resorcinol enantiomer, and the DACH racemate at the 1-phenyloctane/HOPG interface. Molecular models are superimposed on top of the STM images. (g) (R)-resorcinol:(R,R)-DACH:(S,S)-DACH (4:1:1) (I_{set}=0.30 nA, V_{bias}=−1.00 V), (h) (S)-resorcinol:(R,R)-DACH:(S,S)-DACH (4:1:1) (I_{set}=0.30 nA, V_{bias}=−1.00 V). (Adapted with permission from Ref. [47].)*

with respect to the row axis) and at the level of symmetry breaking with the underlying graphite surface. Generally, CCW domains are observed for (*R*)-resorcinol/(*R*,*R*)-DACH and CW domains for (*S*)-resorcinol/(*S*,*S*)-DACH. Examination of the assembly of unlike enantiomers ((*R*)-resorcinol/(*S*,*S*)-DACH or vice versa), revealed no sign of complex formation and the surface was only covered with monolayers of pure resorcinol. Only at higher ratios (1 : 5 or 1 : 10 resorcinol : DACH) a few small domains with complexes were observed. Therefore, it is reasonable to conclude that complexes formed by like enantiomers are more stable than those formed by unlike enantiomers. This result is crucial if one wants to adsorb one particular enantiomer selectively from a racemate. To explore the resolving ability of resorcinol, enantiopure resorcinol was mixed with equal amounts of (*R*,*R*)- and (*S*,*S*)-DACH at different ratios ranging from 4 : 1 : 1 to 1 : 10 : 10 (see Figure 9.10g–h). In the case of (*R*)-resorcinol only two-component patterns with (*R*,*R*)-DACH were visible. The same holds for (*S*)-resorcinol – only CW domains were visualized, indicating (*S*)-resorcinol/(*S*,*S*)-DACH complex formation. These results clearly show the capacity to coadsorb a specific DACH-enantiomer selectively by diastereoselective adsorption.

Similar preferential adsorption experiments have been carried out with OPV3T/thymidine mixtures. Previously, it was mentioned that enantiopure thymidine can be used as a chiral auxiliary to amplify the chirality of achiral OPV3T monolayers. The coadsorption of the enantiomerically pure thymidine resulted in a dynamic transformation from rosettes to dimers due to hydrogen bonding between both units. Thus far, it has been established that stable surface-confined diastereomers can be formed between achiral OPV3T and thymidine. This raises the question of whether chiral OPV3Ts can act as a resolving agent for thymidine by preferential adsorption. First the patterns formed by all possible combinations of OPV3T and thymidine were characterized, followed by optimization of the resolving conditions. Careful analysis revealed that combinations with D-thymidine always produced CCW dimers and those with L-thymidine CW dimers, regardless of the chirality of OPV3T [48]. The dynamic transformation from rosettes to dimers occurred at different solution ratios (thymidine : OPV3T) for the diastereomeric monolayers, implying that, for (*S*)-OPV3T, for instance, DS assemblies are stronger than LS assemblies or DS assemblies have a higher affinity for adsorption. In addition, molecular dynamics simulations supported this observation: D-thymidine can simultaneously interact with (*S*)-OPV3T and the substrate, while L-thymidine can only optimize one interaction. The resolving ability of OPV3T was confirmed by deposition of a mixture of OPV3T with (*rac*)-thymidine. In case of the (*S*)-enantiomer, both CCW and CW dimers were present on the surface but the CCW dimers represented 37% of the surface coverage, while the CW dimers only covered 2%. A similar observation was made for the (*R*)-OPV3T : 42% CW dimers and 5% CCW dimers. Further optimization of the resolving conditions such as the molar OPV3T : thymidine ratio and the OPV3T concentration allowed to obtain an enantiomeric excess of 90%.

9.4.2 Enantioselective Adsorption at the Solution/Solid-State Interface

Nanoporous 2D networks show great potential in recognition processes of chemical, biological and even chiral species at the single molecule level [50, 51]. Currently, the recognition of these systems is limited to compounds of complementary size and/or shape [52–54]. Porous networks can also be used for data storage, site-selective catalysis and surface-confined chemical reactions. To illustrate their potential as separating platform for enantiomers, the molecular recognition between a chiral host network and a chiral DBA

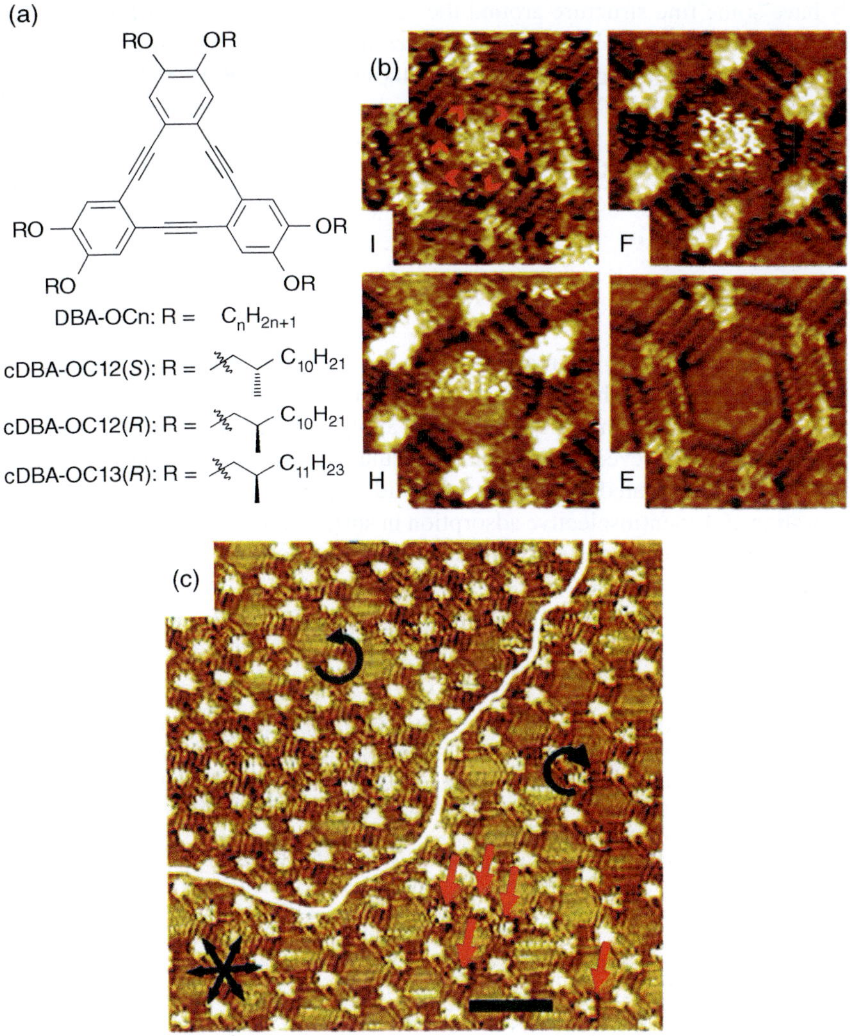

Figure 9.11 *(a) Chemical structure of the achiral and chiral DBAs (b) Examples of the four different occupation levels of the pores: incorporated (I), fuzzy (F), half fuzzy (H) and empty (E) (c) Scanning tunneling microscopy image of a premixed solution containing 30 mol% cDBA-OC12-(S) and 70 mol% DBA-OC12. I_{set} = 200 pA and V_{bias} = −250 mV. The white line marks the boundary between a CCW domain (upper left) and a CW domain (lower right). The black arrows indicate the major symmetry axes of the HOPG surface underneath. The scale bar measures 5 nm. The red arrows indicate cDBA-OC12-(S) sergeant molecules. (Adapted from Ref. [49] with permission of The Royal Society of Chemistry.)*

guest molecules will be considered [49]. A clear difference in occupation level is noted for CW and CCW pores while carrying out sergeants-and-soldiers experiments with a mixture of DBA-OC12/cDBA-OC12-(*S*) (70/30 mol%). In the STM images four different occupation levels of the pores can be distinguished: incorporated (I), fuzzy (F), half-fuzzy (H) and empty (E) as depicted in Figure 9.11b. Incorporated guests appear quite intense and typically have some fine structure around the center spot, which is attributed to the alkyl chains that adopt a windmill-like structure. Pores that appear (half-)fuzzy are believed to be the result of highly mobile molecules. Upon mixing of DBA-OC12 and cDBA-OC12-(*S*), a clear preference towards CCW pores is observed. In total, one out of three CCW nanowells are empty, while for the CW nanowells 85% remain unoccupied. A similar trend is observed for the mixture of DBA-OC12 and cDBA-OC12-(*R*), namely preferential occupation of CW pores, and supports the hypothesis that the pores show enantioselectivity.

9.5 Conclusions

Globally chiral surfaces can be manifested at the solution/solid-state interface by the adsorption of intrinsically chiral molecules. Moreover, amplification methods such as sergeants-and-soldiers and majority rules also apply at interfaces and, as such, global chirality can also be achieved for achiral molecules and racemates. Furthermore, chiral separation has been illustrated at the solution/solid-state interface by preferential diastereomer adsorption and enantioselective adsorption in surface-confined nanoporous networks. This could have profound consequences for the further development of new chiral separation techniques.

References

[1] J. M. Lehn, Supramolecular chemistry – scope and perspectives molecules, supermolecules, and molecular devices, *Angewandte Chemie – International Edition*, **27**(1), 89–112 (1988).
[2] J. M. Lehn, Perspectives in supramolecular chemistry – from molecular recognition towards molecular information-processing and self-organization, *Angewandte Chemie – International Edition*, **29**(11), 1304–1319 (1990).
[3] J. M. Lehn, Supramolecular chemistry, *Science*, **260**(5115), 1762–1763 (1993).
[4] J. C. Love, L. A. Estroff, J. K. Kriebel *et al.*, Self-assembled monolayers of thiolates on metals as a form of nanotechnology, *Chemical Reviews*, **105**(4), 1103–1169 (2005).
[5] S. A. Claridge, W. S. Liao, J. C. Thomas *et al.*, From the bottom up: dimensional control and characterization in molecular monolayers, *Chemical Society Reviews*, **42**(7), 2725–2745 (2013).
[6] D. A. Bonnell and P. N. Ross, *Scanning Tunneling Microscopy and Spectroscopy: Theory, Techniques, and Applications*. Wiley-VCH Verlag GmbH, New York, NY, p. 436 (1993).
[7] F. Zaera, Probing liquid/solid interfaces at the molecular level, *Chemical Reviews*, **112**(5), 2920–2986 (2012).
[8] G. Binnig and H. Rohrer, Scanning tunneling microscopy – from birth to adolescence, *Reviews of Modern Physics*, **59**(3), 615–625 (1987).
[9] G. Binning, H. Rohrer, C. Gerber and E. Weibel, Surface studies by scanning tunneling microscopy, *Physical Review Letters*, **49**(1), 57–61 (1982).
[10] S. De Feyter, A. Gesquiere, P. C. M. Grim *et al.*, Expression of chirality and visualization of stereogenic centers by scanning tunneling microscopy, *Langmuir*, **15**(8), 2817–2822 (1999).
[11] J. Elemans, I. De Cat, H. Xu and S. De Feyter, Two-dimensional chirality at liquid-solid interfaces, *Chemical Society Reviews*, **38**(3), 722–736 (2009).
[12] J. Zhang, A. Gesquière, M. Sieffert *et al.*, Losing the expression of molecular chirality in self-assembled physisorbed monolayers, *Nano Letters*, **5**(7), 1395–1398 (2005).
[13] L. Pérez-García and D. B. Amabilino, Spontaneous resolution under supramolecular, *Chemical Society Reviews*, **31**(6), 342–356 (2002).
[14] T. Balandina, M. W. van der Meijden, O. Ivasenko *et al.*, Self-assembly of an asymmetrically functionalized [6]helicene at liquid/solid interfaces, *Chemical Communications*, **49**(22), 2207–2209 (2013).

[15] R. Fasel, M. Parschau, and K.-H. Ernst, Chirality transfer from single molecules into self-assembled monolayers, *Angewandte Chemie – International Edition*, **42**(42), 5178–5181 (2003).
[16] R. Fasel, M. Parschau and K. Ernst, Amplification of chirality in two-dimensional enantiomorphous lattices, *Nature*, **439**(7075), 449–452 (2006).
[17] M. Stöhr, S. Boz, M. Schär *et al.*, Self-assembly and two-dimensional spontaneous resolution of cyano-functionalized [7]helicenes on Cu111, *Angewandte Chemie – International Edition*, **50**(42), 9982–9986 (2011).
[18] J. Lin, Z. Guo, J. Plas *et al.*, Homochiral and heterochiral assembly preferences at different length scales – conglomerates and racemates in the same assemblies, *Chemical Communications*, **49**(81), 9320–9322 (2013).
[19] N. Katsonis, E. Lacaze and B. L. Feringa, Molecular chirality at fluid / solid interfaces: expression of asymmetry in self-organised monolayers, *Journal of Materials Chemistry*, **18**(18), 2065–2073 (2008).
[20] Y. Wei, K. Kannappan, G. W. Flynn and M. B. Zimmt, Scanning tunneling microscopy of prochiral anthracene derivatives on graphite: chain length effects on monolayer morphology, *Journal of the American Chemical Society*, **126**(16), 5318–5322 (2004).
[21] A. Marchenko, S. Lukyanets and J. Cousty, Adsorption of alkanes on Au(111): Possible origin of STM contrast at the liquid / solid interface, *Physical Review B*, **65**(4), 045414 (2002).
[22] H. Zhou and J. D. Wuest, Crankshafts: Using simple, flat C2h-symmetric molecules to direct the assembly of chiral 2D nanopatterns, *Langmuir*, **29**(24), 7229–7238 (2013).
[23] S. Ahn, J. T. Henssler and A. J. Matzger, Conglomerate with periodic enantiomer inclusion: A mechanism for homochirality erosion, *Chemical Communications.*, **47**(41), 11432–11434 (2011).
[24] S. Ahn and A.J. Matzger, Anatomy of one-dimensional cocrystals: Randomness into order, *Journal of the American Chemical Society*, **131**(38), 13826–13832 (2009).
[25] W. J. Saletra, H. Xu, T. Vosch *et al.*, Chiral non-periodic surface-confined molecular nanopatterns revealed by scanning tunneling microscopy, *CrystEngComm*, **13**(18), 5578–5583 (2011).
[26] A. Mugarza, N. Lorente, P. Ordejón *et al.*, Orbital specific chirality and homochiral self-assembly of achiral molecules induced by charge transfer and spontaneous symmetry breaking, *Physical Review Letters*, **105**(11), 115702 (2010).
[27] J. Liu, T. Chen, X. Deng *et al.*, Chiral hierarchical molecular nanostructures on two-dimensional surface by controllable trinary self-assembly, *Journal of the American Chemical Society*, **133**(51), 21010–21015 (2011).
[28] P. Sleczkowski, N. Katsonis, O. Kapitanchuk *et al.*, Emergence of chirality in hexagonally packed monolayers of hexapentyloxytriphenylene on Au(111): A joint experimental and theoretical study, *Langmuir*, **30**(111), 13275–13282 (2014).
[29] M. M. Green, M. P. Reidy, R. J. Johnson *et al.*, Macromolecular stereochemistry – the out-of-proportion influence of optically-active co-monomers on the conformational characteristics of polyisocyanates – the sergeants and soldiers experiment, *Journal of the American Chemical Society*, **111**(16), 6452–6454 (1989).
[30] S. Lifson, C. E. Felder and M. M. Green, Helical conformations, internal motion, and helix sense reversal in polyisocyanates and the preferred helix sense of an optically active polyisocyanate, *Macromolecules*, **25**(16), 4142–4148 (1992).
[31] K. Tahara, H. Yamaga, E. Ghijsens *et al.*, Control and induction of surface-confined homochiral porous molecular networks, *Nature Chemistry*, **3**(9), 714–719 (2011).
[32] H. Xu, E. Ghijsens, S. J. George *et al.*, Chiral induction and amplification in supramolecular systems at the liquid-solid interface, *European Journal of Chemical Physics and Physical Chemistry*, **14**(8), 1583–1590 (2013).
[33] P. A. Korevaar, S. J. George, A. J. Markvoort *et al.*, Pathway complexity in supramolecular polymerization, *Nature*, **481**(7382), 492–496 (2012).
[34] I. De Cat, Z. Guo, S. J. George *et al.*, Induction of chirality in an achiral monolayer at the liquid / solid interface by a supramolecular chiral auxiliary, *Journal of the American Chemical Society*, **134**(6), 3171–3177 (2012).
[35] Z. Guo, I. De Cat, B. Van Averbeke *et al.*, Nucleoside-assisted self-assembly of oligo (p-phenylenevinylene)s at liquid / solid interface: chirality and nanostructures, *Journal of the American Chemical Society*, **133**(44), 17764–17771 (2011).

[36] N. Katsonis, H. Xu, R. M. Haak *et al.*, Emerging solvent-induced homochirality by the confinement of achiral molecules against a solid surface, *Angewandte Chemie – International Edition*, **47**(27), 4997–5001 (2008).

[37] I. Destoop, E. Ghijsens, K. Katayama *et al.*, Solvent-induced homochirality in surface-confined low-density nanoporous molecular networks, *Journal of the American Chemical Society*, **134**(48), 19568–19571 (2012).

[38] T. Chen, W.-H. Yang, D. Wang and L.-J. Wan, Globally homochiral assembly of two-dimensional molecular networks triggered by co-absorbers, *Nature Communications*, **4**, 1389 (2013).

[39] T. Chen, S. Li, D. Wang *et al.*, Remote chiral communication in coadsorber-induced enantioselective 2D supramolecular assembly at a liquid/solid interface, *Angewandte Chemie – International Edition*, doi: 10.1002/anie.201410927, (2015).

[40] I. Destoop, H. Xu, C. Oliveras-González *et al.*, "Sergeants-and-Corporals" principle in chiral induction at an interface, *Chemical Communications*, **49**(68), 7477–7479 (2013).

[41] M. Linares, P. Iavicoli, K. Psychogyiopoulou *et al.*, Chiral expression at the solid–liquid interface: A joint experimental and theoretical study of the self-assembly of chiral porphyrins on graphite, *Langmuir*, **24**(17), 9566–9574 (2008).

[42] J. Mougous, A. Brackley, K. Foland *et al.*, Formation of uniaxial molecular films by liquid-crystal imprinting in a magnetic field, *Physical Reviews Letters*, **84**(12), 2742–2745 (2000).

[43] A. M. Berg and D. L. Patrick, Preparation of chiral surfaces from achiral molecules by controlled symmetry breaking, *Angewandte Chemie – International Edition*, **44**(12), 1821–1823 (2005).

[44] S. Hickman, A. Hamilton and D. L. Patrick, Controlling molecular alignment in an organic monolayer with a sacrificial liquid crystal template, *Surface Science*, **537**(1–3), 113–122 (2003).

[45] N. E. Gislason, C. Murphy and D. L. Patrick, Mechanistic origins of hierarchical order in organic monolayers deposited from liquid crystal solvents, *Journal of Physical Chemistry C*, **114**(29), 12659–12666 (2010).

[46] C. West, C. Elfakir and M. Lafosse, Porous graphitic carbon: a versatile stationary phase for liquid chromatography, *Journal of Chromatography A*, **1217**(19), 3201–3216 (2010).

[47] H. Xu, W. J. Saletra, P. Iavicol *et al.*, Pasteurian segregation on a surface imaged in situ at the molecular level, *Angewandte Chemie – International Edition*, **51**(48), 11981–11985 (2012).

[48] Z. Guo, I. De Cat, B. Van Averbeke *et al.*, Surface-induced diastereomeric complex formation of a nucleoside at the liquid/solid interface: Stereoselective recognition and preferential adsorption, *Journal of the American Chemical Society*, **135**(26), 9811–9819 (2013).

[49] E. Ghijsens, H. Cao, A. Noguchi *et al.*, Towards enantioselective adsorption in surface-confined nanoporous systems, *Chemical Communications*, doi: 10.1039/C4CC08826C (2015).

[50] S. Stepanow, N. Lin, J. V. Barth and K. Kern, Non-covalent binding of fullerenes and biomolecules at surface-supported metallosupramolecular receptors, *Chemical Communications*, **20**, 2153–2155 (2006).

[51] D. Bonifazi, S. Mohnani and A. Llanes-Pallas, Supramolecular chemistry at interfaces: Molecular recognition on nanopatterned porous surfaces, *European Journal of Chemistry*, **15**(29), 7004–7025 (2009).

[52] Y. T. Shen, K. Deng, X. M. Zhang *et al.*, Switchable ternary nanoporous supramolecular network on photor-regulation, *Nano Letters*, **11**(8), 3245–3250 (2011).

[53] M. Li, K. Deng, S. B. Lei *et al.*, Site-selective fabrication of two-dimensional fullerene arrays by using a supramolecular template at the liquid-solid interface, *Angewandte Chemie – International Edition*, **47**(35), 6717–6721 (2008).

[54] J. Adisoejoso, K. Tahara, S. Okuhata *et al.*, Two-dimensional crystal engineering: A four-component architecture at a liquid-solid interface, *Angewandte Chemie – International Edition*, **48**(40), 7353–7357 (2009).

10

Nanoscale Aspects of Chiral Nucleation and Propagation

Edward G. Latter and Rasmita Raval

10.1 Introduction

Molecular chirality was first discovered by Louis Pasteur in 1848 with the separation of a racemic mixture of sodium ammonium tartrate [1, 2] into two types of crystals. The crystals were mirror images with respect to their morphology – a result of the chirality of the individual molecules of which they were composed. It is known that chiral crystal growth occurs from nucleation points and the nature of the chiral ordering at the beginning lays out a pathway for the growth of the macroscopic crystal. Thus, an understanding of the fundamentals of the nucleation process and the transfer of chiral expression to longer range order is crucial to achieve control over the properties of the crystalline state [3, 4]. Many factors affect chiral nucleation and, as nuclei transform into crystals, the situation becomes complex due to competing pathways of growth versus further nucleation. However, currently, there is a lack of spectroscopic methods for directly observing nanometre size clusters in solution and chiral nucleation remains little understood. This means that knowledge needs to be drawn from analogous systems in order to gain insights into the chiral nucleation process [5].

An important and analogous system that can provide unique insights is the formation of self-assembled chiral molecular organizations and assemblies on well defined metal surfaces. A similar pathway is responsible for the formation of these 2D crystal structures as for the 3D crystal growth: nucleation of a single or small cluster of molecules followed by growth that leads to longer range order. Many of these systems display the chiral resolution familiar to us from 3D systems and provide nanoscale details of the events that drive

Chirality in Supramolecular Assemblies: Causes and Consequences, First Edition. Edited by F. Richard Keene.

system evolution. However, studies in this area are often focussed on the energetic driving forces that determine the final observed structure, while details of the nucleation and growth of chiral assemblies are substantially harder to elucidate. In this chapter, we focus on some of our recent work, which shows that important insights into the beginnings of chiral suprastructures can be obtained by appropriate choice of system and experimental conditions combined with high resolution experimental methods and theoretical calculations.

10.1.1 Chirality at Surfaces

Organic 2D chiral self-assemblies [6–9] can display a hierarchical expression of chirality, as described previously by Raval and coworkers [6, 10–14]. Point (0D) chirality arises because the adsorption of the molecule locally destroys all mirror planes and chirality is observed from both chiral and achiral adsorbates. For chiral molecules, such as amino acids, it is necessary that the chiral centres of the molecules are preserved upon adsorption for all the various chemical forms that are adopted and no racemization or inversion occurs, thus nucleating the first expression of point (0D) chirality [15]. Achiral adsorbates can often produce a chiral motif deriving from the adsorption event alone, or with a polar or azimuthal tilt or the rearrangement of the molecular backbone that breaks all mirror symmetry [16]. Due to the inherent nonchirality of the initial molecule, there is equal probability of nucleating reflectional chiral motifs.

It may also be that chirality is not expressed at the isolated molecule level but requires an ensemble of interacting molecules to induce a chiral manifestation. Understanding the driving forces and mechanisms for the induction and evolution of chirality at this early stage, which may only encompass tens to hundreds of molecules, is a real challenge in the field and is largely unstudied. We shall show that this crucial stage can be influenced by many factors, both energetic and kinetic and even the small data set of information currently available reveals new insights and points to a complex landscape of events. Finally, growth from these small clusters of molecules to form longer-range chiral superstructures, which are bereft of mirror symmetry, can then occur induced by surface-molecule interactions and the relative strengths of intermolecular interactions such as π-π interactions [17], steric repulsion [18] and hydrogen bonding [13, 14].

Some commonly observed routes in the formation of large-scale molecular assemblies at surfaces is shown schematically in Figure 10.1. When a molecule is confronted with a surface, a wide number of phases with different combinations of chemical moieties, orientation and self-organization can potentially exist – a highly polymorphic system can arise. Thus, the depiction below is not an exhaustive route map for all chiral supra-structures but it reflects the importance of the nucleation phase in determining the larger scale structure. Examples used in subsequent sections will highlight some pertinent details of the growth process from a description of the individual molecules, the nucleation stage and the details of the finally observed structures.

10.1.2 Tracking Chiral Nucleation at Surfaces

This chapter draws on data obtained from the adsorption of molecules on a clean well defined metal surface held under ultra-high vacuum (UHV) conditions. An important advantage of surface studies in UHV is that the surface is available for study over a wide

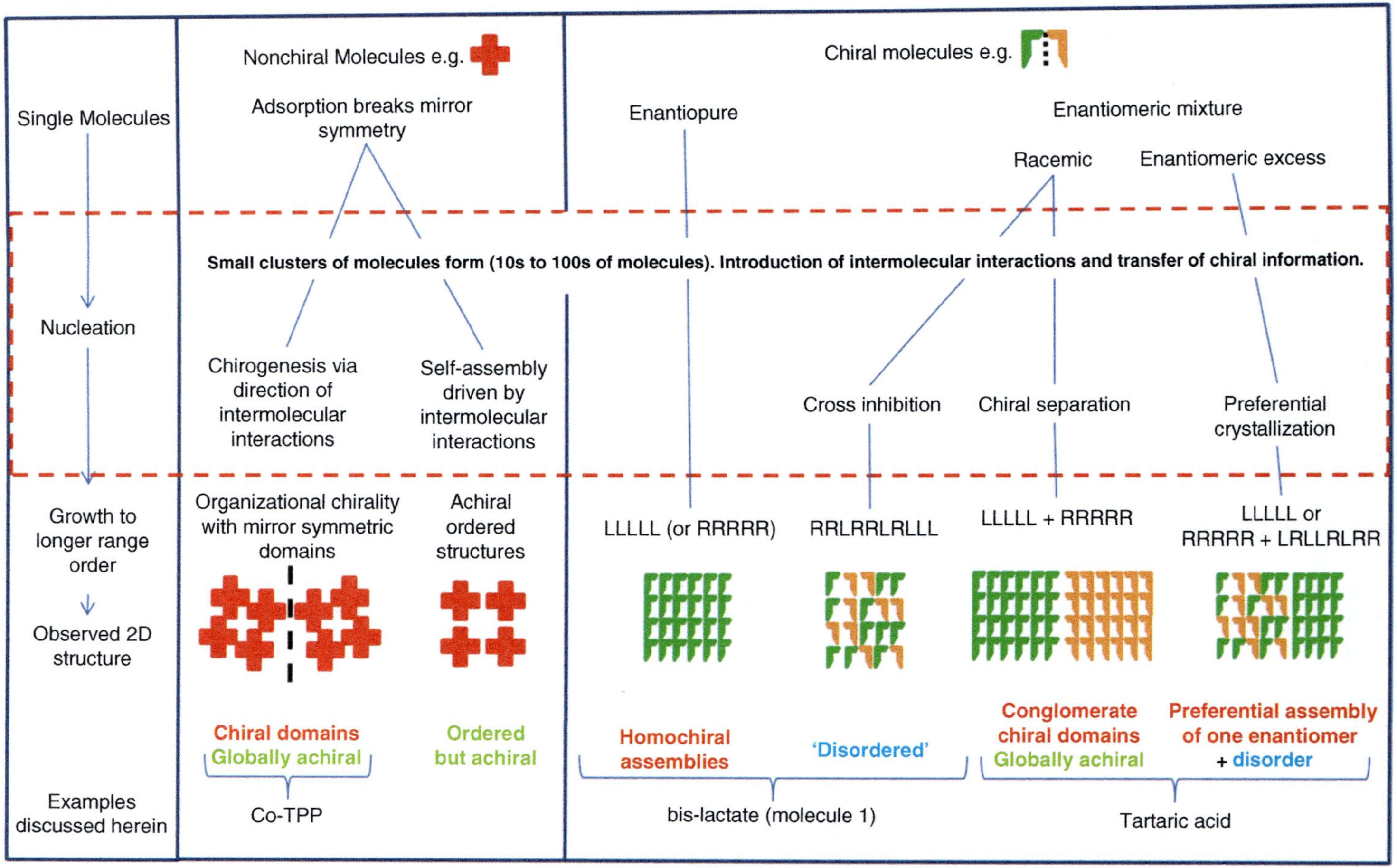

Figure 10.1 Schematic of some pathways for forming chiral structures on surfaces

coverage and temperature range. This allows chiral behaviour to be studied from nucleation to saturation coverage, and often allows the transfer of chiral expression to be mapped through different hierarchical stages. These two accessible experimental variables also allow the system to be studied over a wide and dynamic phase space [19]. For example, the adsorption behaviour of enantiopure (R,R)- tartaric acid on Cu(110) [12, 20, 21] has been shown to be a highly sensitive system where the adsorbed molecules change rapidly between the monotartrate, bitartrate or dimer forms as coverage and temperature are varied. Thus it is important that the chemical nature of the adsorbed molecule is known. In the work described here, surface IR spectroscopy provided key information. Second, it is necessary to follow the behaviour of the system from individual molecules to assemblies. Here, scanning tunnelling microscopy (STM) provides unrivalled access to such information.

Despite high-resolution STM imaging capabilities, it often remains unclear which forces drive the nucleation, molecular deformations and the subsequent chiral assemblies that are experimentally observed. The most effective approach to address this issue is supplementing good-quality experimental data on the adsorbed species (e.g. chemical nature, adsorption site, spatial distribution) with theoretical calculations using periodic density functional theory (DFT), which allows for a detailed analysis of the overall electronic and geometric properties of the adsorbed molecular layer at a surface and, thus, more comprehensive insights into the factors that incubate and propagate chiral expressions.

10.2 Systems of Discussion

This chapter will focus on examples that highlight the importance of the nucleation stage in determining the longer range order of self-assembled structures and the different factors that play a part in the system evolution as summarized below.

- Cobalt-tetraphenylporphyrin (Co-TPP) on Cu(110) demonstrates the role of surface-molecule interactions in enabling chiral motifs to be realized and the subsequent role of intermolecular interactions in favouring chiral over achiral motifs to construct self-assembled nucleation domains in which local chirality is transferred to organizational chirality.
- The second example, a chiral bis-lactate on Cu(110), provides a stark example of how cross- inhibition by opposite enantiomers limits the growth of enantiopure nuclei, thus kinetically inhibiting chiral segregation into a conglomerate.
- The third example, tartaric acid on Cu(110) demonstrates highly nonlinear behaviour when one enantiomer is in excess. Here, entropic factors drive the drastic change in the ability of the majority and minority enantiomers to organise.

10.2.1 System 1: Co-TPP on Cu(110)- Chirogenesis via Intermolecular Interactions

One class of molecules whose self-assembly on well defined metal surfaces has recently elicited great technological and scientific interest is porphyrins. Porphyrins, and associated chlorins and corroles, are invaluable molecular species [22] and are prevalent in biological systems performing numerous tasks. The variety found in nature is based around similar macrocyclic arrangement of four pyrrole rings connected by methine units. This structure allows for the coordination of different metal ions (commonly Fe, Co and Mg) in the interior, great flexibility and provides the potential for a wide variety of functionalization for specific properties. On surfaces they have been shown to self-assemble into a variety of forms such as supramolecules [23, 24]. A number of these systems show chiral self-assembly structures at surfaces [17, 25–32]. Examples of chiral self-assemblies include those reported by Buchner *et al.* for the adsorption of X-TPP (X=H_2, Fe or Co) [26] and by Auwater *et al.* for tetrapyridyl-porphyrin (TPyP) [25], both on Ag(111). Despite the extensive information regarding single-molecule conformations and descriptions of long-range order, insights pertaining to the nucleation stage of growth are currently lacking.

The example presented here represents a comprehensive description from the single molecule level to small clusters of molecules and, finally, the observed long-range order. The nature of the surface molecule and intermolecular interactions during the nucleation stage is key in manifesting chirality, which is then transferred into the longer range organizational chirality [17].

The chemical structure of the achiral Co-TPP molecules and the achiral Cu(110) surface are shown in Figures 10.2a and b respectively. Scanning tunnelling microscopy images (Figure 10.2c) were obtained at the early stages of surface occupation by Co-TPP. These data show that there are actually two types of species present at this nucleation stage, with chirally ordered small ensembles of molecules, which coexist with individual, isolated molecules at the surface. In the context of this chapter it is of interest to note that the size of the ordered ensemble comprises approximately 30 molecules – the early nucleation stage of the growth is being directly observed. The mobility of isolated molecules make them difficult to image but scattered molecules were regularly observed in the vicinity of organized ensembles and they show no significant interactions with other molecules. One such molecule is highlighted in the yellow circle in Figure 10.2c, with a closer view in 2h revealing molecular-level details. The isolated molecules are imaged in a distinctively different way to those within the organized ensemble domain, as can be seen from a comparison with the single molecule in the assembly in Figure 10.2d. This suggests that, at the earliest nucleation stage, two types of isolated 0D motifs are created, but that progression into the early, organized chiral ensembles involves just one of the species.

Periodic DFT calculations were undertaken in order to understand the nucleation of the two different species. These calculations reveal that the surface can indeed accommodate two types of species : one which adopts a saddle shape and appears in the STM experiments as an isolated species, and another that is highly distorted and appears within the ordered ensembles. The difference between the STM images of these two types of adsorbed molecules is a direct result of the different conformations of the molecules. For the isolated molecules, adsorption occurs at the long-bridge site, a significant saddling of the core can be seen and the phenyl rings show slight deformations but lie largely parallel to the surface [17], as shown in Figure 10.2i–k. The saddle conformation is commonly observed [26, 33–35] when porphyrin molecules are adsorbed on a clean metal surface and, therefore, represents an expected outcome for the Co-TPP/Cu(110) system. In contrast, the molecules within the self-assembled ensembles are adsorbed at the short-bridge site and exhibit a substantial chemical interaction of the molecular core with the surface. This causes the porphyrin macrocycle to accommodate close to the surface and in a flat geometry, inducing considerable tilting distortions in the phenyl groups aligned along the [001] rows. In addition, substantial and matching (anti)clockwise twists of the second phenyl pair endow each adsorbed molecule with a distinctive propeller like conformation and lead to a strong

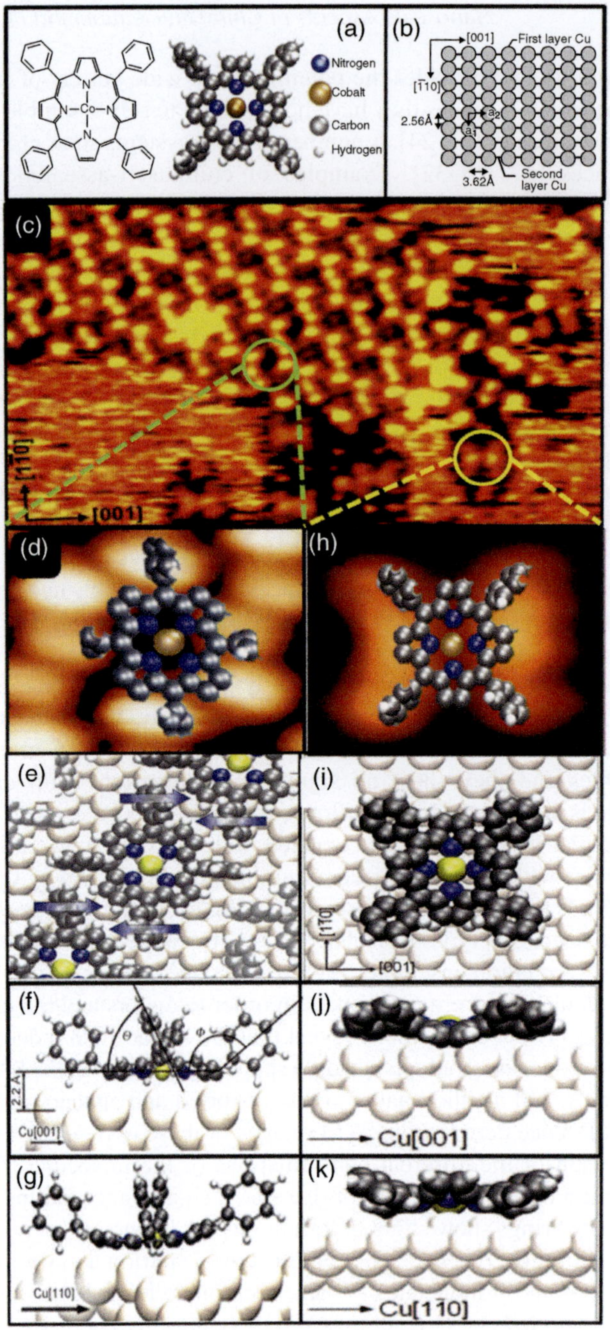

Figure 10.2 *(a) Chemical and structural representation of Co-TPP. (b) The Cu(110) surface. The high-symmetry [001] and [110] directions are shown at the top and the vectors, a_1 and a_2 define the unit cell. (c) Scanning tunnelling microscopy image showing part of the self-assembled structure with a domain border and isolated molecules nearby. Typical examples of single molecules are shown in the green and yellow circles. The single molecule within the assembly is shown in more detail in (d–g) and the isolated molecule in (h–k). Magnified STM images of single molecules in the assembly (d) and the isolated molecule (h) with Co-TPP molecules overlaid. Molecular conformations of single Co-TPP molecules both within assembly (e–g) or isolated (l–k) show the adsorption sites for the minimum energy conformations calculated by DFT. (f) and (j) are the side view along the Cu[001] axis of the Co-TPP. (g) and (k) are side views of Co-TPP viewed along the Cu[1$\bar{1}$0] axis. (Parts of this figure have been reproduced with permission from Ref. [36].)*

expression of chirality at the individual molecule level. The conformations predicted by DFT and STM images for both the isolated and single molecules within the assembly are shown below in Figure 10.2.

Why are two types of species nucleated, and why does only one progress to create an ordered ensemble? To answer these questions it is necessary to consider both the surface-molecule and intermolecular interactions [17]. The overall energy balance for a single adsorbed molecule (E_{gain}) is a combination of the favourable molecule-surface interaction ($E_{mol\text{-}sur}$) and the cost of the deformations of both the molecule ($E_{cost(mol)}$) and the underlying surface ($E_{cost(sur)}$) that occur upon adsorption. These are summarized in Table 10.1 for the two adsorption conformers. For the saddle-shaped molecule adsorbed at the long bridge site a net gain of −96 kJ mol^{-1} (Table 10.1) is calculated. For a single molecule, adopting the chiral propeller conformation, there is a high energy penalty for the molecular distortion but this is compensated by the energy gain from surface-molecule interactions as the macrocycle accommodates close to the surface, leading to $E_{gain} = -97$ kJ mol^{-1}. From these calculations it can be surmised that the two observed conformations are energetically degenerate, thus explaining why both are observed at the nucleation stage.

In order to understand why only the distorted chiral motif progresses to an organized assembly it is important to account for the interaction of neighbouring molecules – that is, an additional term, the intermolecular interaction energy ($E_{m\text{-}m}$), must be added to the energy balance equation to calculate the net gain (E_{net}). Here, a strong deviation in energetics occurs; the distortions of the phenyl groups in the chiral propeller conformer enable highly favourable π-π interactions to be set up with neighbouring molecules in an assembly, which are not possible for the saddled molecule, bringing in an additional energy advantage of −58 kJ mol^{-1}, thus lowering E_{net} for the single molecule in the assembly to −156 kJ mol^{-1}. Intermolecular interactions are generally poorly described by standard DFT but, recently, the inclusion of van der Waals forces has enabled these interactions to be captured [37–39] and provides new insights into the encoding and transfer of chirality in the CoTPP/Cu(110) system [17].

The intermolecular interactions also provide an understanding as to why each domain assembly incorporates molecules possessing the same chirality – propellers that are all clockwise or anticlockwise. During the initial adsorption, and prior to 2D assembly, the phenyls aligned with the [1$\bar{1}$0] axis are presumably free to choose between energetically equivalent twists of about ±65° to accommodate the substrate corrugations. However, upon interaction with a neighbouring molecule, the phenyl pairs are forced to twist in the same sense (clockwise or anti-clockwise) to maximize π-π interactions between the molecules.

Table 10.1 *Summary of costs and gains for a saddle-shaped and chiral propeller conformation of Co-TPP molecules on Cu(110)*

	Chiral propeller conformer	Saddle-shaped conformer
$E_{MOL\text{-}SUR}$ (kJ mol^{-1})	−335	−232
$E_{COST(MOL)}$ (kJ mol^{-1})	+206	+117
$E_{COST(SUR)}$ (kJ mol^{-1})	+31	+19
E_{GAIN} (kJ mol^{-1})	−97	−96
$E_{M\text{-}M}$ (kJ mol^{-1})	−58	–
E_{NET} (kJ mol^{-1})	−156	−96

This traps each individual molecule within an ensemble into a well defined homochiral propeller conformation, thus nucleating the organizational chirality that is observed in the assembled suprastructure. Attractive interactions between phenyl groups occur for T, parallel displaced (PD) and combined PD/T [40–42] configurations, shown in Figure 10.3. They are maximized by forming staggered rows of high interaction that can be visualized as an

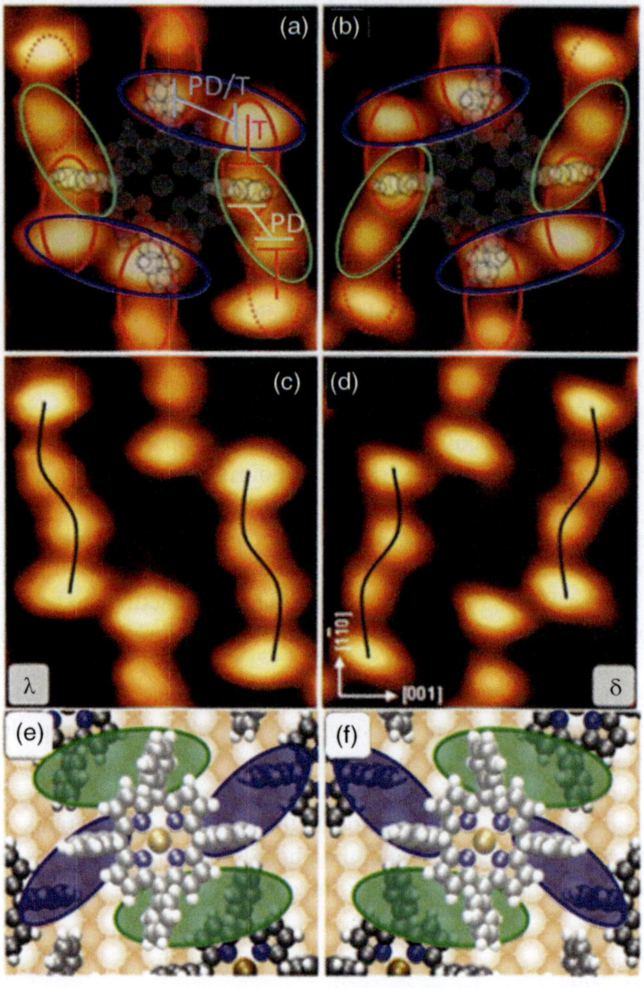

Figure 10.3 *Nucleation of chirality expressed at the single molecule level within the self-assembly. (a and b) Enlarged STM images with Co-TPP molecules overlayed showing the molecular interactions. PD interactions are represented are green ovals, PD/T as blue ovals, T-type as red ovals with dotted representing interactions that are not associated with the central molecules. In (a) the phenyl-phenyl interactions that correspond to the S-like structure are indicated and correspond to the lines drawn to represent the high interaction in (c and d). The intermolecular interactions are shown in the real space model in (e and f). (Reproduced with permission from Ref. [17].)*

S-like structure (Figures 10.3c and d). Clearly, nucleation domains of either chirality are degenerate in energy, and are nucleated with equal probability.

Above is a detailed description of the nucleation phase where clusters of less than 100 molecules are observed. The distinct pattern of phenyl-phenyl interactions is also responsible for the transfer of chiral information from the nucleation phase to the longer range expression of organizational chirality. The directionality of the intermolecular interactions leads to domains being oriented +(20)° or −(20)° with respect to the [001] surface axis, thus the repeat mesh of the organization breaks both mirror-symmetry planes of the underlying Cu(110) surface, as shown in the real space structure in Figure 10.4. The system, therefore, shows organizational chirality with the two domains denoted as λ and δ. These domains can grow to up to 1 μm^2.

The importance of π-π interactions in inducing chirality in porphyrin monolayers is also exemplified by H_2-TPP adsorption on Cu(110)-(2×1)O [29] but in this system, neighbouring molecules are present with alternating chirality to make an overall racemic unit cell comprised of two molecules. A similar nucleation and growth process discussed here has recently been documented by Zhang *et al.* for the adsorption of manganese-phthalocyanine (Mn-PC) on Bi(111) [43]. At low temperatures (4.6 K) Mn-PC was observed in a flat conformation but upon an increase in temperature, intermolecular interactions come into play due to molecular rotations and results in phase separation into two mirror domains.

A number of important conclusions can be drawn from Co-TPP on Cu(110) system. Both the achiral saddle conformation and the twisted conformer are degenerate in energy when intermolecular interactions are discounted. However, as the system progresses from the isolated molecule stage to the nucleation stage, where tens of molecules aggregate, π-π intermolecular interactions come into play and the twisted conformer with a distinct propeller-like shape becomes favoured. This can occur with either a clockwise or anticlockwise sense but the same handedness is adopted by all the molecules within each domain in order to maximize intermolecular interactions. With the two mirror propeller conformers being of equal energy it is expected that the nucleation into one or other of the chiral forms is randomly decided due to fluctuations at the time of the initial interaction. The growth of the observed structure can then be said to be an amplification of chirality through self-assembly.

10.2.2 System 2: Enantiopure and Racemic Mixtures of a Chiral Bis-lactate – Chiral Segregation Nipped in the Bud

Chiral resolution in which a racemic mixture segregates into separate homochiral domains is an important step in generating homochiral crystals of a range of compounds. However, the inhibition of the growth of homochiral domains, be they in fibres, monolayers or bulk crystals, can prevent longer range ordered chiral structures being formed. This can occur at the earliest stages of nucleation as exemplified by the use of tailor-made inhibitors to induce the growth of chiral crystals [44]. However, there is little documentation of the nanoscale events that can inhibit (or promote) transmission of chirality from individual molecules to the crystalline state. The example discussed here is of a 2D self-assembled system, where chiral resolution into a surface conglomerate shows extraordinary cross-inhibition by mirror enantiomers at the growing surface-confined crystallite [36]. This behaviour has not yet been mapped in 3D but such phenomena may well be the reason why less than 10% of molecular chiral crystallization results in conglomerate formation.

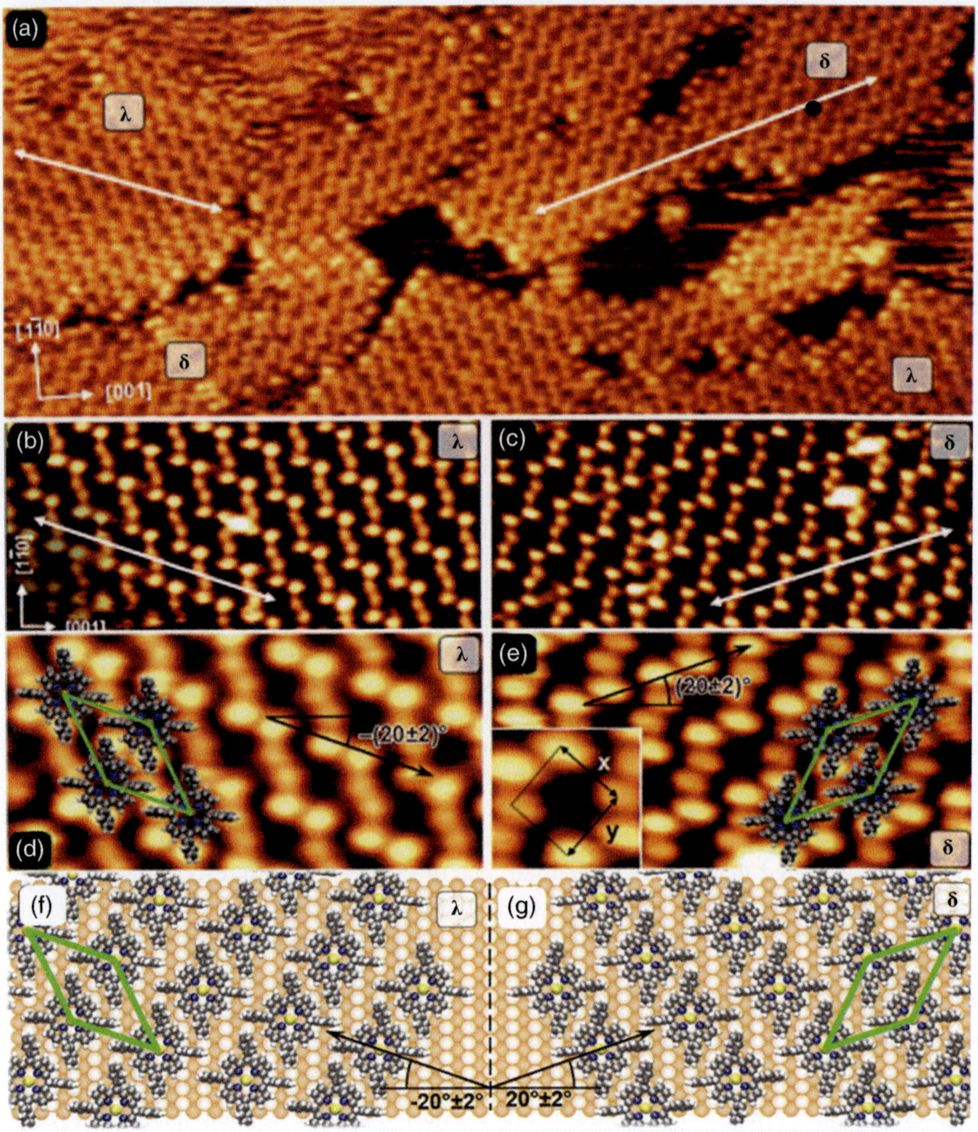

Figure 10.4 *Co-TPP self-assembly on Cu(110) showing the chiral domains with arrows drawn to represent the main growth directions. (a) Large area (542 × 242 Å²) STM image showing coexisting domains with organizational chirality expressed. Detailed images of the two domains are shown in (b) and (c); 140 × 42 Å². (d) and (e) show an enlargement of areas from (b) and (c) with molecules overlaid showing the unit meshes. Chiral directions are given relative to the Cu[110] axis. Inset: the aspect ratio of adsorbed Co-TPP. (f) and (g) The model showing the arrangement of molecules on the surface clearly demonstrates the organizational chirality of the λ and δ domains. (Reproduced with permission from Ref. [17].)*

Using a combination of STM, reflection absorption infrared spectroscopy (RAIRS) and low-energy electron diffraction (LEED), the adsorption of a resorcinol (1,3-dihydroxybenzene) derivative with two chiral lactate groups, molecule **1** (Figure 10.5), on Cu(110) [36] was studied.

STM images (Figures 10.6a and b) obtained for each enantiomer, following deposition at room temperature and annealing up to 370 K, show that each enantiopure system self-assembles into structures with long-range chiral organization. The (R,R) and (S,S)

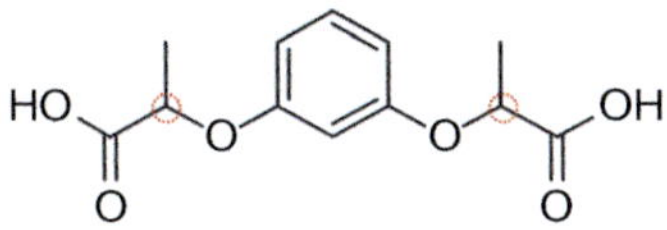

1

Figure 10.5 *Structure of the chiral molecule **1**. The chiral centres are highlighted in the red circles*

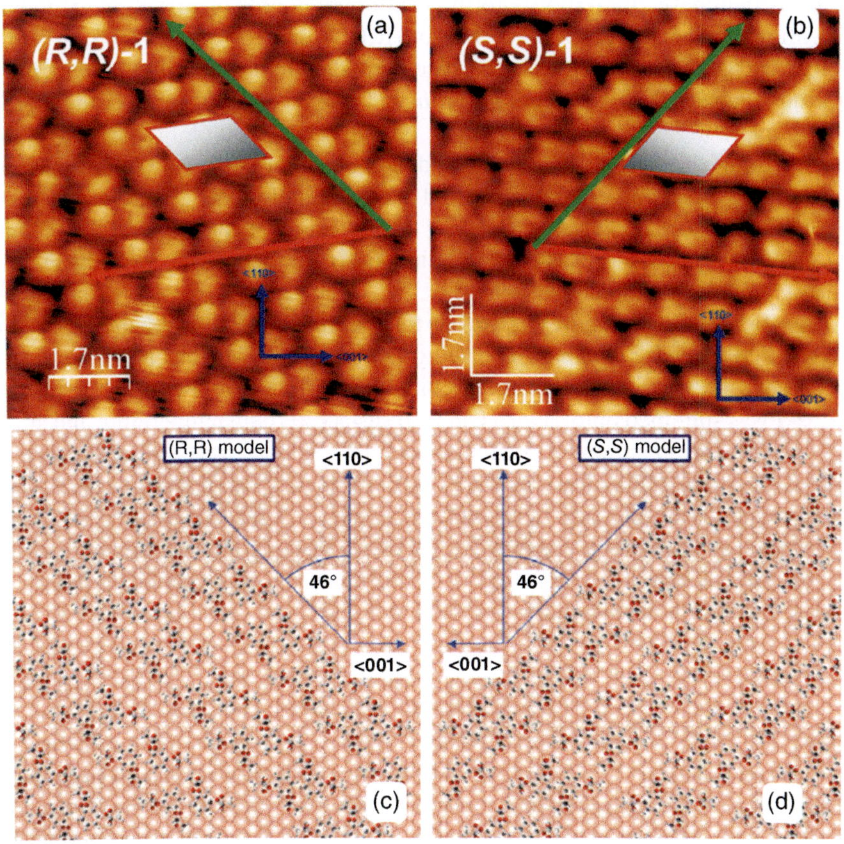

Figure 10.6 *The self-assembled structure of the enantiopure systems on Cu(110) as seen using STM following annealing to 370 K (a, b) for the (R,R)-1 and (S,S)-1 respectively. The models (c, d) and the experimental data clearly show the two structures to be mirror images of each other. (Figure reproduced with permission from Robin* et al. *[36].)*

molecules form mirror enantiomorphous domains in which paired molecule chains grow along nonsymmetry dimensions aligned at ±46° with respect to the $[1\bar{1}0]$ direction of the surface. The extended chiral organization was further echoed in the LEED data, which showed clear diffraction spots, and the unit mesh calculated from the LEED patterns are in excellent agreement with the measured STM distances for the chiral structures [36].

High-resolution STM data provides submolecular information and, when combined with knowledge of the nature of the adsorbed species, it is possible to provide the basis for a local adsorption model of **1** in which the chiral centres of the molecule are preserved and each carboxylate group is bonded to the surface at a short bridge site along the close packed $[1\bar{1}0]$ rows of the metal. This adsorption geometry is found to occur widely for this functionality at a Cu(110) surface [13, 45–48]. Taking into account the measured diameter and asymmetry from the STM images, a diagonal adsorption geometry was concluded to be the most likely scenario. In this configuration the two carboxylate functionalities are adsorbed diagonally to occupy the short bridge sites on adjacent $[1\bar{1}0]$ rows. The long-range order formed shows that these chiral motifs form enantiopure double chains aligned at an angle of −46° with respect to the <110>direction for (R,R)-**1** and +46° for (S,S)-**1**, as shown in Figure 10.6c and d. The adsorption sites for the individual molecules are the same but rotation of the molecules occurs such that close contact between charged carboxylate molecules does not occur, thus avoiding energetically unfavourable Coulombic repulsive forces. Likewise, the double chain formation introduces a spacer area between neighbouring chains, alleviating the Coulomb repulsion and accommodating the compressive stress that is known to accompany carboxylate bonding on the Cu(110) surface [49]. This bonding model directly transmits the chiral information from the nanoscale molecule with two chiral centres to the supramolecular assembly.

Given the ability of enantiopure systems to form stable chiral self-assembled structures, it is of interest to consider whether the driving forces for this self-assembly are sufficient for chiral resolution to occur in the racemic mixture. In order to investigate this, racemic (R,R),(S,S)-**1** was sublimed onto a clean Cu(110) surface at room temperature and annealed to 370 K as for the enantiopure systems, and the adlayer was imaged using STM (Figure 10.7). Distinct dotlike features were observed measuring approximately 7 Å across, which is consistent with the size of a single molecule of **1** on the surface. However, the overall system displays no long-range order. On closer inspection, it becomes clear that the apparent disorder is actually comprised of a multitude of small chiral domains. These display short double-molecule chains oriented along specific directions about ±46° away from the $[1\bar{1}0]$ Cu axis. This pattern of defined growth is remarkably similar to the longer range ordered domains observed for the enantiopure (R,R)-**1** and (S,S)-**1** described above but the lengths in the racemic case generally do not exceed 5 nm. This suggests that they are local aggregations of enantiopure molecules whose growth has been curtailed at the nucleation stage, described by Amabilino and Raval as a 'conglomerate nipped in the bud'.

High-resolution STM images shed more light on the nature of the racemic adlayer (Figure 10.8), with each molecule producing a 'thumbs-up' image with five submolecular features; four give rise to a bulky square part and the fifth is attached to either the left-hand or right-hand edge of the square. These left-handed or right-handed features are attributed to the two enantiomers of molecule **1**. In total, four distinct types of image can be distinguished as each enantiomer is also found in two positions, rotated 180° with respect to each other.

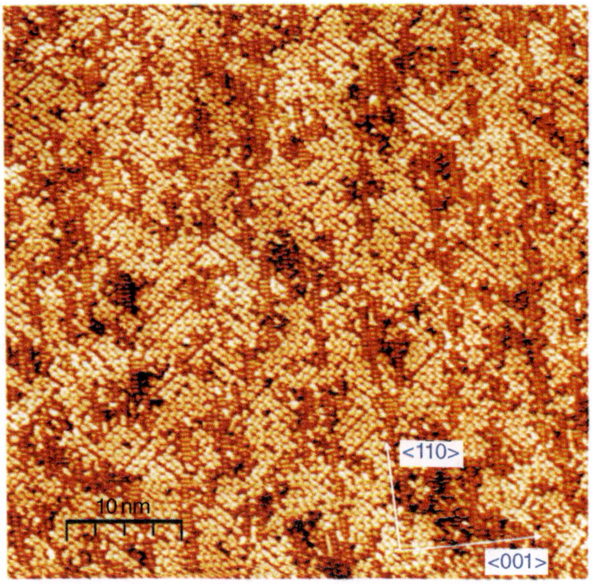

Figure 10.7 *Large area scan ($50 \times 50\,nm^2$) of the global structure observed following the adsorption of a racemic mixture of (R,R)-1 and (S,S)-1 deposited on Cu(110) at room temperature and afterward annealed to 370 K. (Reproduced with permission from Ref. [36].)*

Thus, both the chirality and orientation of individual molecules at the surface can be identified and can be used to understand the details of the local chiral organization observed. It can be seen from Figure 10.8 that the small areas of local order, involving only tens of molecules, adopt the characteristic double-molecule chains that resemble the long range assemblies observed in the enantiopure systems. However, the process of chain elongation has been arrested and molecules of opposite chirality block the growth of the chains. Nevertheless, the preference for conglomerate nucleation is clearly evident, and across the surface the nucleation of nanoscopic enantiopure areas is observed.

The observations above suggest that even at the earliest inception of nucleation, before any recognizable crystal emerges and only tens of molecules are involved, the molecular arrangements take up shapes and habits that resemble those of fully formed crystals. However, the growth into longer range order is effectively halted by the presence of the opposing enantiomer. This is analogous to crystallization, which occurs at boundaries and can be blocked by molecular impurities that act as inhibitors to this process. The example presented here is particularly interesting because the enantiomers act as 'tailor-made' impurities for each other and can be viewed as an extreme example of the tailor-made additives pioneered by Lahav and co-workers, and others [44, 50–52], where structurally similar molecules are effective inhibitors of crystallization.

The conclusions drawn from this example support the hypothesis of Pratt-Brock *et al.* [53] that cross-enantiomer inhibitions exact a very high penalty in the kinetics of conglomerate growth compared to the racemic compound. This can be understood simply as follows: for a racemic compound, an enantiomer at the 'wrong' site only has to diffuse to a neighbouring site to correct the defect. However, for a conglomerate, a large mass

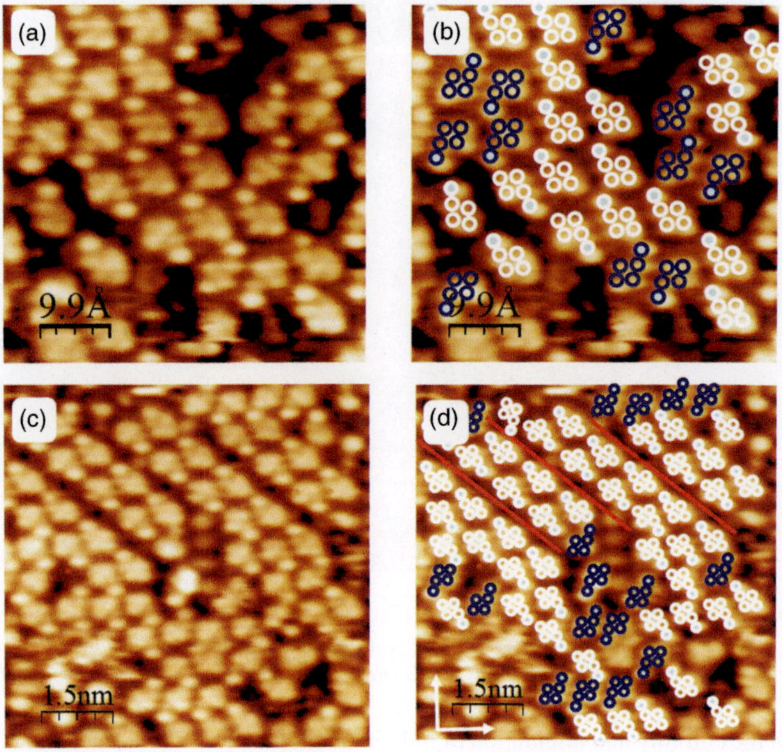

Figure 10.8 *High-resolution STM images (a, c) of a 5×5 nm² area of racemic (R,R)-1 and (S,S)-1 mixture deposited on Cu(110) at room temperature and annealed to 370 K. The image reveals some areas of order but lack of long-range order. Each molecule can be assigned to the different enantiomers, which can be seen more clearly in (b) and (d) where the white and blue overlays represent the enantiomers. The main directions of the Cu axes are shown in (d) – the vertical one is the ‹110› and the horizontal the ‹001› – and apply to all images. (Reproduced with permission from Ref. [36].)*

transport is required for enantiomers to diffuse away from each other and segregate into homochiral domains. Such diffusion-limited kinetics may well be an important factor in determining that less than 10% of 3D molecular crystals form conglomerates. This viewpoint is supported by Monte Carlo simulations of a diffusion-limited surface system where similar organizational patterns to those described here are seen [54].

10.2.3 System 3: Tartaric Acid on Cu(110): Highly Nonlinear Chiral Crystallization

The third system of interest in this chapter provides an example whereby chiral resolution occurs for the racemic mixture but highly nonlinear crystallization behaviour is observed when enantiomeric ratios deviate from the racemic state.

Tartaric acid (TA) on Cu(110) is one of the best understood 2D chiral systems [12, 20, 21, 55, 56]. For the enantiopure systems (R,R-TA and S,S-TA), the behaviour of the doubly dehydrogenated bitartrate species is of interest (Figure 10.9a). The self-assembly of chiral bitartrate motifs into 2D chiral superstructures was imaged using STM and yields an oblique (1 2, −9 0) unit mesh for the R,R-TA and the mirror image (9 0, −1 2) unit mesh for the S,S-TA (Figure 10.9b and d), demonstrating that the chirality at the single molecule level is transferred to the supramolecular level. The bitartrate assembly breaks the mirror

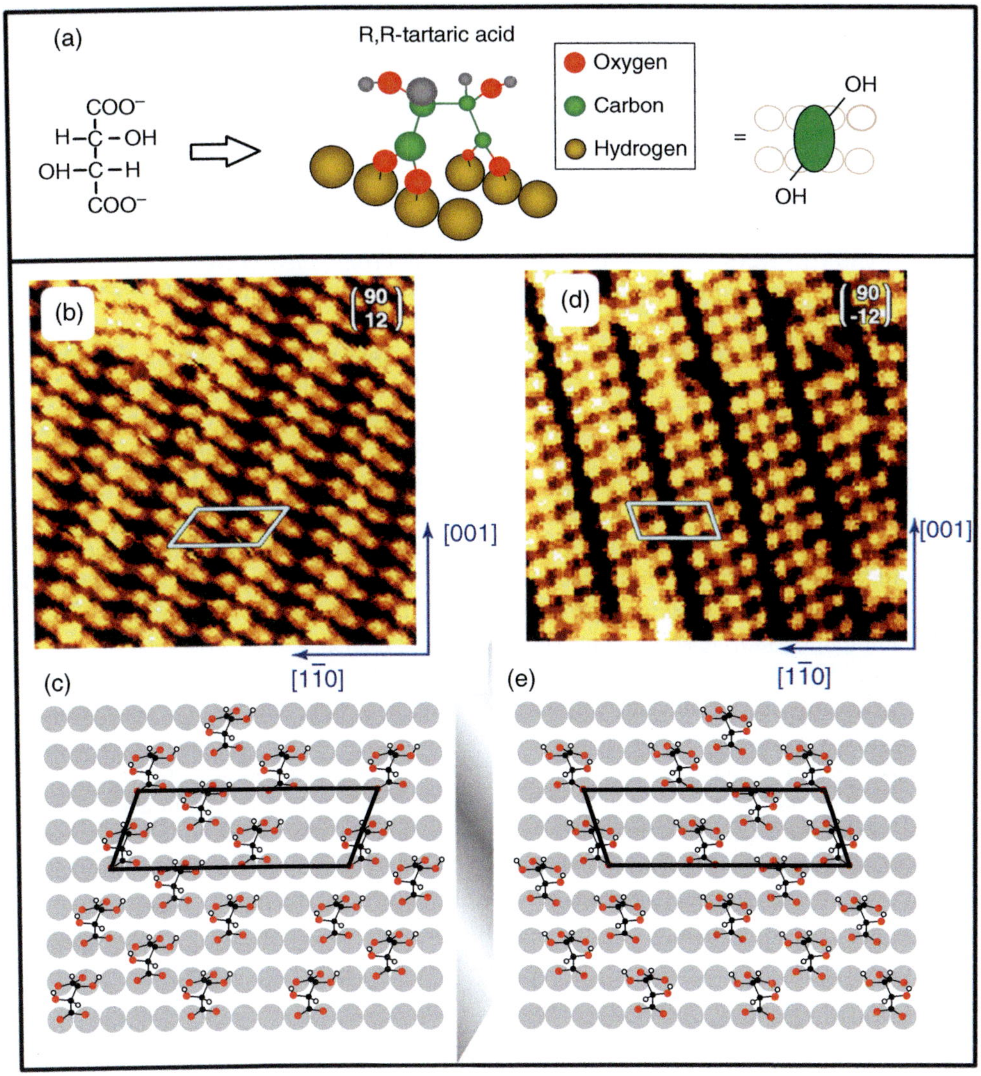

Figure 10.9 *(a) Schematic diagram showing the local bonding and orientation of the adsorbed bitartrate species on a Cu(110) surface. (b)–(e) Scanning tunnelling microscopy images showing the self-assembled structure formed from adsorption of enantiopure (R,R)-TA, (b) and (S,S)-TA (d) on Cu(110) and the respective molecular models shown in (c) (R,R)-TA and (e) (S,S)-TA. It can be seen that the organization of the structure for the enantiomers are mirrored by the other. (Reproduced with permission from Ref. [57].)*

Figure 10.10 *Scanning tunnelling microscopy images of the self-assembled structure formed from the racemic mixture of (R,R)-TA and (S,S)-TA shows the separation of the adsorbed molecules into two types of organized domains on Cu(110), which possess the same (1 2, −9 0) and (9 0, −1 2) as the enantiopure systems. (Reproduced with permission from Ref. [57].)*

symmetry of the metal surface and shows chiral expressions across the lengthscales: 0D at the individual motif level due to the preservation of the chiral centres; and, globally in 2D as revealed by LEED [12, 20] patterns that possess no mirror symmetry.

Turning to the racemic mixture, 50 : 50 (R,R)-TA : (S,S)-TA, the STM data in Figure 10.10 shows that the enantiomers separate into large mirror symmetric domains. The organization adopted is the same as that observed for the enantiopure systems and both superstructures are represented equally in terms of surface coverage. This leads to the conclusion that the racemic mixture has successfully separated into the homochiral domains, as a direct 2D analogy of the classical Pasteur experiments [1, 2] that resolved a racemic mixture of sodium ammonium tartrate solution into enantiopure crystals.

Haq *et al.* [57] extended studies of this system to enantiomerically unbalanced systems. From considering the racemic mixture case one might predict that such a perturbation would simply result in both enantiopure domains being formed, but with their proportional coverage dependent on the enantiomeric excess (e.e.). LEED and STM data (Figure 10.11), however, reveal that even a small chiral perturbation of the system results in a surprising deviation from the expected behaviour, with the system demonstrating highly nonlinear behaviour in the organization as a function of the e.e. The majority enantiomer readily organizes into its favoured self-assembled structure; however, the minority enantiomer is significantly impeded from realizing its superstructure. A monolayer with a compositional e.e. of just 0.2 resulted in 75% of the organized matter displaying the majority structure. The STM data provides a remarkable visualization of this deviation, with the domains formed by the major component clearly identifiable but any ordering of the minority component is restricted to small patches within a complex and disordered structure. It would appear that the result of the small e.e. is to all but destroy the organized domains of the minority component.

In order to gain further insight into this unexpected behaviour, the system was modelled [57] using kinetic Monte Carlo (KMC) simulations based on periodic DFT calculations [55]. The time-resolved simulations, Figure 10.12, reveal important details of the nucleation and growth process. During the initial nucleation stages for the racemic mixture, the system

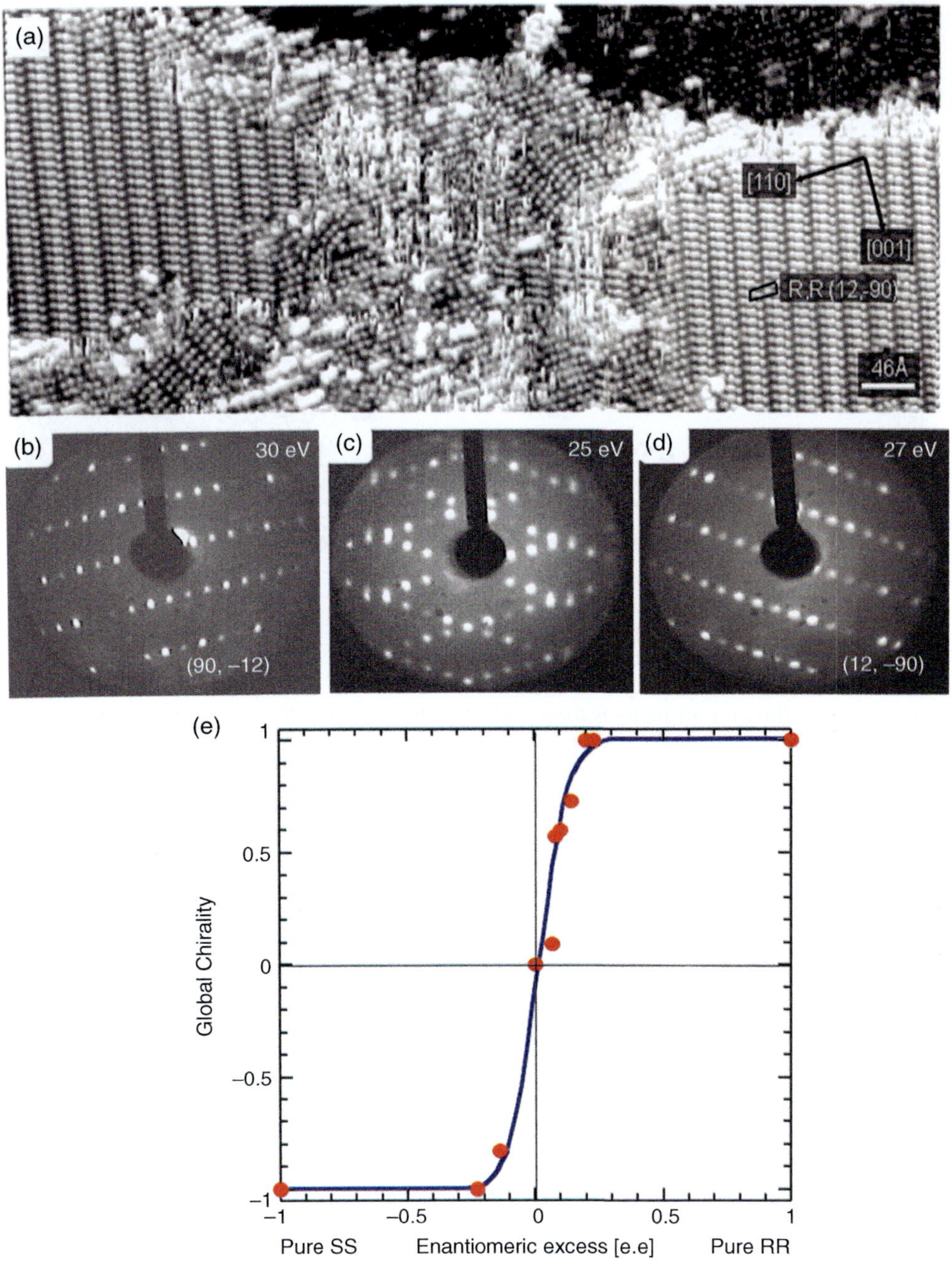

Figure 10.11 *(a) STM images of a 60:40 mixture of (R, R)-TA and (S,S)-TA on Cu(110) showing areas of disorder and organized assemblies of (R,R)-TA. LEED patterns observed following the adsorption of (b) a small excess of (S,S), which shows only the (9 0, −1 2) structure, (c) a racemic mixture showing a superposition of diffraction spots from the (9 0, −1 2) and (1 2, −9 0) structures and (d) a small excess of (R,R) which shows only the (1 2, −9 0) structure. (e) The extent of the globally organized chiral structures derived from LEED pattern intensities versus enantiomeric excess. Experimental data (red points) demonstrates strong nonlinear behaviour. (Reproduced with permission from Ref. [57].)*

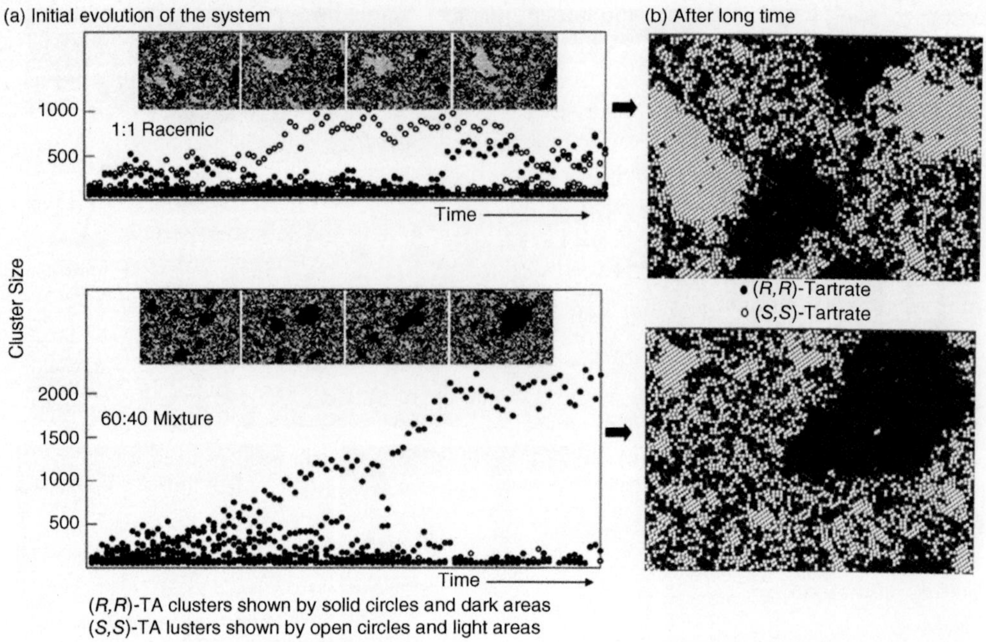

Figure 10.12 *Based on KCM and Monte Carlo simulations, (a) shows the organizational behaviour during the nucleation phase of the system evolution. The top panel is for a racemic mixture whereas the bottom panel is for a system with e.e. = 0.2. (b) After sufficient time has elapsed, well defined organised areas form. For the racemic mixture (top panel) there is even coverage between the two enantiomers but for the 60 : 40 mix (bottom panel) the (R,R)-TA forms large organized areas whereas the (S,S)-TA does not. (Reproduced with permission from Ref. [57].)*

shows large stochastic fluctuations but soon the cluster size for both enantiomers is of the order of 1000 molecules. As the system evolves with longer times the predicted organization is in good agreement with the STM and LEED experimental data, with both enantiomers organizing in equal measure. In contrast, the 60 : 40 mixture shows that even during the initial nucleation stage, the majority enantiomer is preferentially organized and cluster sizes are 2–3 times greater than observed for the racemic mixture at the same time points. Similarly, the minority enantiomer is adversely affected and the cluster sizes rarely exceed 250 molecules, markedly lower than the racemic scenario. With time evolution, a very skewed organizational outcome emerges that agrees with the experimental observations.

This nonlinear behaviour in crystallization can be attributed to the configurational entropy of the system. At low coverages, the whole system is disordered but as the coverage increases the system is forced to organize due to the higher density of the organized domains. Three phases are then present: R,R-TA, S,S-TA and a disorganized mixture. The adsorbates that are involved in the disorganized regions contribute primarily to the entropy of the system and this is maximized if the proportions of R,R-TA and S,S-TA are equal within the disorganized areas. To achieve this, the majority enantiomer will nucleate first. With further time evolution, small patches of the minority enantiomer appear, but these are always significantly smaller than the large domains occupied by the majority enantiomer. This can also be understood in terms of maximizing the configurational entropy. Hence, to retain a racemic mixture in the disordered regions, an equivalence has to maintained for the RR and SS enantiomers entering or leaving the organized phase via the domain boundaries. This can only be achieved if the total boundary lengths of the ordered domains are equal for the majority and minority components and, for an enantiomerically unbalanced system, this can only occur when the majority component occupies a few large domains and the minority component forms many small domains within the disorder. Our STM data show that the majority enantiomer creates large domains, containing thousands of molecules, while the minority domain is effectively constrained to small domains of tens to hundreds of molecules – essentially restricted to domain sizes typically expected at the nucleation stage. This work clearly shows entropy-driven effects in nucleation and chiral ordering, and may provide some insights into the nanoscale origins of the analogous 3D phenomenon of preferred crystallization in mixed enantiomer systems [5, 58–60].

10.3 Conclusions

The formation of chiral assemblies on well defined metal surfaces is hierarchical in nature; chirality is first expressed at the single molecule level followed by the transfer of chiral information into small nucleation domains and then into longer range ordered structures. By appropriate choice of experimental conditions and system, the crucial nucleation phase whereby chiral information is transferred from single adsorbed molecules to small clusters, which lay out a pathway for further growth, can be studied. The examples discussed above highlight different aspects of nanoscale nucleation, captured with detailed STM images and DFT calculations. They provide unique insights into the complex nature of the nucleation stage and the transfer of chirality from single molecules through to long-range ordered structures.

The three systems discussed show that multiple effects can determine nucleation and chiral growth including the effect of the surface in nucleating chiral deformations at the single molecule level but selection of molecules to form nuclei is determined by intermolecular interactions as clearly demonstrated with our Co-tetraphenylporphyrin/Cu(110) system.

Another remarkable insight that emerges is the critical role that enantiomers play in the organization of racemic mixtures with opposite enantiomers drastically affecting the growth of homochiral domains. Our work has delineated two completely different ways in which this occurs; for the bis-lactate/Cu(110) system, limited kinetic diffusion lead to the 'wrong' enantiomer being trapped at the boundaries of homochiral nuclei and suppressing further growth, thus leading to 'nipped in the bud' conglomerates. This was only uncovered by STM imaging. In the tartaric acid/Cu(110) system, the racemic mixture readily creates a conglomerate but slight enantiomeric excess leads to drastic reduction in the ability of the minority enantiomer to organize, driven by entropic factors.

In all cases, the nanoscale characterization that is afforded by surface systems is pivotal in providing information on nucleation events. The separation of molecules into their enantiopure forms is of great practical use in many fields and insights contributing to new understanding of the driving forces and mechanisms involved in the nucleation process are critical in advancing this field for technological benefit.

References

[1] Pasteur L. Mémoire sur la relation qui peut exister entre la forme cristalline et la composition chimique, et sur la cause de la polarisation rotatoire. *Comptes rendus hebdomadaires des séances de l'Académie des Sciences* 1848; **26**: 538.

[2] Pasteur L. Recherches sur les relations qui peuvent exister entre la forme cristalline, la composition chimique et le sens de la polarisation rotatoire. *Annales de chimie et de physique* 1848; **24**: 442–459.

[3] Coquerel G, Amabilino DB. The Nanoscale Aspects of Chirality in Crystal Growth: Structure and Heterogeneous Equilibria. Chirality at the Nanoscale. Wiley-VCH Verlag GmbH & Co. KGaA; 2009. pp. 305–348.

[4] Erdemir D, Lee AY, Myerson AS. Nucleation of crystals from solution: classical and two-step models. *Accounts of Chemical Research* 2009; **42**(5): 621–629.

[5] Kellogg R, Kaptein B, Vries T. Dutch Resolution of Racemates and the Roles of Solid Solution Formation and Nucleation Inhibition. In: Sakai K, Hirayama N, Tamura R, editors. Novel Optical Resolution Technologies. Springer, Berlin; 2007. pp. 159–197.

[6] Barlow SM, Raval R. Complex organic molecules at metal surfaces: bonding, organisation and chirality. *Surface Science Reports* 2003; **50**(6–8): 201–341.

[7] Raval R. Chiral expressions at metal surfaces. *Current Opinion in Solid State and Materials Science* 2003; **7**(1): 67–74.

[8] Hazen RM, Sholl DS. Chiral selection on inorganic crystalline surfaces. *Nature Materials* 2003; **2**(6): 367–374.

[9] Guo Z, De Cat I, Van Averbeke B *et al.* Nucleoside-assisted self-assembly of oligo(p-phenylenevinylene)s at liquid / solid interface: chirality and nanostructures. *Journal of the American Chemical Society* 2011; **133**(44): 17764–17771.

[10] Barlow SM, Raval R. Nanoscale insights in the creation and transfer of chirality in amino acid monolayers at defined metal surfaces. *Current Opinion in Colloid and Interface Science* 2008; **13**(1–2): 65–73.

[11] Haq S, Hanke F, Sharp J *et al.* Versatile bottom-up construction of diverse macromolecules on a surface observed by scanning tunneling microscopy. *ACS Nano* 2014; **8**(9): 8856–8870.

[12] Ortega Lorenzo M, Baddeley CJ, Muryn C, Raval R. Extended surface chirality from supramolecular assemblies of adsorbed chiral molecules. *Nature* 2000; **404**(6776): 376–379.

[13] Forster M, Dyer MS, Persson M, Raval R. Probing conformers and adsorption footprints at the single-molecule level in a highly organized amino acid assembly of (S)-proline on Cu(110). *Journal of the American Chemical Society* 2009; **131**(29): 10173–10181.

[14] Forster M, Dyer MS, Persson M, Raval R. 2D Random organization of racemic amino acid monolayers driven by nanoscale adsorption footprints: proline on Cu(110). *Angewandte Chemie International Edition* 2010; **49**(13): 2344–2348.

[15] Raval R. Nanoscale insights into the creation of chiral surfaces. *Journal of Molecular Catalysis A: Chemical* 2009; **305**(1–2): 112–116.

[16] Ernst K-H. Molecular chirality at surfaces. *Physica Status Solidi B* 2012; **249**(11): 2057–2088.

[17] Donovan P, Robin A, Dyer MS *et al.* Unexpected deformations induced by surface interaction and chiral self-assembly of coII-tetraphenylporphyrin (Co-TPP) adsorbed on Cu(110): a combined STM and periodic DFT study. *Chemistry – A European Journal* 2010; **16**(38): 11641–11652.

[18] Fasel R, Parschau M, Ernst K-H. Chirality transfer from single molecules into self-assembled monolayers. *Angewandte Chemie International Edition* 2003; **42**(42): 5178–5181.

[19] Raval R. Chiral expression from molecular assemblies at metal surfaces: insights from surface science techniques. *Chemical Society Reviews*. 2009; **38**(3): 707–721.

[20] Ortega Lorenzo M, Haq S, Bertrams T *et al.* Creating chiral surfaces for enantioselective heterogeneous catalysis: R,R-Tartaric Acid on Cu(110). *The Journal of Physical Chemistry B* 1999; **103**(48): 10661–10669.

[21] Ortega Lorenzo M, Humblot V, Murray P *et al.* Chemical transformations, molecular transport, and kinetic barriers in creating the chiral phase of (R, R)-tartaric acid on Cu(110). *Journal of Catalysis* 2002; **205**(1): 123–134.

[22] Auwarter W, Ecija D, Klappenberger F, Barth JV. Porphyrins at interfaces. *Nature Chemistry* 2015; **7**(2): 105–120.
[23] Fendt L-A, Stöhr M, Wintjes N *et al.* Modification of supramolecular binding motifs induced by substrate registry: formation of self-assembled macrocycles and chain-like patterns. *Chemistry – A European Journal.* 2009; **15**(42): 11139–11150.
[24] Heim D, Seufert K, Auwärter W *et al.* Surface-assisted assembly of discrete porphyrin-based cyclic supramolecules. *Nano Letters* 2010; **10**(1): 122–128.
[25] Auwärter W, Weber-Bargioni A, Riemann A *et al.* Self-assembly and conformation of tetrapyridyl-porphyrin molecules on Ag(111). *The Journal of Chemical Physics* 2006; **124**(19): 194708.
[26] Buchner F, Kellner I, Hieringer W *et al.* Ordering aspects and intramolecular conformation of tetraphenylporphyrins on Ag(111). *Physical Chemistry Chemical Physics* 2010; **12**(40): 13082–13090.
[27] Écija D, Trelka M, Urban C *et al.* Molecular conformation, organizational cvhirality, and iron metalation of meso-tetramesitylporphyrins on copper(100). *The Journal of Physical Chemistry C* 2008; **112**(24): 8988–94.
[28] Mao XF, Lin T, Adisoejoso J, Shi Z *et al.* Coordination self-assembly of bromo-phenyl and pyridyl functionalized porphyrins with Fe on an Au(111) surface. *Physical Chemistry Chemical Physics* 2013; **15**(30): 12447–50.
[29] Wagner M, Puschnig P, Berkebile S *et al.* Alternating chirality in the monolayer H2TPP on Cu(110)-(2 x 1)O. *Physical Chemistry Chemical Physics* 2013; **15**(13): 4691–4698.
[30] Yang Z-Y, Durkan C. Edge and terrace structure of CoTPP on Au(111) investigated by ultra-high vacuum scanning tunnelling microscopy at room temperature. *Surface Science* 2010; **604**(7–8): 660–665.
[31] Écija D, Seufert K, Heim D *et al.* Hierarchic self-assembly of nanoporous chiral networks with conformationally flexible porphyrins. *ACS Nano* 2010; **4**(8): 4936–4942.
[32] Otsuki J. STM studies on porphyrins. *Coordination Chemistry Reviews* 2010; **254**(19–20): 2311–2341.
[33] Seufert K, Bocquet M-L, Auwärter W *et al.* Cis-dicarbonyl binding at cobalt and iron porphyrins with saddle-shape conformation. *Nature Chemistry* 2011; **3**(2): 114–119.
[34] Diller K, Klappenberger F, Marschall M *et al.* Self-metalation of 2H-tetraphenylporphyrin on Cu(111): An x-ray spectroscopy study. *The Journal of Chemical Physics* 2012; **136**(1): 014705.
[35] Wolfle T, Gorling A, Hieringer W. Conformational flexibility of metalloporphyrins studied by density-functional calculations. *Physical Chemistry Chemical Physics* 2008; **10**(37): 5739–5742.
[36] Robin A, Iavicoli P, Wurst K *et al.* A racemic conglomerate nipped in the bud: a molecular view of enantiomer cross-inhibition of conglomerate nucleation at a durface. *Crystal Growth and Design* 2010; **10**(10): 4516–4525.
[37] Puzder A, Dion M, Langreth DC. Binding energies in benzene dimers: nonlocal density functional calculations. *The Journal of Chemical Physics* 2006; **124**(16): 164105.
[38] Thonhauser T, Puzder A, Langreth DC. Interaction energies of monosubstituted benzene dimers via nonlocal density functional theory. *The Journal of Chemical Physics* 2006; **124**(16): 164106.
[39] Temprano I, Thomas G, Haq S *et al.* 1D self-assembly of chemisorbed thymine on Cu(110) driven by dispersion forces. *The Journal of Chemical Physics* 2015; **142**(10): 101916.
[40] Grimme S, Muck-Lichtenfeld C, Antony J. Analysis of non-covalent interactions in (bio) organic molecules using orbital-partitioned localized MP2. *Physical Chemistry Chemical Physics* 2008; **10**(23): 3327–3334.
[41] Sinnokrot MO, Sherrill CD. High-accuracy quantum mechanical studies of $\pi-\pi$ Interactions in Benzene dimers. *The Journal of Physical Chemistry A.* 2006; **110**(37): 10656–10668.
[42] Tsuzuki S, Honda K, Uchimaru T *et al.* Origin of attraction and directionality of the π/π Interaction: model chemistry calculations of benzene dimer interaction. *Journal of the American Chemical Society* 2002; **124**(1): 104–112.
[43] Zhang T-T, Wang C-J, Sun K, Yuan H-K, Wang J-Z. Step-by-step self-assembly of manganese phthalocyanine on Bi(111) surface: from single molecule to two-dimensional domains. *Applied Surface Science* 2014; **317**: 1047–1051.

[44] Addadi L, Berkovitch-Yellin Z, Weissbuch I *et al.* A Link Between Macroscopic Phenomena and Molecular Chirality: Crystals as Probes for the Direct Assignment of Absolute Configuration of Chiral Molecules. John Wiley & Sons, Inc., New York; 2007. pp. 1–85.
[45] James JN, Sholl DS. Theoretical studies of chiral adsorption on solid surfaces. *Current Opinion in Colloid and Interface Science* 2008; **13**(1–2): 60–64.
[46] Jones G, Jones LB, Thibault-Starzyk F *et al.* The local adsorption geometry and electronic structure of alanine on Cu{110}. *Surface Science* 2006; **600**(9): 1924–1935.
[47] Rankin RB, Sholl DS. Assessment of heterochiral and homochiral glycine adlayers on Cu(110) using density functional theory. *Surface Science* 2004; **548**(1–3): 301–308.
[48] Sayago DI, Polcik M, Nisbet G *et al.* Local structure determination of a chiral adsorbate: Alanine on Cu(110). *Surface Science.* 2005; **590**(1): 76–87.
[49] Hermse CGM, van Bavel AP, Jansen APJ *et al.* Formation of chiral domains for tartaric acid on Cu(110): a combined DFT and kinetic Monte Carlo study. *The Journal of Physical Chemistry B* 2004; **108**(30): 11035–11043.
[50] Plasson R, Kondepudi DK, Asakura K. Three-dimensional description of the spontaneous onset of homochirality on the surface of a conglomerate crystal phase. *The Journal of Physical Chemistry B.* 2006; **110**(16): 8481–8487.
[51] Weissbuch I, Kuzmenko I, Berfeld M *et al.* Stereochemical studies on chiral systems in two dimensions. *Journal of Physical Organic Chemistry* 2000; **13**(7): 426–434.
[52] Weissbuch I, Leiserowitz L, Lahav M. Stochastic 'mirror symmetry breaking' via self-assembly, reactivity and amplification of chirality: Relevance to abiotic conditions. *Topics in Current Chemistry* 2005; **259**: 123–165.
[53] Pratt-Brock C, Schweizer WB, Dunitz JD. On the validity of Wallach's rule: on the density and stability of racemic crystals compared with their chiral counterparts. *Journal of the American Chemical Society* 1991; **113**(26): 9811–9820.
[54] Uñac R, Vidales A, Zgrablich G. Effect of interaction energies on the adsorption of glycine onto a Cu(110) surface: a Monte Carlo simulation. *Adsorption Science and Technology* 2009; **27**(7): 633–642.
[55] Barbosa LAMM, Sautet P. Stability of chiral domains produced by adsorption of tartaric acid isomers on the Cu(110) surface: a periodic density functional theory study. *Journal of the American Chemical Society* 2001; **123**(27): 6639–6648.
[56] Fasel R, Wider J, Quitmann C *et al.* Determination of the absolute chirality of adsorbed molecules. *Angewandte Chemie – International Edition* 2004; **43**(21): 2853–2856.
[57] Haq S, Liu N, Humblot V *et al.* Drastic symmetry breaking in supramolecular organization of enantiomerically unbalanced monolayers at surfaces. *Nature Chemistry* 2009; **1**(5): 409–414.
[58] Avalos M, Babiano R, Cintas P *et al.* Symmetry breaking by spontaneous crystallization – Is it the most plausible source of terrestrial handedness we have long been looking for? – A reappraisal. *Origins of Life and Evolution of the Biosphere* 2004; **34**(4): 391–405.
[59] Coquerel G. Preferential crystallization. *Topics in Current Chemistry* 2006; **269**: 1–51.
[60] Kondepudi DK, Crook KE. Theory of conglomerate crystallization in the presence of chiral impurities. *Crystal Growth and Design* 2005; **5**(6): 2173–2179.

11

Chirality in Organic Hosts

Daniel Fankhauser and Christopher J. Easton

11.1 Introduction

Organic receptors have long held the attention of chemists, induced by the ground-breaking work of Charles J. Pederson [1, 2], Jean-Marie Lehn [3] and Donald J. Cram [4, 5], who were awarded the 1987 Nobel Prize in chemistry for their development and use of such molecules [6–8]. Nowadays there is a plethora of literature about organic host systems, describing their properties and applications. In particular, studies of chiral hosts are a key research field with great potential for implementation in many areas. Interaction of a chiral host with a racemic guest leads to the formation of diastereomeric host–guest inclusion complexes, which differ in thermodynamic stability. This feature can be used in analytical chemistry to separate racemic guest mixtures or simply to distinguish enantiomers. Another important application of chiral hosts is their use as catalysts in asymmetric reactions, where they may be regarded as enzyme mimics, often providing high yields and enantiomeric excesses (*ee*s) of chiral products.

In this chapter, we review particular features and applications of chiral host systems based primarily on cyclodextrins (CD) **1–3**, calixarenes and crown ethers (Figure 11.1). Our selected examples include past milestones as well as recent developments in this area. We provide a brief coverage of some analytical applications in section 11.2, then focus on applications in asymmetric reactions in section 11.3.

11.2 Chiral Hosts in Analytical Applications

Cyclodextrins dominate reports of molecular recognition using chiral host structures [9–19] but studies with chiral calixarenes [20–23] and crown ethers [24] have also been

Chirality in Supramolecular Assemblies: Causes and Consequences, First Edition. Edited by F. Richard Keene.

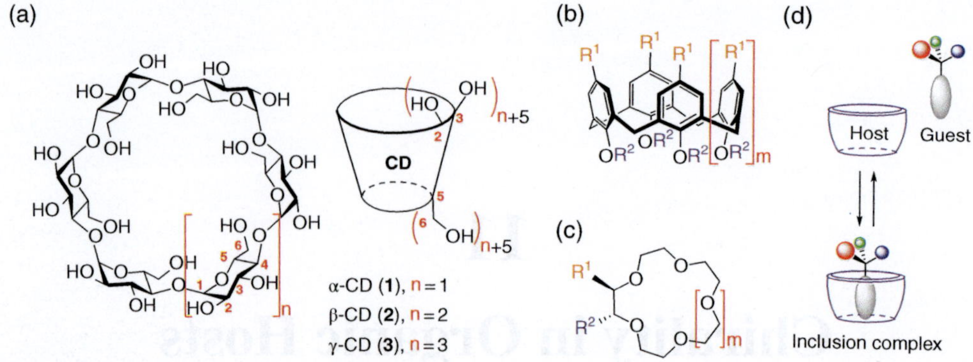

Figure 11.1 *(a) The natural cyclodextrins **1–3**, (b) a calixarene scaffold, (c) a crown ether scaffold and (d) schematic formation of a host–guest inclusion complex*

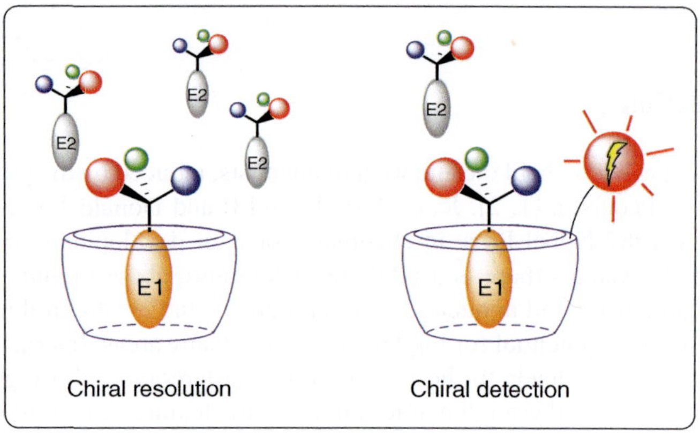

Figure 11.2 *Analytical applications involving molecular recognition by chiral hosts*

described. Here we report selected examples of some frequently used and inspiring applications in analytical chemistry (Figure 11.2). Our intention is to provide only a broad introduction to this very diverse field, noting that other more specialist and comprehensive reviews are available [24–27].

(A) Chiral Resolution A simple method for separation of enantiomers is liquid-liquid extraction using chiral hosts as phase transporters [24]. Peacock and Cram [28] applied the chiral crown ether **4** to the separation of racemic α-amino acids and their esters, such as compounds **5** and **6**, in a biphasic system of $CDCl_3/CD_3CN$ and D_2O, achieving enantiomer distributions up to 52 : 1 (Figure 11.3).

In a typical procedure, the racemic perchlorate salt **5** was treated with the host **4** in a biphasic system of $CDCl_3/CD_3CN$ (77/23, w/w) and 4 M $LiClO_4$ in D_2O at pH 1 ($HClO_4$) and 25 °C. The mixture was shaken and cooled to 0 °C before the layers were separated, yielding a 52 : 1 ratio of the *R*- and *S*-enantiomers of phenylglycine **5** from the organic phase. The proportion of CD_3CN had a large impact on the chiral recognition, in that more or less than 23% lowered the efficiency of the enantiomeric separation. It is possible that

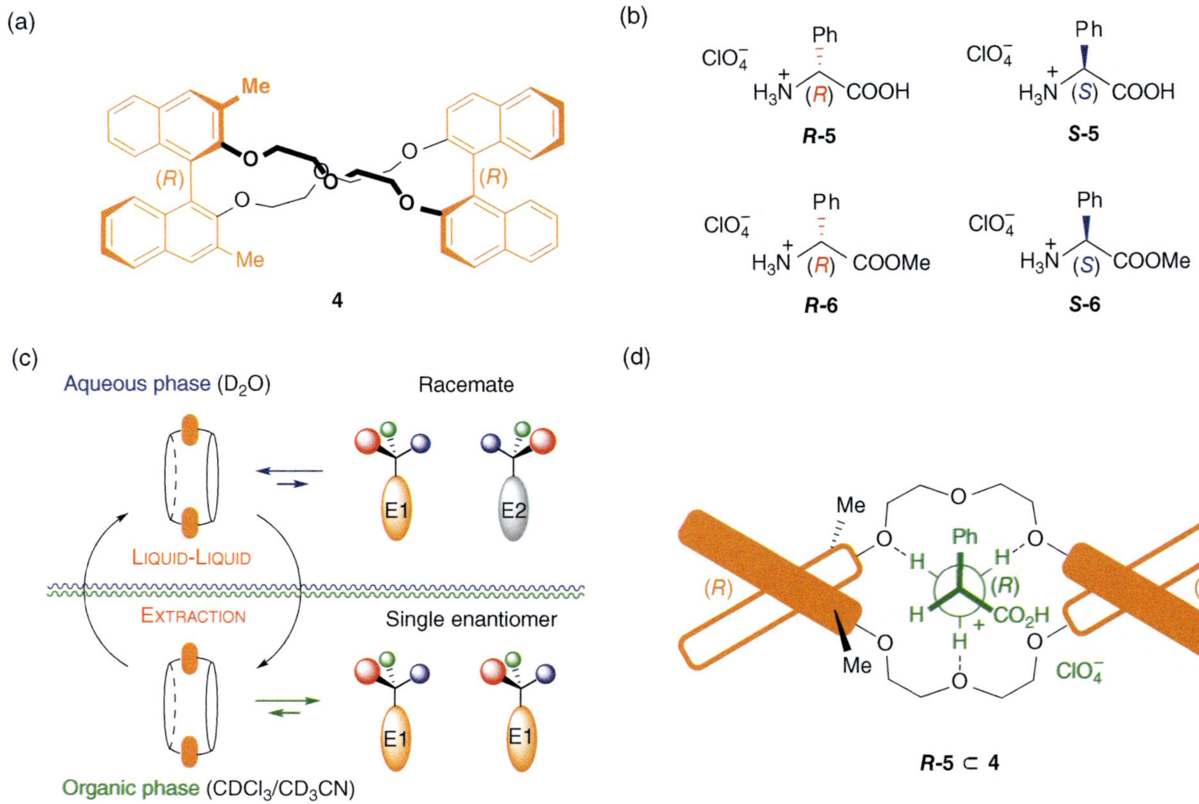

Figure 11.3 *(a) The host **4** used for chiral resolution, of (b) compounds **5** and **6**, by (c) liquid-liquid extraction and (d) the proposed structure of a favoured host–guest complex [28]*

the organic medium has to be sufficiently polar to facilitate host–guest complexation and dissociation, but not too polar so that it competes with the host's binding sites. Smaller substituents at the guest's chiral centre decreased the enantioselectivity. The preference for the complexation and thus transport of *R*-amino acids was rationalized by proposing a favoured host–guest complex (Figure 11.3). The cavity of the host **4** is more suited to *R*-amino acids, where there is enough space for the guest's phenyl substituent between the host's naphthalene walls while placing the small α-hydrogen of the guest next to the host's sterically demanding methyl groups. Effective resolution of the racemic ester **6** was also achieved, giving a 21 : 1 ratio of the *R*- and *S*-enantiomers.

Today, high-performance liquid chromatography (HPLC) is probably the most widely applied technique for separation of enantiomers, where cyclodextrins feature quite prominently in the production of chiral stationary phases [25]. For example, Han *et al.* [29] achieved notable results with the immobilized per-2, 3-dimethyl-β-cyclodextrin **7** (Cyclobond I 2000 DM) as the stationary phase, in the separation of a large number of furan derivatives such as compounds **8–10** (Figure 11.4).

The separations were accomplished in the reverse-phase mode using MeOH/H_2O as the mobile phase. The enantioselectivity (α) and resolution factor (R_S) were calculated according to the equation $\alpha = k_{E2}/k_{E1}$, where $k_E = (t_E - t_0)/t_0$, and $R_S = 2 \times (t_{E2} - t_{E1})/(\omega_{E2} + \omega_{E1})$. Enantioseparation requires $\alpha > 1.02$ and baseline separation $R_S > 1.5$. The data obtained for

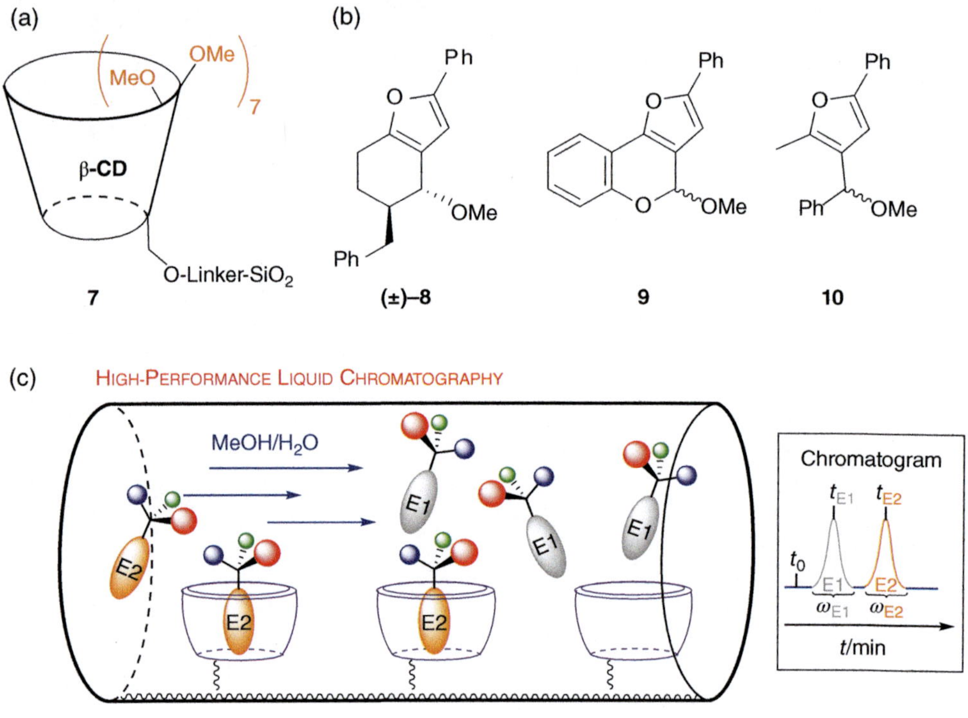

Figure 11.4 *(a) The host **7** stationary phase (Cyclobond I 2000 DM) used for HPLC chiral resolution of (b) the furans **8–10**, and (c) a schematic representation of the separation [29].* t_0 *= void time,* t_E *= retention time,* ω_E *= base peak width*

the racemic tetrahydrobenzofuran derivative **8** were $\alpha = 1.55$ and $R_S = 3.1$, for the furochromene **9** $\alpha = 1.15$ and $R_S = 1.2$, and for the furan **10** $\alpha = 1.10$ and $R_S = 0.8$. Additional investigations with other compounds revealed that adjusting the pH with 0.1% triethylamine / acetic acid enhanced enantioseparation in particular cases, while the addition of 0.5% methyl *tert*-butyl ether improved peak shape and hence separation efficiency while decreasing retention times.

Apart from the use of cyclodextrins and chiral calixarenes as stationary phases in HPLC, numerous other applications have been described in gas chromatography (GC), capillary electrophoresis (CE), ion chromatography (IC) and supercritical fluid chromatography (SFC) [30, 31].

(B) Chiral Detection Techniques have also been developed to sense quantitatively one enantiomer of a mixture in solution, mainly based on cyclodextrins [26]. For example, Parker and coworkers [32, 33] designed a potentiometric sensor with the peroctylated-α-cyclodextrin **11** to measure the enantiomeric purity of ephedrine hydrochloride (**12**) (Figure 11.5). The best results were achieved by embedding the host **11** in a polyvinyl chloride (PVC) membrane on a potentiometric ion-sensitive electrode filled with an aqueous solution of NH_4Cl. The electrode is calibrated using solutions with various ratios of the enantiomers of ephedrine hydrochloride (**12**), after which the enantiomeric purity of an arbitrary mixture may be measured with a remarkable limit of detection down to 10^{-6} mol/L. The sensitivity is only slightly affected by background Na^+, K^+ or Ca^{2+} ions. The functionality of the sensor is based on the complexation of the ephedrine's aryl moiety in the cavity of the host embedded in the membrane and the enantiomeric purity could be measured even in presence of the pseudo-ephedrine diastereoisomers.

Aswathy and Sony [34] developed a fluorescence sensor based on thio-β-cyclodextrin **13**, which chiroselectively recognizes aromatic amino acids **14–16** (Figure 11.6).

The host **13** is immobilized on Au nanoparticles and saturated with fluorescein included within the cyclodextrin annulus. Upon excitation of complexed fluorescein, the fluorescence signal is quenched through surface energy transfer (SET) to the Au. However, complexation of an aromatic amino acid releases bound fluorescein, thus switching off the SET process and generating a fluorescence signal. The experiments were performed in a Tris-HCl buffer pH 7.6 at 37 °C using phenylalanine (**14**), tryptophan (**15**) and tyrosine (**16**) as analytes because their aromatic units drive inclusion complexation with the host **13**. In each case the L-enantiomers selectively complex with the host **13**, changing the fluorescence spectra and facilitating a discrimination from their D-enantiomers. The enantioselectivity is highest for tryptophan (**15**), followed by phenylalanine (**14**) then tyrosine (**16**), reflecting better complexation with more hydrophobic substrates. The detection limit for the L-enantiomers was 10^{-6} mol/L in each investigated case.

Organic hosts have also been used to differentiate enantiomers through the formation of diastereomeric complexes that give rise to distinct signals in nuclear magnetic resonance (NMR) spectroscopy. The hosts are then being used as chiral shift reagents and systems based on cyclodextrins, calixarenes and resorcinarenes have been described [27, 35]. For example, Pham and Wenzel [36] developed the water-soluble resorcin[4]arene **17** as a chiral shift reagent for the differentiation of enantiomers such as those of compounds **18–20** (Figure 11.7).

The host **17** is equipped with L-pipecolinic acid groups on one rim and sulfonated substituents that render it water soluble. The guests form complexes by insertion of their

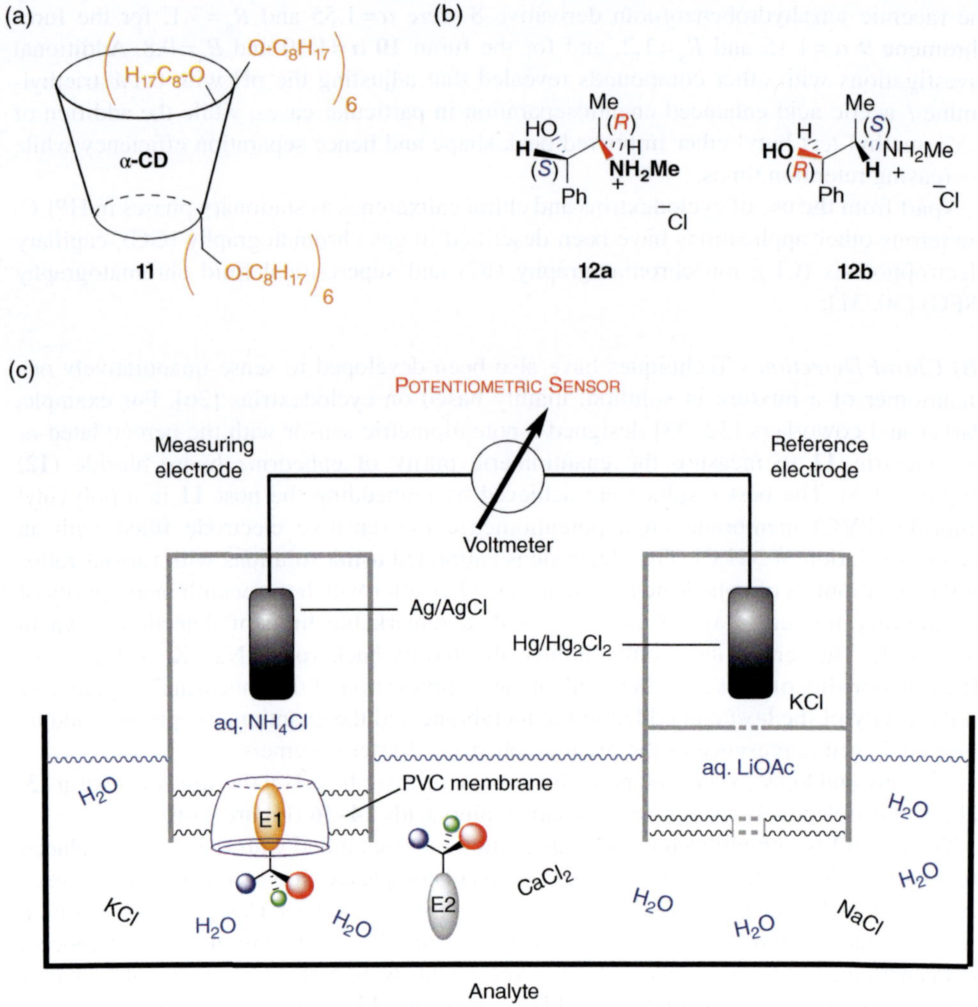

Figure 11.5 *(a) The cyclodextrin **11** that is embedded in a membrane for potentiometric chiral sensing of (b) ephedrine hydrochloride (**12**), and (c) a schematic representation of the ion-selective electrode sensor [32, 33]*

aromatic units in the host cavity, which shields the aromatic and nearby aliphatic hydrogens, thus causing the chemical shifts that vary in extent for each enantiomer. All experiments were performed using at least an equimolar amount of host to guest. For compounds **18–20**, large enantiomeric discriminations were observed, with chemical shift differences up to 0.34 ppm.

As an extension of this work, Wenzel *et al.* [37] designed a method for analysing the enantiomeric purity of aliphatic amines and amino alcohols, such as compounds **21–23**, by derivatization with an aromatic anhydride to facilitate host–guest complexation (Figure 11.8).

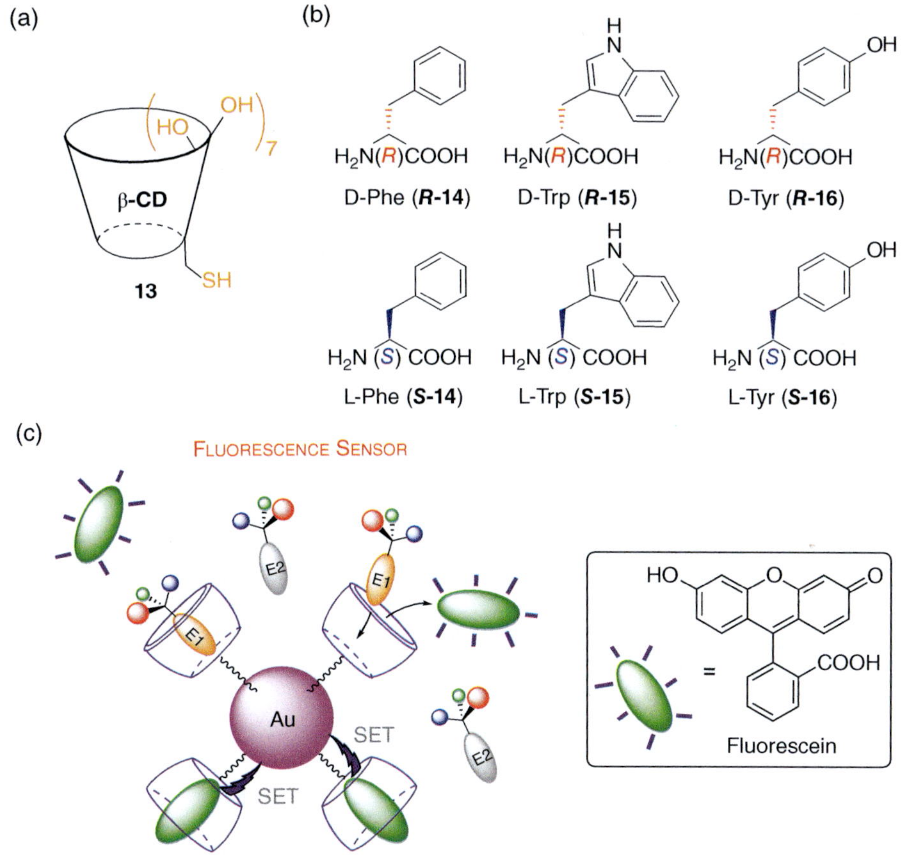

Figure 11.6 *The cyclodextrin **13** that is coupled to Au nanoparticles for fluorescence chiral sensing of (b) phenylalanine (**14**), tryptophan (**15**) and tyrosine (**16**), and (c) a schematic representation of fluorescence sensing using fluorescein as the fluorophore [34.] SET = Surface energy transfer*

The derivatization process has several advantages: (i) the configuration of the substrates is not modified; (ii) naphthyl protons 1 and 4 display sharp singlets in the ^{1}H NMR spectra which are distinct from other signals; and (iii) the relative extents of the complexation-induced shifts of the *R*- and *S*-enantiomers are consistent within series of amino alkanes and 2-amino alcohols, hence facilitating assignment of absolute configuration.

11.3 Chiral Hosts in Asymmetric Reactions

The previous section covered stereoselective binding by organic hosts but not asymmetric reactions of the bound species. Such reactions occur frequently, with the hosts behaving as catalysts. Native and modified cyclodextrins are most commonly used due to the intrinsic chirality of their sugar subunits [9, 38, 39]. However, other examples have been reported

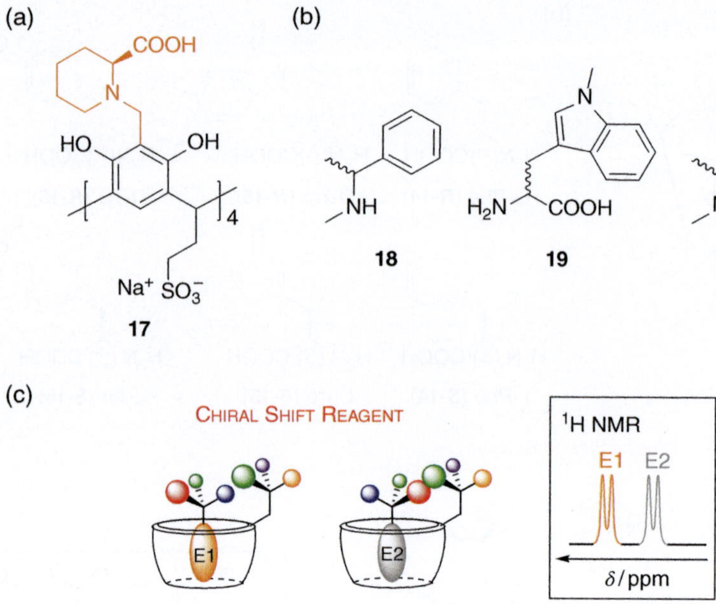

Figure 11.7 The resorcinarene **17** used as a chiral shift reagent in ^1H NMR spectroscopy to distinguish (b) the enantiomers of compounds **18–20**, and (c) a schematic representation of the use of a host chiral shift reagent [36]

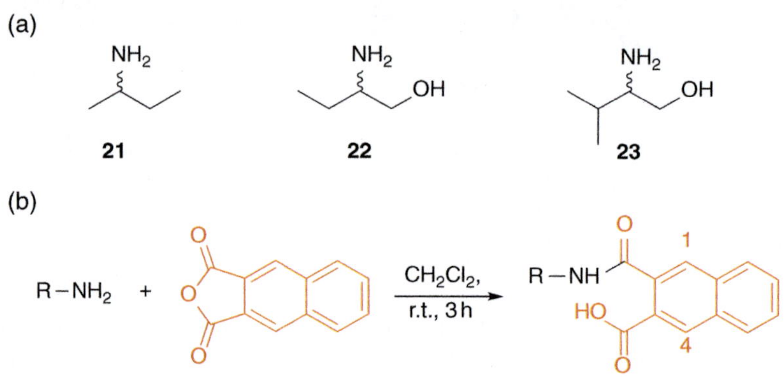

Figure 11.8 (a) The amines **21–23** (b) derivatized with an aromatic anhydride providing an aromatic unit to enable chiral analysis through ^1H NMR spectroscopy using a host as a chiral shift reagent [37]

with chiral calixarenes [40] or chiral crown ethers in phase-transfer catalysis [41, 42]. Many of these catalytically active host systems have been described as artificial enzymes, chemzymes or nanoreactors [43–46].

In this section, we review selected examples of cyclodextrins, calixarenes and crown ethers being used as chiral hosts in asymmetric reactions. This section is divided into four subsections (Figure 11.9): (i) native chiral hosts, which comprise the simplest catalysts; (ii) hosts modified with achiral substituents, which enhance the chiral properties of the

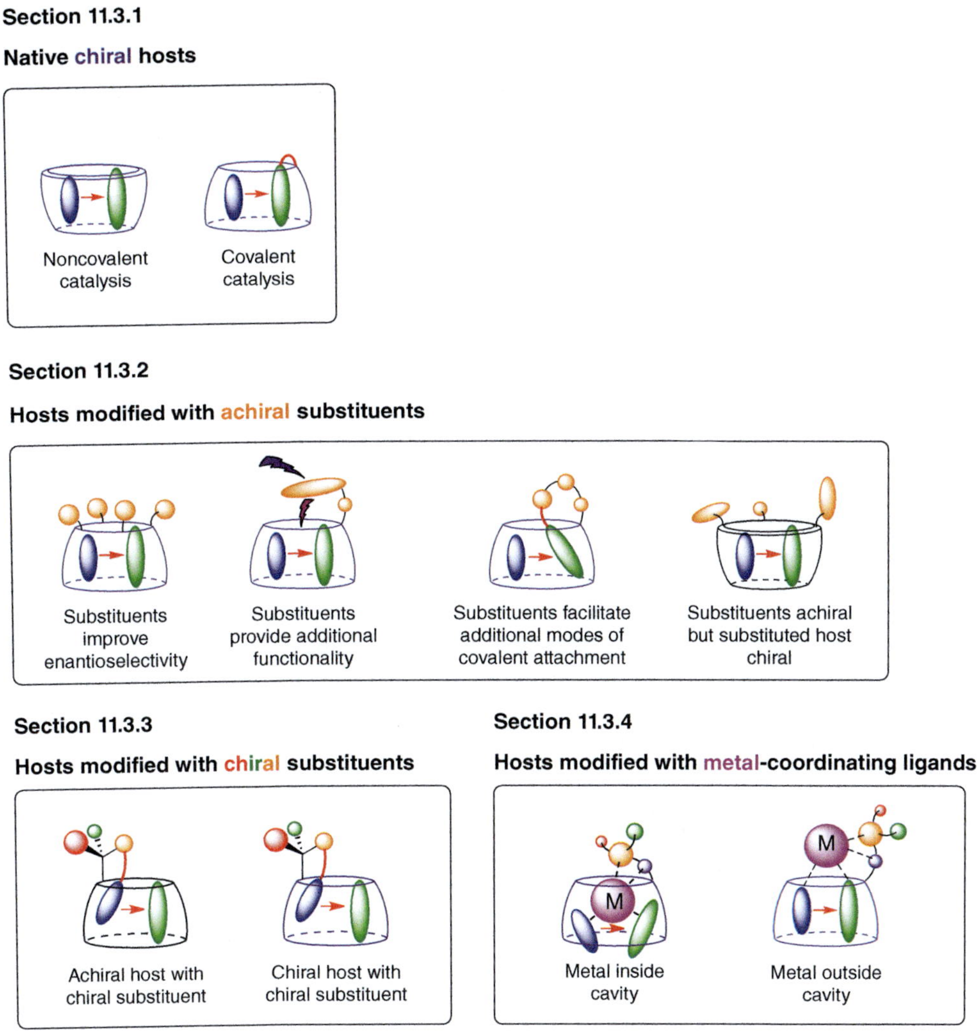

Figure 11.9 *Section 11.3 overview – chiral hosts in asymmetric reactions*

basic structures; (iii) hosts modified with chiral substituents, which provide an external asymmetric unit; and (iv) hosts modified with metal-coordinating ligands, which allow their use in transition-metal catalysis.

11.3.1 Native Chiral Hosts

In the late 1950s, Cramer [47, 48] published the first studies regarding asymmetric reactions using cyclodextrin inclusion complexes. Based on his experience with cyclodextrins in the separation of racemic mixtures as well as in catalysis he concluded that it should be possible to perform asymmetric reactions of included species. He successfully demonstrated the properties of cyclodextrins to promote asymmetry in oxidation and hydrolysis, and the formation of a cyanohydrin. Comprehensive kinetic studies of the hydrolysis reaction lead him to the conclusion that

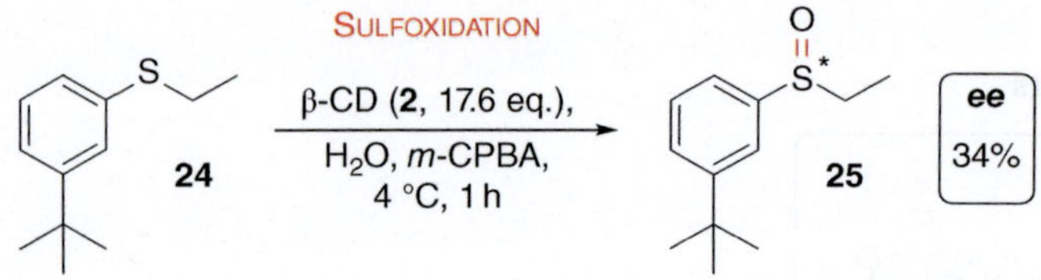

Figure 11.10 *Sulfoxidation using β-CD (**2**) as the catalyst [49]*

the transformation occurs asymmetrically as a result of the influence of the cyclodextrin on the reaction rate, through formation of an inclusion complex. Although only low *ee*s were obtained, Cramer established the principles and hence laid the foundation for future work. As native hosts with intrinsic chirality, as well as being inexpensive and readily available, cyclodextrins offered a general approach for studies of catalytic asymmetric reactions.

(A) Noncovalent Catalysis Several studies investigated the influence of native cyclodextrins on the enantioselectivity of reactions in solution. For example, enantioselective sulfoxidation of aryl alkyl sulfides was published by Czarnik [49], using β-CD (**2**) as the catalyst. He reported an *ee* of 34% for conversion of the sulfide **24** with *m*-chloroperbenzoic acid (*m*-CPBA) to the chiral sulfoxide **25** (Figure 11.10).

Related sulfoxidation studies with other aryl alkyl sulfides and hydrogen peroxide (H_2O_2) as the oxidant yielded lower *ee* values but disclosed, by comparison, the general importance of both substrate inclusion by the cyclodextrin and complexation with the oxidant for enantioselectivity. While *m*-chlorobenzene derivatives are known to form cyclodextrin-inclusion complexes and it is reasonable to assume that *m*-CPBA does so, H_2O_2 does not. Hence, the oxidant that binds to the host enhances the chiral induction to provide higher *ee* values.

Abreu *et al.* [50] used α-CD (**1**) as a phase transfer catalyst (PTC) in the direct hydration of alkenes. They reported an *ee* of 32% for the conversion of the piperidine derivative **26** with $Hg(OAc)_2$ to give the alcohol **27** (Figure 11.11).

The direct alkene hydration was achieved through application of the known two-step procedure comprising oxymercuration by the mercuric salt, followed by demercuration with sodium borohydride ($NaBH_4$), yielding the Markovnikov adduct. The cyclodextrin plays a key role in this reaction, facilitating the transport of the substrate from the organic layer into the aqueous phase where the hydration occurs in the catalyst's asymmetric environment. A control experiment without α-CD (**1**) gave a much lower yield and no enantioselectivity, revealing the importance of the catalyst. A comparison of several aromatic and nonaromatic substrates indicated that an aromatic ring is needed to achieve enantioselectivity, presumably because it is required for host–substrate complexation. Higher reaction temperatures in the direct hydrations decreased the *ee* values.

The influence of β-CD (**2**) on diastereo- and enantio-selectivity in Diels–Alder reactions has been investigated by Schneider and Sangwan [51]. They reported an *ee* of 21% for the addition of cyclopentadiene (**28**) to diethyl fumarate (**29**) giving the enantiomeric adduct **30** (Figure 11.12).

Kinetic analysis indicated >90% complexation of the participating molecules as well as an increase of the reaction rate by a factor of 70. Additional investigations with maleic acid derivatives showed a preference for the *endo* products, which is probably the effect of a more compact transition state within the cyclodextrin cavity.

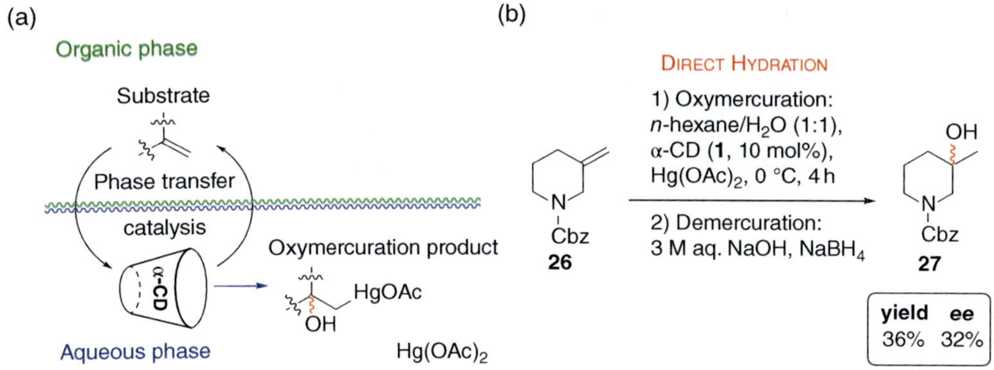

Figure 11.11 *Direct hydration of alkenes using α-CD (**1**) as a PTC [50]: (a) simplified model for phase transfer catalysis and (b) enantioselective direct hydration yielding the alcohol **27***

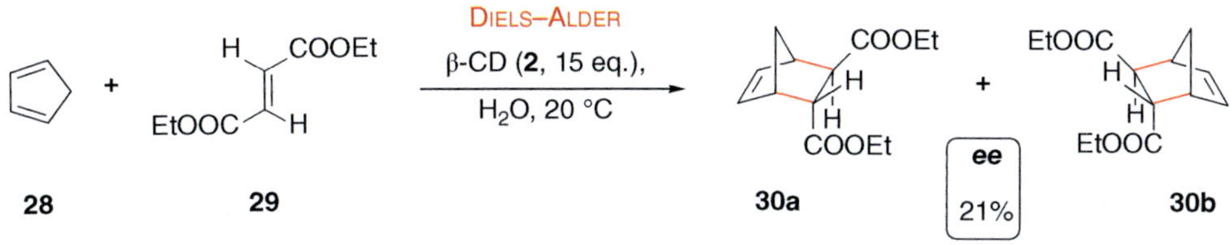

Figure 11.12 *Diels–Alder reaction using β-CD (**2**) as the catalyst [51]*

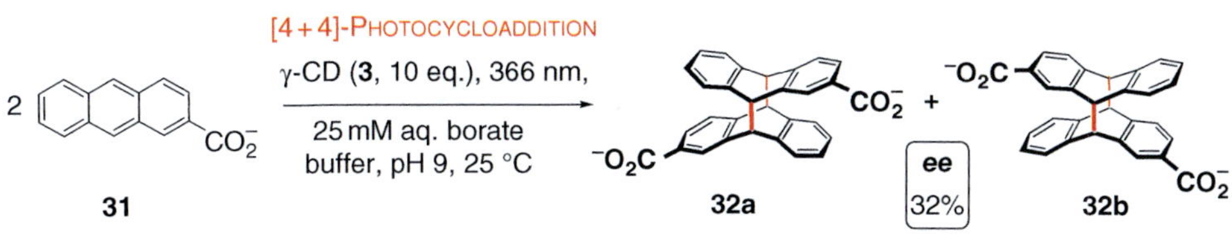

Figure 11.13 *[4 + 4]-Photocycloaddition using γ-CD (**3**) as the catalyst [52]*

Another example of noncovalent asymmetric catalysis in solution by a native cyclodextrin comprises a [4+4]-photocycloaddition with γ-CD (**3**), reported by Nakamura and Inoue [52]. Photodimerization of 2-anthracenecarboxylic acid (**31**) gave the enantiomeric cyclodimer **32** in *ee*s up to 41% (Figure 11.13).

[4+4]-Photocycloaddition of the anthracene derivative **31** yields four possible regioisomers, being the two chiral cyclodimers **32** and **33**, as well as the two achiral adducts **34** and **35** (Figure 11.14). Chromatographic resolution of all six isomers was achieved with tandem achiral/chiral HPLC. Hence the photocyclodimerization ratio with 10 eq. γ-CD (**3**) at 25 °C was found to be 45% **32** : 7% **33** : 43% **34** : 5% **35**, with an *ee* of 32% for **32** and 3% for **33**. The *ee* of the cyclodimer **32** improved to 41% on lowering the temperature to 0 °C. A study without γ-CD (**3**) yielded similar ratios of regioisomers but negligible enantioselectivities, revealing the impact of the cyclodextrin **3**.

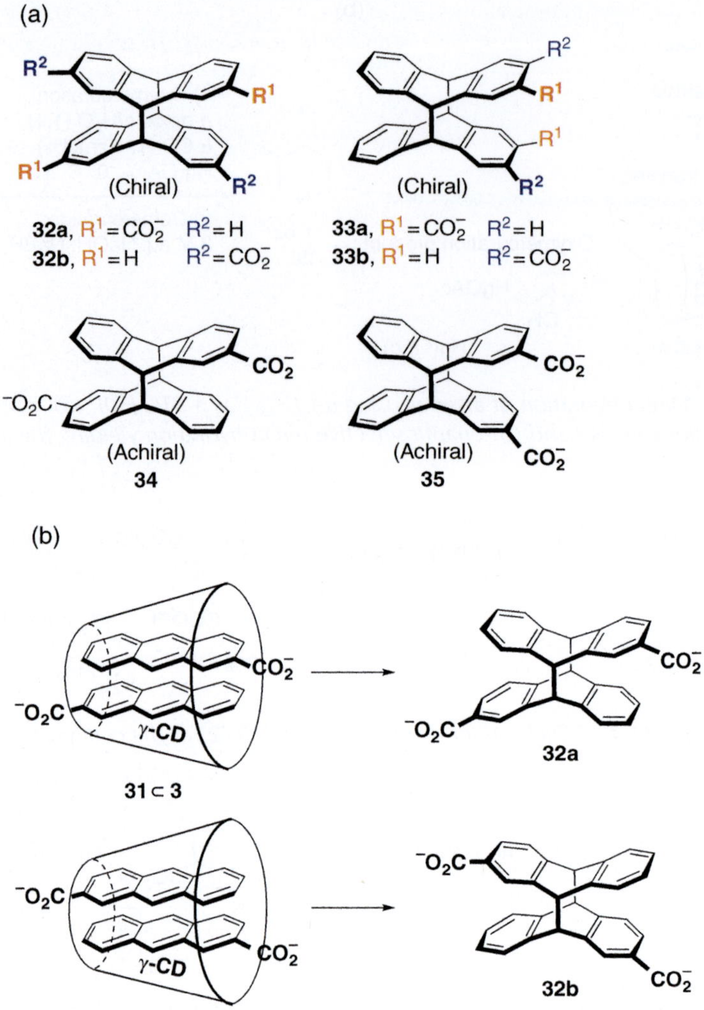

Figure 11.14 *(a) Possible cyclodimerization products being the chiral species* **32** *and* **33**, *as well as the achiral adducts* **34** *and* **35**, *and (b) diastereomeric 2:1 complexes of* **31** *in γ-CD (***3***) (left) leading to the corresponding enantiomers of the dimerization product* **32** *(right) [52]*

UV/Vis, circular dichroism and ^{1}H NMR spectroscopic studies indicated an enthalpically driven 2:1 inclusion complexation of compound **31** in γ-CD (**3**) with relatively high complex stability, shown in the K_a value of $3.9 \times 10^4 M^{-1}$ at 25 °C. The complexation proceeds *via* initial formation of a 1:1 complex with a K_a value of $1.6 \times 10^2 M^{-1}$ at 25 °C. The distribution of all possible cyclodimers primarily depends on the populations of the corresponding precursor complexes having different longitudinal orientations of two molecules of the anthracene derivative **31** inside the cavity of γ-CD (**3**). Hence, the two enantiomers of the regioisomer **32** are obtained from the corresponding diastereomeric 2:1 complexes (Figure 11.14), with the enantioselectivity depending mainly on the populations of the precursor complexes and not significantly on their relative photoreactivities.

Increased asymmetric induction and higher *ee*s are observed in reactions with aqueous suspensions of cyclodextrins, presumably because the cyclodextrin-reactant complexes are more rigidly held with fixed geometries in the solid phase. For example, Kawajiri and Motohashi [53] investigated a reduction using an aqueous suspension of β-CD (**2**) as the catalyst. They reported *ee*s up to 84% for conversion of the benzoylferrocene (**36**) with $NaBH_4$ to the optically active alcohol **37** (Figure 11.15).

Ferrocene and its derivatives are known to bind strongly in the cavity of β-CD (**2**), forming 1 : 1 inclusion complexes. The enantioselective reduction of the ferrocene derivative **36** is attributable to formation of an additional hydrogen bond between the carbonyl of the acyl group and a C(2)-hydroxyl group of the cyclodextrin. $NaBH_4$ then attacks selectively from the more exposed face of the carbonyl group, yielding the ferrocenyl alcohol **37** having the *R*-configuration. Consistent with a salting-out effect by the high concentration of KCl, reduction in its absence showed a drop in enantioselectivity by around 40%.

Compared to the oxidation of aryl alkyl sulfides in solution [49] (see Figure 11.10), which yielded *ee*s up to 34%, Sakuraba *et al.* [54] reported sulfoxidation with *ee*s up to 81% using an aqueous suspension of β-CD (**2**) as the catalyst. They oxidized methyl 1-naphthyl sulfide (**38**) with peracetic acid (AcOOH) to the chiral sulfoxide **39** (Figure 11.16).

The sulfide **38** forms a highly crystalline 1 : 1 complex with β-CD (**2**), which accounts for the observed high enantioselectivity. Using the too small α-CD (**1**) or the too large γ-CD (**3**) as host resulted in a significant decrease in *ee*, to 1% and 19%, respectively. These results demonstrate the importance of proper size-complementarity between host and guest. Based on studies with several sulfides it is evident that not only does the choice of oxidant have an influence on the enantioselectivity but also the choice of solvent under the heterogeneous conditions. Most importantly, the crystalline state must be maintained during the procedure. Highest yields and *ee*s were only achieved using water as solvent. This observation is attributable to two factors: (i) the substrates generally not being very soluble in water, and hence staying inside the cyclodextrin cavity; and (ii) the hydrophobic effect driving the encapsulation, thus further stabilizing the inclusion complexes. Variable temperature studies disclosed better *ee*s at lower temperatures, which might reflect increased rigidity of the host–guest complexes. Interestingly, the molar ratio of oxidant to substrate had no significant impact on either the yield or the enantioselectivity.

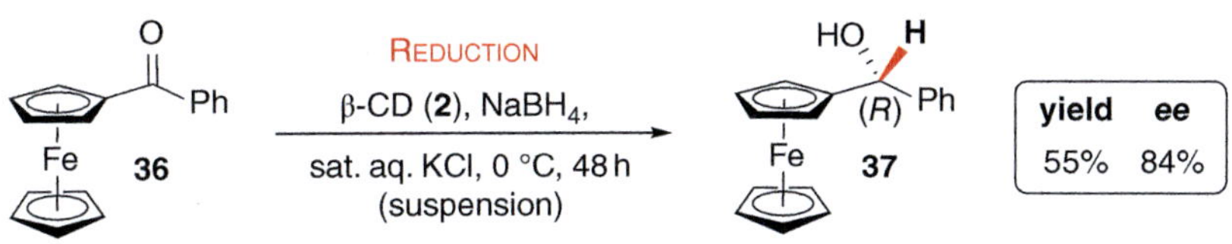

Figure 11.15 *Reduction using an aqueous suspension of β-CD (**2**) as the catalyst [53]*

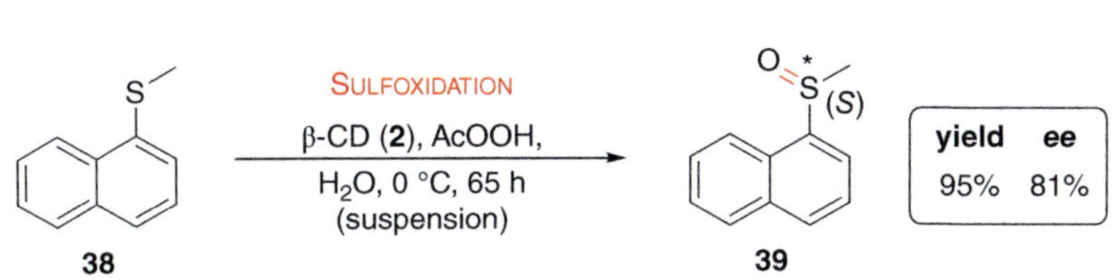

Figure 11.16 *Sulfoxidation using an aqueous suspension of β-CD (**2**) as the catalyst [54]*

Asymmetric catalysis with cyclodextrins is also observed in unimolecular reactions. Harano *et al.* [55] performed a [3,3]-sigmatropic rearrangement using an aqueous suspension of β-CD (**2**). They reported a 46% *ee* for rearrangement of the xanthate **40** to the dithiocarbonate **41** (Figure 11.17).

The xanthate **40** forms a 1 : 1 inclusion complex with β-CD (**2**) in which the hydroxyl groups of the host are understood to hydrogen bond to the dithiocarbonate moiety, to control the reaction geometry. The use of α-CD (**1**) or γ-CD (**3**) as host yielded a much lower *ee*, of 3% or 12%, respectively. This reveals again the importance of proper host–guest complementarity. Related studies using substrates without an aryl group also resulted in low enantioselectivities, indicating that an aromatic moiety is required for substrate inclusion, to facilitate the important hydrogen bonding.

Asymmetric catalysis of aldol addition with β-CD (**2**), in an aqueous suspension (Figure 11.18), as well as in the solid-state, was studied by Li *et al.* [56]. They reported *ees* up to 65% for the reaction of acetyl ferrocene (**42**) with the aromatic aldehydes **43–45** to give the aldol adducts **46–48**.

The ferrocene unit of compound **42** fits neatly inside the cavity of β-CD (**2**), forming a 1 : 1 inclusion complex where reaction of the acyl group is likely to be restricted, causing the progress of the reaction to be asymmetric. The use of alkali-metal halides (KCl or NaCl) as additives showed a positive effect on the enantioselectivity as did a lower reaction temperature, leading to an *ee* of 47% for the production of the chloride **48** at 0 °C in aqueous suspension.

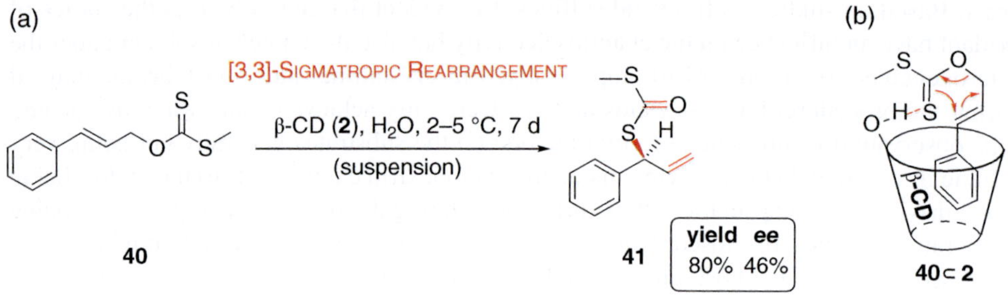

Figure 11.17 *(a) [3,3]-Sigmatropic rearrangement using an aqueous suspension of β-CD (**2**) as the catalyst and (b) a schematic representation of the proposed reaction geometry [55]*

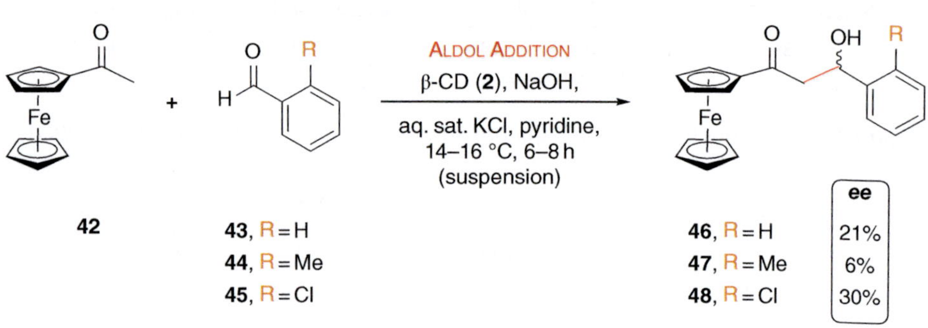

Figure 11.18 *Aldol additions using an aqueous suspension of β-CD (**2**) as the catalyst [56]*

The reaction of acetyl ferrocene (**42**) with benzaldehyde (**43**) was also investigated in the solid state, by grinding the dried complex of the ferrocene **42** and β-CD (**2**) with NaOH and the aldehyde **43** for 20 min, resulting in a 65% *ee*, the highest value obtained in this study. This outcome also suggests higher enantioselectivities result from more rigid assemblies of reactants.

Tanaka *et al.* [57, 58] have also reported very high enantioselectivities in the hydrohalogenation of *trans*-2-butenoic acid (**49**), using the crystalline 1 : 1 complex with α-CD (**1**). Reaction with gaseous HBr afforded *S*-3-bromobutanoic acid (**50**) in 58% *ee* (Figure 11.19). Even more remarkably, halogenation of compound **49** with chlorine gas at 25 °C for 20 h produced the corresponding adduct in 100% *ee*, albeit in only 17% yield.

(B) Covalent Catalysis The hydroxyl groups of the native cyclodextrins also provide the opportunity to form transient covalent bonds as part of the catalysis of reactions of included species, thereby introducing another specific interaction between the host and reactant that may affect the stereoselectivity. A thoroughly investigated aspect of covalent catalysis by cyclodextrins is the hydrolysis of esters, which serves further as a model for covalent catalysis by enzymes. The hydrolysis of esters involves the steps of (i) host–guest complexation; (ii) transesterification and (iii) ester hydrolysis and product release.

A detailed kinetic analysis of enantioselectivity in the hydrolysis of esters by native cyclodextrins was performed by Fornasier *et al.* [59]. For example, the β-CD (**2**) catalysed hydrolysis of the ester **51** to the acid **52** proceeds with an overall enantioselectivity of 19 : 1, taking into account both the enantioselectivity of complexation by the cyclodextrin (K_R and K_S) as well as the stereoselectivity of the transesterifications of the complexed species (k_{cR} and k_{cS}, Figure 11.20).

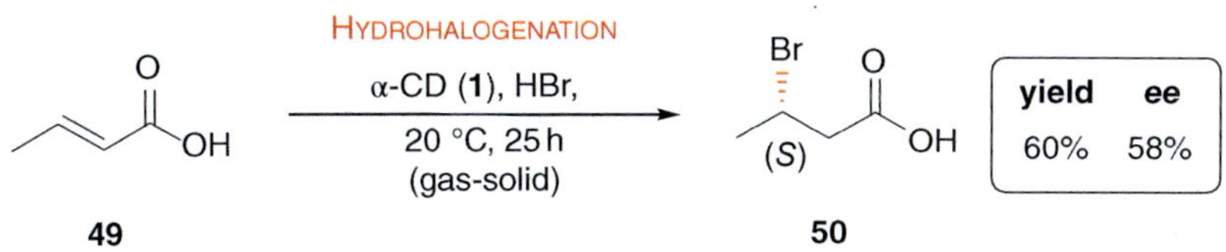

Figure 11.19 *Hydrohalogenation using α-CD (**1**) as the catalyst [57, 58]*

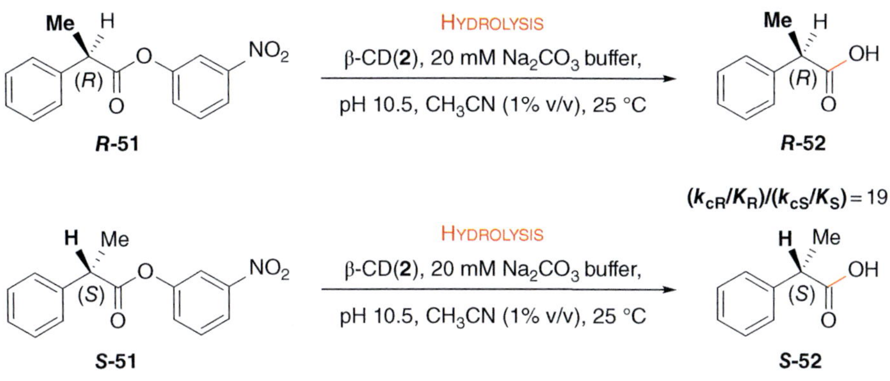

Figure 11.20 *Enantioselective hydrolysis using β-CD (**2**) as a covalent catalyst [59]*

***Table 11.1** Thermodynamic parameters for hydrolysis of the ester **51** to the acid **52** catalysed by α-CD (**1**) or β-CD (**2**) in 20 mM Na_2CO_3 buffer pH 10.5 with CH_3CN (1% v/v) at 25 °C [59]*

Host	Guest	k_c/k_{un} [a]	K_R/K_S [b]	k_{cR}/k_{cS} [c]	$(k_{cR}/K_R)/(k_{cS}/K_S)$ [d]	ee
α-CD (**1**)	***R*-51**	175	1.1	8.7	9.3	81% (*R*)
α-CD (**1**)	***S*-51**	20				
β-CD (**2**)	***R*-51**	77.5	1.2	15.5	19.0	90% (*R*)
β-CD (**2**)	***S*-51**	5.0				

Notes:
[a] k_c = rate constant for transesterification of the complexed species, k_{un} = rate constant for uncatalysed hydrolysis.
[b] Ratio of the complexation constants for the enantiomers.
[c] Ratio of the rate constants for the transesterifications of the complexed species.
[d] Overall enantioselectivity.

A comprehensive study with several esters showed that the cyclodextrin-catalysed hydrolysis rates are generally larger for the *R*-enantiomers. Many insights are gained by comparing the thermodynamic parameters for hydrolysis of the ester **51** with α-CD (**1**) or β-CD (**2**) (Table 11.1). Firstly and most importantly, substantial enantioselectivity arises from the different reactivity of the complexed species (k_{cR}/k_{cS}) rather than from different complexation constants (K_R/K_S). Secondly, the extent of acceleration of the reactions by the cyclodextrins (k_c/k_{un}) is rather modest compared to the efficiencies of enzymes. Finally, better enantioselectivity is obtained in this particular case with β-CD (**2**). The overall enantioselectivity $[(k_{cR}/K_R)/(k_{cS}/K_S)]$ of 19 : 1 corresponds to an *ee* of 90%.

The above analysis considers only the acylation of the cyclodextrin by complexed esters but not the subsequent deacylation of the transesterified catalyst. Coates *et al.* [60, 61] investigated both the acylation and deacylation in more detail, showing that both processes may display complementary stereoselectivity (Figure 11.21). Treating the racemic ibuprofen acid chloride **53** with β-CD (**2**) yielded the CD-bound species **54** in a diastereomeric ratio (*dr*) of 5 : 1 in favour of the *R*-form. Further studies disclosed that hydrolysis of the diastereomer **54a** to (*R*)-ibuprofen (**55**) proceeded an order of magnitude faster than hydrolysis of **54b** to (*S*)-ibuprofen (**55**). Thus the overall stereoselectivity of the acylation and deacylation is a factor of 50 : 1 in favour of (*R*)-ibuprofen (**55**), corresponding to an *ee* of 96%.

The enantioselectivity of hydrolysis of esters with cyclodextrins as covalent catalysts can be fine tuned by further restricting the geometry of substrate binding and transesterification. Breslow and coworkers [62, 63] demonstrated this by limiting rotational degrees of freedom in the acylation transition state through rigidifying the structure of several ferrocene esters. In this manner, they achieved an overall enantioselectivity of 62 : 1, corresponding to an *ee* of 97%.

11.3.2 Hosts Modified with Achiral Substituents

(A) Substituents Improve Enantioselectivity Through modification, the stereoselectivity displayed by the native cyclodextrins has been enhanced, even in cases where the alteration involves only the introduction of one or more achiral substituents [64, 65]. Kida *et al.* [66]

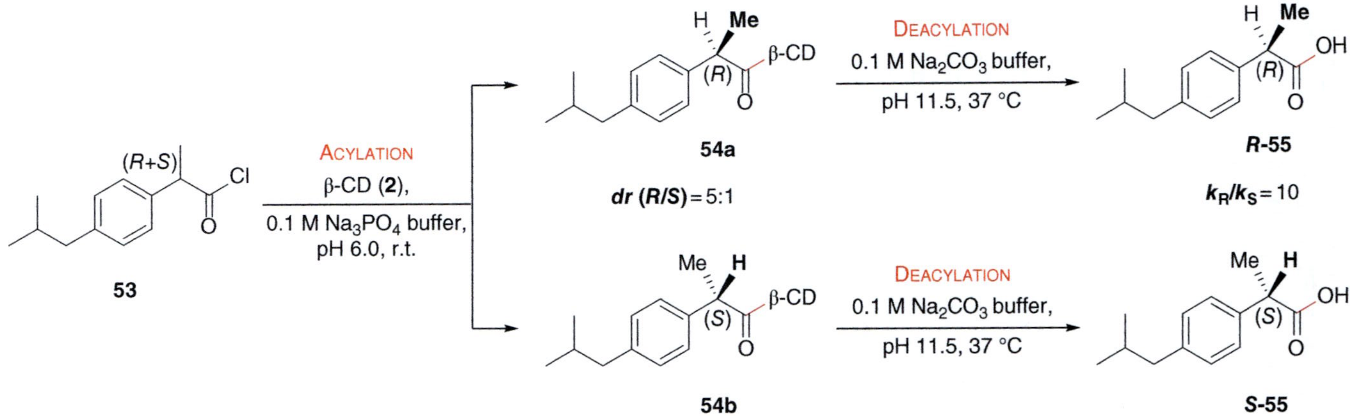

Figure 11.21 *Diastereoselective acylation and deacylation of β-CD (**2**) [60, 61]*

reported enantioselective *N*-acylation using the triisopropylsilyl- (TIPS)-modified β-CD **56**, which forms a dimeric nanocapsule with guests in nonpolar solvents (Figure 11.22). Kinetic resolution of the racemic aromatic amine **57** through enantioselective encapsulation to limit reaction with the anhydride **58** produces the amide **59** in up to 91% *ee*.

The dimeric nanocapsule of the TIPS-β-CD **56** with the encapsulated guest **57** is stabilized by multipoint interactions. An X-ray crystal structure of the inclusion complex formed in benzene revealed hydrogen bonding between the host units as well as between the amino group of ***S*-57** and a C(3)-hydroxyl group of **56**. Within the cavity, the guest forms a sandwich-type complex with two benzene molecules *via* π···π interactions and sits in the centre of the capsule. Further NMR studies in benzene and cyclohexane indicated that the complexation is enthalpy driven and that the naphthalene ring of the *S*-form of the amine **57** penetrates deeper into the cavity than that of the *R*-form. The K_a values in cyclohexane at 10 °C for the 2 : 1 complexes were $6.1 \times 10^9\ M^{-1}$ for ***S*-57** and $1.5 \times 10^8\ M^{-1}$ for the *R*-enantiomer, corresponding to a remarkable selectivity of 41 : 1. This explains the chiral recognition of the nanocapsule and the enantiodifferentiation of the kinetic resolution. While ***S*-57** is selectively encapsulated by the nanocapsule and thus shielded from the acylating reagent, the *R*-enantiomer is *N*-acylated in the bulk solution. The *ee* of 91% is based on 70% conversion of 0.1 eq. of the anhydride **58**. The use of more of the anhydride **58** resulted in lower *ee*s, but even with 98% conversion of 0.5 eq. the *ee* remained high at 75%.

Inoue and coworkers [67, 68] have used a modified cyclodextrin to affect the enantioselectivity of the [4+4]-photocycloaddition discussed above (Figures 11.13 and 11.14).

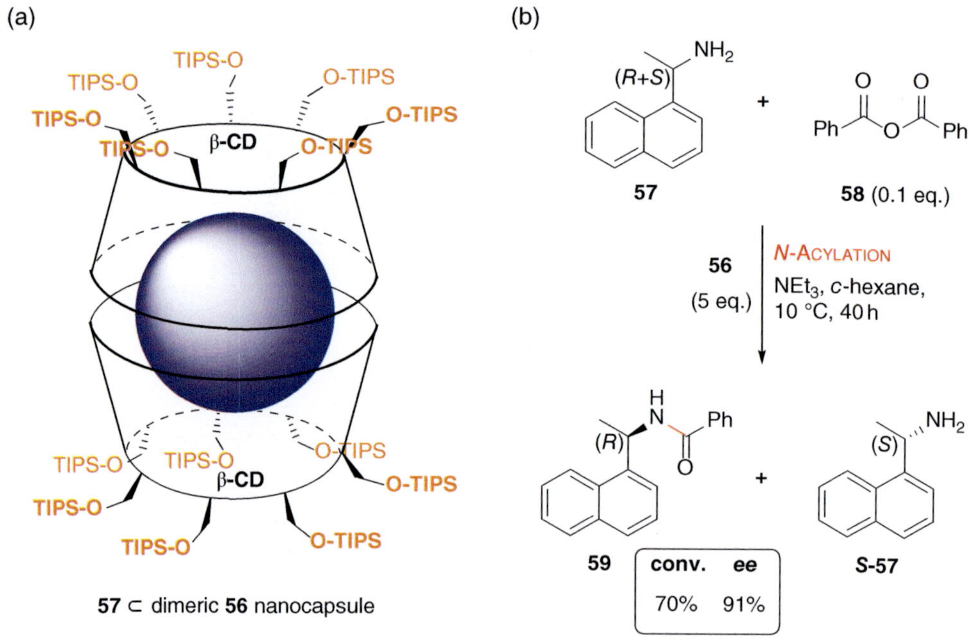

Figure 11.22 *(a) The dimeric TIPS-β-CD **56** nanocapsule which (b) induces kinetic resolution of the racemic aromatic amine **57** through* N*-acylation [66]*

Using γ-CD (**3**) as the catalyst, the cyclodimer **32** was produced in at most 41% *ee* through photodimerization of 2-anthracenecarboxylic acid (**31**), whereas applying the capped γ-CD **60** (Figure 11.23) resulted in the opposite enantioselectivity and up to 58% *ee*.

The ratios of the photodimers **32–35** produced using the capped γ-CD **60** and native γ-CD (**3**) but under otherwise identical conditions are compared in Table 11.2. Most noticeably, use of the host **60** reverses the enantioselectivity of formation of **32**, improves the *ee*s of both of the chiral cyclodimers **32** and **33**, and significantly increases the proportion of the dimers **33** and **35** having *head-to-head* orientation of the carboxyl groups. The latter is consistent with the hydrophobic area below the cap of the γ-CD **60** hindering accommodation of the polar carboxyl group of the anthracene **31**, hence destabilizing the *head-to-tail* orientation.

UV/Vis, circular dichroism and ^{1}H NMR spectroscopic studies indicated an enthalpically driven stepwise formation of a 2:1 inclusion complex of **31** in the capped-γ-CD **60** with a first K_a value of $1.3 \times 10^3\,M^{-1}$ and a second of $1.3 \times 10^4\,M^{-1}$ at 20 °C. By comparison, the corresponding values with native γ-CD (**3**) are $1.8 \times 10^2\,M^{-1}$ and $5.7 \times 10^4\,M^{-1}$. The second association constant is higher than the first in both systems because of more efficient π···π and van der Waals interactions upon inclusion of the second molecule of the anthracene derivative **31**.

(B) Substituents Provide Additional Functionality Cyclodextrins have also been modified with more directly functional substituents that enable catalysis of different types of chemical processes. An effective system for enantioselective photoisomerization was

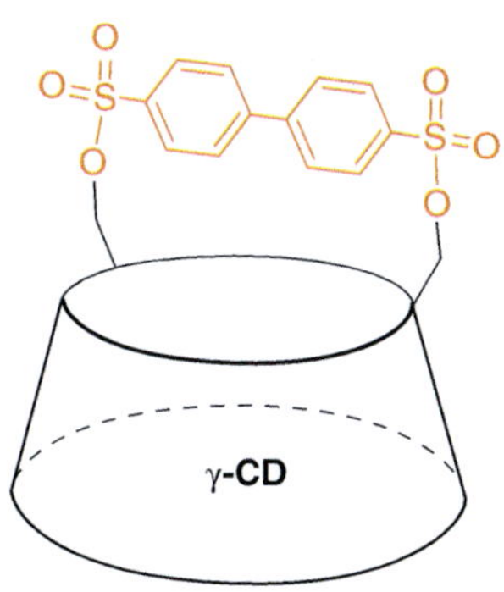

60

Figure 11.23 *The capped γ-CD **60** used to affect the enantioselectivity of [4+4]-photocycloaddition of 2-anthracenecarboxylic acid (**31**) [67, 68]*

Table 11.2 *Photodimerization of the anthracene derivative **31** using 2.5 eq. of the capped γ-CD **60** or the native γ-CD (**3**) as the catalyst [67, 68]*

	Relative yields (and ees)			
Catalyst	Chiral **32** (%)	Chiral **33** (%)	Achiral **34** (%)	Achiral **35** (%)
Capped γ-CD **60**	28 (58)	20 (14)	39	13
γ-CD (**3**)	46 (37)	7 (1)	43	5

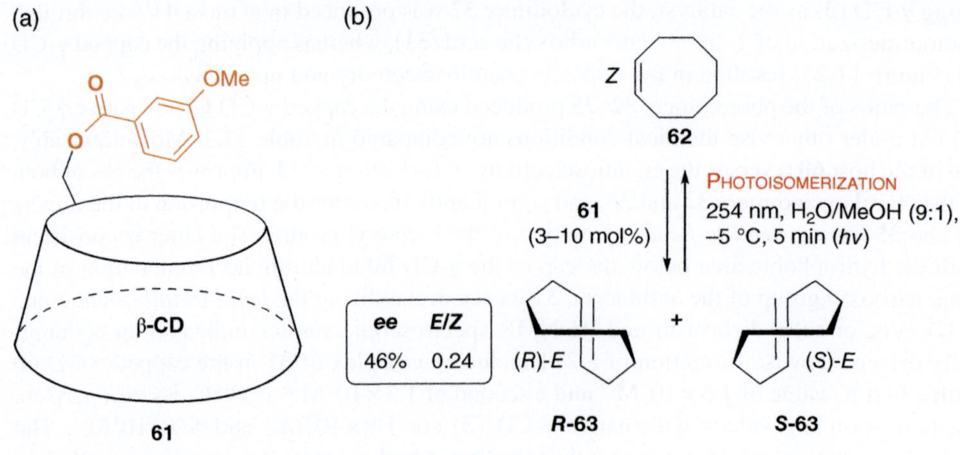

Figure 11.24 *(a) The modified β-cyclodextrin **61** used as a catalyst of (b) the enantioselective photoisomerization of the octene **62** [69, 70]*

designed by Lu *et al.* [69, 70] involving equipping a cyclodextrin with a photosensitizer. Accordingly, using the sensitizer-substituted β-CD **61** as a catalyst, the *Z*-octene (**62**) converted to the enantiomeric *E*-octene (**63**) in 46% *ee* in favour of the *R*-form (Figure 11.24).

Circular dichroism studies revealed that the cyclodextrin **61** and the octene **62** form a 1 : 1 host–guest complex with a K_a value of $2.3 \times 10^4\,M^{-1}$ at 18 °C in H_2O/MeOH (9 : 1). In the complex, the chromophoric substituent partially covers the enclosed guest, reducing the surface area of guest exposed to the aqueous media, enhancing van der Waals interactions in the complex, and facilitating efficient sensitizer-guest energy transfer upon irradiation. Small variations to the sensitizer, such as changing or replacing its methoxy group, greatly influence its orientation and impact on the guest complexation, energy transfer to the guest, the resulting *E*/*Z*-isomer product ratio, *ee* and absolute configuration of the products. While changing the temperature has negligible effect on the enantioselectivity, the solvent composition influences the position of the sensitizer, impacting on the reaction outcome.

Kanagaraj *et al.* [71] have exploited the amino groups of the modified β-CD **64** in base catalysis of an enantioselective Henry reaction. Quite remarkable *ee*s were achieved in the reaction of nitromethane (**65**) with the nitrobenzaldehydes **66–68** to give the Henry adducts **69–71** (Figure 11.25).

The Henry reaction is understood to take place within a ternary complex of host and reactants. Encapsulated inside the cavity of the catalyst **64**, the aldehydes **66–68** form hydrogen bonds of their carbonyl groups with the amino groups of the host. Additionally, encapsulated nitromethane (**65**) is activated by the host's amino groups acting as an internal base. The importance of the amino groups was verified through a comparative study using native β-CD (**2**), which showed no reaction of nitromethane (**65**) with *p*-nitrobenzaldehyde (**68**) after 7 h at room temperature. Adding triethylamine as an external base in another control experiment with β-CD (**2**) gave an adduct yield of 78% but with only a very low 1.2% *ee*, under otherwise identical conditions to those where the catalyst **64** realized a yield of 99% with 65% *ee*. The use of DMF or methanol as cosolvent instead of CH_3CN decreased the enantioselectivity, while the use of less than 1 eq. of the catalyst **64** lowered

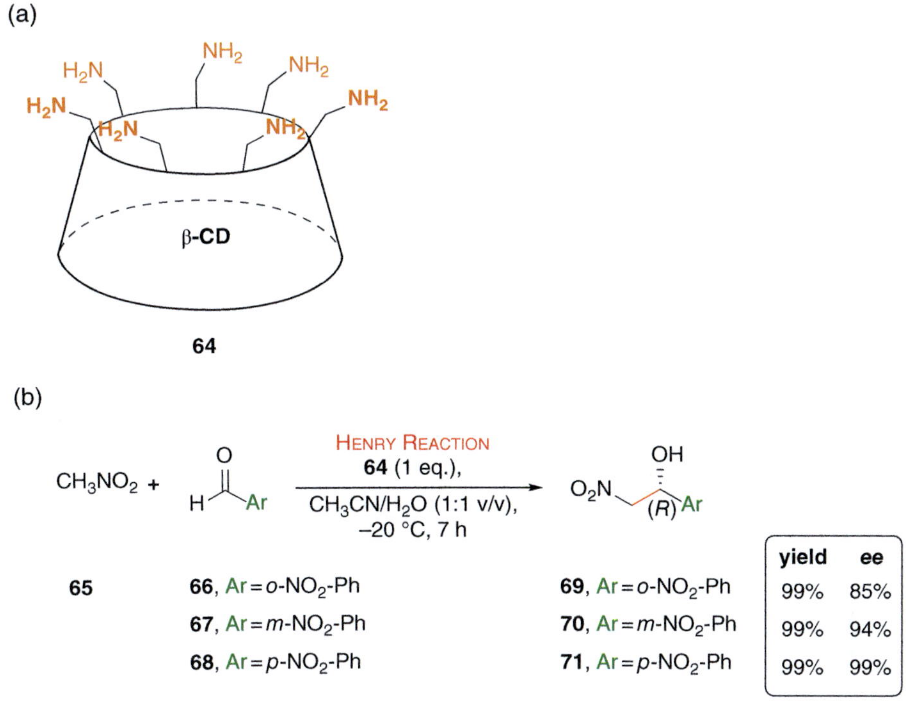

Figure 11.25 *(a) The functionalized β-cyclodextrin* ***64*** *used as a catalyst of (b) enantioselective Henry reactions [71]*

both the enantioselectivity and product yield. Lower reaction temperatures had a positive impact on the *ee*. The catalyst **64** may be recycled without significant loss of activity. It has also been used in asymmetric Michael additions, yielding products with up to 87% *ee* [72].

(C) Substituents Facilitate Additional Modes of Covalent Attachment While the hydroxyl groups of the cyclodextrins enable enzyme-like covalent catalysis as discussed above, other introduced substituents capable of forming a covalent bond with an encapsulated substrate provide alternative reaction pathways. A synthetic model for the enzyme glyoxalase was built by Tamagaki *et al.* [73]. They achieved a 46% *ee* using the (2-mercaptoethyl)amino-β-CD **72** to catalyse the transformation of phenylglyoxal (**73**) to (*S*)-mandelic acid (**74**) (Figure 11.26).

The catalyst **72** combines the binding site and reactive group in the same molecule, which is crucial for good enantiocontrol. By comparison, native α-CD (**1**), β-CD (**2**) and γ-CD (**3**) in combination with 2-(dimethylamino)ethanethiol yielded only 26%, 11% and 2% *ee*, respectively. This trend originates from the decreasing rigidity of the substrate complexes of the larger hosts. The smallest α-CD (**1**) facilitates the greatest van der Waals interactions and hence gives the highest *ee* in this series. However, in the covalently bonded system **72**, the substrate's movement is additionally constrained, further enhancing the enantioselectivity. The proposed structure for the covalently bound hemithioacetal-cyclodextrin intermediate **75** is shown in Figure 11.26. The pH has a substantial effect on the enantioselectivity, which decreases under more basic conditions.

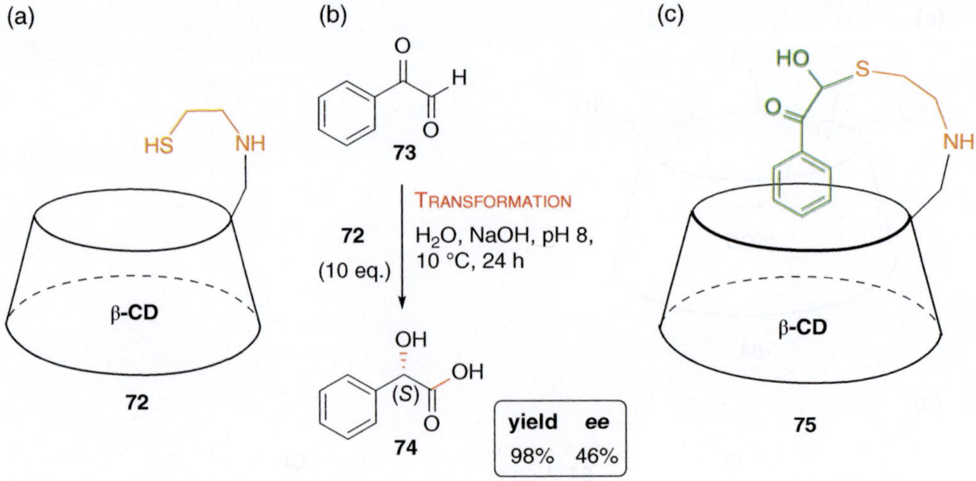

*Figure 11.26 (a) The customized β-cyclodextrin **72** used in (b) the enantioselective conversion of phenylglyoxal (**73**) to (S)-mandelic acid (**74**), and (c) the proposed reaction intermediate **75** [73]*

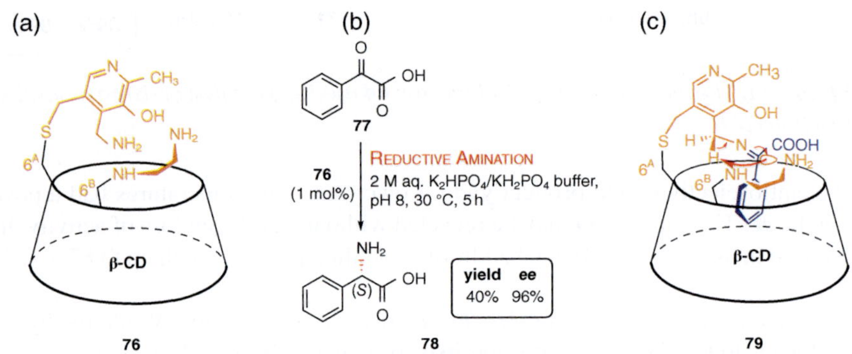

*Figure 11.27 (a) The customized β-cyclodextrin **76** used as a catalyst for (b) reductive amination, and (c) the proposed reaction intermediate **79** [14, 74]*

Another example of covalent catalysis with a customized cyclodextrin comprises the synthetic aminotransferase of Tabushi *et al.* [14, 74]. They reported a remarkable 96% *ee* in the reductive amination of phenylglyoxylic acid (**77**) to (*S*)-phenylglycine (**78**), applying the disubstituted β-CD **76** as the catalyst (Figure 11.27).

The catalyst **76** comprises two cooperating functional groups that work synergistically, with the pyridoxamine moiety promoting substrate binding *via* imine formation and the flexible amino chain facilitating proton transfer and imine-enamine tautomerization, before hydrolysis releases the product. Enantiocontrol results from the proton transfer step to the covalently bound imine-cyclodextrin intermediate **79** (Figure 11.27). The active catalyst **76** is then regenerated through reaction with a second amino acid.

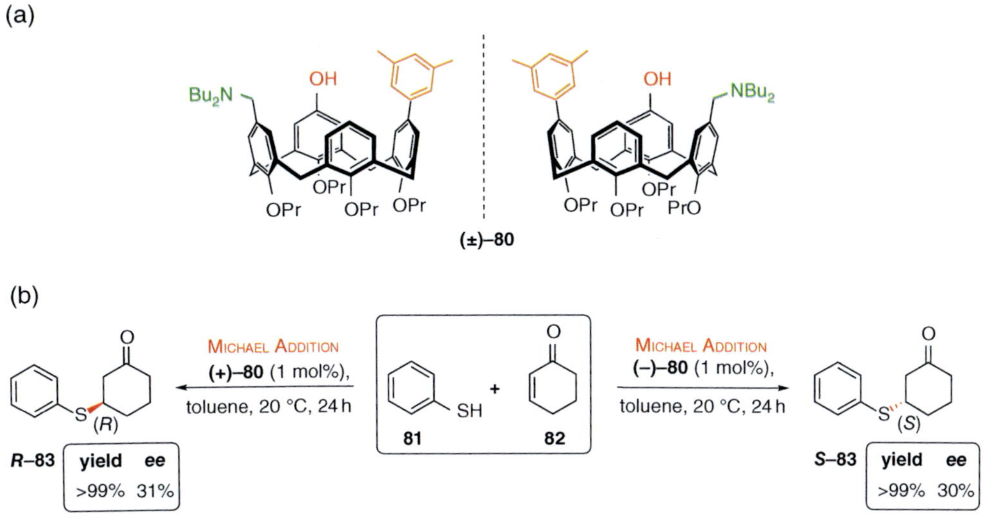

Figure 11.28 *(a) The catalyst* ***80*** *used for (b) enantioselective Michael addition [78]*

(D) Substituents Achiral but Substituted Host Chiral Up to this point, the focus of the articles reviewed has been on cyclodextrins due to their inherent chirality, which is lacking in other organic hosts such as the cucurbiturils and calixarenes. While unfunctionalized calixarenes are achiral, through substitution even with an achiral group they may become chiral if the substituents are attached in an asymmetric manner [75–77]. Shirakawa *et al.* [78] synthesized chiral calix[4]arenes and applied them as catalysts in an enantioselective Michael addition. After optical resolution of the enantiomers of the catalyst **80**, they achieved up to 31% *ees* in the reaction of thiophenol (**81**) with 2-cyclohexen-1-one (**82**). Each of the enantiomers of the adduct **83** was selectively produced through appropriate use of the enantiomers of catalyst **80** (Figure 11.28).

The catalyst **80** is fixed in a cone conformation, exhibiting an ABCD substitution pattern on one rim, including amino, hydroxyl and 3,5-dimethylphenyl groups. All three extensions are understood to contribute to the reactivity and selectivity in the reaction, with the amino group activating thiophenol (**81**) by forming an ammonium thiolate complex, the hydroxyl group activating 2-cyclohexen-1-one (**82**) *via* hydrogen bonding to the carbonyl group, and the 3,5-dimethylphenyl group selectively blocking one enantioface of the electrophile **82** such that the nucleophile **81** can only attack from the opposite face. Related catalyst versions without the bulky phenyl group achieved only up to 15% *ees*, and the methyl ether of the catalyst **80** showed negligible enantioselectivity.

11.3.3 Hosts Modified with Chiral Substituents

(A) Achiral Host with Chiral Substituent A chiral calixarene may also be produced through introduction of a single chiral substituent. An enantioselective aldol addition using the proline-extended calix[4]arene **84** as the catalyst was reported by Li *et al.* [79]. They achieved up to >99% *ees* in the reactions of cyclohexanone (**85**) with the *o*-benzaldehydes **43**, **66** and **86** to give the adducts **87–89** (Figure 11.29).

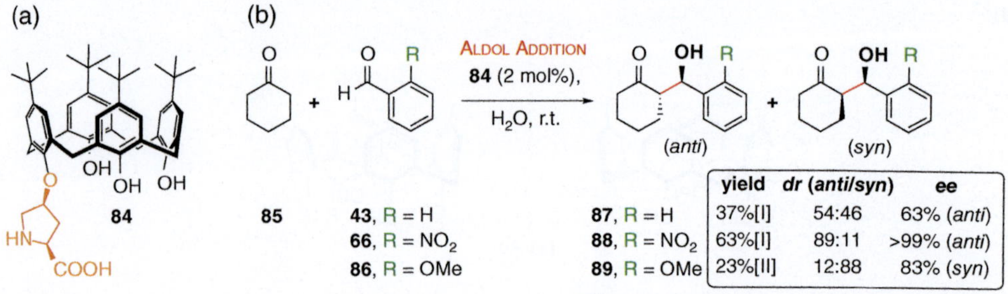

	yield	dr (anti/syn)	ee
87, R = H	37%[I]	54:46	63% (anti)
88, R = NO_2	63%[I]	89:11	>99% (anti)
89, R = OMe	23%[II]	12:88	83% (syn)

[I] Reaction time of 48 h. [II] Reaction time of 96 h.

Figure 11.29 *(a) The catalyst **84** used in (b) enantioselective aldol additions yielding the diastereomeric compounds **87–89** [79]*

The activity of the catalyst **84** is related to its hydrophobic and hydrophilic regions. Water has a marked effect on the outcome. In aldol additions tested in solvent-free conditions using neat cyclohexanone (**85**), or when using DMSO as the solvent, both the yield and enantioselectivity were reduced by 50%. In order to evaluate the impact of the calix[4]arene scaffold, aldol additions using L-proline or L-3-hydroxyproline as the catalyst were performed. Only traces of products were observed. The best result with the catalyst **84** was achieved in the aldol addition of *p*-nitrobenzaldehyde (**68**) with cyclopentanone, with a quantitative yield, good diastereoisomeric ratio (*anti*/*syn* 32 : 68), and excellent enantioselectivities of >99% for both the *anti* and *syn* adducts.

Interestingly, electronic effects of the aromatic rings significantly influence the diastereo- and enantio-selective outcomes. While *o*-nitrobenzaldehyde (**66**) promotes mainly formation of the *anti*-isomer of the adduct **88** with >99% *ee*, *o*-methoxybenzaldehyde (**86**) selectively generates the opposite *syn*-isomer of the adduct **89** with 83% *ee*. Further investigations using *meta*- and *para*-substituted benzaldehydes showed the same trend. Hence, incorporating electron-withdrawing or donating groups on the aromatic substrate tunes the ratio of *anti*/*syn* adducts.

Buckley *et al.* [80] reported enantioselective nucleophilic addition using the resorcin[4]arene catalysts **90** and **91** in the addition of Et_2Zn to benzaldehyde (**43**), to produce (*S*)-1-phenylpropan-1-ol (**92**) in up to 83% *ee* (Figure 11.30).

The dissymmetry of the catalysts **90** and **91** arises from their three stereogenic centres: (i) the inherent chiral axis of the C_n-symmetric resorcin[4]arene resulting from the asymmetric substitution pattern and designated *P* or *M*; (ii) the auxiliary substituent chirality designated *R* or *S*; and (iii) the interring stereogenic centre again designated *R* or *S* (Figure 11.30). The stereochemical assignment of the chiral axis and auxiliary substituent may be matched (both clockwise or both anticlockwise priorities) or mismatched (one clockwise, the other anticlockwise). Addition of Et_2Zn to benzaldehyde (**43**) produced the *S*-enantiomer of the adduct **92** in 83% *ee* using the matched system **90**, but in only 5% *ee* with the mismatched host **91** (Figure 11.30). A host system that lacked axial chirality produced racemic material while using a simple chiral benzoxazine as the catalyst resulted in a reduced yield of 57% and only 18% *ee*, revealing the importance of the combined features of the catalysts **90** and **91**.

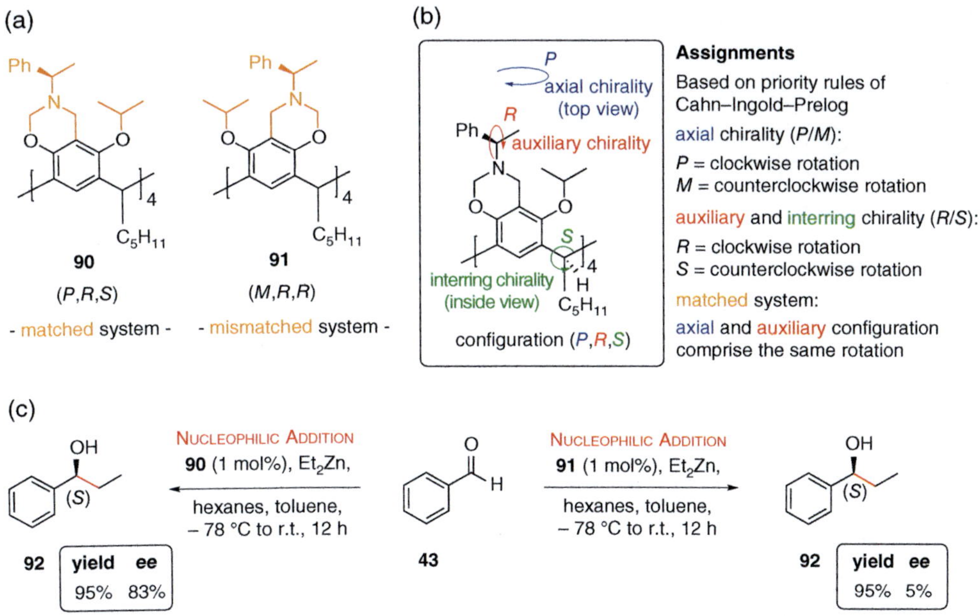

Figure 11.30 *(a) Inside view and stereochemical assignments of the catalysts* ***90*** *and* ***91****, (b) explanation of the assignments based on Cahn–Ingold–Prelog rules, and (c) use of the catalysts* ***90*** *and* ***91*** *for enantioselective nucleophilic addition of* Et_2Zn *to the aldehyde* ***43*** *[80]*

(B) Chiral Host with Chiral Substituent Chiral substituents have also been introduced to cyclodextrins to enhance the asymmetry. Another enantioselective aldol addition, using a diamine-extended β-cyclodextrin as catalyst, was reported by Hu *et al.* [81]. They achieved a 97% *ee* in the conversion of acetone (**94**) with *p*-nitrobenzaldehyde (**68**) to the *R*-adduct **95**, using the diamino-β-CD **93** (Figure 11.31).

The catalyst **93** is a good mimic of the enzyme aldolase. Fluorescence and circular dichroism spectroscopy with 2-naphthaldehyde as a substrate revealed the formation of a 1 : 1 complex with a K_a value of 403 M^{-1} in 50 mM acetate buffer containing 2% glycol at pH 4.8. This is twice the association constant of the corresponding aldol adduct with acetone (**94**) and twice the value with native β-CD (**2**) as the host. The former comparison indicates that the catalyst **93** is not strongly inhibited by product formation, which is the basis for good catalytic turnover. The latter indicates the modest effect of the substituent of the catalyst **93** on substrate complexation. Other, simple cyclohexadiamines showed low activities and poor enantioselectivities in catalysis of aldol additions. Furthermore, addition of 1-bromoadamantane, which is a good cyclodextrin guest, inhibits the catalyst **93**.

NMR spectroscopy investigations showed that, in water, the catalyst **93** forms an intramolecular self-inclusion complex of its diamino extension. However, in acetate buffer at pH 4.8, protonation of the amino groups decreases the hydrophobicity of the substituent and drives it out of the cavity. The diamino extension is displaced even more upon aldehyde complexation.

Kinetic studies of the aldol addition with acetone (**94**) and *p*-nitrobenzaldehyde (**68**) indicated Michaelis–Menten kinetics at pH 4.8 and 25 °C, showing a K_m of 6.3 mM for the

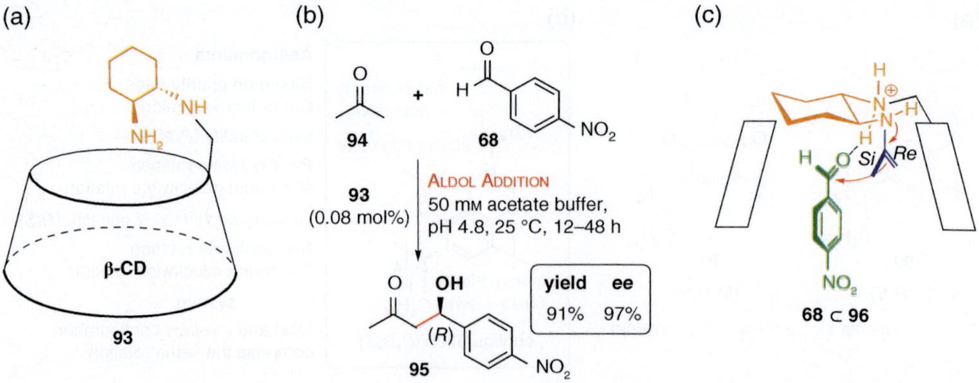

Figure 11.31 (a) The catalyst **93** used for (b) enantioselective aldol addition yielding the adduct **95** and (c) the proposed reaction intermediate **96** with the aldehyde **68** encapsulated and positioned by a hydrogen-bond for nucleophilic attack from the Si-face [81]

aldehyde **68** and a k_{cat} of 6.1×10^{-3} min^{-1}. Studies with a variety of substrates indicate that a proper geometry and orientation of the aldehyde in the cavity is essential for effective catalysis. Furthermore, only small ketones were effective substrates. The highest *ee* of >99% in the aldol additions with the catalyst **93** was achieved with acetone (**94**) and 4-phenylbenzaldehyde. Diastereoselective reactions with cyclic ketones were also conducted, yielding mainly the *anti*-conformers with high enantioselectivities.

The proposed catalytic cycle is thought to involve three steps: (i) assembly of the ternary complex of catalyst, aldehyde and ketone, followed by formation of an enamine-intermediate; (ii) nucleophilic addition of the enamine to the aldehyde; and (iii) imine hydrolysis and release of the product to regenerate the catalyst. Enamine formation was identified as rate limiting. The observed enantioselectivity is explained by the greater accessibility of the *Si*-face of the enamine to the aldehyde, which is positioned by a hydrogen bond between its carbonyl group and a host amino group (Figure 11.31).

11.3.4 Hosts Modified with Metal-Coordinating Ligands

Another class of modified organic hosts are those substituted with metal-coordinating ligands. These are used in asymmetric phase-transfer and transition-metal catalysis. It may not necessarily be the case that catalysis always occurs through host–guest encapsulation, rather than simply by metal coordination, but interesting systems have been reported and this is a growing area in asymmetric synthesis.

(A) Metal Inside Cavity Crown ethers are probably the simplest systems that complex metals. Functionalization of this scaffold with chiral extensions facilitates their application in asymmetric synthesis. As an example, an asymmetric phase-transfer Darzens reaction was reported by Rapi *et al.* [82] using a D-glucose-azacrown ether derivative as catalyst. They achieved a 96% *ee* in the conversion of 4-phenyl-α-chloroacetophenone (**98**) with benzaldehyde (**43**) to the *trans*-epoxy ketone **99** using the phase-transfer catalyst **97** (Figure 11.32).

Darzens reactions with substituted benzaldehydes were also investigated. All *trans*-epoxy ketones were produced in >98% diastereomeric excess, suggesting the selectivity of

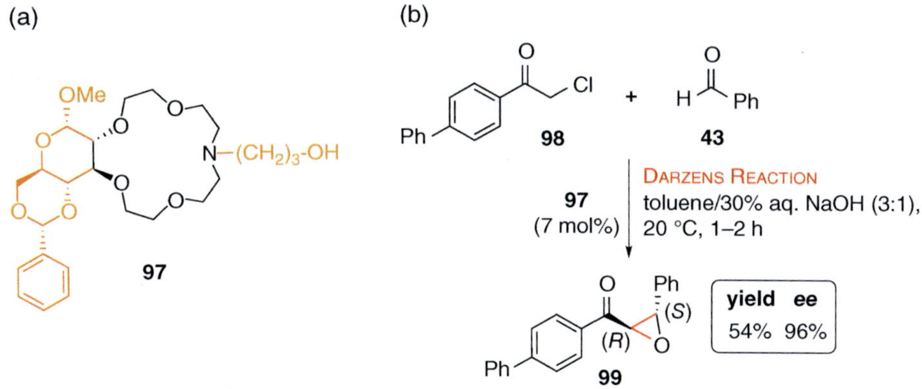

Figure 11.32 *(a) The D-glucose-azacrown ether derivative **97** used as a phase-transfer catalyst of (b) the enantioselective Darzens reaction yielding the adduct **99** [82]*

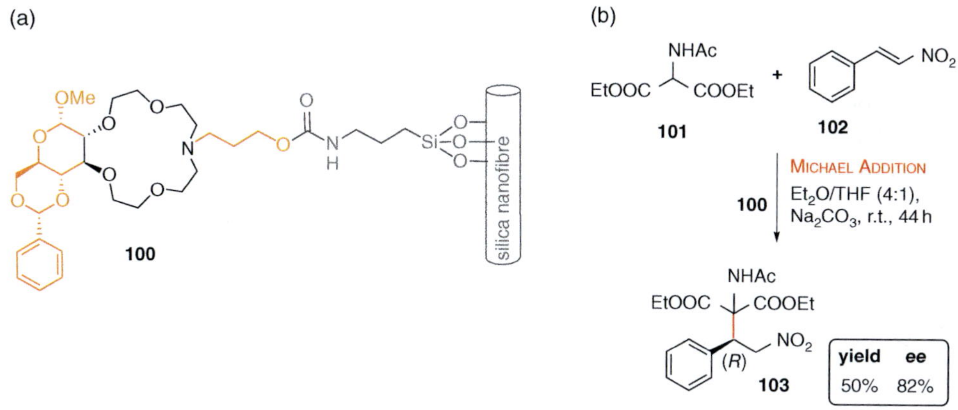

Figure 11.33 *(a) The immobilized D-glucose-azacrown ether derivative **100** used as a catalyst of (b) the enantioselective Michael addition yielding the adduct **103** [84]*

these reactions is due to the formation of a *Z*-enolate of the chloroacetophenone **98**. The yield and enantioselectivity were found to be dependent on the benzaldehyde's ring substituent and its position. While benzaldehyde (**43**) clearly demonstrated the highest 96% *ee*, electron-withdrawing groups tended to increase yield but decrease the *ee*, while electron-donating groups had a negative impact on yield but improved the enantioselectivity. This observation is attributable to the more electrophilic aldehydes being more reactive and giving higher yields but at the expense of reacting less stereoselectively. Among *ortho*-, *meta*- and *para*-substituted benzaldehydes, the latter revealed the best enantiocontrol.

Catalytic properties of D-glucose-azacrown ethers are also demonstrated in asymmetric epoxidations and Michael additions [83]. An interesting feature of the enantioselective Michael addition developed by Soti *et al.* [84] was to immobilize the catalyst on silica nanofibres. They reported an 82% *ee* for the conversion of diethyl acetamidomalonate (**101**) with *trans*-β-nitrostyrene (**102**) to the adduct **103** using the catalyst **100** (Figure 11.33).

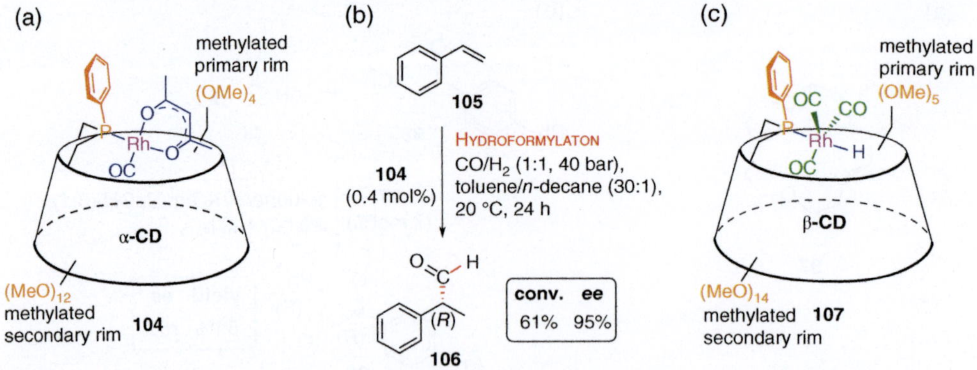

Figure 11.34 *(a) The rhodium monophosphane-α-cyclodextrin complex* ***104*** *used as a catalyst for (b) enantioselective hydroformylation yielding compound* ***106****, and (c) the active species* ***107*** *formed under hydroformylation conditions examined using the β-cyclodextrin analogue [85, 86]*

The reaction rate and *ee* were both lower than those obtained using the non-immobilized catalyst **97**, but the immobilized form has the advantage of being easily removed from the reaction mixture by filtration and it is then reusable without regeneration.

Cyclodextrins modified to complex a transition metal have also been reported to exhibit interesting behaviour in asymmetric reactions. Jouffroy *et al.* [85, 86] used a phosphane derivative for an enantioselective hydroformylation. They achieved a 95% *ee* in the conversion of styrene (**105**) to the branched aldehyde **106** using the square-planar rhodium monophosphane-α-cyclodextrin complex **104** as catalyst (Figure 11.34).

The phosphane-α-cyclodextrin precursor of the catalyst **104** adopts the conformation in which the phosphorus lone pair points into the cavity. This facilitates a tight embrace of the introduced transition metal, promotes chirality transfer and, most desirably, restricts the phosphane coordination to a single ligand, hence influencing the regioselectivity of reaction. Under the applied reaction conditions, the observed selectivity of 98% of the branched aldehyde **106** to 2% of the linear regioisomer clearly revealed the impact of the catalyst. The selectivity for the branched aldehyde **106** was enhanced to almost 100% with a 93% *ee* by lowering the reaction temperature to 4 °C; however, this dropped the conversion to 34%. In contrast, raising the reaction temperature to 40 °C resulted in a conversion of 99%, while the regioselectivity reduced to 96% and the enantioselectivity fell to 90%. A particular feature of this catalytic system is that, unlike classical diphosphane-rhodium complexes, in this case higher CO partial pressures enhanced both catalytic activity and enantioselectivity.

While investigating the catalytically active species formed under hydroformylation conditions, activation of a β-cyclodextrin analogue in toluene at 80 °C with CO/H_2 (1 : 1, 40 bar) resulted in formation of the trigonal bipyramidal complex **107**, showing a rare linear P–Rh–H arrangement which may be formed in order to minimize steric interactions between the cyclodextrin cavity wall and the carbonyl ligands. Interestingly, hydroformylations with the α-cyclodextrin complex **104** or the β-cyclodextrin analogue provided similar results, indicating that the reaction is insensitive to cavity size.

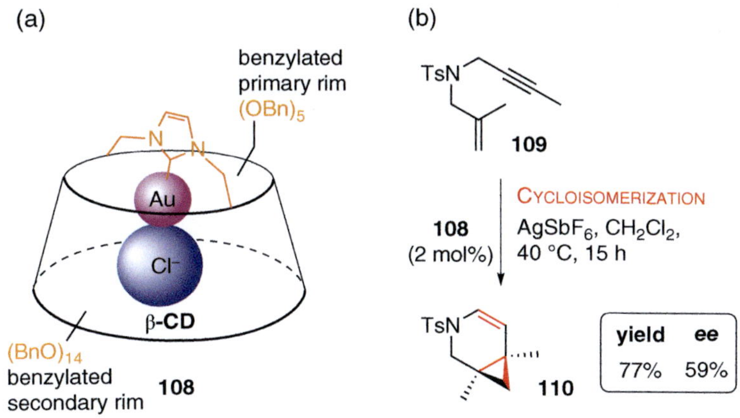

Figure 11.35 (a) The carbene-capped β-cyclodextrin AuCl complex **108** used as a catalyst of (b) the enantioselective cycloisomerization yielding the bicyclic product **110** [87]. Ts = p-toluenesulfonyl

Another cyclodextrin-based metal-complex system was reported by Guitet *et al.* [87] who applied a heterocyclic carbene-capped β-cyclodextrin-AuCl complex in an enantioselective cycloisomerization. They achieved a 59% *ee* in the conversion of the alkyne **109** to the bicyclic product **110** with the catalyst **108** (Figure 11.35).

The substituent of the catalyst **108** comprises the carbene unit, which is partially included and oriented towards the centre of the host cavity. This provides an inner coordination site for metals, hence burying the Au-ion deep inside the cavity. The AuCl complex is stabilized by various interactions with the surrounding host, including C–H···M, C–H···X–M and π···X–M interactions. The latter occur between the catalyst benzyl substituents and the halide, inducing the closure of the cyclodextrin cavity at the secondary rim by two benzyl substituents. However, the cycloisomerization reveals that the encapsulated metal is still accessible and reactive. The asymmetric induction is linked to the structure of the host, which is slightly distorted due to the asymmetric capping with the carbene unit at the primary rim, leaving two primary benzyl substituents on one side and three on the other.

(B) Metal Outside Cavity Schlatter *et al.* [88] reported a ruthenium-amino alcohol-β-cyclodextrin complex for enantioselective reduction of aromatic and aliphatic ketones. They achieved up to 97% *ee* in the conversion of 4′-*tert*-butylacetophenone (**113**) to the *S*-enantiomer of the alcohol **114** using the catalyst **111**, while the *R*-enantiomer was produced using the alternative catalyst **112** (Figure 11.36).

The cyclodextrin unit of the catalyst **112** is the only chiral element within the structure, nevertheless the *R*-alcohol **114** is formed with 47% *ee*. Further studies with related ketones showed that the *ee*s obtained with this catalyst correlate with the K_a values of the ketone–β-cyclodextrin (**2**) complexes, which reflect the extent of preorganization of the substrates for stereospecific hydride addition to the *Si*-face. The catalyst **111** comprises a second chiral element on the substituent, resulting in a higher *ee* and in favour of the opposite enantiomer of the alcohol **114**. This demonstrates that the effect of the chirality of the substituent on the enantioselectivity of the reduction overrides that of the cyclodextrin unit. Even so, the cyclodextrin component of the catalyst **111** was shown to be important because reduction

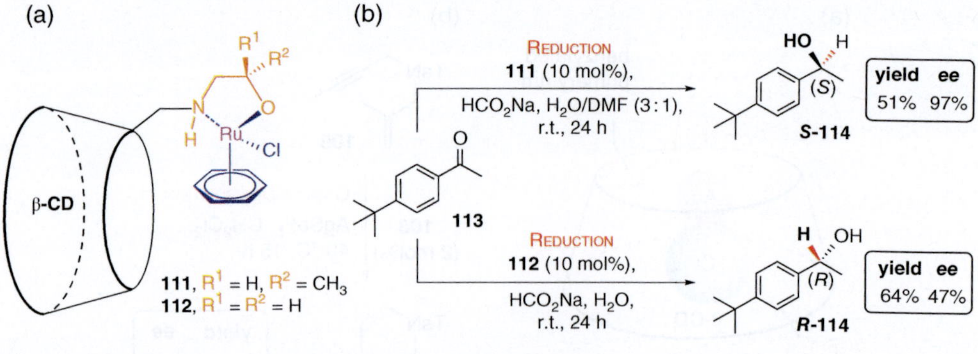

Figure 11.36 *(a) The ruthenium-amino alcohol-β-cyclodextrin complexes **111** and **112** used as catalysts for (b) enantioselective reductions yielding the alcohol **114** [88]*

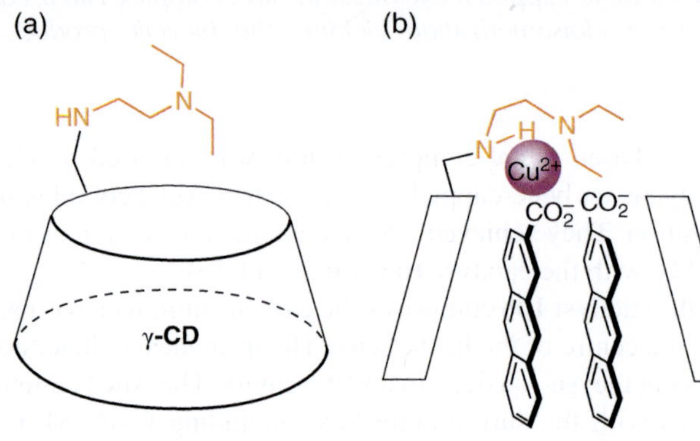

Figure 11.37 *(a) The diamino-substituted γ-cyclodextrin **115** and (b) the complex of the cyclodextrin **115**, Cu^{2+} and the anthracene derivative **31** proposed to lead to enantioselective production of the photodimer **33** [90]*

of acetophenone with (*S*)-1-amino-2-propanol afforded the corresponding alcohol in only 50% *ee*, compared to 77% *ee* with the catalyst **111**. An α-keto ester and several aliphatic ketones were also reduced in yields and with *ee*s both up to 95%. Schlatter and Woggon [89] additionally developed similarly modified cyclodextrins for ruthenium-catalysed hydrogenations, giving product yields up to 99% and with up to 98% *ee*.

Inoue and coworkers [90] used a cyclodextrin substituted with a metal-coordinating ligand to extend their studies of the enantioselective [4+4]-photocycloaddition of anthracene derivatives discussed above (Figures 11.13, 11.14 and 11.23). They applied the diamino-substituted γ-cyclodextrin **115** with $Cu(ClO_4)_2$ to the photodimerization of 2-anthracenecarboxylic acid (**31**) and obtained the cyclodimer **33** in up to 70% *ee* (Figure 11.37).

At –50 °C, the photodimers **32–35** are produced in the ratios summarized in Table 11.3, using the substituted cyclodextrin **115** with or without added Cu^{2+}, or native γ-CD (**3**), as the catalyst. The obvious effect of the modified cyclodextrin **115** in combination with the copper is to significantly increase the proportions of the otherwise sterically and

Table 11.3 *Photodimerization of the anthracene derivative* ***31*** *using 10 mol% of the diamino-substituted γ-CD* ***115*** *with or without added* Cu^{2+} *(5 eq.), or native γ-CD (**3**), as the catalyst in aq. phosphate buffer/MeOH (1 : 1) pH 5, at >320 nm and –50 °C for 1 h [90]*

Catalyst	**Conversion** (%)	Relative yields (and ees)			
		Chiral **32** (%)	Chiral **33** (%)	Achiral **34** (%)	Achiral **35** (%)
Diamino γ-CD **115** + Cu^{2+}	12	7 (13)	52 (70)	13	28
Diamino γ-CD **115**	51	15 (0)	34 (45)	21	25
γ-CD (**3**)	63	26 (32)	26 (16)	41	7

electrostatically unfavoured *head-to-head* dimers **33** and **35**, and the *ee* of the chiral one **33**. This reveals the effectiveness of the copper complex of the catalyst **115**, although the requirement for a fivefold excess of Cu^{2+} compared to the cyclodextrin **115** indicates weak chelation of the metal. The extent of reaction after 1 h is less with the copper complex of the cyclodextrin **115**, probably because the rigid geometric constraints within that complex restrict the reaction rate. Accordingly, extending the reaction time to 40 h resulted in 83% conversion of the reactant **31** and then produced the cyclodimer **33** in 64% *ee*.

Evidence of formation of the complex of the cyclodextrin **115**, Cu^{2+} and the anthracene derivative **31** (Figure 11.37) was obtained using UV/Vis spectroscopy, which showed stepwise formation of a 2 : 1 inclusion complex of the anthracene **31** in the diamino γ-CD **115** with added Cu^{2+}, having $K_{a,1} = 5.7 \times 10^2\,M^{-1}$ and $K_{a,2} = 3.2 \times 10^3\,M^{-1}$ at –50 °C in phosphate buffer/MeOH (1 : 1). By comparison, without Cu^{2+} the values were found to be $K_{a,1} = 8.4 \times 10^2\,M^{-1}$ and $K_{a,2} = 2.0 \times 10^3\,M^{-1}$.

Using a calix[4]arene-based system, Bonaccorso *et al.* [91] studied a manganese complex for enantioselective epoxidation of aryl alkenes. They achieved a 93% *ee* in the conversion of chromene **117** with NaOCl to the epoxide **118** using the Mn-salen calix[4]arene **116** as the catalyst (Figure 11.38). The diimine bridge of the catalyst **116** exhibits a dissymmetry that favours the *Si*-attack on the alkene **117** by the catalytically active oxo-Mn(V) species. The epoxidation of styrene and other compounds was also investigated; however, the extent of reaction remained generally below 50% and the maximum *ee* was 56%, seen in the reaction of *cis*-β-methylstyrene.

11.4 Conclusion

Chiral host systems based mainly on cyclodextrins, calixarenes and crown ethers have been explored frequently in enantioselective molecular recognition studies. A major role is played by the inherently chiral cyclodextrins and, more recently, chiral calixarenes, while the application of chiral crown ethers is generally limited to phase transfer processes. Analytical applications have been described, using chiral host systems as phase transporters in extraction, stationary phases in chromatography, specific sensors, or chiral shift reagents for detecting enantiomers. Chiral host systems have further shown their superior properties

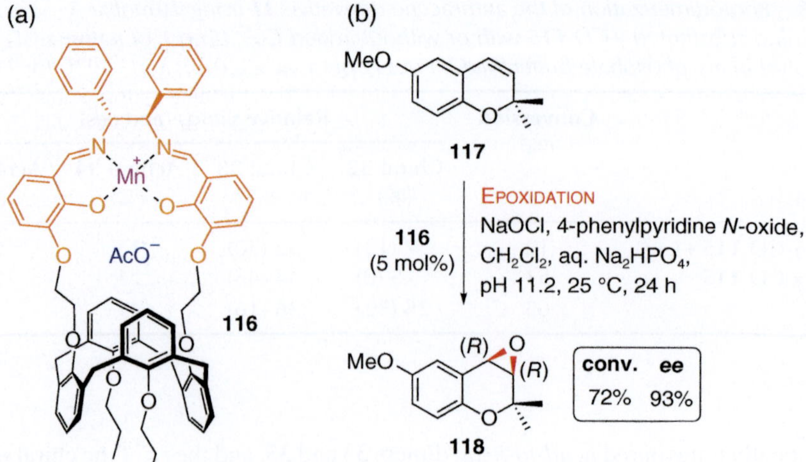

Figure 11.38 *(a) The manganese-salen calix[4]arene complex **116** used as a catalyst for (b) enantioselective epoxidation yielding compound **118** [91]*

in many asymmetric reactions, mimicking the catalysis shown by natural enzymes. While older publications mainly explored the use of cyclodextrins as native hosts, recent literature concentrates more on functionalized cyclodextrins, calixarenes and crown ethers. To achieve good enantioselectivities, the formation of a rigid host–guest complex is required, which is generally supported by the application of lower temperatures. The advantage of encapsulation catalysis is evident from comparing reactions involving native cyclodextrins in solution and solid state. While the *ee* values achieved in the liquid phase generally lie far below 50%, the enhanced rigidity of the assembly in the solid state improves the *ee*s up to nearly 100%. Alternative methods for enhanced enantiocontrol are the use of covalent catalysis and the specific functionalization of the host-systems with interacting functional groups. A wide range of forces have been reported, which stabilize the host–guest complexes, to thus facilitate the chiral induction in reactions of enclosed substrates. Important in the design of new chiral host systems is a match of chirality between the host and any introduced substituent, facilitating best results in enantioselectivity.

Acknowledgements

We gratefully acknowledge the award an Early Postdoc.Mobility Fellowship to Dr Daniel Fankhauser, by the Swiss National Science Foundation (P2EZP2_159061), to fund his position as research fellow at the Australian National University.

References

[1] C. J. Pedersen, *J. Am. Chem. Soc.* 1967, **89**, 2495–2496. Cyclic Polyethers and their Complexes with Metal Salts.
[2] C. J. Pedersen, *J. Am. Chem. Soc.* 1967, **89**, 7017–7036. Cyclic Polyethers and their Complexes with Metal Salts.

[3] J.-M. Lehn, *Acc. Chem. Res.* 1978, **11**, 49–57. Cryptates: The Chemistry of Macropolycyclic Inclusion Complexes.

[4] D. J. Cram, T. Kaneda, R. C. Helgeson *et al.*, *J. Am. Chem. Soc.* 1985, **107**, 3645–3657. Host–Guest Complexation. 35. Spherands, the First Completely Preorganized Ligand Systems.

[5] D. J. Cram, G. M. Lein, *J. Am. Chem. Soc.* 1985, **107**, 3657–3668. Host–Guest Complexation. 36. Spherand and Lithium and Sodium Ion Complexation Rates and Equilibria.

[6] C. J. Pedersen, *Angew. Chem.* 1988, **100**, 1053–1059; *Angew. Chem. Int. Ed. Engl.* 1988, **27**, 1021–1027. The Discovery of Crown Ethers (Noble Lecture).

[7] J.-M. Lehn, *Angew. Chem.* 1988, **100**, 91–116; *Angew. Chem. Int. Ed. Engl.* 1988, **27**, 89–112. Supramolecular Chemistry – Scope and Perspectives: Molecules, Supermolecules, and Molecular Devices (Nobel Lecture).

[8] D. J. Cram, *Angew. Chem.* 1988, **100**, 1041–1052; *Angew. Chem. Int. Ed. Engl.* 1988, **27**, 1009–1020. The Design of Molecular Hosts, Guests, and their Complexes (Nobel Lecture).

[9] C. J. Easton, S. F. Lincoln, *Chem. Soc. Rev.* 1996, **25**, 163–170. Chiral Discrimination by Modified Cyclodextrins.

[10] B. Chankvetadze, *Chem. Soc. Rev.* 2004, **33**, 337–347. Combined Approach Using Capillary Electrophoresis and NMR Spectroscopy for an Understanding of Enantioselective Recognition Mechanisms by Cyclodextrins.

[11] F. Bellia, D. La Mendola, C. Pedone *et al.*, *Chem. Soc. Rev.* 2009, **38**, 2756–2781. Selectively Functionalized Cyclodextrins and their Metal Complexes.

[12] A. Cooper, D. D. MacNicol, *J. Chem. Soc., Perkin Trans. 2* 1978, 760–763. Chiral Host–Guest Complexes: Interaction of α-Cyclodextrin with Optically Active Benzene Derivatives.

[13] I. Tabushi, Y. Kuroda, T. Mizutani, *J. Am. Chem. Soc.* 1986, **108**, 4514–4518. Artificial Receptors for Amino Acids in Water. Local Environmental Effect on Polar Recognition by 6*A*-Amino-6*B*-carboxy- and 6*B*-Amino-6*A*-carboxy-*β*-cyclodextrins.

[14] I. Tabushi, *Pure Appl. Chem.* 1986, **58**, 1529–1534. Chiral Selection and Chiral Induction by the Use of Regiospecifically Di-(or Poly)substituted Cyclodextrins.

[15] K. Takahashi, Y. Ohtsuka, K. Hattori, *Chem. Lett.* 1990, **19**, 2227–2230. Enantioselective Inclusion Behavior of Monosubstituted Cyclodextrins with Guest-Induced Substituent Movement.

[16] K. Takahashi, *Bull. Chem. Soc. Jpn.* 1993, **66**, 550–554. Preparation and Molecular Recognition Ability of Mono[6-deoxy-6-(L-tyrosylamino)]-*β*-cyclodextrin. Formation and Guest Binding Properties of Distorted Cavity Formed Outside the Parent Cyclodextrin.

[17] S. E. Brown, J. H. Coates, C. J. Easton *et al.*, *J. Chem. Soc., Chem. Commun.* 1994, **47**. Tryptophan Anion Complexes of β-Cyclodextrin (Cyclomaltaheptaose), an Aminopropylamino-β-cyclodextrin and its Enantioselective Nickel(II) Complex.

[18] S. E. Brown, J. H. Coates, C. J. Easton, S. F. Lincoln, *J. Chem. Soc., Faraday Trans.* 1994, **90**, 739–743. Metallocyclodextrins of 6^A-(3-Aminopropylamino)-6^A-deoxy-*β*-cyclodextrin: Their Formation and Enantioselective Complexation of (*R*)- and (*S*)-Tryptophan Anions in Aqueous Solution.

[19] S. E. Brown, C. A. Haskard, C. J. Easton, S. F. Lincoln, *J. Chem. Soc., Faraday Trans.* 1995, **91**, 1013–1018. Complexation of Phenylalanine and Histidine by *β*-Cyclodextrin, 6^A-(3-Aminopropylamino)-6^A-deoxy-*β*-cyclodextrin and its Metallocyclodextrins in Aqueous Solution.

[20] Y.-S. Zheng, J. Luo, *J. Incl. Phenom. Macrocycl. Chem.* 2011, **71**, 35–56. Inherently Chiral Calixarenes: A Decade's Review.

[21] L. Mutihac, J. H. Lee, J. S. Kim, J. Vicens, *Chem. Soc. Rev.* 2011, **40**, 2777–2796. Recognition of Amino Acids by Functionalized Calixarenes.

[22] S. Shirakawa, A. Moriyama, S. Shimizu, *Org. Lett.* 2007, **9**, 3117–3119. Design of a Novel Inherently Chiral Calix[4]arene for Chiral Molecular Recognition.

[23] U. Darbost, X. Zeng, M. Giorgi, I. Jabin, *J. Org. Chem.* 2005, **70**, 10552–10560. Optically Pure Calix[6]tris-ammoniums: Syntheses and Host–Guest Properties toward Neutral Guests.

[24] B. Schuur, B. J. V. Verkuijl, A. J. Minnaard *et al.*, *Org. Biomol. Chem.* 2011, **9**, 36–51. Chiral Separation by Enantioselective Liquid–Liquid Extraction.

[25] S. M. Han, *Biomed. Chromatogr.* 1997, **11**, 259–271. Direct Enantiomeric Separations by High Performance Liquid Chromatography using Cyclodextrins.

[26] P. Shahgaldian, U. Pieles, *Sensors* 2006, **6**, 593–615. Cyclodextrin Derivatives as Chiral Supramolecular Receptors for Enantioselective Sensing.

[27] T. J. Wenzel, *J. Incl. Phenom. Macrocycl. Chem.* 2014, **78**, 1–14. Calixarenes and Calix[4]resorcinarenes as Chiral NMR Solvating Agents.
[28] S. C. Peacock, D. J. Cram, *J. Chem. Soc., Chem. Commun.* 1976, 282–284. High Chiral Recognition in α-Amino-acid and -ester Complexation.
[29] X. Han, T. Yao, Y. Liu *et al.*, *J. Chromatogr. A* 2005, **1063**, 111–120. Separation of Chiral Furan Derivatives by Liquid Chromatography using Cyclodextrin-Based Chiral Stationary Phases.
[30] Y. Xiao, S.-C. Ng, T. T. Y. Tan, Y. Wang, *J. Chromatogr. A* 2012, **1269**, 52–68. Recent Development of Cyclodextrin Chiral Stationary Phases and their Applications in Chromatography.
[31] N. Li, R. G. Harrison, J. D. Lamb, *J. Incl. Phenom. Macrocycl. Chem.* 2014, **78**, 39–60. Application of Resorcinarene Derivatives in Chemical Separations.
[32] P. S. Bates, R. Kataky, D. Parker, *J. Chem. Soc., Chem. Commun.* 1992, 153–155. A Chiral Sensor Based on a Peroctylated α-Cyclodextrin.
[33] R. Kataky, P. S. Bates, D. Parker, *Analyst* 1992, **117**, 1313–1317. Functionalized α-Cyclodextrins as Potentiometric Chiral Sensors.
[34] B. Aswathy, G. Sony, *J. Lumin.* 2014, **154**, 541–548. Fluorescence Turn-on Recognition of Chiral Amino Acids using Dye Incorporated β-CD Functionalized AuNPs Assembly.
[35] S. E. Brown, J. H. Coates, S. F. Lincoln *et al.*, *J. Chem. Soc., Faraday Trans.* 1991, **87**, 2699–2703. Chiral Molecular Recognition: A ^{19}F Nuclear Magnetic Resonance Study of the Diastereoisomer Inclusion Complexes formed between Fluorinated Amino Acid Derivatives and α-Cyclodextrin in Aqueous Solution.
[36] N. H. Pham, T. J. Wenzel, *Chirality* 2012, **24**, 193–200. A Water-Soluble Calix[4]resorcinarene with L-Pipecolinic Acid Groups as a Chiral NMR Solvating Agent.
[37] T. J. Wenzel, R. D. Rollo, R. L. Clark, *Magn. Reson. Chem.* 2012, **50**, 261–265. Chiral Discrimination of Aliphatic Amines and Amino Alcohols using NMR Spectroscopy.
[38] R. Breslow, S. D. Dong, *Chem. Rev.* 1998, **98**, 1997–2011. Biomimetic Reactions Catalyzed by Cyclodextrins and their Derivatives.
[39] K. Takahashi, *Chem. Rev.* 1998, **98**, 2013–2033. Organic Reactions Mediated by Cyclodextrins.
[40] D. M. Homden, C. Redshaw, *Chem. Rev.* 2008, **108**, 5086–5130. The Use of Calixarenes in Metal-Based Catalysis.
[41] T. Ooi, K. Maruoka, *Angew. Chem.* 2007, **119**, 4300–4345; *Angew. Chem. Int. Ed.* 2007, **46**, 4222–4266. Recent Advances in Asymmetric Phase-Transfer Catalysis.
[42] S. Shirakawa, K. Maruoka, *Angew. Chem.* 2013, **125**, 4408–4445; *Angew. Chem. Int. Ed.* 2013, **52**, 4312–4348. Recent Developments in Asymmetric Phase-Transfer Reactions.
[43] Y. Murakami, J.-i. Kikuchi, Y. Hisaeda, O. Hayashida, *Chem. Rev.* 1996, **96**, 721–758. Artificial Enzymes.
[44] D. M. Vriezema, M. Comellas Aragonès, J. A. A. W. Elemans *et al.*, *Chem. Rev.* 2005, **105**, 1445–1489. Self-Assembled Nanoreactors.
[45] J. Bjerre, C. Rousseau, L. Marinescu, M. Bols, *Appl. Microbiol. Biotechnol.* 2008, **81**, 1–11. Artificial Enzymes, "Chemzymes": Current State and Perspectives.
[46] Z. Dong, Q. Luo, J. Liu, *Chem. Soc. Rev.* 2012, **41**, 7890–7908. Artificial Enzymes based on Supramolecular Scaffolds.
[47] F. Cramer, *Chem. Ind. (London)* 1958, 892–893. Asymmetric Catalysis by Inclusion Compounds.
[48] F. Cramer, *Chem. Ber.* 1959, **92**, 1739–1747. Stereospezifische Reaktionen mit Einschlussverbindungen.
[49] A. W. Czarnik, *J. Org. Chem.* 1984, **49**, 924–927. Cyclodextrin-Mediated Chiral Sulfoxidations.
[50] A. R. Abreu, I. Costa, C. Rosa *et al.*, *Tetrahedron* 2005, **61**, 11986–11990. New Enantioselective Method for Hydration of Alkenes using Cyclodextrins as Phase Transfer Catalyst.
[51] H.-J. Schneider, N. K. Sangwan, *Angew. Chem.* 1987, **99**, 924–925; *Angew. Chem. Int. Ed. Engl.* 1987, **26**, 896–897. Changes of Stereoselectivity in Diels–Alder Reactions by Hydrophobic Solvent Effects and by β-Cyclodextrin.
[52] A. Nakamura, Y. Inoue, *J. Am. Chem. Soc.* 2003, **125**, 966–972. Supramolecular Catalysis of the Enantiodifferentiating [4+4] Photocyclodimerization of 2-Anthracenecarboxylate by γ-Cyclodextrin.
[53] Y. Kawajiri, N. Motohashi, *J. Chem. Soc., Chem. Commun.* 1989, 1336–1337. Strong Asymmetric Induction without Covalent Bond Connection.

[54] H. Sakuraba, K. Natori, Y. Tanaka, *J. Org. Chem.* 1991, **56**, 4124–4129. Asymmetric Oxidation of Alkyl Aryl Sulfides in Crystalline Cyclodextrin Complexes.
[55] K. Harano, H. Kiyonaga, T. Hisano, *Tetrahedron Lett.* 1991, **32**, 7557–7558. [3,3]-Sigmatropic Rearrangement of Allylic Xanthates in β-Cyclodextrin Complexes.
[56] Y.-M. Li, X.-J. Yao, X. Feng *et al.*, *J. Organomet. Chem.* 1996, **509**, 221–224. Asymmetric Aldol Condensation of Acetyl Ferrocene with Aromatic Aldehydes via the Formation of the β-Cyclodextrin Inclusion Complexes.
[57] Y. Tanaka, H. Sakuraba, H. Nakanishi, *J. Chem. Soc., Chem. Commun.* 1983, 947–948. Asymmetric Halogenation and Hydrohalogenation of Olefins in Crystalline Cyclodextrin Complexes.
[58] Y. Tanaka, H. Sakuraba, H. Nakanishi, *J. Org. Chem.* 1990, **55**, 564–567. Asymmetric Halogenation and Hydrohalogenation of *trans*-2-Butenoic Acid in a Crystalline α-Cyclodextrin Complex.
[59] R. Fornasier, F. Reniero, P. Scrimin, U. Tonellato, *J. Chem. Soc., Perkin Trans.* **2** 1987, 193–196. Enantioselectivity Effects in the Hydrolytic Cleavage of Activated Substrates with α- and β-Cyclodextrins.
[60] J. H. Coates, C. J. Easton, S. J. van Eyk *et al.*, *J. Chem. Soc., Chem. Commun.* 1991, 759–760. Chiral Differentiation in the Deacylation of 6^A-*O*-{2-[4-(2-Methylpropyl)phenyl]propanoyl}-β-cyclodextrin.
[61] J. H. Coates, C. J. Easton, N. L. Fryer, S. F. Lincoln, *Chem. Lett.* 1994, **23**, 1153–1156. Complementary Diastereoselectivity in the Synthesis and Hydrolysis of Acylated Cyclodextrins.
[62] G. L. Trainor, R. Breslow, *J. Am. Chem. Soc.* 1981, **103**, 154–158. High Acylation Rates and Enantioselectivity with Cyclodextrin Complexes of Rigid Substrates.
[63] R. Breslow, G. Trainor, A. Ueno, *J. Am. Chem. Soc.* 1983, **105**, 2739–2744. Optimization of Metallocene Substrates for β-Cyclodextrin Reactions.
[64] G. Wenz, *Angew. Chem.* 1994, **106**, 851–870; *Angew. Chem. Int. Ed. Engl.* 1994, **33**, 803–822. Cyclodextrins as Building Blocks for Supramolecular Structures and Functional Units.
[65] A. R. Khan, P. Forgo, K. J. Stine, V. T. D'Souza, *Chem. Rev.* 1998, **98**, 1977–1996. Methods for Selective Modifications of Cyclodextrins.
[66] T. Kida, T. Iwamoto, H. Asahara *et al.*, *J. Am. Chem. Soc.* 2013, **135**, 3371–3374. Chiral Recognition and Kinetic Resolution of Aromatic Amines via Supramolecular Chiral Nanocapsules in Nonpolar Solvents.
[67] C. Yang, A. Nakamura, T. Wada, Y. Inoue, *Org. Lett.* 2006, **8**, 3005–3008. Enantiodifferentiating Photocyclodimerization of 2-Anthracenecarboxylic Acid Mediated by γ-Cyclodextrins with a Flexible or Rigid Cap.
[68] C. Yang, T. Mori, Y. Inoue, *J. Org. Chem.* 2008, **73**, 5786–5794. Supramolecular Enantiodifferentiating Photocyclodimerization of 2-Anthracenecarboxylate Mediated by Capped γ-Cyclodextrins: Critical Control of Enantioselectivity by Cap Rigidity.
[69] R. Lu, C. Yang, Y. Cao *et al.*, *Chem. Commun.* 2008, 374–376. Supramolecular Enantiodifferentiating Photoisomerization of Cyclooctene with Modified β-Cyclodextrins: Critical Control by a Host Structure.
[70] R. Lu, C. Yang, Y. Cao *et al.*, *J. Org. Chem.* 2008, **73**, 7695–7701. Enantiodifferentiating Photoisomerization of Cyclooctene Included and Sensitized by Aroyl-β-cyclodextrins: A Critical Enantioselectivity Control by Substituents.
[71] K. Kanagaraj, P. Suresh, K. Pitchumani, *Org. Lett.* 2010, **12**, 4070–4073. Per-6-amino-β-cyclodextrin as a Reusable Promoter and Chiral Host for Enantioselective Henry Reaction.
[72] P. Suresh, K. Pitchumani, *Tetrahedron-Asymmetr.* 2008, **19**, 2037–2044. Per-6-amino-β-cyclodextrin Catalyzed Asymmetric Michael Addition of Nitromethane and Thiols to Chalcones in Water.
[73] S. Tamagaki, J. Narikawa, A. Katayama, *Bull. Chem. Soc. Jpn.* 1996, **69**, 2265–2268. Asymmetric Transformation of Phenylglyoxal into Mandelic Acid Catalyzed by Cyclodextrin-Based Glyoxalase Models.
[74] I. Tabushi, Y. Kuroda, M. Yamada *et al.*, *J. Am. Chem. Soc.* 1985, **107**, 5545–5546. A-(Modified B_6)-B-[ω-amino(ethylamino)]-β-cyclodextrin as an Artificial B_6 Enzyme for Chiral Aminotransfer Reaction.

[75] V. Böhmer, D. Kraft, M. Tabatabai, *J. Inclus. Phenom. Mol.* 1994, **19**, 17–39. Inherently Chiral Calixarenes.

[76] A. Dalla Cort, L. Mandolini, C. Pasquini, L. Schiaffino, *New. J. Chem.* 2004, **28**, 1198–1199. 'Inherent Chirality' and Curvature.

[77] M. J. McIldowie, M. Mocerino, M. I. Ogden, *Supramol. Chem.* 2010, **22**, 13–39. A Brief Review of C_n-Symmetric Calixarenes and Resorcinarenes.

[78] S. Shirakawa, T. Kimura, S. Murata, S. Shimizu, *J. Org. Chem.* 2009, **74**, 1288–1296. Synthesis and Resolution of a Multifunctional Inherently Chiral Calix[4]arene with an ABCD Substitution Pattern at the Wide Rim: The Effect of a Multifunctional Structure in the Organocatalyst on Enantioselectivity in Asymmetric Reactions.

[79] Z.-Y. Li, J.-W. Chen, L. Wang, Y. Pan, *Synlett* 2009, 2356–2360. Highly Enantioselective Direct Aldol Reactions Catalyzed by Proline Derivatives Based on a Calix[4]arene Scaffold in the Presence of Water.

[80] B. R. Buckley, J. Y. Boxhall, P. C. B. Page *et al.*, *Eur. J. Org. Chem.* 2006, **2006**, 5117–5134. Mannich and *O*-Alkylation Reactions of Tetraalkoxyresorcin[4]arenes – The Use of Some Products in Ligand-Assisted Reactions.

[81] S. Hu, J. Li, J. Xiang *et al.*, *J. Am. Chem. Soc.* 2010, **132**, 7216–7228. Asymmetric Supramolecular Primary Amine Catalysis in Aqueous Buffer: Connections of Selective Recognition and Asymmetric Catalysis.

[82] Z. Rapi, P. Bakó, G. Keglevich *et al.*, *Tetrahedron-Asymmetr.* 2012, **23**, 489–496. Asymmetric Phase Transfer Darzens Reactions Catalyzed by D-glucose- and D-Mannose-Based Chiral Crown Ethers.

[83] A. Makó, Á. Szöllősy, G. Keglevich *et al.*, *Monatsh. Chem.* 2008, **139**, 525–535. Synthesis of Methyl-α-D-Glucopyranoside-Based Azacrown Ethers and their Application in Enantioselective Reactions.

[84] P. Sóti, L. Telkes, Z. Rapi *et al.*, *J. Inorg. Organomet. Polym.* 2014, **24**, 713–721. Synthesis of an Aza Chiral Crown Ether Grafted to Nanofibrous Silica Support and Application in Asymmetric Michael Addition.

[85] M. Jouffroy, R. Gramage-Doria, D. Armspach *et al.*, *Angew. Chem.* 2014, **126**, 4018–4021; *Angew. Chem. Int. Ed.* 2014, **53**, 3937–3940. Confining Phosphanes Derived from Cyclodextrins for Efficient Regio- and Enantioselective Hydroformylation.

[86] M. Jouffroy, R. Gramage-Doria, D. Sémeril *et al.*, *Beilstein J. Org. Chem.* 2014, **10**, 2388–2405. Phosphinocyclodextrins as Confining Units for Catalytic Metal Centres. Applications to Carbon–Carbon Bond Forming Reactions.

[87] M. Guitet, P. Zhang, F. Marcelo *et al.*, *Angew. Chem.* 2013, **125**, 7354–7359; *Angew. Chem. Int. Ed.* 2013, **52**, 7213–7218. NHC-Capped Cyclodextrins (ICyDs): Insulated Metal Complexes, Commutable Multicoordination Sphere, and Cavity-Dependent Catalysis.

[88] A. Schlatter, M. K. Kundu, W.-D. Woggon, *Angew. Chem.* 2004, **116**, 6899–6902; *Angew. Chem. Int. Ed.* 2004, **43**, 6731–6734. Enantioselective Reduction of Aromatic and Aliphatic Ketones Catalyzed by Ruthenium Complexes Attached to β-Cyclodextrin.

[89] A. Schlatter, W.-D. Woggon, *Adv. Synth. Catal.* 2008, **350**, 995–1000. Enantioselective Transfer Hydrogenation of Aliphatic Ketones Catalyzed by Ruthenium Complexes Linked to the Secondary Face of β-Cyclodextrin.

[90] C. Ke, C. Yang, T. Mori *et al.*, *Angew. Chem.* 2009, **121**, 6803–6805; *Angew. Chem. Int. Ed.* 2009, **48**, 6675–6677. Catalytic Enantiodifferentiating Photocyclodimerization of 2-Anthracenecarboxylic Acid Mediated by a Non-Sensitizing Chiral Metallosupramolecular Host.

[91] C. Bonaccorso, G. Brancatelli, F. P. Ballistreri *et al.*, *Dalton Trans.* 2014, **43**, 2183–2193. Novel Chiral (salen)Mn(III) Complexes Containing a Calix[4]arene Unit in 1,3-alternate Conformation as Catalysts for Enantioselective Epoxidation Reactions of (Z)-Aryl Alkenes.

12

Chirality Related to Biocatalysis and Enzymes in Organic Synthesis

Declan P. Gavin and Anita R. Maguire

12.1 Introduction

The production of chiral compounds as single enantiomers in the synthesis of drugs and intermediates is extremely important to the pharmaceutical industry. In 1992 the US Food and Drug Administration (FDA) issued a policy on stereoisomeric drugs encouraging the commercialization of clinical drugs as single enantiomers [1]; thus chirality is now a crucial aspect of the design and discovery of new drugs. As a result, methods to produce these chiral compounds in an enantioselective manner have become proportionally important.

Due to the stereoselective interactions of a chiral drug with optically active biological macromolecules, two stereoisomeric compounds tend to differ in their pharmacokinetic/pharmacodynamic properties [2]. Indeed, enantiomeric discrimination displayed by metabolizing enzymes often results in a preference for one enantiomer of a chiral drug [2, 3]. Therefore, one stereoisomer may be responsible for the desired activity of the drug, but its paired enantiomer could have a completely different activity, be an antagonist of the active compound, or may even be inactive [2]. Thus the production of optically active drugs as single enantiomers can be economically desirable.

Amongst the strategies available to the synthetic organic chemist for controlling the stereochemical outcome of a reaction, catalysis has become the option of choice since the mid-1980s. In this context the enormous potential of microorganisms and enzymes in the field of asymmetric synthesis has been recognized and the use of whole cells and isolated enzymes is an attractive option for the chemical industry [4, 5].

Chirality in Supramolecular Assemblies: Causes and Consequences, First Edition. Edited by F. Richard Keene.

12.2 Biocatalysis

Biocatalysis involves the use of enzymes or whole cells that contain the desired enzyme or enzyme system as catalysts for chemical reactions. Biocatalysis is a broad and growing area of asymmetric synthesis and it is now considered an extremely useful tool for the organic chemist [5–9]. In such a dynamic and burgeoning area, it is not possible to include every aspect of the field and it is not the objective of this chapter to cover in detail the intricacies of every reaction type; the objective is to introduce the concept of biocatalysis and acquaint the reader with some of the many elegant transformations that biocatalysts can perform. Several excellent, general reviews and books exist on the topic for those interested in further reading [5–12].

It is important to highlight from the outset that enzymes can be highly specific for a single substrate (or a small number of structurally similar substrates) – enzymes involved in anabolic pathways tend to fit this description. However, some enzymes do display impressive substrate 'promiscuity'. When applied as biocatalysts these enzymes can have a broad substrate scope coupled with high stereoselectivity and they have been extensively used for this purpose [13]. This is particularly true of enzymes that are involved in metabolic or detoxification pathways. In contrast, the use of enzymes which display narrow substrate scope on non-natural substrates can be challenging, although there are now several options available to the biochemist to overcome this obstacle.

12.2.1 Historical Context

Historically the use of enzymes as catalysts in synthetic organic chemistry suffered from two major drawbacks. Firstly, access to sufficient quantities of many enzymes for practical applications was not possible as it was necessary to extract them from natural sources. However, molecular cloning and heterologous protein expression strategies developed in the 1970s, whereby a given enzyme occurring in one organism can be overexpressed in another host organism such as *E. coli* [14] facilitated production of large quantities of recombinant enzymes. Secondly, the exquisite selectivity displayed by biocatalysts was somewhat of a double-edged sword as natural enzymes could suffer from narrow substrate scope and/or poor stability in a synthetic chemistry context. This can sometimes be circumvented through clever engineering of reaction conditions but the advent of directed evolution in the 1990s – described in section 12.5 – enabled the use of tailored biocatalysts through random, rational and semi-rational protein engineering. Furthermore, the introduction of immobilized enzymes allowed for the development of biocatalysts with improved stability and reusability. Also noteworthy is the increasing trend for the mining of enzymes from extremophiles [15–17] that have evolved to perform under ostensibly harsh reaction conditions. These enzymes can also provide new templates for further evolution of the biocatalyst, starting from a point in which the enzyme is already relatively robust.

12.2.2 Importance of Biocatalysis

Although it finds plentiful application in the laundry, agriculture and food industries, arguably the most important role for biocatalysis is in the pharmaceutical sector, where the wonderful stereoselectivity of enzymes can be exploited [18]. Reviews documenting industrial applications are plentiful [19–31]. Indeed, biocatalysis is considered to be the principle green chemistry technology adopted by the fine chemicals and pharmaceutical industries for the manufacture of chemicals [32, 33]. Nowadays, many enzymatic transformations are possible and meet the standards of established asymmetric reactions. Allied to their environmental advantages, the impressive regioselectivity and stereoselectivity of enzymes can enable difficult syntheses (often requiring more circuitous routes through functional group-protection and/or use of chiral auxiliaries) to be circumvented [34]. Isolated enzymes or whole cells can also allow access to a diverse range of enantioselective transformations which are often difficult to achieve via chemical catalysis. A broad range of biocatalytic transformations can now be applied to synthetic chemistry and this has brought to the fore the concept of 'biocatalytic retrosynthesis', whereby chemists incorporate biocatalytic methodologies into synthetic routes [35, 36]. In this way biocatalysis could enable the development of new synthetic routes to high-value, enantiomerically pure molecules.

12.2.3 Biocatalytic Methodologies

From a synthetic perspective, many biocatalytic methodologies are now accessible to the organic chemist. Turner and O'Reilly have described some of the commercially available biocatalysts and the synthetic transformations they catalyze (Scheme 12.1) [36].

12.2.4 Enzyme Classes

In the earlier literature, the naming of enzymes was extremely complex as they were named by individual workers. This led to a lack of clarity in a period when many new enzymes were being discovered, with the same enzyme known by several different names, or, conversely, the same name was given to different enzymes. In the 1950s, Malcolm Dixon and Edwin Webb introduced some order to enzyme nomenclature. They had previously noticed that, despite the many enzymes that had been discovered, the number of reactions that could be catalyzed was limited. The International Commission on Enzymes was established in 1956 by the International Union of Biochemistry in order to resolve the issue of nomenclature [37]. As a result, the EC classification system was established, whereby every enzyme is given four numbers after the abbreviation EC. The first indicates the reaction type (of which there are six). The second number describes the structural changes which occur during the enzyme catalysis (e.g. the type of substrate which is acted upon); the third number outlines particular enzyme characteristics involved in the catalytic reaction (e.g. the nature of the cofactor), while the fourth is a running number identifying the specific enzyme. For example, an oxidoreductase enzyme such as an alcohol dehydrogenase, which oxidizes a substrate by reducing $NAD(P)^+$, is classified as EC 1.x.1.- [38]. Here, 'x' refers to the group oxidized (1 for alcohol, 2 for ketone or aldehyde, etc.). If an acyl group is transferred the (transferase) enzyme is classified as EC 2.3.y -, where 'y' specifies the type of acceptor, and so forth. This system has been adopted by the International Union of Biochemistry and Molecular Biology (IUBMB) in conjunction with the International Union of Pure and Applied Chemistry (IUPAC) into the *Enzyme Nomenclature* list of enzymes [39]. The IUBMB enzyme data are publicly available on the Web [37]. A summary of the enzyme classes is provided below (Table 12.1). The history of enzyme nomenclature and classification has been addressed previously in the literature [39, 40] and further useful information on enzyme naming and classification is available on the ExplorEnz web site [41].

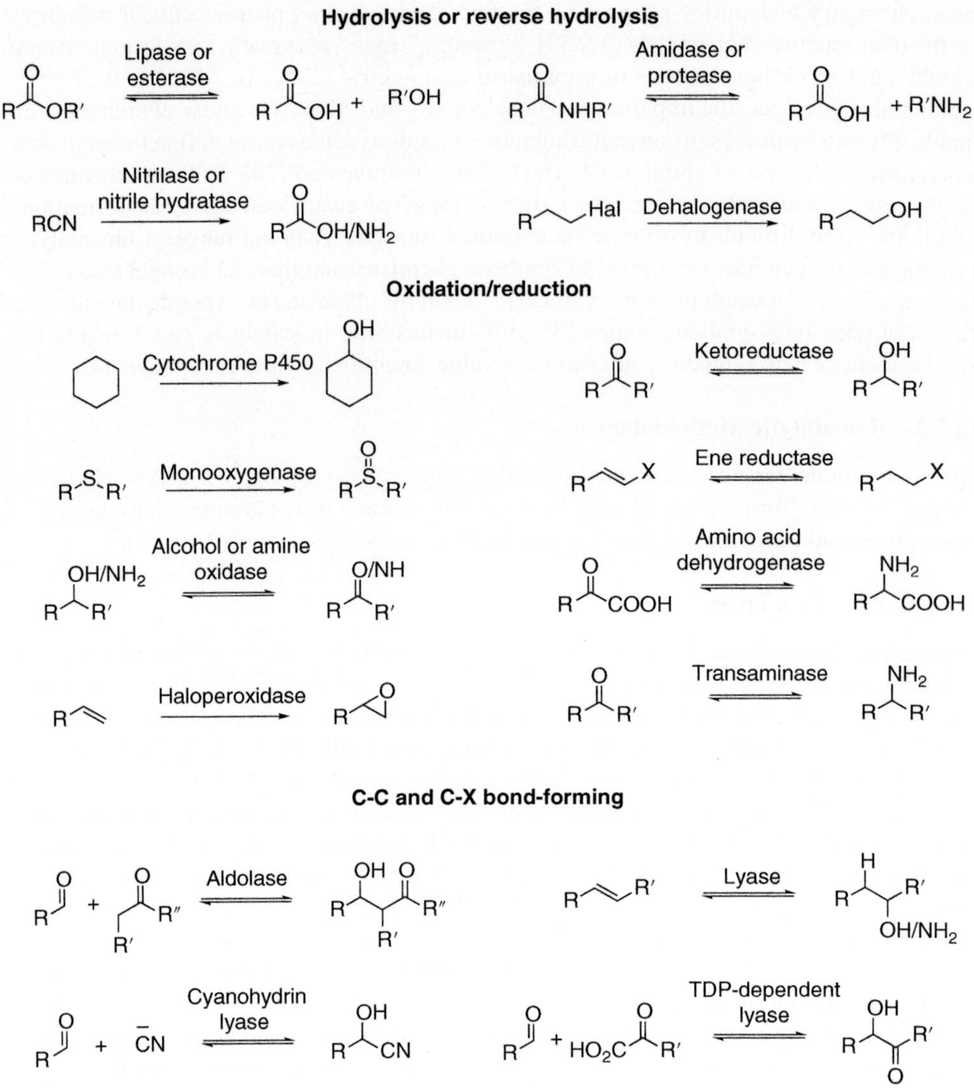

Scheme 12.1 *Examples of commercially available biocatalysts and the chemical transformations they perform [36]*

12.2.5 Advantages and Disadvantages of Biocatalysis

Biocatalysis satisfies many of the 12 principles of green chemistry [42, 43] and the reactions are typically run at moderate or low temperatures. Allied to their high (stereo)selectivity, enzyme-catalyzed reactions usually proceed with characteristically high turnover numbers, with rate accelerations of 10^8–10^{10} when compared to the corresponding non-catalyzed reactions. These turnover rates are generally far superior to those associated with chemical

Table 12.1 *Summary of enzyme classes (adapted from reference [10])*

Enzyme	Examples	Reaction catalyzed	EC class
Oxidoreductases	Dehydrogenase, oxidase, oxygenase, peroxidase	Oxidation or reduction	1
Transferases	Transaminase, glycosyltransferase, transaldolase	Transfer of a group from one molecule to another	2
Hydrolases	Lipase, protease, esterase, nitrilase, nitrile hydratase, glycosidase, phosphatase	Hydrolysis reaction in water	3
Lyases	Decarboxylase, dehydratase, deoxyribose-phosphate aldolase	Non-hydrolytic bond cleavage	4
Isomerases	Racemase, mutase	Intramolecular rearrangement	5
Ligases	DNA ligase	Bond formation requiring triphosphate	6

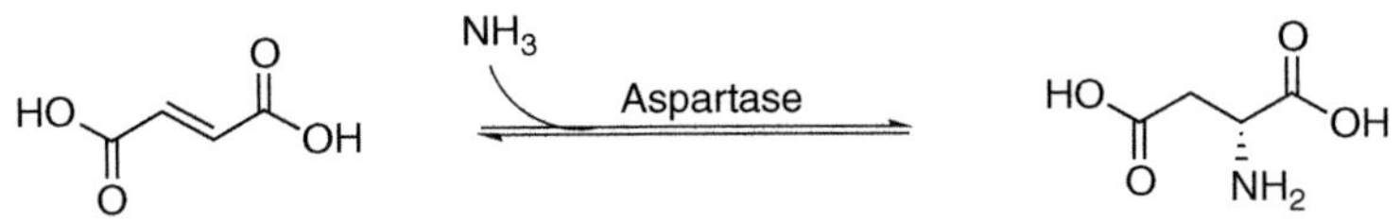

Scheme 12.2 *Aspartame produced on a commercial scale using aspartase as the biocatalyst*

catalysts [44–46]. Thus many enzymatic reactions can be performed with a mole percentage of 10^{-3} – 10^{-4}% of catalyst [8]. An example of the massive catalytic power of biocatalysts is shown in Scheme 12.2. In the commercial production of L-aspartic acid, a synthetic intermediate in the production of the artificial sweetener Aspartame®, immobilized aspartase (aspartate ammonia lyase) catalyzes the stereospecific transfer of ammonia to the unsaturated diacid. In this reaction 1 kg of immobilized biocatalyst produces more than 100 000 kg of product [9, 47].

The principal disadvantages associated with biocatalysis are that enzymes are sensitive and generally are not compatible with extreme reaction conditions such as excessive heat or low/high pH. They can also have a limited substrate scope and usually function well only at physiological pH, in dilute solutions of the substrate. Some enzymes are inactive or unstable in organic solvents and this can be problematic from the point of view of solubility of the substrate(s) and product(s). Further process complications can arise when using water as a solvent, such as the need for an extractive reaction work up to isolate the product from the enzyme catalyst, and so forth. However, since the early 1990s many of these disadvantages have been overcome.

Both substrate and product inhibition of the biocatalyst can be problematic in many biotransformations and enzyme engineering can often improve this. Alternatively, engineering of the reaction conditions (e.g. *in situ* removal of (by)product) can be employed to drive the reaction to completion [48].

Table 12.2 *Associated advantages and disadvantages of biocatalysis [12]*

Advantages	Disadvantages
High regio- and stereo-selectivity	Limited substrate scope
Non-toxic	Poor catalyst stability
Can be recycled	Limited enzyme availability
Biodegradable waste	Can require cosubstrates and/or cofactors
Large rate enhancements	Inactivation may occur: • at high temperatures • at extreme pH • in organic solvents • through substrate/product inhibition

12.2.6 Whole Cells/Isolated Enzymes

Biotransformations can be conducted using either whole cells (containing the enzyme/enzyme system of interest) or the isolated enzyme. The use of whole cells is attractive as no costly purification or cofactor recycling is necessary. Microorganisms from natural environments can possess more than 4000 genes and this can result in unwanted competing enzyme activities on the substrate. This can (usually) be avoided by using recombinant cells in which the required protein is (over) expressed, as the protein of interest for the reaction is the dominant fraction in the cell's proteome [8]. Thus, side reactions are minimized. Generally speaking, isolated enzymes often display superior activity to their whole cell counterparts and better productivity can be obtained due to higher substrate concentration tolerance. The excess biomass and the associated complications in the workup when using whole cells can also be a drawback. This has resulted in a general preference for the use of purified, isolated enzymes in organic synthesis, particularly in large-scale syntheses [8, 49].

12.3 Biocatalytic Methodologies: Kinetic/Dynamic Kinetic Resolution and Asymmetric Transformations/Chemoselective Desymmetrizations

Biocatalysts can be efficient promoters of many achiral transformations, finding widespread use in the fine chemicals and materials industry [50–52]. However, it is in the resolution or synthesis of chiral molecules that their remarkable power is truly evident. Exquisite stereo/chemo/regio-selectivity can be displayed by enzymes.

Although much progress has been made in asymmetric synthesis in recent decades in both transition metal- and latterly organo-catalyzed reactions [53, 54], the resolution of a racemic mixture continues to be one of the major weapons in the artillery of the organic chemist for the production of single enantiomers [55, 56]. From a biocatalytic perspective, the resolution of enantiomers can be performed via a kinetic resolution or a dynamic kinetic resolution. The fundamental principle for any resolution is that the reaction rate of one enantiomer must be faster than that of the other, i.e. $k_{\text{Fast}} > k_{\text{Slow}}$ [57, 58]. The resolution of racemates through enzymatic catalysis is one of the principal tools used by organic chemists for the production of chiral building blocks and enantiopure biologically active compounds [8, 20, 58, 59]. Biocatalysts can also perform asymmetric transformations – the production of a stereogenic centre from a transformation involving a prochiral substrate.

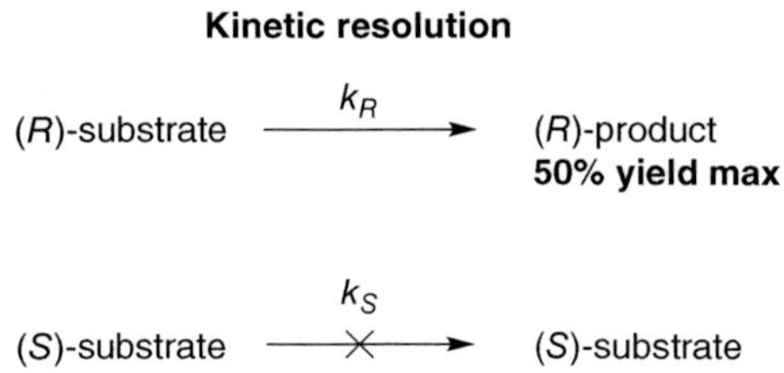

Figure 12.1 *Kinetic resolution of a racemic mixture*

12.3.1 Kinetic Resolution

In a kinetic resolution, one of the enantiomers in a racemic mixture is more readily transformed into product than its mirror image [10]. In such a process, the prerequisite is that the rate of reaction of the (*R*)-enantiomer is not equal to the rate of reaction of the (*S*)-enantiomer (Figure 12.1).

In an enzymatic kinetic resolution, chiral discrimination by the biocatalyst occurs and only one enantiomer in the racemic mixture is converted to product. The principal drawback of a kinetic resolution is that a maximum yield of 50% can be achieved and therefore it is usually accompanied by additional processing such as separation and recycling of unwanted enantiomers. Despite this, kinetic resolutions remain a reliable method for the production of enantiopure compounds and an impressive body of work on enzymatic methodologies has been published and this has been reviewed in the literature [60–62].

To conveniently compare kinetic resolutions, a series of equations were developed by Sih *et al.* to calculate the enantioselectivity of an enzyme in a given process [63]. This allows synthetic chemists to make highly useful predictions on the efficacy of the reaction and the performance of the enzyme since the enantiomeric purity of the substrate and product are dependent on the extent of conversion in the resolution.

The enantiomeric ratio, E, is a measure of the ability of an enzyme to distinguish between enantiomers. In practical terms, the higher the E-value, the more selective the bioresolution, resulting in a higher yield of product with better enantiomeric excess (*ee*). A non-selective reaction has an E-value of 1, while resolutions with an E-value below 20 warrant further development. A resolution that gives an E value of 200 or above requires little optimization.

To calculate E, three variables are measured; enantiomeric purity of the starting material (ee_s), enantiomeric purity of the product (ee_p), and extent of conversion (c) and one of the three equations in Figure 12.2 are used [64].

Kroutil *et al.* have developed a program that allows the enantiomeric ratio to be calculated using these equations [65, 66]. Although they include assumptions such as an irreversible reaction, one substrate and product, and no product inhibition, they have been widely used as a measure of enzyme performance in biocatalytic kinetic resolutions [64].

12.3.2 Dynamic Kinetic Resolution

For a dynamic kinetic resolution (DKR) to occur, a kinetic resolution must be accompanied by racemization of the substrate. In a dynamic kinetic resolution, there is an increase in

$$E = \frac{\ln[1-c(1+ee_p)]}{\ln[1-c(1-ee_p)]} \qquad E = \frac{\ln[(1-c)(1-ee_s)]}{\ln[(1-c)(1+ee_s)]} \qquad E = \frac{\ln\left[\dfrac{1-ee_s}{1+\dfrac{ee_s}{ee_p}}\right]}{\ln\left[\dfrac{1+ee_s}{1+\dfrac{ee_s}{ee_p}}\right]}$$

Figure 12.2 *Equations used in the calculation of E [64]*

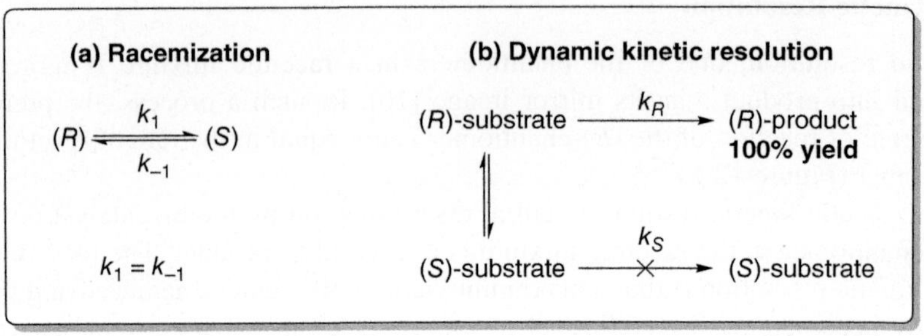

Figure 12.3 *Dynamic kinetic resolution allows a theoretical yield of 100%*

entropy in the presence of both enantiomers and this is the driving force for the process [67]. Dynamic kinetic resolution is a more desirable methodology than a simple kinetic resolution as, theoretically, a 100% yield is achievable (Figure 12.3). This has made DKR an attractive process and indeed chemoenzymatic methods of DKR have been thoroughly reviewed in book and journal form [10, 57, 61, 68, 69].

Due to the requirement for racemization of the substrate, a number of elements must coalesce for a successful DKR process. Firstly, the kinetic resolution step must be irreversible, with an E-value of at least 30. The rate of racemization must also be greater than the rate of reaction of the slow-reacting enantiomer. From a biocatalytic perspective, racemization can be induced through both metal and non-metal catalysis, although it is more common to employ a transition metal catalyst to perform the racemization step [70].

12.3.3 Asymmetric Transformations

In an asymmetric transformation, a new stereogenic centre is formed from a prochiral substrate. Thus a theoretical yield of 100% is possible. Depending on the particular transformation, one or two new stereogenic centres can be introduced. An alcohol dehydrogenase-mediated reduction of a ketone exemplifies this type of reaction, where a ketone is (enantiospecifically) reduced to a chiral alcohol (Scheme 12.3).

12.3.4 Chemoselective Desymmetrizations

In a chemoselective desymmetrization, a prochiral/meso substrate is transformed to a chiral product through the selective reaction of one functional group, ultimately eliminating the symmetrical element of the substrate which precludes chirality [71]. An example is the

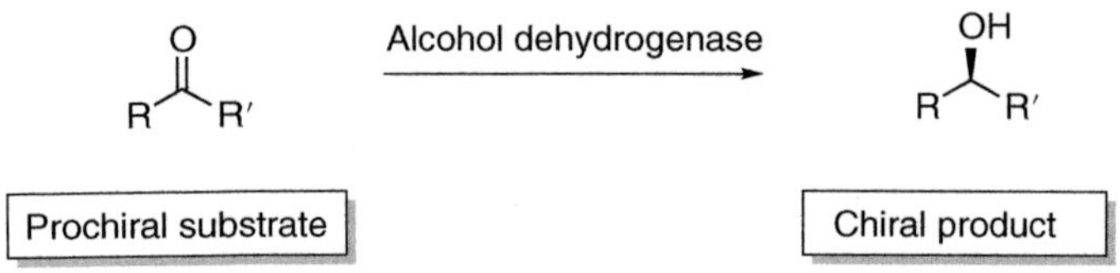

Scheme 12.3 A biocatalytic asymmetric reduction of a prochiral ketone to a chiral alcohol

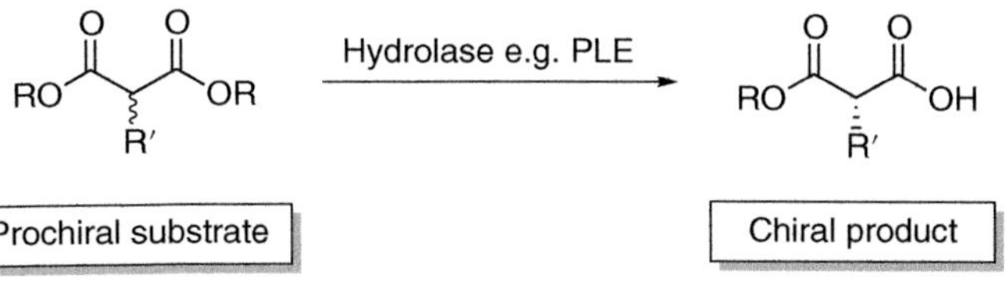

Scheme 12.4 PLE-mediated chemoselective desymmetrization

selective hydrolysis of one of two enantiotropic ester groups by a hydrolase such as pig-liver esterase (PLE) to give a chiral acid product (Scheme 12.4).

12.4 Optimization of Biocatalyst Performance

Evidently enzymes have evolved over millions of years to act in a physiological environment, on a narrow range of natural substrates at low concentrations. These parameters often do not fit the required conditions from a synthetic chemistry point of view, as many transformations need to be conducted at elevated temperature, in organic solvents, and so forth. In theory, water is an excellent medium in which to run an organic transformation; it is non-toxic, non-flammable, cheap and widely available. Practically, however, many substrates are poorly soluble and product extraction can be difficult due to the high boiling point/heat of vaporization. Furthermore, unwanted side reactions such as hydrolysis, polymerization and racemization can occur. In some instances these issues can be overcome by using alternative reaction media such as organic solvents, supercritical fluids and ionic liquids [72]. Other options for improving biocatalyst stability/solvent tolerance include immobilization and protein engineering (discussed in section 12.5) [73].

12.4.1 Organic Solvents

Historically, one of the commonly held prejudices against enzymes was that they are not stable in non-aqueous media. Certainly, in aqueous-organic mixtures proteins can denature. However, in anhydrous solvents they are less likely to do so [74] and various crystalline enzymes can retain their native structures in non-aqueous media [75]. Countless studies over the last few decades have shown that many enzymes are stable and active in organic solvent (in some cases the presence of an organic solvent will increase enzyme activity/selectivity) and these have been reviewed [76–78].

12.4.2 Immobilization

Immobilization involves attachment or dispersion of an enzyme or cell to an insoluble support material, creating a heterogeneous biocatalyst. The main advantages associated with immobilization are ease of handling/recycling, ease of product recovery and increased temperature and pH stability of the biocatalyst [79]. Biocatalysis in organic media can cause the formation of enzyme aggregates, which leads to decreased catalyst performance due to inaccessibility of the substrate. In some cases, where enzyme aggregates are problematic, the activity of the enzyme can be significantly improved through immobilization [80–83]. Due to the associated advantages, enzyme immobilization has found application in several areas, including the pharmaceutical, food and textile industries [81, 84].

The most frequently employed immobilization methods are adsorption/deposition, covalent attachment, cross-linking and entrapment [85]. Despite many recent advancements in immobilization technologies, a broadly applicable methodology has not yet been discovered, with the challenge of achieving high catalytic activity in conjunction with the advantages above very much being a case-by-case undertaking. Immobilization techniques for biocatalysis have received much attention and more detailed descriptions are available in the literature [86–88].

12.4.3 Ionic Liquids

Ionic liquids are salts that do not crystallize at room temperature [89]. They are non-volatile and non-flammable. The use of ionic liquids for organic synthesis has been associated with several other environmental and safety advantages, such as their excellent chemical and thermal stability. Furthermore, they have been described as environmentally benign [90]. As the combination of ions is variable, properties such as viscosity, density, hydrophobicity and solubility are tunable. The employment of ionic liquids as solvents for biocatalysis has received a lot of attention in the literature and this work has been reviewed in great detail [91–94]. Drawbacks of ionic liquids include their relative expense and, in some cases, their toxicity [95–97].

12.5 Protein Engineering

Due to advances in molecular biology and more specifically recombinant DNA technology, many naturally occurring enzymes have been discovered and applied as biocatalysts in asymmetric synthesis. These natural enzymes (known as the wild type) are not optimized for practical conditions on non-natural substrates and the likelihood is that a given enzyme will not be optimal for a specific chemical process. Although in many cases careful engineering of the reaction conditions can improve efficiency/enantioselectivity to a point where wild-type enzymatic resolutions are synthetically very useful [98, 99], it is not always possible to overcome the constraints of natural enzymes satisfactorily with this approach. From a biocatalytic perspective, protein engineering is the premier tool in circumventing the limitations associated with using enzymes as catalysts. The use of tailor-made enzymes has had a profound effect on the field of biocatalysis. The creation of mutant libraries of enzymes allows the screening of several thousand non-natural enzymes against a given target substrate [100]. This, in tandem with high throughput screening methodologies [101], allows for the rapid development of an enzyme (or several enzymes in a single system) which is designed for a specific chemical reaction. Not only can protein engineering improve the activity/selectivity of an enzyme for a certain substrate – it can also positively affect other parameters which can be critical for a successful chemical process [102], such as (thermo) stability and/or tolerance of organic solvents [78, 79, 103, 104].

Indeed, protein engineering can be used to improve several enzyme characteristics in tandem to make the biocatalyst more compatible with the reaction conditions. An excellent example of this is shown in Scheme 12.5, whereby researchers at Merck and Codexis synthesized the antidiabetes drug sitagliptin from the diketone intermediate shown via a transaminase mutant [105]. ω-Transaminases are discussed in more detail in section 12.7.4. The example below is particularly noteworthy as it illustrates the full and breathtaking power of enzyme engineering – an enzyme that displayed poor activity towards a non-natural substrate was created by rational design and subjected to 11 rounds of evolution, with the final variant, carrying 27 mutations, delivering the ketoamine product in 92% yield and 99% *ee*. Moreover, this mutant significantly outperformed its predecessor(s) in terms of the other parameters which were essential to the overall process, such as substrate concentration and tolerance of organic solvent. The biotransformation replaced an earlier synthesis which used a rhodium catalyst for the enantioselective reduction of an intermediate enamine. A recrystallization step to improve the enantiopurity of the final product and additional measures for removal of the transition metal were eliminated from the process through the use of the biocatalytic method.

The improvements in yield, efficiency and waste generation in this process resulted in the awarding of 2010 Presidential Green Chemistry Award to the researchers at Merck and Codexis.

In many ways, protein engineering has brought about an explosion of interest in the use of biocatalysis as a synthetic tool – greater knowledge of enzyme structures and catalytic mechanisms, coupled with the ability to express and test libraries of mutant enzymes in a relatively short time period, has allowed the creation of enzymes with increased activity, selectivity and stability for use in asymmetric synthesis. The two approaches for the use of protein engineering as a biochemical tool are directed evolution and rational design.

	1st round selection	11th round selection
Substrate concentrations		
$^{i}PrNH_2$	0.5M	1M
Prositagliptin ketone	2 g/L	50 g/L
Reaction conditions		
DMSO	5%	50%
Temperature	22 °C	45 °C
pH	7.5	8.5

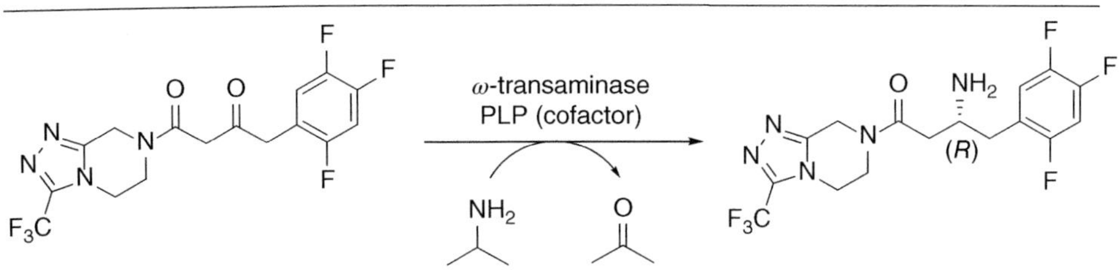

Scheme 12.5 *Use of an engineered ω-transaminase for the synthesis of sitagliptin. Here protein engineering enabled access to an enzyme which delivered an important pharmaceutical intermediate in 92% yield and >99% ee [105]*

12.5.1 Directed Evolution and Semi-Rational Design

Directed evolution mimics the process that has bestowed the complex biological structures and function of natural enzymes – that of natural evolution. Several excellent books and reviews have been published on the use of directed evolution for biocatalysis [106–114]. In this technique (Figure 12.4), an enzyme (known as the parent protein) is chosen based upon its suitability for a desired transformation and/or substrate. The parent will typically not be optimized for the transformation. The parent gene is mutated to produce a library of mutant genes through a series of random point mutations. Each gene encodes a variant protein. The genes are then transformed to a suitable host and following protein expression, these mutants are then tested against the substrate. This process can be repeated several times using a parent from the previous round until the enzyme with the desired properties is achieved or no further improvement is possible. Several techniques can be used to create mutant libraries, of which the error-prone polymerase chain reaction (PCR) is the most common [115]. Other techniques include site-saturated mutagenesis, chemical mutagenesis, mutator strains and DNA shuffling [116–119]. The major milestones in directed evolution technologies have been well documented [107, 111] and the various methods for mutant library creation have also been reviewed [120]. More recently, approaches that require minimal researcher intervention, such as phage-assisted continuous evolution (PACE) [121], have been developed to complement the above methods [122]. Due to the creation of (up to) several thousand mutant proteins, the use of a high-throughput screening method is desirable in tandem with these techniques [123–126].

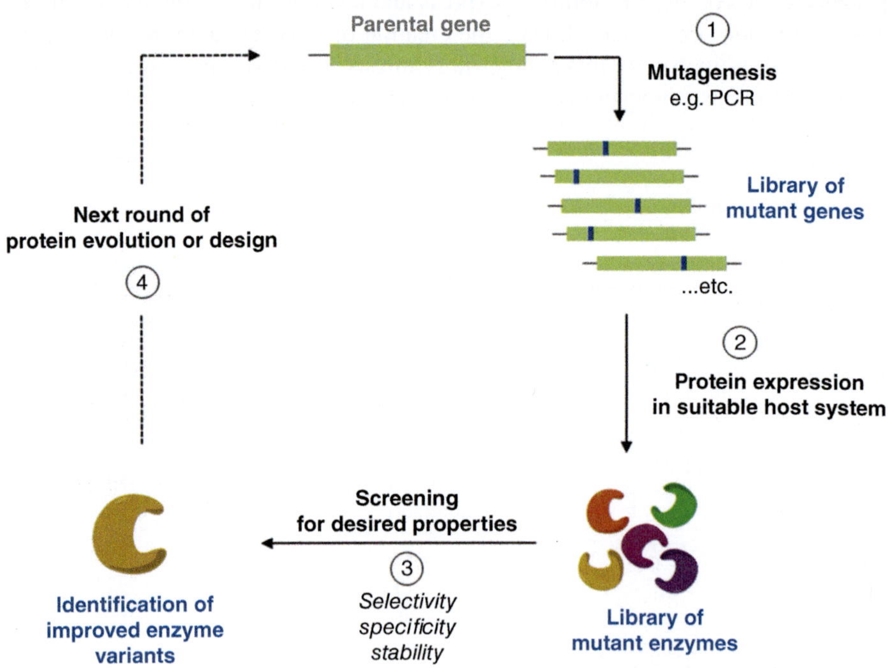

Figure 12.4 General scheme for directed evolution

Due to the so-called 'numbers problem' [127], directed evolution can require careful experimental design. For example for a protein composed of 350 amino acids 400 possible sequences exist [20]. A single mutation of an amino acid would lead to 7600 variants and a double mutation would lead to 144 000 variants [12]. It is therefore desirable to control the number of possible mutations for a given enzyme and this has given rise to semirational design of proteins (focused directed evolution), whereby the active site of the biocatalyst is identified. The position of the mutations is then chosen based on this and is saturated with all possible mutations resulting in the creation of smaller, focused libraries. For this approach, a fundamental understanding of both protein structure and mechanistic aspects of the biotransformation is necessary [128, 129]. Site-saturation methods such as ISM or CASTing are commonly used in a semirational approach [119]. When no high throughput screening assay is available a semirational method is particularly advantageous.

12.5.2 Rational Design

Rational design of biocatalysts differs from directed evolution in that the mutants are designed based on their protein structure [130]. Specific amino acid residues suitable for mutation are identified and, importantly, the mutations are carried out based on the predicted improvements for the particular enzyme. The mutants are prepared by site-directed mutagenesis. In contrast to semirational engineering, whereby only the position of the mutation is controlled, both the position **and** nature of the mutation is designed [109]. As with directed evolution, they are then transformed into the host expressing organism (e.g. *E. coli*), and the variants are expressed and analyzed for the desired traits. An overview of rational design is shown in Figure 12.5.

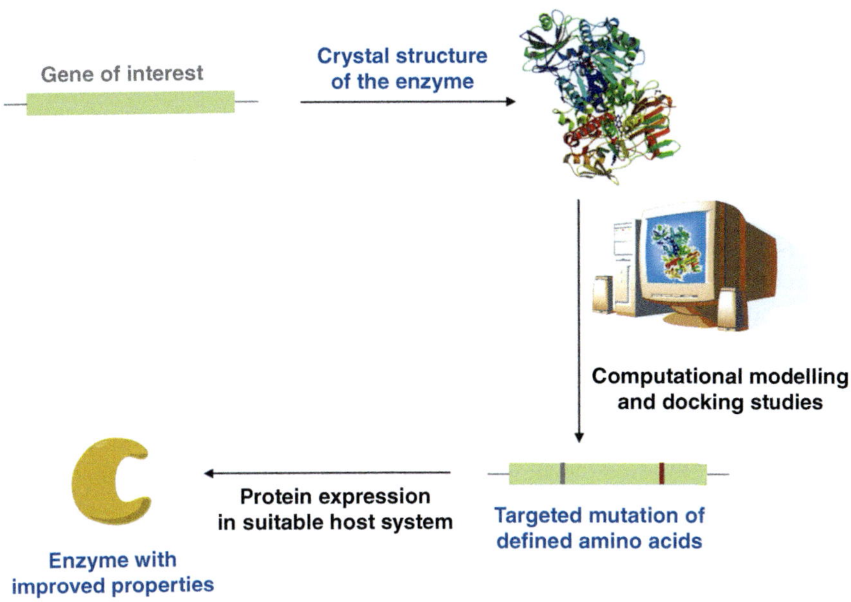

Figure 12.5 General scheme for rational design of proteins

Several computational tools exist for analyzing enzyme structure and predicting the effects of mutations on its properties [131–134]. Using rational design for biocatalysis can allow the redesign of the active site in an enzyme [135] and even alter its specific reactivity. Some excellent reviews are to be found in the literature [136, 137].

The most obvious drawback of a rational design approach is that a detailed knowledge of the protein structure is required (i.e. a crystal structure) and this is not always possible. Furthermore, the identification of mutations that would result in the desired enzymatic properties necessitates an understanding of the catalytic mechanism, substrate-enzyme interactions and reaction dynamics [138]. Of course, directed evolution and rational design of enzymes are not mutually exclusive; they are commonly used in tandem very successfully [109] and specific examples will be discussed in this chapter.

12.6 Hydrolysis/Reverse Hydrolysis

Hydrolases are catabolic enzymes that catalyze bond cleavage by reaction with water [64]. This class of enzyme displays several characteristics that make them useful to the organic chemist. Hydrolases often have broad substrate specificity and therefore accept many synthetic intermediates, showing high selectivity even toward non-natural substrates. Furthermore, besides hydrolysis, hydrolases also catalyze some related reactions such transesterification and alcoholysis (cleavage using an alcohol instead of water). Another feature that makes hydrolases convenient biocatalysts is that they do not require external cofactors. Their reactivity can alter depending on the reaction conditions – typically they perform hydrolysis in water and transesterification/alcoholysis (or related reactions) in organic solvents. In particular, lipases and esterases are stable and active in neat organic solvents [78]. This is also true of some proteases.

12.6.1 Hydrolases in Biocatalysis – An Overview

Owing to their catalytic promiscuity and ease of use, amongst the biocatalysts in organic synthesis, hydrolases are unequivocally the most studied and an enormous body of work exists detailing their application in the field [44, 61, 64, 139–144]. They are used frequently in both organic synthesis and in the production of bulk pharmaceuticals and chemicals [139, 145, 146]. Almost 75% of enzymes used industrially are hydrolases [147]. Lipases (triacylglycerol ester hydrolases), which catalyze the hydrolysis of fats and oils, have found particularly widespread use [61, 64, 139, 140, 148–151] and are therefore the most comprehensively examined of all the biocatalysts [8]. Proteases and esterases have also been used extensively in asymmetric synthesis [152, 153].

Table 12.3 depicts some of the well-known biocatalytic, stereoselective transformations, which are catalyzed by hydrolases. These biocatalysts are commonly used in (dynamic) kinetic resolutions of amides and esters but they are also efficient promoters of non-chiral transformations. The vast majority of hydrolase-mediated reactions use C-heteroatom hydrolases, such as lipases, esterases or proteases; however, C-C hydrolases are present in nature, catalysing many interesting reactions and a limited number have been applied to biocatalytic reactions [144, 154]. A wide variety of immobilized

Table 12.3 *Selected well-known hydrolase-mediated enantioselective reactions*

Biocatalytic transformation	Selected references
Hydrolysis	
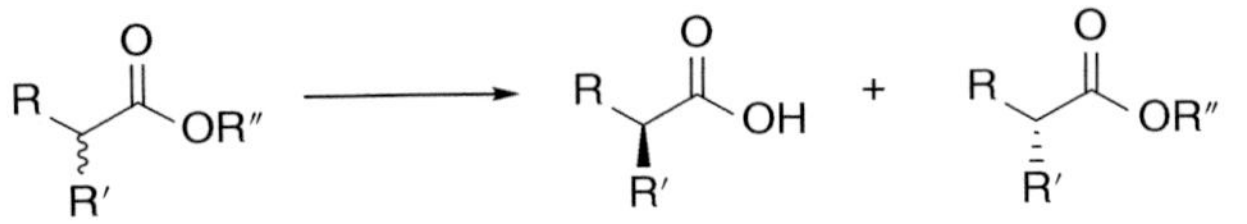	[98, 99, 156–159]
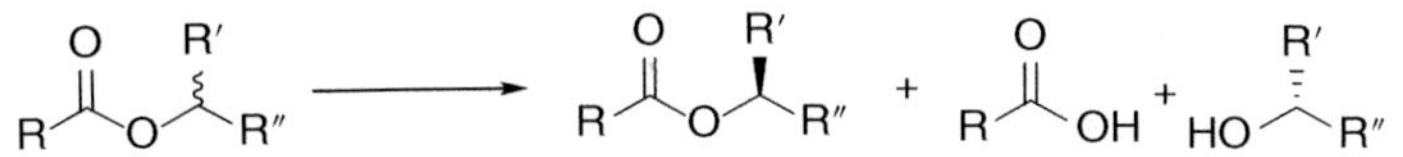	[160–163]
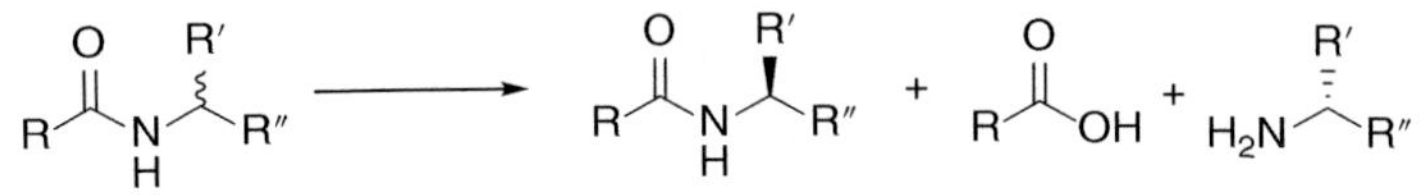	[164–167]
Esterification/amidification	
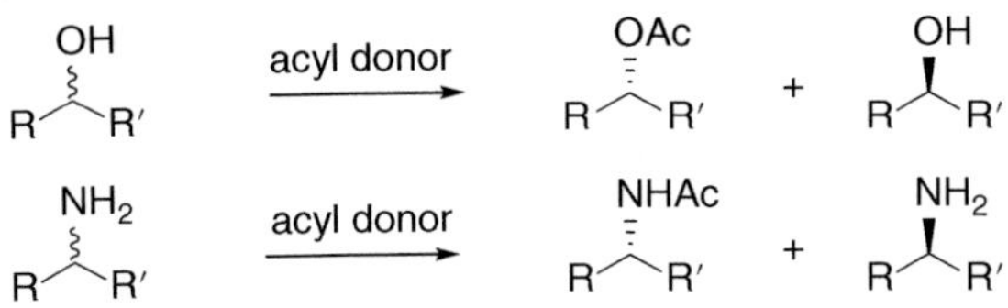	[151, 168–172]
Chemoselective desymmetrization	
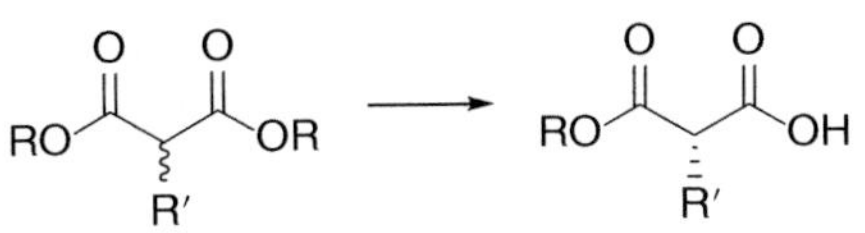	[173–176]
Epoxide hydrolysis	
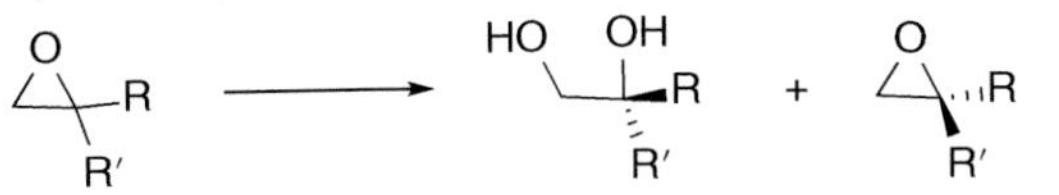	[177, 178]
Nitrile hydrolysis	
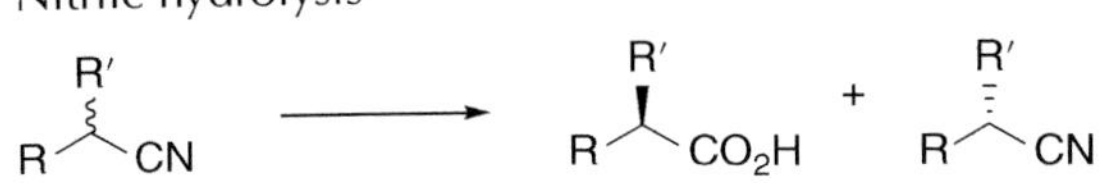	[179, 180]

hydrolases are also commercially available and these have been applied to several key biotransformations [155].

Detailed discussions on the reaction mechanisms and kinetics of hydrolase-mediated transformations are available in the literature [64, 148]. As an in-depth discussion on the breadth of hydrolase-mediated transformations is not possible in a review of this size, a general overview of the common reactions types along with some exemplars will be presented.

12.6.2 Esterification/Hydrolysis of Esters

Esterification reactions and hydrolysis of esters are (for the most-part) carried out by lipases and esterases [64]. As many lipases are derived from bacteria or fungi and are excreted as extracellular enzymes, they can be conveniently produced on a large scale [8]. A soluble substrate is required for proteases but it is not necessary for lipases and esterases. In the case of lipases, a second phase is actually advantageous as in many instances it 'activates' the enzyme and can increase its activity by tenfold to a hundredfold [64].

As mentioned in section 12.4, hydrolases catalyze esterification reactions in organic solvent [181–183]. Indeed, the synthetic utility of these enzymes can be greatly enhanced by using them in non-aqueous media [75]. Here, several parameters can influence the selectivity and efficiency of the transformation, particularly choice of solvent and acetylating agent [75, 140, 143, 168]. The reversibility of the reaction certainly poses one problem. This can sometimes be circumvented by using vinyl acetate as the acetylating source, since acetaldehyde is the side product and acyl transfer cannot be reversed [59, 184]. However, some hydrolases are known to be deactivated by acetaldehyde [140]. Further drawbacks of producing acetaldehyde as the side product (particularly on a production scale) are its low flash and boiling points, its toxicity [185] and its proclivity for polymerization [186]. Hence several alternatives, including carboxylic acids [187], methoxyacetates [188], anhydrides [189], other enol esters [169] (such as isopropyl acetate) and even oxime esters [190] have all been applied as acyl donors to hydrolase-mediated esterification reactions [191, 192]. In some cases reaction design and additives such as organic solvents can have a positive effect on enzyme activity and selectivity [193].

Alcohols are an extremely versatile class of substrate for synthetic applications and accordingly have received the most attention in the context of hydrolase-catalyzed kinetic resolution [8, 64]. Lipases, while showing high selectivity towards a wide range of substrates, are particularly selective towards alcohols – indeed, hundreds of secondary alcohols have been resolved using lipases [64]. Chiral primary alcohols are less studied in asymmetric biocatalysis and are more challenging substrates as the stereocentre is not adjacent to the hydroxyl functionality. Enantiopure alcohols can be resolved through acetylation of the racemic mixture (esterification) or hydrolysis of the corresponding ester (Table 12.4). The structures of some pharmaceutically important chiral synthons, which have been resolved using these methods, are also shown.

In terms of the resolution of both primary and secondary alcohols, lipase B from *Candida antarctica* (CAL-B) and lipase from *Pseudomonas cepacia* (PCL) play predominant roles. Indeed, lipase A and particularly lipase B from *Candida antarctica* are ubiquitous biocatalysts for esterification reactions and because of this are recognizable to non-biochemists [199]. Both are active in many organic solvents, which makes them perfect for acylations [200], whilst simultaneously allowing the solubilization of hydrophobic reactants/products. CAL-B and immobilized CAL-B (sold commercially as Novozym 435) are used in the synthesis of several chiral building blocks [201, 202] including intermediates in the synthesis of Lotrafiban [158], Saxagliptin [203] and profen esters [171, 202]. Dynamic kinetic resolutions have also been used for the quantitative synthesis of enantiomerically pure alcohols [204]. Conveniently, hydrolases can also resolve compounds which have a second stereoisomer; usually a *cis*- or *trans*-substrate is used (Table 12.4).

Table 12.4 *Some examples of lipase-catalyzed hydrolysis/esterification reactions yielding enantiopure alcohols*

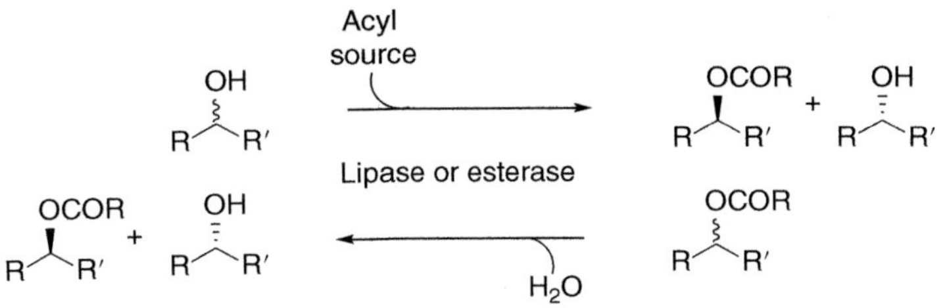

Substrate	Acyl source/ H_2O	Enantiomerically pure products	Notes
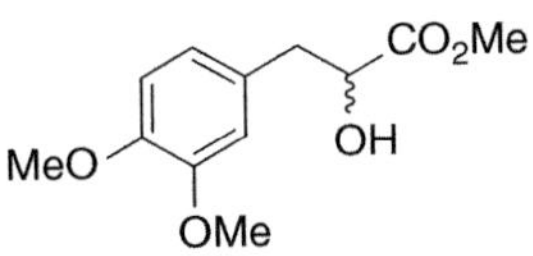	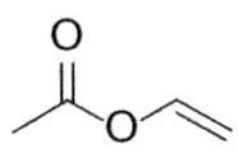	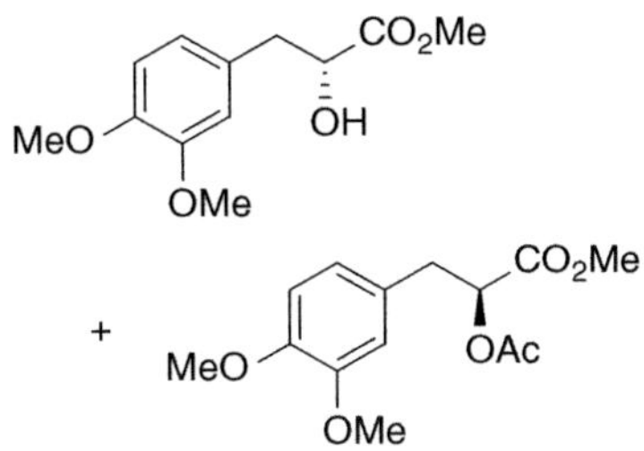	Lipase from *Burkholderia cepacia* used to resolve hydroxyester [194]
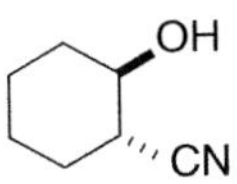 (*Trans*)	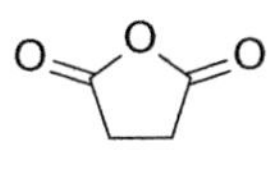	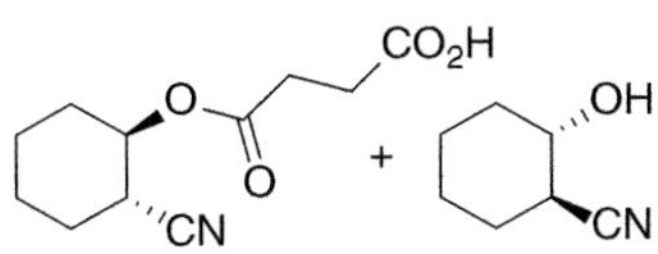	*Trans*-1,2-hydroxynitrile resolved using Novozym 435 [195]
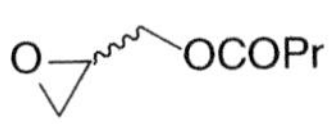	H_2O	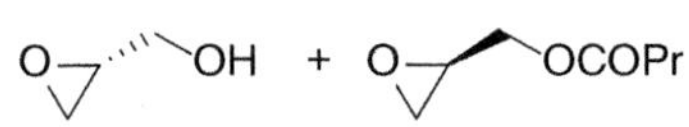	Lipase from *Thermomyces lanuginosa* resolved important chiral building block [196]
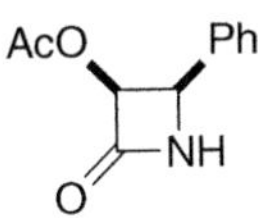 (*cis*)	H_2O	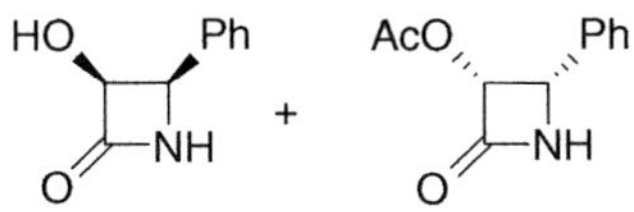	Lipase from *Pseudomonas cepacia* (PCL) used to resolve racemic *cis*-acetate for intermediate in anti-cancer drug Taxol [197]
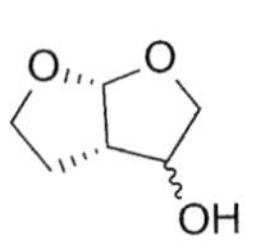	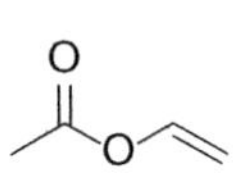	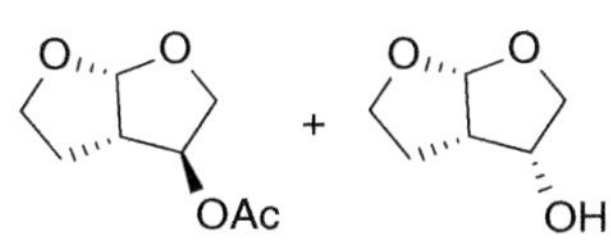	Bisfuran intermediate in synthesis of HIV protease inhibitors resolved with immobilized lipase PS [198]

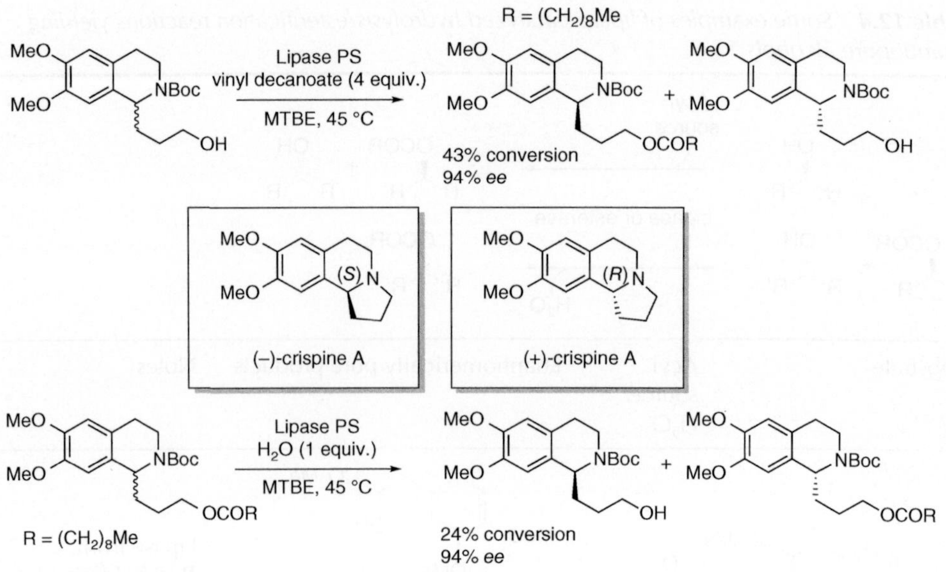

Scheme 12.6 *Esterification of a primary alcohol and hydrolysis of the corresponding ester, bearing a remote stereocentre, for the synthesis of crispine-A [205]. MTBE=methyl tert-butyl ether*

An interesting resolution of a primary alcohol with a remote stereocentre was investigated by Fülöp *et al.* Here, another advantageous property of the hydrolase family was exploited – opposite selectivities are typically seen for hydrolysis and esterification reactions [64]. This allowed the synthesis of both enantiomers of the antitumour alkaloid crispine A (Scheme 12.6) using the same lipase, from *Burkholderia cepacia* (lipase PS) [205]. Significant screening of conditions (temperature variation, structure of acyl donor, use of organic solvent additives) took place in order to deliver a highly enantioselective, preparative-scale reaction in both cases.

Hydrolase-catalyzed kinetic bioresolution has been used to provide access to enantioenriched carboxylic acids; indeed, resolution of a racemic mixture where the chiral centre is adjacent to the reacting carbonyl centre is a common transformation [64]. Here, the chiral moiety is adjacent to the carbonyl group of the ester. 2-Aryl carboxylic acids (such as profen acids) [206] have been resolved with excellent enantioselectivity [156, 207]. Reports of resolution where the stereocentre is not adjacent to the reacting site are less prevalent; however, hydrolase recognition of compounds bearing remote stereocentres has been reported. Through engineering of reaction conditions such as addition of organic cosolvents and temperature variation, 3-aryl alkanoic acids can also be resolved in excellent enantioselectivity (Scheme 12.7) [98, 99].

An indole ester intermediate (again with a stereocentre at the 3-position of the ester) for a prostaglandin D2 (DP) receptor antagonist targeting allergic rhinitis has also been resolved in excellent enantioselectivity using a lipase catalyst (Scheme 12.7) [208]. This process was optimized to deliver the product acid in >99% *ee* with 50% conversion at scale. Again, the addition of an organic cosolvent (in this case DMF) was used to optimize conversion and *ee*.

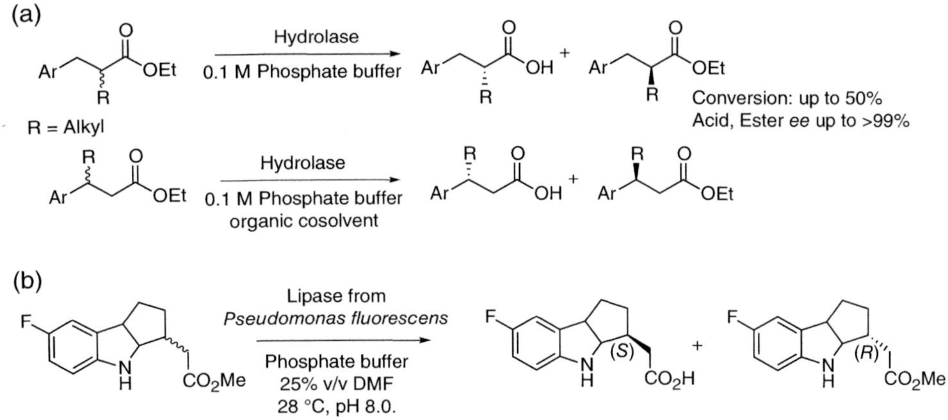

Scheme 12.7 *(a) Resolution of 2- and 3-substituted alkanoic acids [98, 99] and (b) the resolution of a pharmaceutically-important indole ester [208]*

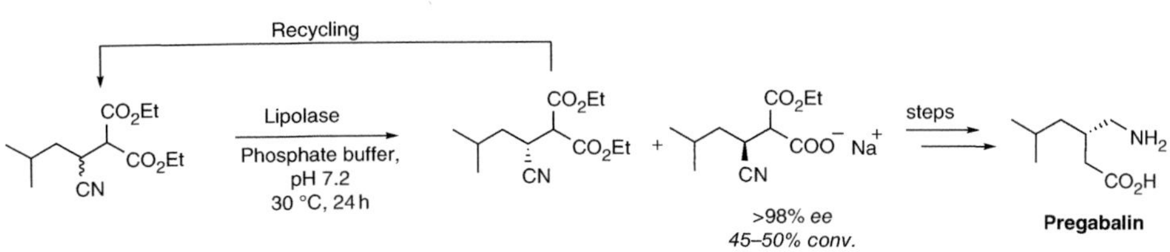

Scheme 12.8 *Biocatalytic route for the synthesis of pregabalin [173]*

Unsurprisingly, lipases have found widespread application in the pharmaceutical industry [28, 34, 64, 146]. Desymmetrization reactions are another elegant transformation catalyzed by hydrolases. Researchers at Pfizer successfully implemented a lipase-mediated biocatalytic desymmetrization in the synthesis of (*S*)-(+)-3-aminomethyl-5-methylhexanoic acid (pregabalin), a treatment for central nervous system disorders. In the first-generation manufacturing process for pregabalin, a three-step synthesis was performed with a single isolation step [209]. The crude racemic pregabalin was then subjected to a classical resolution using (*S*)-(+)-mandelic acid in a three-step crystallization process. This resulted in an overall yield of about 20%. Furthermore, the undesired χ-amino acid could not be recycled, which increased both manufacturing cost and waste material. A biocatalytic route has been devised to address these issues (Scheme 12.8), whereby Lipolase, a commercially available lipase from *Thermomyces lanuginosus* (incidentally also used in the detergent industry) was used to produce the desired (*S*)-mono acid enantiomer in excellent enantioselectivity (>98% *ee*) and high-resolution yield (>45% conversion) [173]. Subsequent decarboxylation, hydrolysis and hydrogenation afforded the desired product.

Convenient racemization of the (*R*)-enantiomer [210] increased the efficiency of the process further – a yield of 40% was achieved after one recycling. In terms of a green-chemistry approach this route was a tremendous success – it was projected that annually over 10 million gallons of organic solvents and nearly 2000 tonnes of raw materials were eliminated through this biocatalytic synthesis [211].

Inevitably, hydrolase engineering has been studied extensively [153, 184]. Indeed the seminal work in the area of directed evolution for enantioselective biocatalysis was conducted by the group of Reetz in 1997, using lipase from *Pseudomonas aeruginosa* (PAL) [212]. The chiral racemic ester, shown in Scheme 12.9, was subjected to hydrolysis using this enzyme. The wild-type enzyme was slightly (*S*)-selective ($E = 1.2$). After four rounds of evolution the enantioselectivity was significantly improved ($E = 11$). Hydrolysis gives the corresponding carboxylic acid and *p*-nitrophenol, which can be spectrophotometrically detected for high throughput screening [213]. Of course, this initial demonstration paved the way for many other studies that used this technique for transformations employing a vast array of enzymes.

Later, further improved mutants were expressed [214, 215] and this methodology has also been extended to other lipases and substrates [157, 216] – and, indeed, directed evolution has been extensively applied to lipases and esterases for organic synthesis [153, 217, 218].

Kinetic resolutions of tertiary alcohols have received less attention than their secondary counterparts and present a more difficult challenge [219–221] due to the sterically demanding structure of these important pharmaceutical building blocks. Two examples of this resolution are shown in Scheme 12.10. An esterase from *Bacillus subtilis* (BS2) was initially expressed

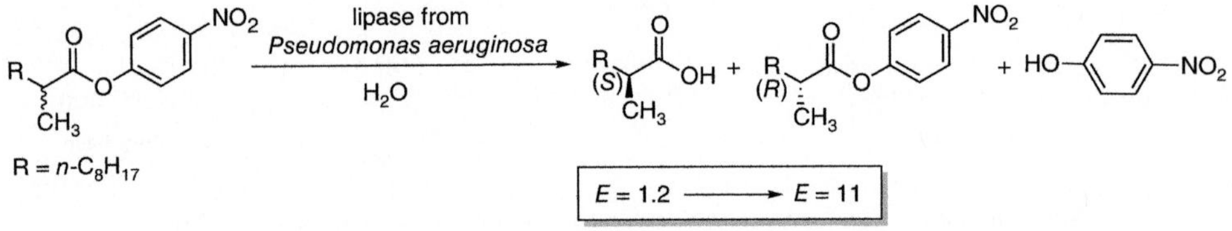

Scheme 12.9 *Seminal work on directed evolution for biocatalysis from the group of Reetz [212]*

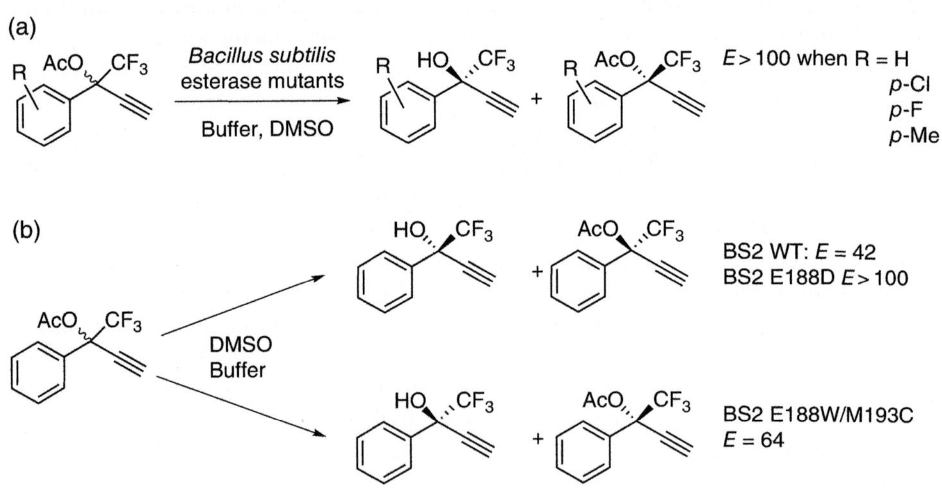

Scheme 12.10 *Mutants of esterase from Bacillus subtilis used in the kinetic resolution of tertiary alcohols [222, 223]. (a) Shows the substrate scope of the mutant enzymes while (b) is an example of inversion of stereo-preference through directed evolution*

and showed activity towards acetates of tertiary alcohols [222]. Molecular modelling studies were used to identify mutants that were more enantioselective than the natural enzyme. Although the substrate range of this family of mutant esterases was quite limited, a number of tertiary alcohols were resolved in excellent enantioselectivity (example (a)).

Example (b) shows complete inversion of stereo-preference from the mutant esterase [223]. BS2 mutant E188D exhibited improved enantioselectivity when compared to the wild-type enzyme, while two other mutants, E188W and M193C, showed the inversion of enantioselectivity towards the same substrate. Other examples of directed evolution-mediated switches in enantio-preference have also been reported [157].

12.6.3 Epoxide Hydrolases

The ring-opening of racemic 2, 2-disubstituted epoxides (using water as the nucleophile) is another hydrolase-catalyzed route to enantiopure alcohols and epoxide hydrolases have been used for this purpose in kinetic resolutions (see Table 12.3) [224–226]. This is an effective method for the synthesis of (enantiopure) diols and epoxides; however, it is limited to the use of water as a nucleophile.

12.6.4 Hydrolases in the Resolution of Chiral Amines

Optically active amines and amino acids play a prominent role in the pharmaceutical industry and are frequently used as building blocks in the synthesis of APIs [227]. Hydrolases can also catalyze the enantioselective acylation of amines and have been used for this purpose [169]. An excellent example of this is the BASF process (Scheme 12.11) [228]. An amine such as α-methylbenzylamine can be resolved using methoxyacetic acid ethyl ester as the acyl source and typically these reactions are highly selective ($E>200$).

In general, less reactive acetylating sources (such as esters or carbonates) are used with amines to minimize competing non-catalytic nucleophilic attack from the heteroatom [8].

Along with kinetic resolutions [64, 188, 227, 229], chiral amines have also been resolved via dynamic processes [62, 68, 69, 230]. As with alcohol substrates, several metal and nonmetal-based racemization catalysts have been employed [230]. A highly efficient, enantioselective example of this is shown in Scheme 12.12, whereby Bäckvall *et al.* deployed a ruthenium-based catalyst for the purpose of substrate racemization [231]. A series of primary amines was synthesized in good to excellent yield and excellent enantioselectivity (99% *ee* in many cases). This methodology has been applied to the synthesis of an important chiral intermediate for the preparation of the Pfizer antidepressant norsertraline [232].

Hydrolysis of amides is another time-honoured biocatalytic methodology for the production of enantiopure materials. Here, amidases, aminoacylases and hydantoinases

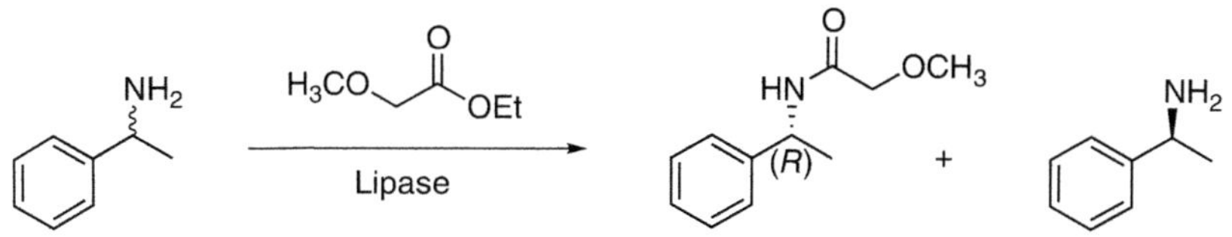

Scheme 12.11 *The BASF process for the production of enantiomerically pure amines [228]*

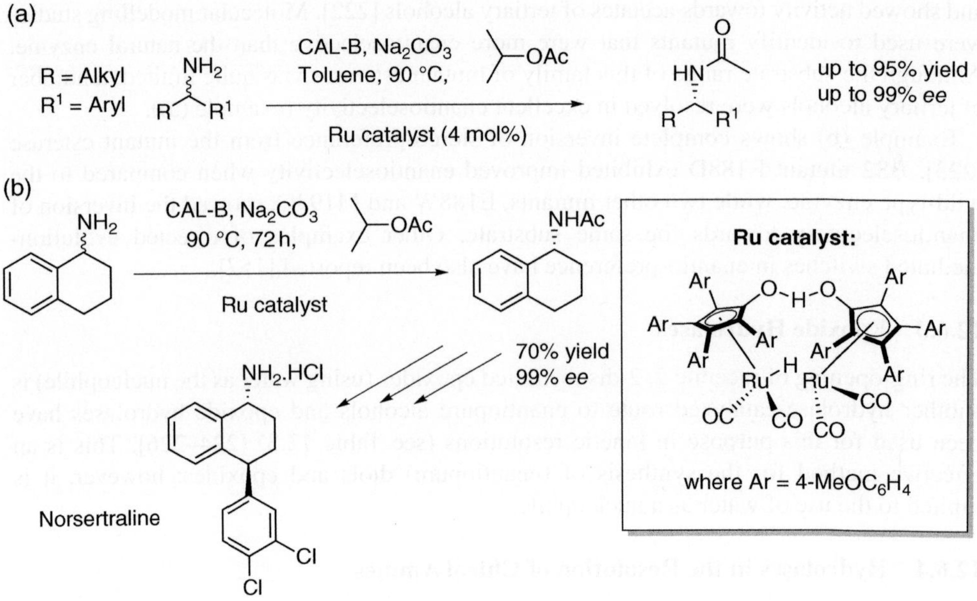

Scheme 12.12 (a) Dynamic kinetic resolution of racemic amines employing a ruthenium racemization catalyst [231] and (b) this methodology applied to the synthesis of norsertraline [232]

are the catalysts of choice, particularly for the synthesis of enantiomerically pure amino acids and their derivatives [64, 233, 234]. Generally, lipases do not catalyze the hydrolysis of amides, although there are a number of exceptions, including the improvement of enzyme activity through directed evolution [235]. Cyclic amides (α- and β-lactams) are chemically more stable than their acyclic counterparts but can be hydrolyzed by lactamases (a special group of proteases) to give the corresponding amino acid and the non-converted lactam. Selected examples of amide hydrolysis to yield enantiomerically pure amines are shown in Table 12.5.

Although the nitrile functional group can be chemically hydrolyzed to the corresponding carboxylic acid using strong acid or base, enzymatic catalysis of this transformation can be more advantageous as it is conducted under mild conditions [241]. Nitrilases (EC 3.5.5.1) directly hydrolyse the cyano functionality to carboxylic acid plus ammonia [242]. These enzymes can be employed for their regioselectivity; however, they can be enantioselective and have been used in (dynamic) kinetic resolutions [64, 243]. An example that has been exploited industrially is shown in Scheme 12.13 [244, 245]. Here, the reversibility of hydrogen cyanide addition to benzaldehyde is used in a dynamic kinetic resolution. The (*R*)-selective nitrilase from *Alcaligenes faecalis* furnishes enantiopure (*R*)-mandelic acid in 91% yield. This methodology can be applied to the enantioselective synthesis of a broad range of 2-hydroxy carboxylic acids [179].

The enantioselectivity of nitrilases has also been utilized in chemoselective desymmetrizations [180, 246, 247] and in the synthesis of enantiopure building blocks [248, 249], particularly α-hydroxy acids and α-amino acids [250].

Table 12.5 *Selected examples of amide hydrolysis to yield enantiomerically pure amines/amino acids*

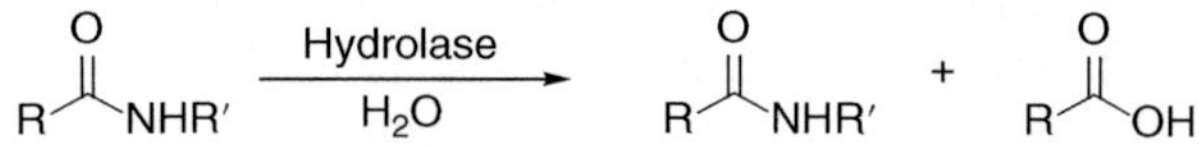

Substrate(s)	Product(s)	Enzyme [ref]
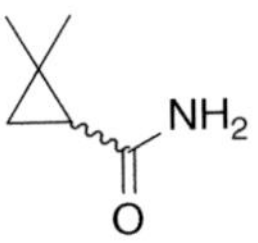	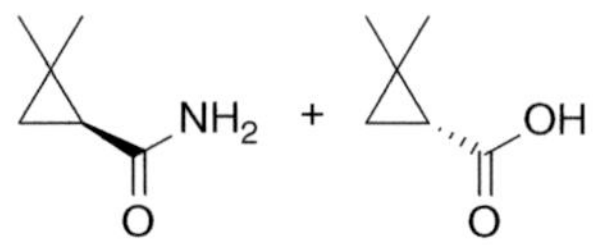	Amidase from *Delftia tsuruhatensis* [236, 237]
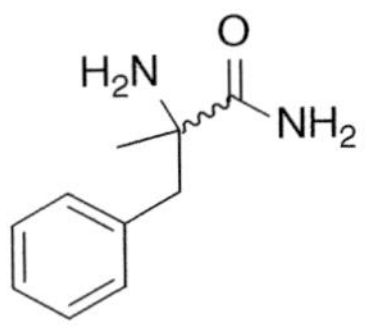	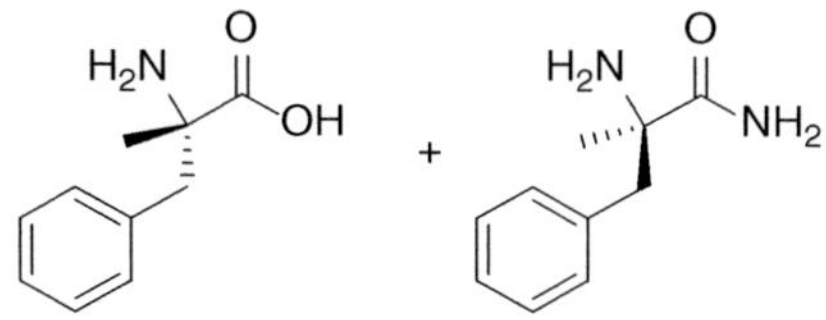	Amidase from *Mycobacterium neoaurum* [238]
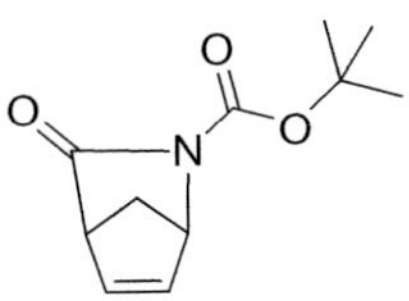	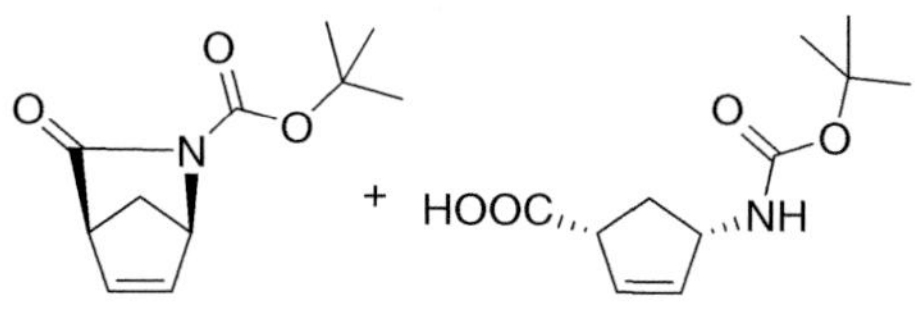	Savinase [239]
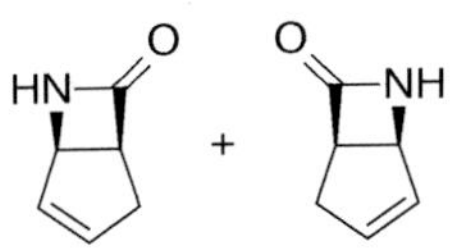	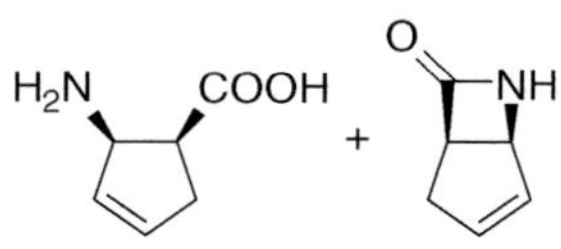	Lactamase from *Rhodococcus sp.* [240]
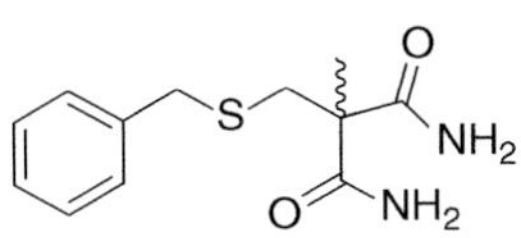	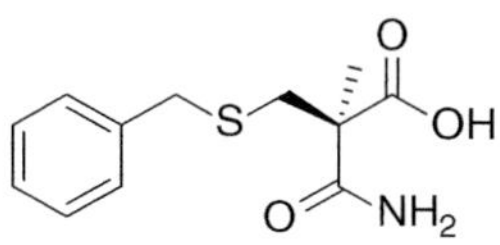	Amidase from *Rhodococcus erythropolis* [49]

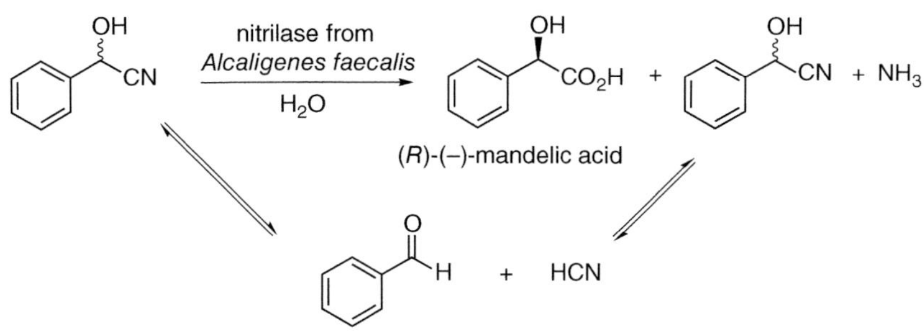

Scheme 12.13 *Nitrilase-mediated DKR to yield enantiopure (R)-mandelic acid [244]*

12.7 Redox Reactions

In terms of enzymes known in nature, oxidoreductases are the most plentiful group. This class comprises several groups of biocatalysts which are useful in asymmetric synthesis, including dehydrogenases, monooxygenases, dioxygenases, oxidases and peroxidases [251]. Oxidoreductases catalyze an impressively diverse range of chemical transformations [252]. The oxidation of alcohols, aldehydes, ketones, and primary and secondary amines, hydroxylation of aromatic carbon atoms and dehydrogenation of carbon-carbon single bonds are all performed by oxidoreductases [251]. Baeyer–Villiger oxidation, double bond epoxidation and heteroatom oxidation also fall within the scope of this class of enzyme. As they catalyze redox processes, many of these reactions are freely reversible, and indeed oxidoreductases can also promote the reduction of aldehydes, ketones, carboxylic acids, and double and triple carbon-carbon bonds as well [253].

12.7.1 Cofactors

Oxidoreductases require redox cofactors, which donate or accept the chemical equivalents that are needed for reduction (or oxidation) [8]. In this context, nicotinamide coenzymes are the crucially important, as 90% of known oxidoreductases need either NAD(H) or NADP(H) as a coenzyme [254]. Amino acid dehydrogenases (AADH) and alcohol dehydrogenases (ADH) have been used extensively for the generation of enantiopure materials and the vast majority of these are NAD(P)H-dependent enzymes [255].

NAD(P)H cofactors have two main drawbacks – they are relatively unstable molecules and are very expensive, particularly in the context of large-scale reactions. Due to this, initial research into redox reactions focused on the use of whole cells, where the cofactor regeneration was in-built in the cell [8]. Due to the preference for purified enzymes in these systems mentioned in section 12.2.6, intensive research has more recently led to the use of isolated enzymes with cofactor regeneration systems. The coenzyme is used in a substoichiometric quantity and a cofactor recycling system is used to regenerate the species whose oxidation state has changed during the reaction. Owing to the importance of regenerating these cofactors, much effort has been invested in investigating various systems, each of which has its advantages and disagdvantages [256–260]. Substrate-coupled and enzyme-coupled systems are used for cofactor regeneration. In the substrate-coupled system, the same dehydrogenase enzyme is used to simultaneously transform substrate and cosubstrate. The cosubstrate must be used in large excess to drive the equilibrium towards the desired direction. In the enzyme-coupled system, a second, independent enzymatic transformation is used for cofactor recycling. In this case, the two enzymatic reactions must proceed independently and therefore the biocatalysts employed must have different substrate specificities to allow this to happen.

Some examples of prevalent cofactor recycling systems are shown in Table 12.6 [6, 261]. In general only the natural cofactors (NAD(H) or NADP(H)) can be used – this is especially true for alcohol dehydrogenases – but there have been reports showing the use and application of synthetic nicotinamide cofactor analogues as a potential method for addressing the cost and stability problems associated with the natural species [262–264].

Discussions on the mechanistic aspects of hydride transfer to a prochiral ketone [265, 266] can be found in the literature [12, 267]. Of course, in such a broad and ever growing field, it is not possible to cover every reaction/enzyme type and the interested reader is pointed to some excellent reviews on this impressive class of enzyme [251, 268–270]. Some selected examples are described below.

Table 12.6 *Prevalent NAD(P)H regeneration systems [6]*

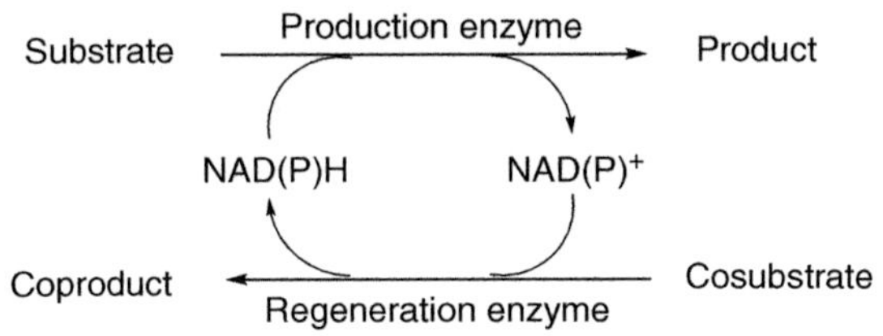

Regeneration enzyme	Cofactor	Cosubstrate	Coproduct
Formate dehydrogenase (FDH)	NADH	Formic acid (salts)	CO_2
Glucose dehydrogenase (GDH)	NADH/NADPH	Glucose	D-Glucono-1, 5-lactone
Alcohol dehydrogenase (ADH)	NADH/NADPH	Alcohol	Aldehyde/ketone
Alcohol dehydrogenase (ADH)	NADH/NADPH	Diol	Lactone
Phosphite dehydrogenase (PDH)	NADH	Phosphorous acids (salts)	Phosphates
Hydrogenase (Hase)	NADH/NADPH	H_2	–

12.7.2 Reduction of Ketones

Alcohol dehydrogenases (also called ketoreductases or KREDs), in both isolated and whole cell form, have been used in the synthesis of enantiopure materials. The commercial availability of KRED kits has led to widespread screening of these enzymes against a broad range of substrates [229, 269].

The biocatalytic reduction of ketones to yield secondary alcohols is a particularly powerful tool for the synthesis of enantiomerically pure secondary alcohols and thus it has been widely investigated [8, 261, 271, 272]. Baker's yeast is the most commonly used microorganism for this purpose and reports of yeast-catalyzed reductions of non-natural substrates date back almost 100 years [273]. Apart from carbonyl groups, this cheap and readily available microorganism has been used for the reduction of C=C bonds [274] and in redox reactions of heteroatom-containing compounds [275]. Due to its extensive use in organic synthesis, baker's yeast-mediated transformations have been widely reviewed in the literature [12, 276–279].

The first industrial application of baker's yeast was in the synthesis of trimegestone, a norsteroidal progestomimetic compound (Scheme 12.14) in 1997 [280].

Isolated alcohol dehydrogenase (ADH) enzymes have also been used in the enantioselective synthesis of pharmaceutical intermediates [201]. An excellent representative example is shown in Scheme 12.15. Researchers at Codexis evolved a ketoreductase enzyme to reduce oxacarbazepine to give (*S*)-lipacarbazepine, an intermediate in the synthesis of lipacarbazepine acetate [281], a treatment for epilepsy. Following four rounds of directed evolution, 15 000 KRED variants were tested. The optimal enzyme was identified and the chiral alcohol product was isolated in 96% yield and >99% *ee*. Isopropanol was used as the reductant for cofactor recycling and the reaction was successfully scaled up to a 500 mL reaction vessel with a substrate : catalyst ratio of 100 : 1.

This reaction was found to reach equilibrium at 90% conversion with this particular system. In a paradigmatic example of reaction engineering to optimize the yield, the reaction was driven to completion by the removal of acetone by employing a nitrogen sweep and parallel addition of a premixed solution of IPA and buffer to maintain a constant reaction volume.

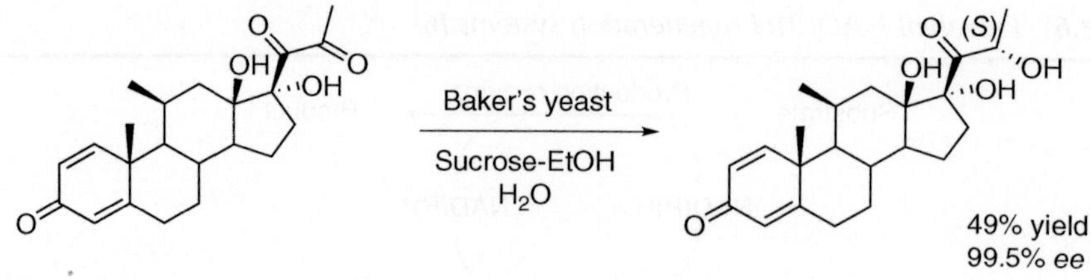

Scheme 12.14 *Baker's yeast-mediated bioreduction for the synthesis of trimegestone [280]*

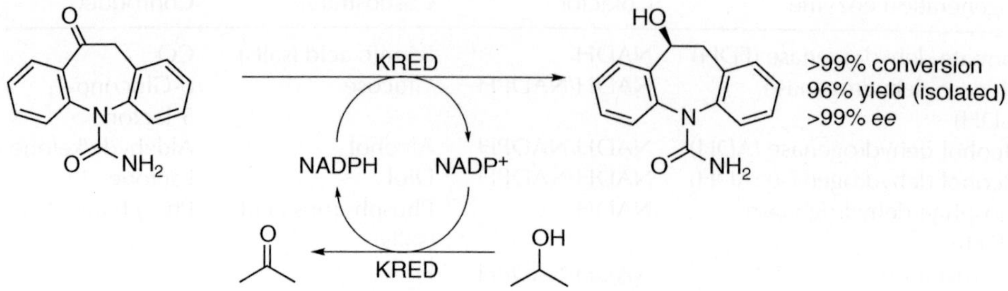

Scheme 12.15 *The use of a KRED variant in the synthesis of (S)-lipacarbazepine [281]*

Several other eloquent systems have been published and selected examples are shown in Table 12.7.

Tetrahydrothiophene-3-one (Table 12.7, entry (ii)) is a near-symmetrical compound and is thus considered a particularly challenging substrate. A highly evolved variant of *Lactobacillus kefir* was used to provide access to the desired enantiopure alcohol. Eight rounds of directed evolution were required to engineer the highly selective biocatalyst.

A key intermediate in the production of atorvastatin is the chiral 3, 5-dihydroxy carboxylate side chain. The first step in this synthesis is a biocatalytic reduction of ethyl-4-chloroacetoacetate (Table 12.7, entry (iii)). A ketoreducatse was employed for this reduction, in tandem with a NADP-dependent glucose dehydrogenase (GDH) for cofactor regeneration [282]. The desired (*S*)-hydroxybutyrate product was obtained in 96% yield and >99.5% *ee*. Many other pharmaceutically important chiral alcohols have been enantioselectively synthesized via biocatalytic reduction of the corresponding ketone and some examples are shown (Table 12.7, entries (v)–(vii)).

Hollmann and coworkers have reported the use of a 'smart-cosubstrate', 1,4-butanediol, in substrate-coupled NADH-dependent biotransformations. Here, the diol is (irreversibly) transformed to a lactone coproduct which is thermodynamically and kinetically inert [289]. This may be an excellent means of overcoming the reversibility and poor thermodynamic driving force associated with reductions using traditional cosubstrates (e.g. IPA) whilst simultaneously negating the requirement for molar surpluses. More recently, this was applied to the scaled-up asymmetric reduction of ethyl 4, 4, 4-trifluoroacetoacetate (Table 12.7, entry (iv)) [285].

When compared to the use of hydrolases on solid supports, very few examples of ketoreductase immobilization have been reported [290, 291]. This is possibly due to the lower tolerance for organic solvents that many KREDs exhibit. Despite this, the use of an immobilized KRED that is stable and active in organic solvents has been reported [292]. This was prepared from the commercially available ketoreductase, P1B2, from Codexis and

Table 12.7 *Selected examples of ADH-mediated reductions yielding enantiopure alcohols*

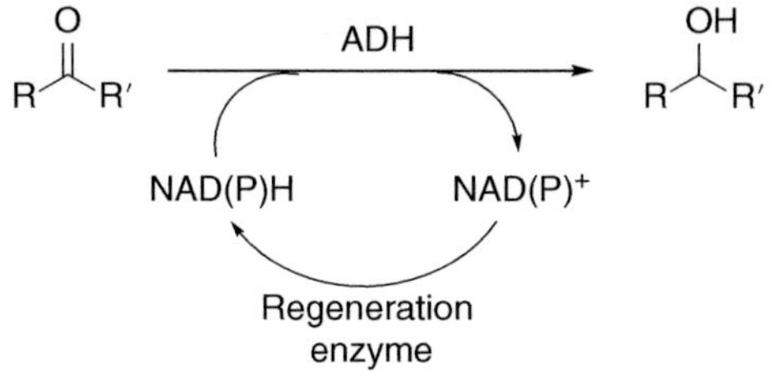

Entry	Substrate	Product	Comments
(i)	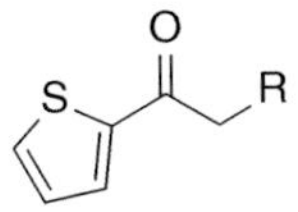	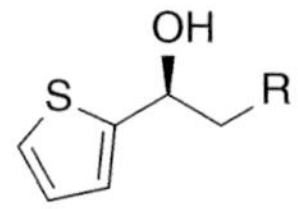	Intermediate in the synthesis of duloxetine [283]
(ii)	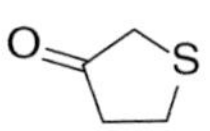	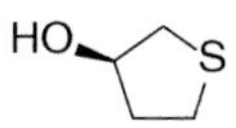	Almost symmetrical tetrahydrothiophene-3-one converted to product in 99% *ee* [284]
(iii)	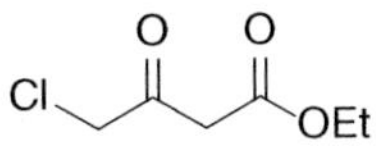	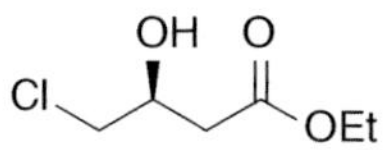	Intermediate in the synthesis of atorvastatin side chain [282]
(iv)	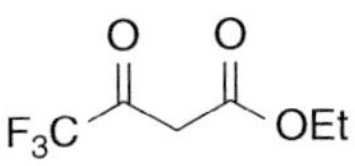	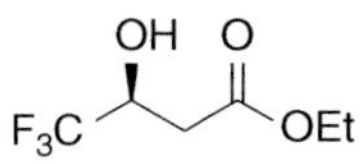	Scale up of 'smart-cosubstrate' approach [285]
(v)	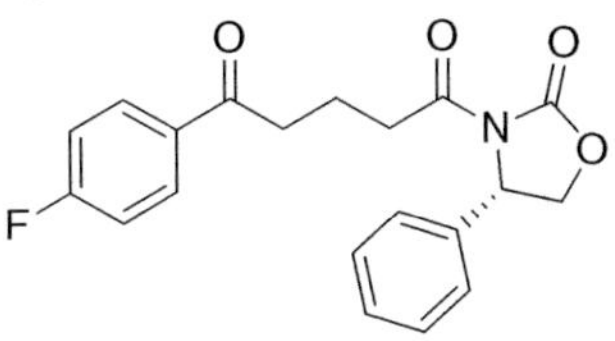	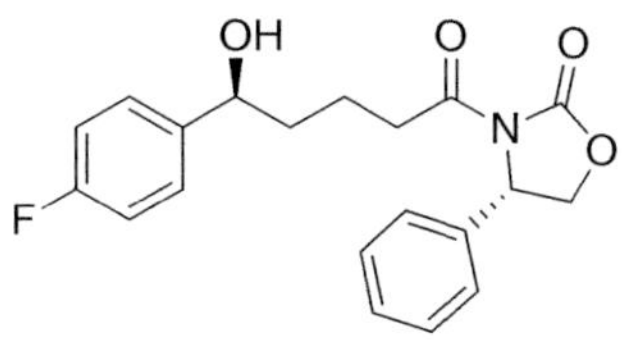	Chiral precursor to ezetimibe, engineered KRED furnished product in 99% *ee* [286]
(vi)	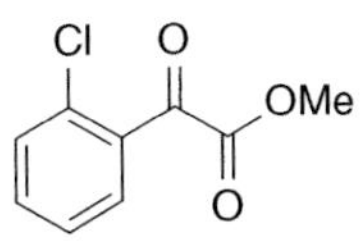	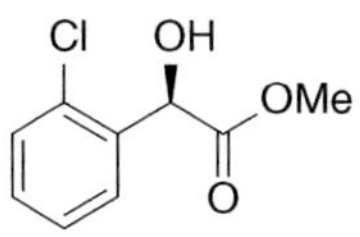	Carbonyl reductase mined from *Candida glabrata* – intermediate in synthesis of clopidogrel (Plavix) [287]
(vii)	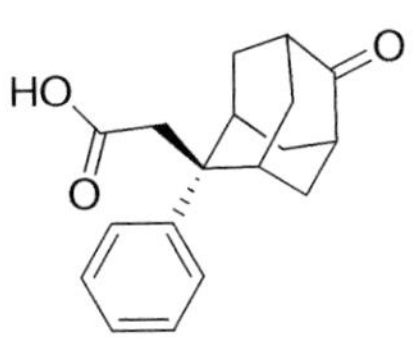	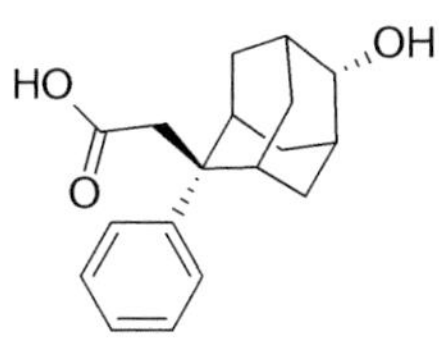	Potential intermediate in 11*β*-hydroxysteroid dehydrogenase inhibitors, KRED from *Candida utilis* used as biocatalyst [288]

was applied to the synthesis of (*R*)-1-(3, 5-bis(trifluoromethyl)phenyl)ethanol, a key intermediate in the synthesis of emend, an antiemetic. The immobilized KRED was stable in a range of organic solvents and the optimized conditions were found to be 50 g/L ketone and 100 g/L immobilized P1B2 in 50/40/10 hexanes/IPA/water at 40 °C. The reaction gave the desired chiral alcohol with 99% conversion, 98% isolated yield and >99% *ee*

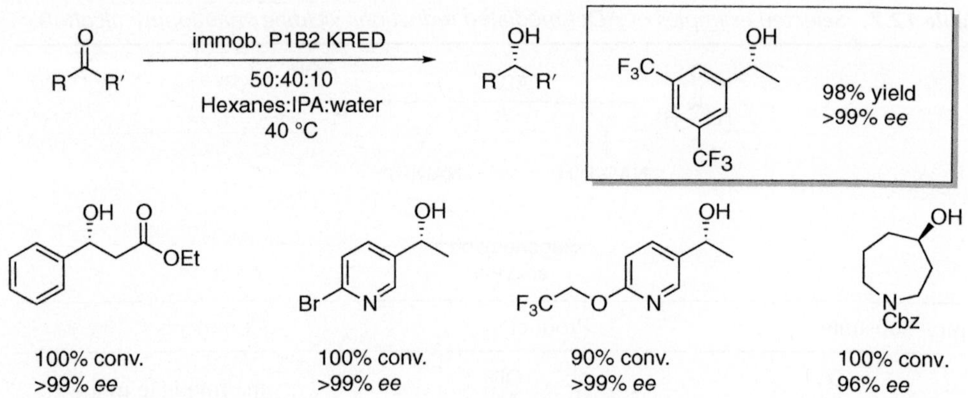

Scheme 12.16 *Synthesis of (R)-1-(3, 5- bis(trifluoromethyl)phenyl)ethanol and a range of other enantiopure alcohols via an immobilized KRED [292]*

(Scheme 12.16). Several other (*R*)-alcohols were prepared with excellent enantioselectivity (this time using a solvent system of 90:10 IPA:water at a temperature of 30 °C) and the structures of some are shown below. Surprisingly, the of use buffer was unnecessary for this transformation as water was equally effective under the same operating conditions.

12.7.3 Aldehyde Reductions

The reduction of an aldehyde does not yield a new stereocentre at the reacting site because a primary alcohol is formed; however, α-substituted aldehydes can be converted to enantiomerically pure β-substituted alcohols via an ADH-mediated (dynamic) kinetic resolution. This approach has been used in the enantiospecific reduction of 2-methylvaleraldehyde, an important chiral intermediate (Scheme 12.17 (a)) [293]. Initially a commercially available KRED isolated from *Lactobacillus kefir* was identified as a viable starting point. Directed evolution was then used to engineer an enzyme to resolve kinetically the racemate in an enantiospecific manner. Following identification of the optimal mutant, the reaction was also performed on a preparative scale, resulting in 97.5% yield and 98.5% *ee*. α-Substituted aldehydes readily racemize at high pH and (*R*)-methylvaleraldehyde was found to do so at pH 12. In this case the mutant enzyme was not stable above pH 9 and therefore a dynamic kinetic resolution was not possible. Dynamic kinetic resolutions of this type have been reported however [294, 295] and one such example is shown in Scheme 12.17 (b) [296]. An ADH was used to access enantioenriched (*S*)-profenols from the corresponding aldehyde. Interestingly, the enzyme was derived from the hyperthermophile *Sulfolobus solfactaricus*, which allowed the reaction to be conducted at 80 °C. Just 5% ethanol was used as the reductant and the enzyme exhibited broad substrate scope, often with excellent enantioselectivity. Selected structures are shown.

12.7.4 Reductive Aminations

Amino acid dehydrogenases, such as leucine dehydrogenase, have been used to synthesize amino acids from the corresponding α-ketoacid [297]. The main drawback of this methodology is that it is limited to α-keto acid substrates. Recently the potential of ω-tramsaminase enzymes in the synthesis of enantiomerically pure amines has been realized [298–306]. ω-Transaminases (ωTAs, E.C. 2.6.1.X) are pyridoxal-5-phosphate (PLP) dependent transferase

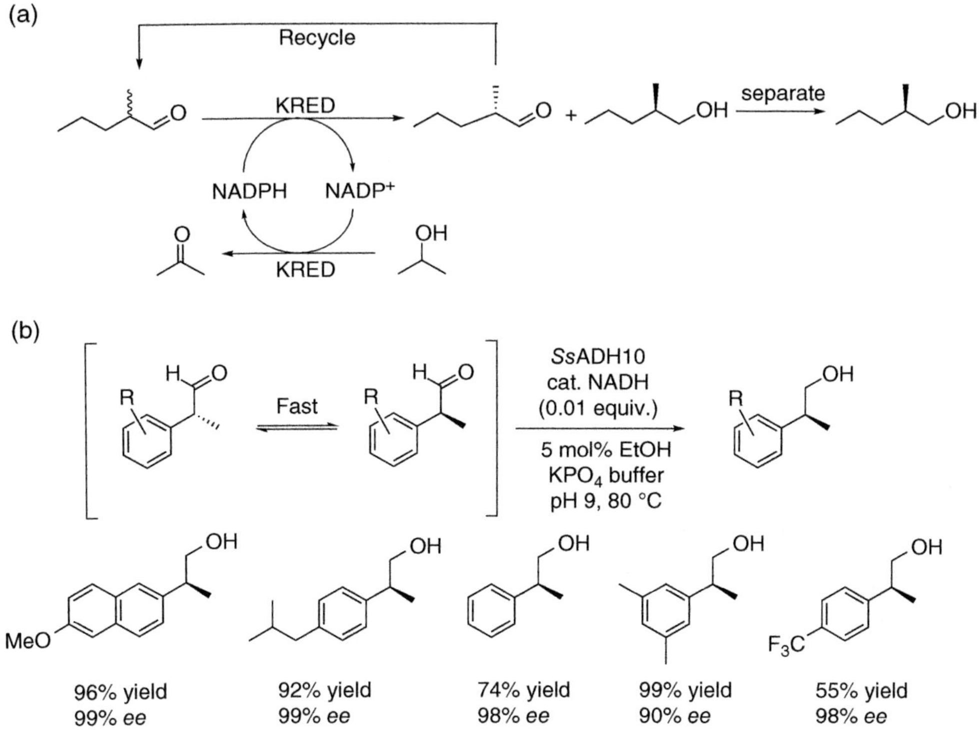

Scheme 12.17 *The kinetic resolution of 2-methylvaleraldehyde (a), [293] and (b) the dynamic kinetic resolution of profenols via an ADH from the thermophile Sulfolobus solfactaricus [296]*

enzymes that catalyze the reversible transfer of an amine group from a donor molecule (e.g. isopropylamine or alanine) to a carbonyl acceptor, without the requirement for an adjacent carboxyl moiety. The cofactor simply facilitates the transfer of ammonia and electrons. Pyridoxal-5-phosphate is required only in catalytic amounts as it is recycled during the reaction. Again, the reversibility of transformation is important as ω-transaminases can be used in kinetic resolutions [307], whereby the substrate is a racemic amine and one enantiomer selectively reacts (an oxidative deamination), or an asymmetric transformation via a reductive amination (Scheme 12.18).

Desymmetrization reactions are possible through the use of concurrent, stereo-complementary ω-transaminases (Scheme 12.19) [309]. The (*S*)-selective ωTA from *Polaromonas species* (PS-ωTA) was used in oxidative mode in tandem with (*R*)-selective ωTAs from *Neorsatoya fischeri* (NF-ωTA) or *Mycobacterium vanbaadenii* (MV-ωTA) in reductive mode by Yun and coworkers. A series of racemic amines with various substitutions on the aromatic ring were deracemized using this approach.

The (*R*)-amines were isolated in excellent yield and *ee* after 24 hours. Stereo-complementary approaches have also been taken for the synthesis of pharmaceutically important intermediates [310].

Recently research has been undertaken in the development of biocatalytic cascade syntheses of enantiomerically pure amines [299], particularly in the area of cascade reactions involving ω-transaminases [308]. The conversion of an alcohol to a chiral amine is as interesting as

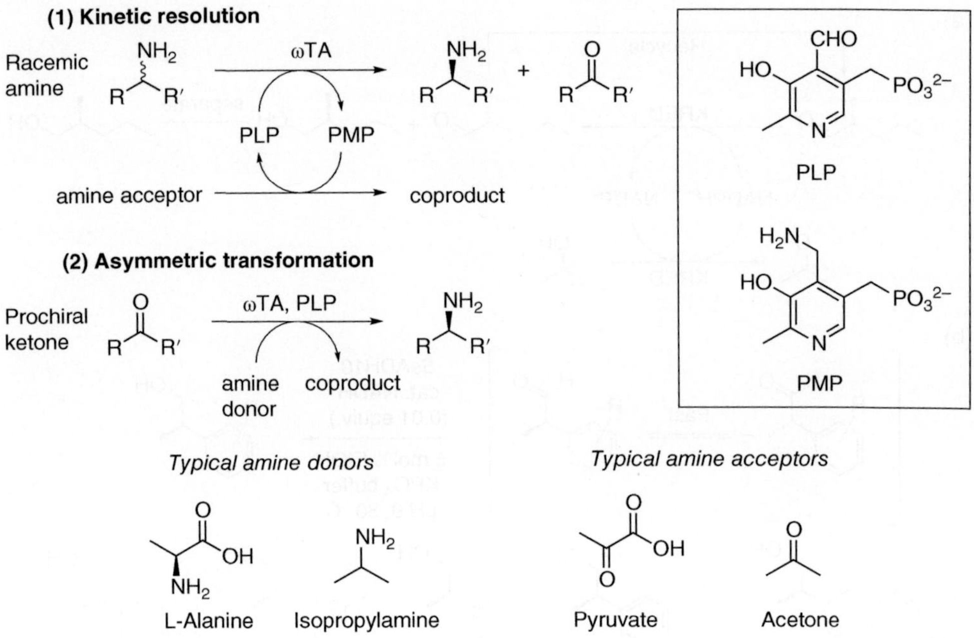

Scheme 12.18 *Routes to chiral amines via ω-transaminases [308]*

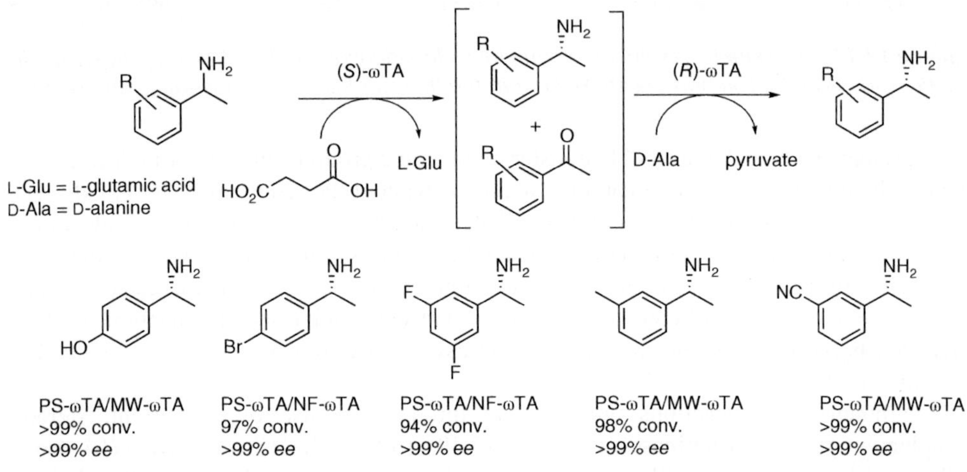

Scheme 12.19 *Concurrent one-pot deracemization cascade for aromatic amines afforded by stereo-complementary ω-transaminases and a selection of amines obtained in the reaction. ω-Transaminases originating from Polarinus species (PS), Neorsatoya fischeri (NF) and Mycobacterium vanbaadenii (MV) were used [309]*

it is synthetically desirable. It is traditionally a challenging chemical transformation, often using stoichiometric amounts of toxic or explosive reagents [311, 312]. Building on previous elegant work with primary alcohol substrates [313, 314], Kroutil and coworkers have developed a synthetic 'redox-self-sufficient' cascade, which combines an alcohol oxidation with a transaminase-mediated reductive amination in a single pot (Scheme 12.20) [315].

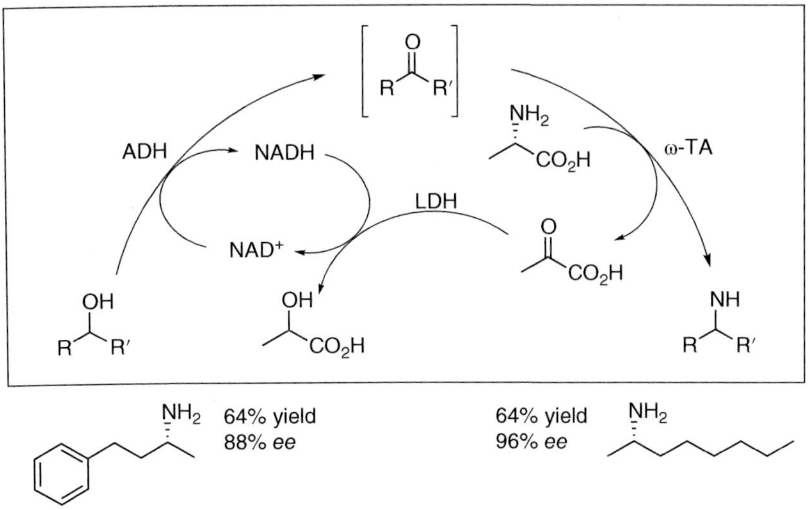

Scheme 12.20 *Redox-neutral network using Lactate Dehydrogenase (LDH) to remove pyruvate [315]. Optimized conditions – ω-TA (lyophilized whole cells), ADH-A from Rhodococcus ruber, LDH from rabbit muscle, phosphate buffer (pH 7.5), NADH, NAD+, PLP, L-alanine (5 equivalents). The structures of two amine products which were synthesized in high ee are also shown*

Direct amination of secondary alcohols is a more challenging transformation than that of the primary alcohol because stereo-preference/selectivity of the ADH must be considered – either a nonspecific ADH or two enantiocomplementary ADHs need to be employed for this step. Initial experiments with this system showed poor conversions due to the equilibrating network and *in situ* lactate dehydrogenase (LDH)-mediated pyruvate removal was then used to increase the yield (Scheme 12.20). When the substrate was an enantiomerically pure alcohol, good yields (up to 64%) and excellent *ee* (up to 96%) were reported.

12.7.5 Reduction of C=C Bonds

Although biocatalytic reduction of conjugated C=C bonds has been known since the 1930s using members of the old yellow enzyme family (OYE) [274, 316], references should be in square brackets with comma afterwards it is only recently that their synthetic potential as biocatalysts has been established [274, 317–321]. Ene-reductases (EREDs) catalyze the reduction of activated olefins via stereoselective addition of hydrogen to a C=C bond (Scheme 12.21). Historically asymmetric bioreductions of activated C=C bonds were performed using whole cells, particularly baker's yeast [277, 278] but today a range of OYEs is commercially available. Ene-reductases are particularly exciting biocatalysts due to the possibility of creating up to two new (adjacent) stereocentres and have been developed for industrial use by BASF, in recent process development work at Pfizer [322] and also in the preparation of some pharmaceutically important chiral synthons [323, 324]. Selected compounds which have been synthesized in a highly enantioselective manner by EREDs from the corresponding α, β-unsaturated substrates are shown in Scheme 12.21.

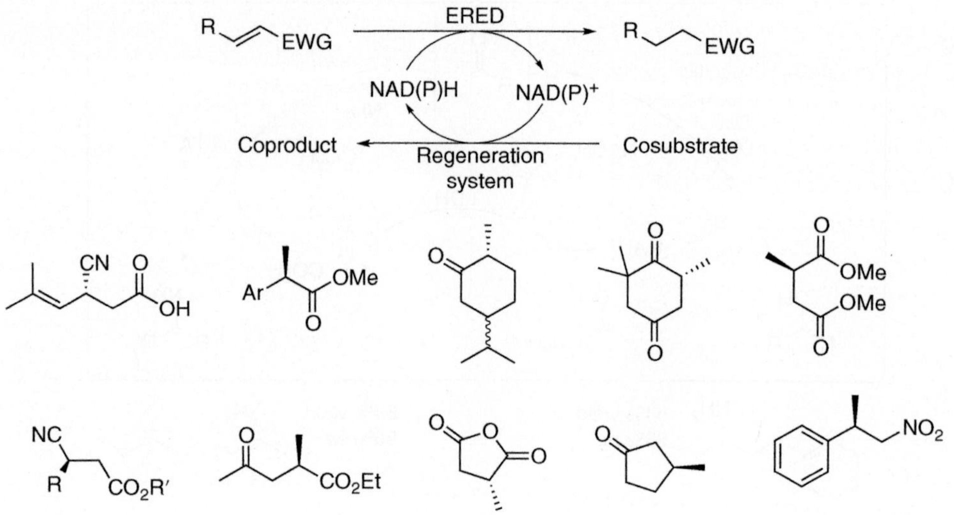

Scheme 12.21 *Selection of enantiopure products obtained from ERED-mediated reduction of the corresponding α, β-unsaturated substrate [274, 319, 325, 326]*

12.7.6 Enantioselective Oxidation/Reduction Cascade Reactions

The reversibility of ADH-catalyzed processes can also be exploited in asymmetric synthesis, although examples are less prevalent in the literature [255]. In this case $NAD(P)^+$ regeneration systems are necessary. For example, the oxidation of secondary alcohols is less studied than its reductive counterpart (ketone to alcohol) as chirality is destroyed in the reaction, i.e. sacrificing an sp^3 carbon for an sp^2 one. Nevertheless, ADH-mediated oxidative kinetic resolutions have been reported [327–332]. Dynamic kinetic resolutions are also possible through *in situ* reduction of the ketone byproduct. Through thoughtful reaction design, Kroutil *et al.* have outlined a one-pot reaction that combines two stereocomplementary ADHs, in oxidative and reductive modes respectively, to deracemize various alcohols in excellent yield and *ee*, as a green alternative to Mitsunobu stereoinversion of alcohols (Scheme 12.22) [333, 334].

In the example shown, the enantiopure (*R*)-alcohols were isolated in quantitative yield. A similar tandem, (*S*)-selective system (involving (*R*)-selective ADH for the oxidative step and an (*S*)-selective ADH for the reductive step), was also designed, again furnishing the alcohols in quantitative yield in enantiopure form.

12.7.7 Oxidases

Alcohol or amine oxidases offer an alternative to the use of NAD(P)H-dependent dehydrogenases for biocatalytic oxidations. Oxidases use molecular oxygen as the oxidant and are being increasingly used as biocatalysts in organic synthesis [255, 268, 270]. A range of different metals and cofactors is used by oxidases to transfer electrons from the substrate to the acceptor (O_2), including metal- (e.g. copper-dependent galactose oxidase) or flavin-dependent (e.g. glycolate oxidase) [335, 336]. Recently they have been applied as biocatalysts in asymmetric synthesis, in the oxidation of C-O and C-N bonds [270].

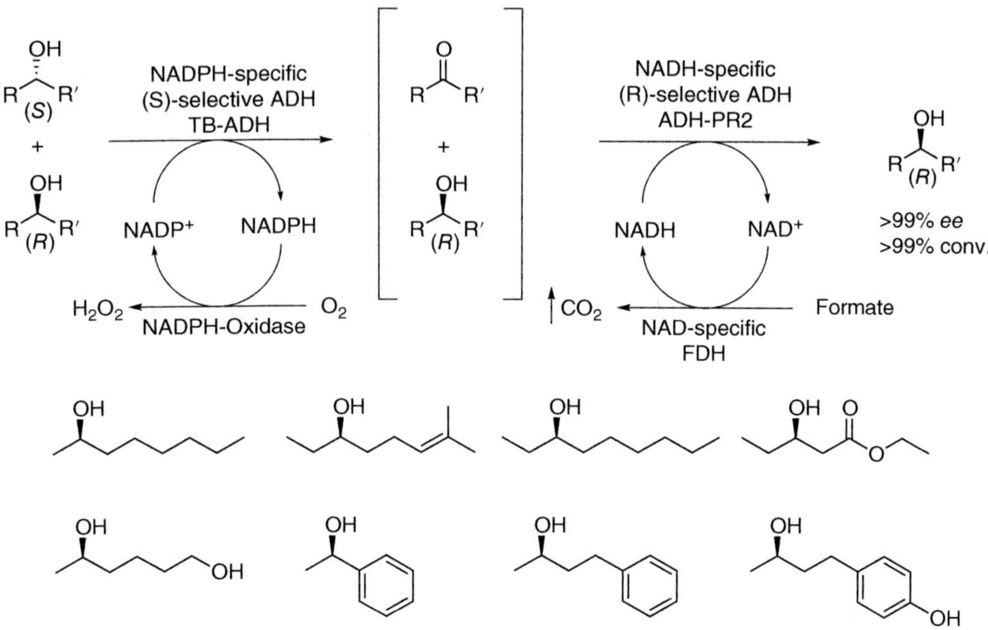

Scheme 12.22 *Concurrent oxidation and reduction to yield enantiopure (R)-secondary alcohols. The structures shown were synthesized in >99% conversion and >99% ee. A corresponding tandem reaction involving a (R)-selective oxidation and (S)-selective reduction furnished these products in excellent conversion to the enantiopure (S)-alcohol [334]. TB-ADH: ADH from Thermoanaerobium brocki; ADH-PR2: commercially available ADH*

12.7.7.1 Monoamine Oxidases

Both L-amino oxidase (EC 1.4.3.2) and D-amino oxidase (EC 1.4.3.3) have broad substrate scope in compounds bearing amino acid side chains and have been used for the deracemization of amino acids [337, 338]. Monoamine oxidase is a flavin-dependent amine oxidase (along with putrescine oxidase, diamine oxidase, etc.) and is in some cases a more useful biocatalyst as it can catalyze the oxidation of amines in substrates that do not have the adjacent α-carbonyl group [270]. Monoamine oxidase N from *Aspergillus niger* (MAO-N; EC 1.4.3.4) was first identified in 1995 [339, 340]. The wild-type enzyme catalyzes the oxidation of simple amines such as benzylamine to the corresponding imine [341].

Through seminal work by the Turner group, the substrate scope of the wild-type enzyme has been significantly broadened and the deracemization or selective oxidative deamination of primary [341], secondary [343] and tertiary amines is now possible [344]. This was achieved through a combination of directed evolution, rational design and high-throughput screening. The synthetic utility of MAO-N mutants is evidenced by the number of important substrates which have been successfully deracemized (examples in Scheme 12.23) [342]. A non-selective chemical reducing agent (BH_3-NH_3) for reduction of the imine (or iminium ion in the case of tertiary alcohols) intermediate allows the accumulation of a single enantiomer of the amine. This deracemization protocol was also implemented in the synthesis of several API building blocks, for example antihistamine levocetirizine and

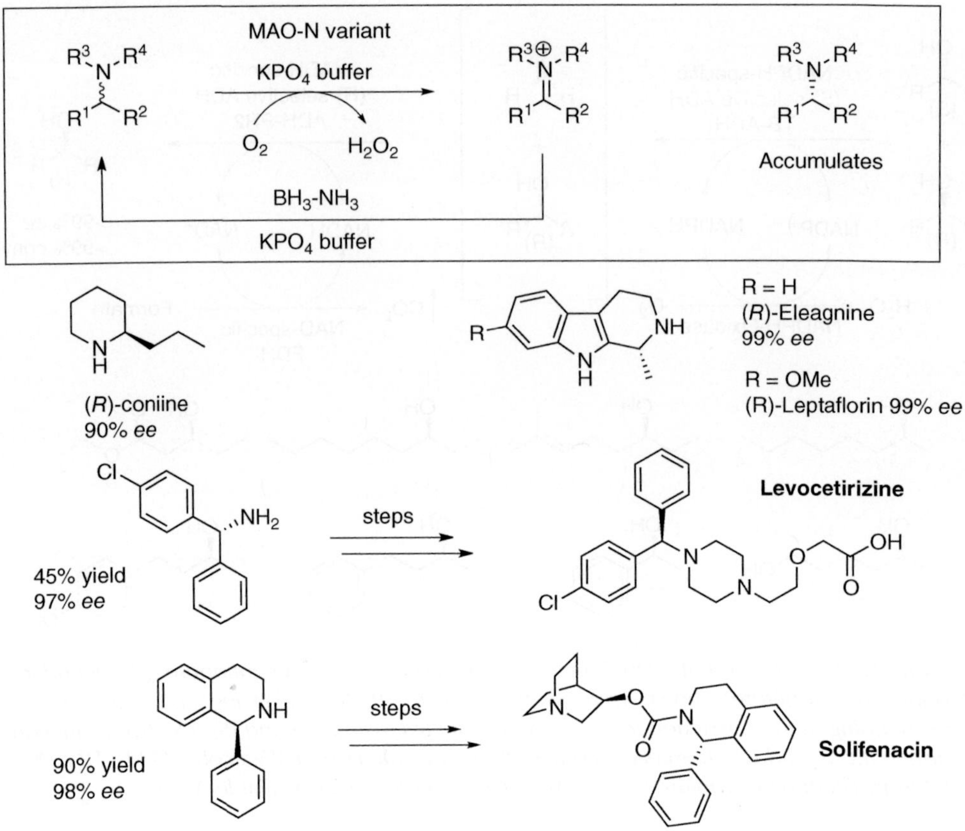

Scheme 12.23 *Application of MAO-N variants for the deracemization of natural products and important chiral building blocks [342]*

solifenacin (an antimuscarinic) – Scheme 12.23. MAO-N-catalyzed biotransformations have also been used for the resolution of intermediates in the synthesis of telaprevir and boceprevir, treatments for hepatitis C [345, 346].

12.7.7.2 C-O Oxidations using Oxidases

C-O oxidases have also been used in biocatalysis [270]. From a chirality point of view, they have been used in kinetic resolutions of chiral alcohols/hydroxyacids [330, 347, 348] or in deracemization reactions that involve a second, reductive enzyme [349, 350]. In particular, galactose oxidase mutants have shown high selectivity in kinetic resolutions involving aromatic secondary alcohols [351].

12.7.8 Other Oxidations

Oxygenases are another important and well studied class of enzyme, catalysing the selective incorporation of oxygen into C-H, C-C and C=C bonds, as well as oxyfunctionalization of several heteroatoms using molecular oxygen or hydrogen peroxide as the oxygen source [6, 352].

The Baeyer–Villiger reaction, which incorporates an oxygen atom into the position adjacent to a carbonyl group, is an important transformation in synthetic chemistry [353–356].

Baeyer–Villiger monooxygenases (BVMOs) are flavin-dependent enzymes [371], which have been applied in the oxidative kinetic resolution of chiral acyclic and cyclic ketones to give esters and lactones respectively [353]. Prominent examples include the kinetic resolution of benzyl ketones, esters and cycloalkanones with various substituents and also desymmetrizations of symmetrical substrates (representative examples in Table 12.8) [372]. Protein engineering of stereoselective Baeyer–Villiger monooxygenases has resulted in the resolution of quite an impressive range of compounds [373, 374].

Dynamic kinetic resolutions are also possible [354, 375] as exemplified by Gotor and coworkers. Here, the acidity of the α-proton in a 1, 3-dicarbonyl compound was utilized to deliver enantiomerically pure, functional group-rich products in quantitative yield in many cases (Scheme 12.24) [376]. Wild-type phenylacetone monooxygenase from *Thermobifida fusa* and 4-hydroxyacetophenone monooxygenase from *Pseudomonas fluorescens* were used for this transformation.

Baeyer–Villiger monooxygenases are also useful catalysts in asymmetric sulfoxidation [377, 378], which can be a challenging transformation via chemical synthesis due to overoxidation. Recently they have been applied for this purpose in the synthesis of esomeprazole (a proton pump inhibitor) and armodanafil (an analeptic drug for the treatment of narcolepsy) – Scheme 12.25 [379, 380].

Cytochrome P450 enzymes are a monooxygenase superfamily and are ubiquitous in nature. They catalyze a number of different chemical reactions [382], of which the most interesting is the selective oxidation of (unactivated) C-H bonds (using molecular oxygen) as this synthetically challenging transformation is very difficult to achieve via chemical synthesis. Unlike the BVMOs, they are heme-dependent. The selective hydroxylation of steroids has been in use industrially since the 1970s [383, 384] and is an excellent example of the high selectivity of these enzymes (Scheme 12.26) [385, 386].

Unsurprisingly, due to their ability to stereoselectively hydroxylate unactivated C-H bonds under mild conditions, P450 monooxygenases have received a large amount of attention in the literature [387–390]. Unfortunately, due to their tendency for low activity, their limited stability and poor solvent tolerance, P450s have not (as yet) been widely exploited in industrial biocatalysis [391–393]. Evolved P450 biocatalysts have also been studied for the purpose of improving the stability and substrate tolerance of this class of enzyme [394, 395].

Other exciting oxidative enzymes worthy of mention here are the dioxygenases, which catalyze selective *cis*-dihydroxylation of arenes [396, 397] and the peroxygenases, which are structurally related to the P450 monooxygenases and therefore could effect similar transformations [398–401]. In contrast to the P450s (which require reduced nicotinamide cofactors), peroxygenases only use hydrogen peroxide as the cosubstrate. Peroxygenases can catalyze achiral transformations but from an asymmetric viewpoint have been used in stereoselective sulfoxidations, epoxidations of olefins and in hydroxylation reactions of sp^3-hybridized C-H bonds [396]. Some representative highly stereoselective peroxygenase-mediated reactions are shown in Table 12.9.

Dioxygenases are also exciting biocatalysts as they offer a route to *cis*-diols from arenes and there is no known chemical equivalent to this reaction. Many *cis*-diols have been

Table 12.8 *Examples of reactions catalyzed by BVMOs in asymmetric synthesis*

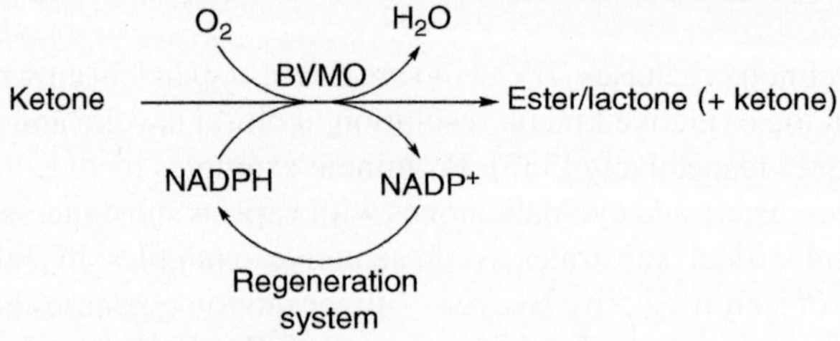

Substrate	Product(s)	Ref.
Kinetic resolution		
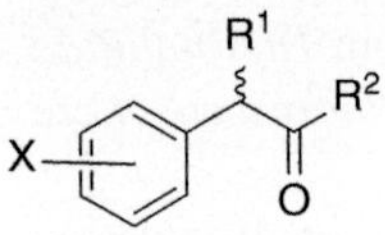 R^1, R^2 = alk X = H, alk, hal	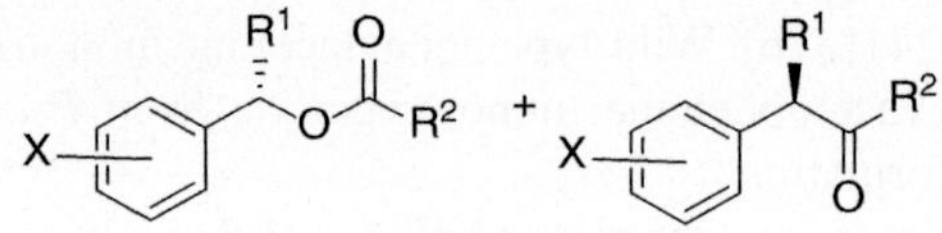 *E*: up to >200	[357–359]
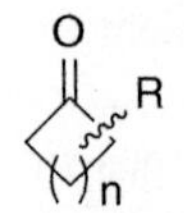 n = 1,2,3 R = Alk, Ar	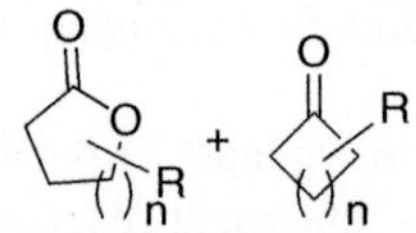 *E*: up to >200	[360–364]
Desymmetrization		
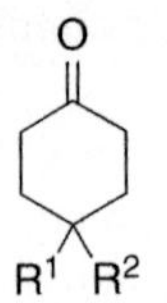 R^1 = Alk, Ar R^2 = H, Alk, Ar	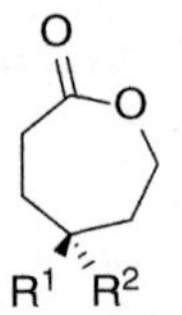 up to 99% *ee*	[365–367]
Regiodivergent oxidation		
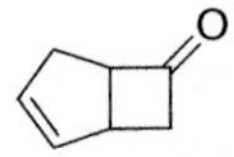	'normal' lactone 'abnormal' lactone 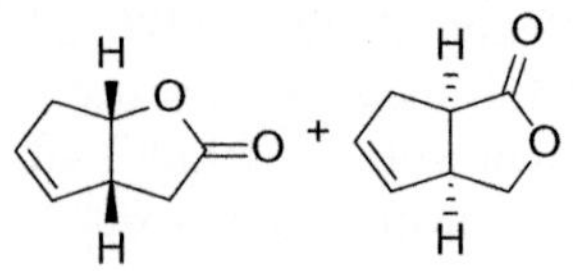 >99% *ee* both lactones	[368–370]

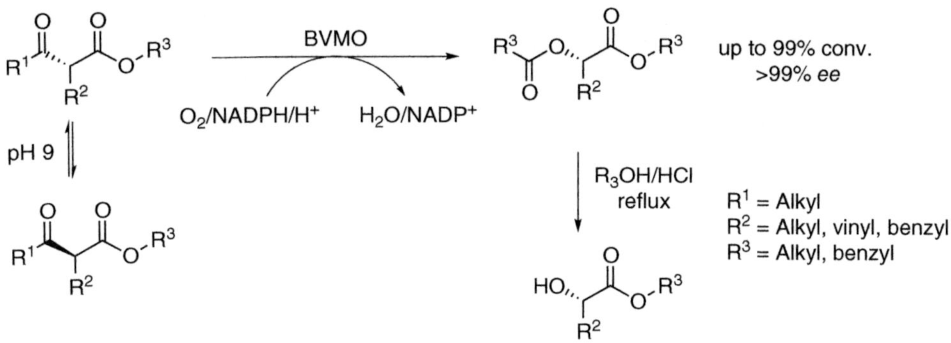

Scheme 12.24 *BVMO-catalyzed dynamic kinetic resolution of β-ketoesters to yield enantioenriched α-hydroyxesters [376]*

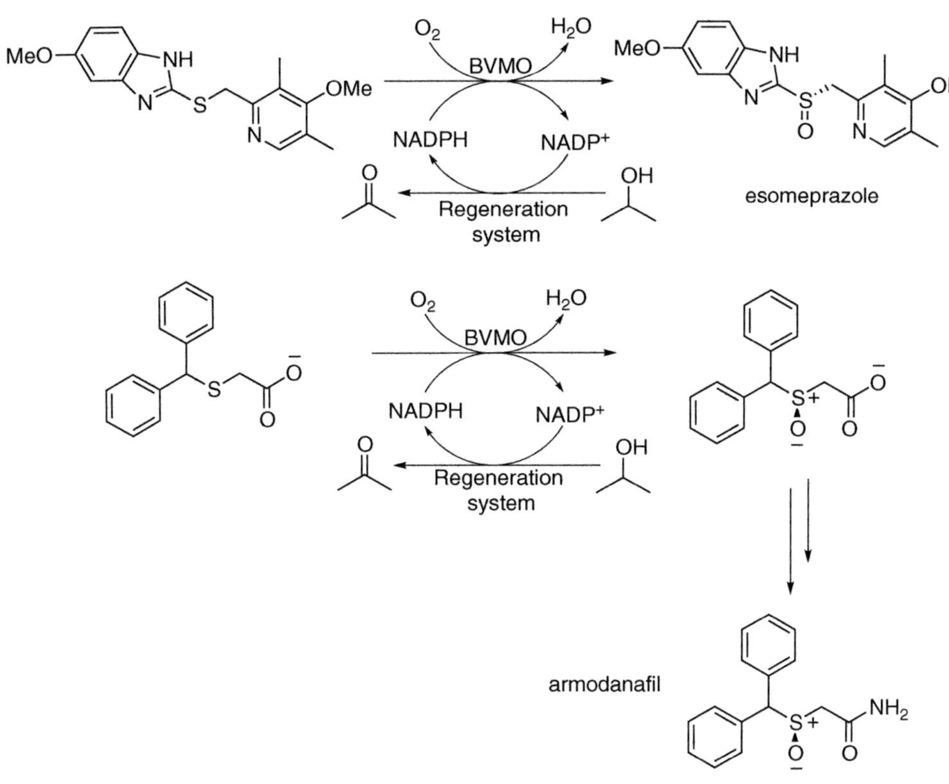

Scheme 12.25 *BVMO-mediated asymmetric sulfoxidations applied to the synthesis of two pharmaceutically important compounds [379–381]*

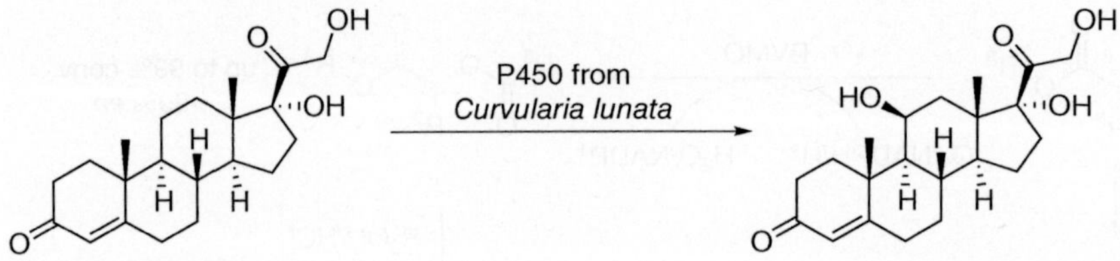

Scheme 12.26 *An example of a P450-mediated stereoselective hydroxylation used in the pharmaceutical industry [386]*

Table 12.9 *Representative highly enantioselective peroxygenase-mediated reactions*

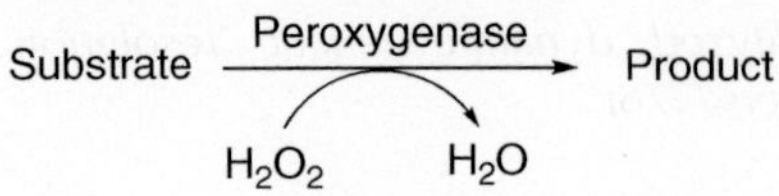

Substrate	Product	Ref.
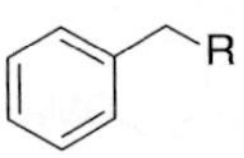	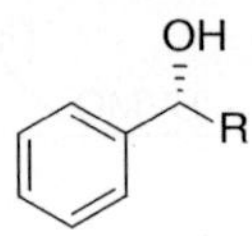	[402]
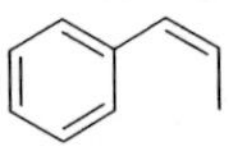	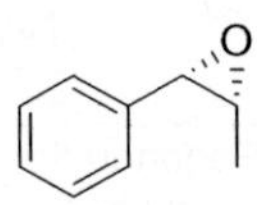	[402]
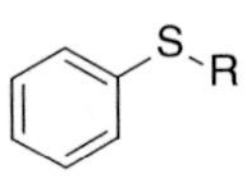	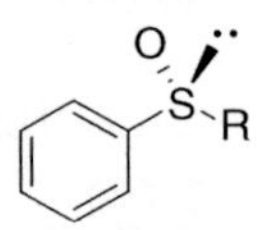	[403]
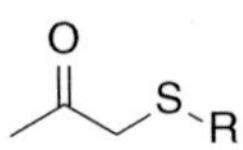	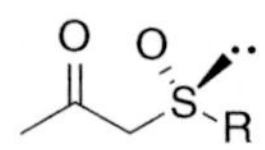	[404]

synthesized via dioxygenase-mediated dihydroxylation of the corresponding aromatic compound [255, 405–407]. Despite this, preparative-scale syntheses are rare. A selection of compounds which have been synthesized using dioxygenases as biocatalysts is shown in Scheme 12.27 [408–413].

12.8 C-C and Other C-X Bond Formation

Biocatalysts are also efficient promoters of synthetically useful carbon-carbon and carbon-heteroatom bond-forming reactions. Some examples will be discussed in the following pages.

12.8.1 C-C Bond Formation

Asymmetric C-C bond formation is an integral part of organic synthesis. Non-enzymatic, stereospecific methods tend to use metal or organocatalytic enolate complexes, which often

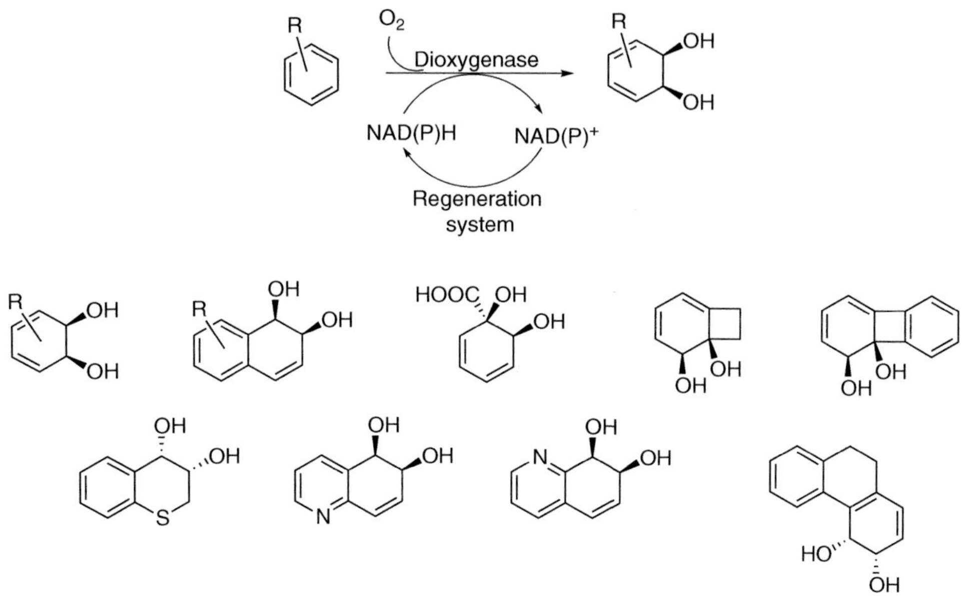

Scheme 12.27 *Selected cis-diols that have been synthesized using dioxygenase biocatalysts [408–413]*

require low temperature and careful design of reaction conditions. Because of this, enzymes that can catalyze this type of transformation have been investigated extensively since the mid-1990s [414–417]. While hydrolases have been found to catalyze C-C bond formations [418–421], aldolases and hydroxynitrile lyases are more frequently used.

12.8.1.1 Hydroxynitrile Lyases

It was a hydroxynitrile lyase (HNL) that was used as a catalyzt in the first enantioselective synthesis described, over 100 years ago, in the formation of enantiomerically pure mandelic acid [422]. The enzyme used, from almonds – *Prunus amygdalus* (PaHNL) – was investigated further many years later and was found to be (*R*)-selective; today, however, both (*R*)- and (*S*)-selective HNLs are available commercially and a wide range of substrates can be used [423, 424]. This class of enzyme catalyze the (stereoselective) addition of cyanide to aldehydes (and to some extent methyl ketones) [425] to give synthetically useful cyanohydrins (Scheme 12.28) [426].

Two issues often encountered in this transformation are the spontaneous non-enzymatic (and thus non-stereoselective) cyanohydrin formation and racemization of the product due to reaction equilibration [430]. Common techniques employed to circumvent these unwanted complications are lowering the pH of the solution and minimizing the water activity of the medium through addition of a water-miscible cosolvent [8, 423]. Immobilization and evolution of HNLs have been used to improve enzyme properties, particularly for the purpose of increasing stability at lower pH [425, 430–434]. Selected compounds that have recently been preparatively synthesized via HNLs are shown in Scheme 12.28.

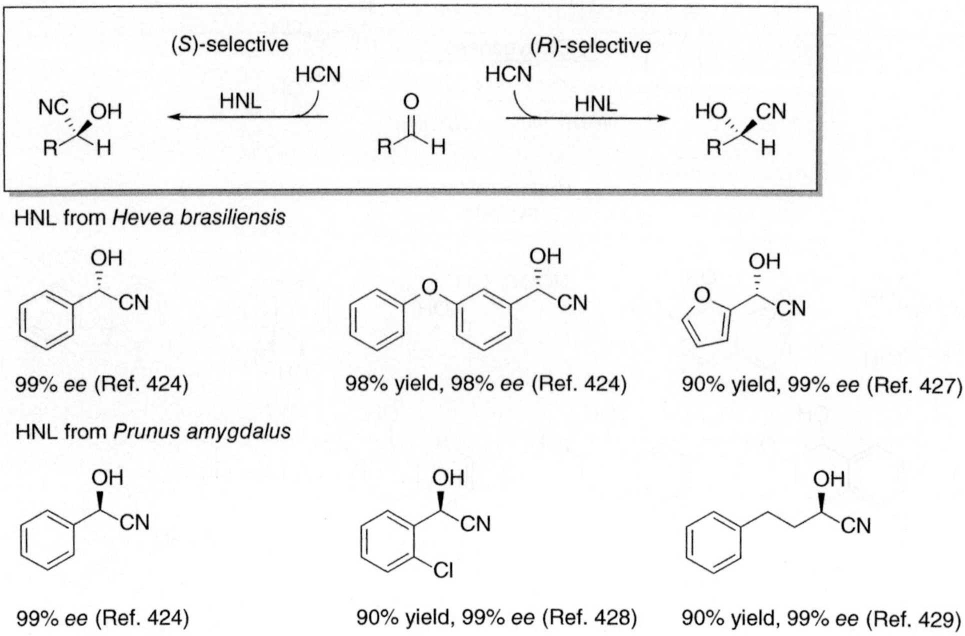

Scheme 12.28 *Selected compounds prepared using hydroxynitrile lyase biocatalysts*

12.8.1.2 Aldolases

In nature, aldolases (aldehyde lyases) catalyze the reverse-aldol reaction and are involved in carbohydrate metabolism; however, in biocatalysis they are employed for the forward reaction as it is more synthetically useful [6]. A donor molecule (usually a ketone) is stereoselectively added to an acceptor (an aldehyde). Aldolases can be split into two groups; class I aldolases activate the donor via Schiff base (imine) formation and class II use a metal (Zn^{2+})-enolate to activate. As a general rule, these enzymes tolerate a broad substrate scope for the acceptor but are much more selective toward the structure of the donor. Therefore, more specifically, they have been classified according to their donor preference into four groups (Table 12.10) [6, 8, 415].

Aldolase reactions are usually characterized by high enantioselectivity, and where applicable, diastereoselectivity in the products that are formed. Engineered aldolases have also been explored as biocatalysts [435] and an impressive range of chiral compounds are accessible [8, 436, 437].

12.8.2 Halohydrin Dehalogenases

Ring opening of epoxides is a useful tool in the synthesis of β-functionalized alcohols. Epoxide hydrolases have been used for the synthesis of enantiopure diols and the non-converted epoxides (Section 12.6.3). Halohydrin dehalogenases (EC 4.5.1.X) can also be used for this purpose [438, 439]. Unlike epoxide hydrolases, they are not limited to the use of water as a nucleophile and therefore they offer a route to a wider range of chiral building blocks [438]. Halohydrin dehalogenase (HheC) from *Agrobacterium radiobacter* AD1 has

Table 12.10 *Classification of aldolases according to their donor preference [6,8,415]*

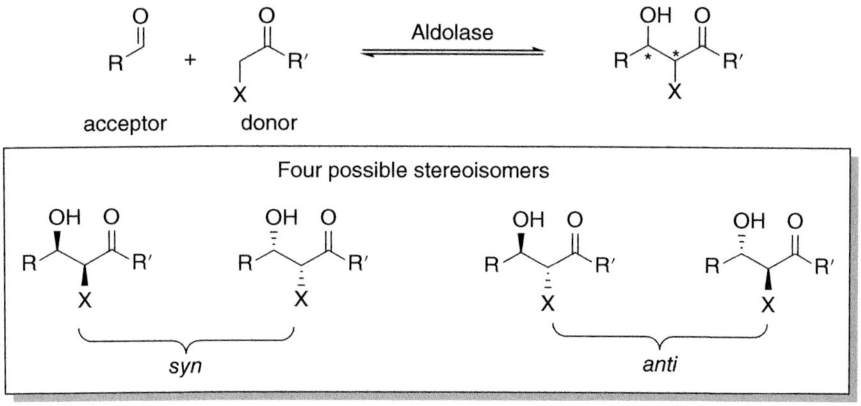

Name	Donor	Acceptor	Product
DHAP-dependent aldolases	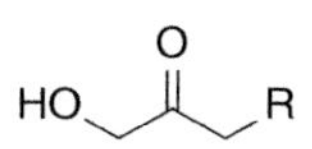 $R = PO_3^{2-}$ or H	Aldehyde	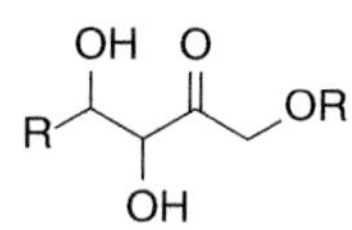
Glycine-dependent aldolases	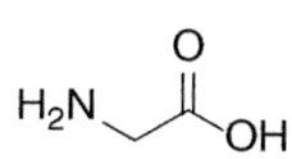	Aldehyde	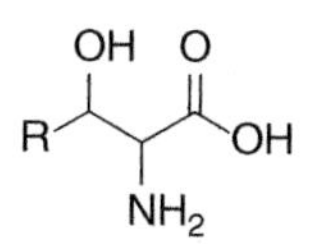
Acetaldehyde-dependent aldolases	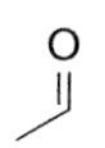	Aldehyde	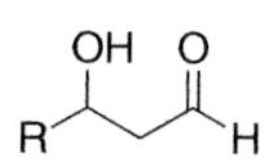
Pyruvate-dependent aldolases	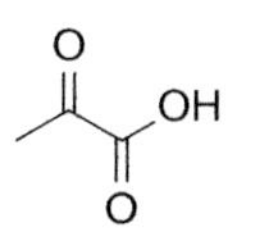	Aldehyde	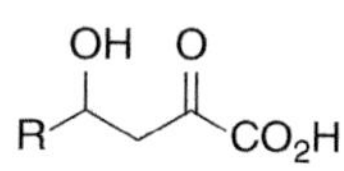

been used in kinetic resolutions involving alkyl epoxide substrates, showing excellent activity and enantioselectivity for a range of different nucleophiles [440, 441]. Later, the scope of the resolution was expanded to include arylaliphatic epoxides, using an azide nucleophile (Scheme 12.29) [442]. Halohydrin dehalogenases have also found application in the pharmaceutical industry in the synthesis of atorvastatin [282].

12.8.3 Nitrile Hydratases

Nitrile hydratases (EC 4.2.1.84) are lyases that catalyze the hydration of a nitrile functionality to an amide [443]. Unlike nitrilases (Section 12.6.5), which hydrolyse the cyano group all the way to the carboxylic acid, nitrile hydratases form a carboxamide, which can be further hydrolyzed by an amidase to furnish the acid (Scheme 12.30), typically in a whole

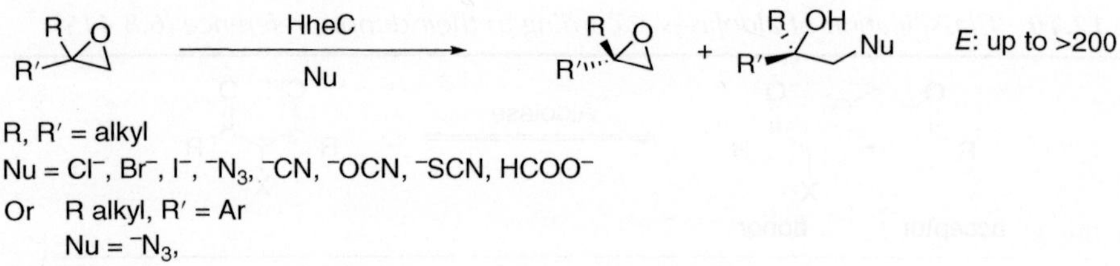

Scheme 12.29 *Halohydrin dehalogenase-mediated kinetic resolutions using a range of nucleophiles for the ring opening of 2, 2-disubstituted epoxides [440, 442]*

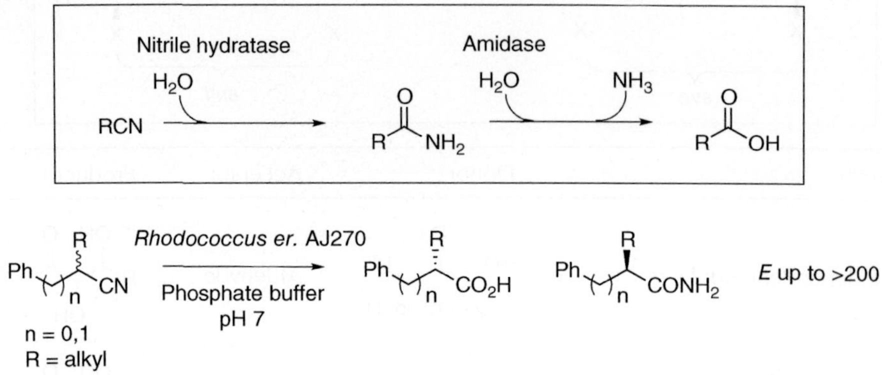

Scheme 12.30 *Nitrile hydratase-mediated hydration of an amide followed by amidase-catalyzed hydrolysis, and their use in the synthesis of enantiopure amides and carboxylic acids [445–447]*

cell system. Nitrile hydratases can be enantioselective in their reactivity and have recently been exploited for this purpose [443, 444]. An example of the enantioselectivity derived a nitrile hydratase/amidase system is shown in Scheme 12.25, whereby *Rhodococcus erythropolis* AJ270 (a microorganism containing both enzymes) was used to furnish enantiopure amides and acids from the corresponding cyano compound [445–447]. This type of system has also been used in the synthesis of optically active α-amino amides and also in desymmetrization reactions [448–451].

12.8.4 Addition of H_2O/NH_3 to C=C Bonds

Nature has provided several hydratases (or hydro-lyases) to promote the stereoselective addition of water to both electron-rich and electron-poor C=C bonds [452, 453]. This is another illustration of the remarkable power of enzymes, given that water is a very small molecule and is a poor nucleophile (without activation) [452]. The most prominent example is the fumarase-catalyzed production of L-malate from fumarate, which is conducted on a multi-ton scale by many companies (Scheme 12.31) [227]. The reaction is performed using solid calcium fumarate and yields solid (*S*)-calcium malate.

Ammonia can also be asymmetrically added to a C=C-double bond of an unsaturated acid (catalyzed by several lyases), allowing access to chiral α-amino acids [454, 455]. Aspartases or aspartate ammonia lyases (EC 4.3.1.1) are industrially important, catalysing the reversible amination of fumarate to L-aspartic acid (shown in Scheme 12.2) [456].

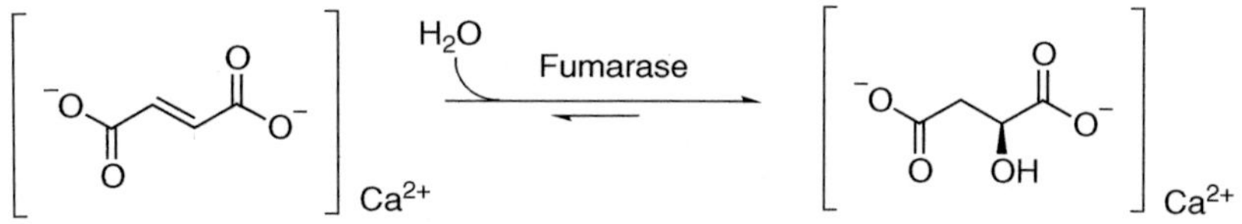

Scheme 12.31 *Industrial production of L-malate from fumarate using fumarase biocatalysts*

Another interesting enzyme in this family is 3-methylaspartate ammonia lyase (MAL; EC 4.3.1.2). The wild-type organism exhibits a narrow substrate scope [457]; however, engineered MAL accepts a broader spectrum of nucleophiles, including linear and cyclic amines, enabling the synthesis of enantiomerically pure *N*-substituted amino acids [458].

12.9 Future and Outlook

Biocatalysis certainly has become established as a powerful synthetic tool for the preparation of enantiopure compounds and this approach continues to play an ever increasing role in synthetic chemistry. This has been especially enabled through the varying approaches to enzyme engineering and stabilization. Owing to the potential environmental and economic advantages associated with biocatalytic methodologies it will continue to grow, particularly as new emerging technologies take a foothold [459]. It is predicted that new functional group transformations will be possible through biocatalysis in the near future [6, 7, 36, 460] and these newly emerging domains will doubtlessly pave the way for further development in the field, while the more established areas will continue to play a pivotal part in the production of chiral chemicals.

References

[1] FDA's policy statement for the development of new stereoisomeric drugs (1992). *Chirality*, **4** (5), 338–340.
[2] Jamali, F., Mehvar, R., Pasutto, F.M. (1989) Enantioselective aspects of drug action and disposition: therapeutic pitfalls. *J. Pharm. Sci.*, **78** (9), 695–715.
[3] Lu, H. (2007) Stereoselectivity in drug metabolism. *Expert Opin. Drug Metab. Toxicol.*, **3** (2), 149–158.
[4] Patel, R.N. (2008) Chemo-enzymatic synthesis of pharmaceutical intermediates. *Expert Opin. Drug Discov.*, **3** (2), 187–245.
[5] Reetz, M.T. (2013) Biocatalysis in organic chemistry and biotechnology: past, present, and future. *J. Am. Chem. Soc.*, **135** (34), 12480–12496.
[6] Torrelo, G., Hanefeld, U., Hollmann, F. (2015) Biocatalysis. *Catal. Lett.*, **145** (1), 309–345.
[7] Nestl, B.M., Hammer, S.C., Nebel, B.A., Hauer, B. (2014) New generation of biocatalysts for organic synthesis. *Angew. Chem. Int. Ed.*, **53** (12), 3070–3095.
[8] Faber, K. (2011) *Biotransformations in Organic Chemistry*, 6th edn, Springer, Heidelberg.
[9] Liese, A., Seebach, K., Wandrey, C. (eds) (2006) *Industrial Biotransformations*, 2nd edn, Wiley-VCH Verlag GmbH & Co. KGaA, Weinheim.
[10] Bommarius, A.S., Riebel, B. R. (2004) *Biocatalysis: Fundamentals and Applications*, Wiley-VCH Verlag-GmBH & KGaA, Weinheim.
[11] Davis, B.G., Boyer, V. (2001) Biocatalysis and enzymes in organic synthesis. *Nat. Prod. Rep.*, **18** (6), 618–640.
[12] Milner, S.E., Maguire, A. R. (2012) Recent trends in whole cell and isolated enzymes in enantioselective synthesis. *Arkivoc*, (**i**), 321–382.

[13] Bornscheuer, U.T., Kazlauskas, R.J. (2004) Catalytic promiscuity in biocatalysis: using old enzymes to form new bonds and follow new pathways. *Angew. Chem. Int. Ed.*, **43** (45), 6032–6040.
[14] Glick, B.R., Pasternak, J.J., Patten, C.L. (2010) *Molecular Biotechnology: Principles and Applications of Recombinant DNA*, 4th edn, ASM Press, Washington DC.
[15] Bergquist, P.L., Morgan, H. W., Saul, D. (2014) Selected enzymes from extreme thermophiles with applications in biotechnology. *Curr. Biotechnol.*, **3** (1), 45–59.
[16] Antranikian, G., Vorgias, C.E., Bertoldo, C. (2005) Extreme environments as a resource for microorganisms and novel biocatalysts, in *Marine biotechnology I* (eds R. Ulber and Y. Le Gal), Springer, Heidelberg, pp. 219–262.
[17] Oren, A. (2015) Halophilic microbial communities and their environments. *Curr. Opin. Biotechnol.*, **33**, 119–124.
[18] Aldridge, S. (2013) Industry backs biocatalysis for greener manufacturing. *Nat. Biotech.*, **31** (2), 95–96.
[19] Pollard, D.J., Woodley, J.M. (2007) Biocatalysis for pharmaceutical intermediates: the future is now. *Trends Biotechnol.*, **25** (2), 66–73.
[20] Tao, J., Xu, J.-H. (2009) Biocatalysis in development of green pharmaceutical processes. *Curr. Opin. Chem. Biol.*, **13** (1), 43–50.
[21] Hoyos, P., Pace, V., Hernaiz, M. J., Alcantara, A.R. (2014) Biocatalysis in the pharmaceutical industry. A Greener Future. *Curr. Green Chem.*, **1** (2), 155–181.
[22] Patel, R.N. (2006) Biocatalysis: synthesis of chiral intermediates for drugs. *Curr. Opin. Drug Discov. Devel.*, **9** (6), 741–764.
[23] Patel, R.N. (2003) Microbial/enzymatic synthesis of chiral pharmaceutical intermediates. *Curr. Opin. Drug Discov. Devel.*, **6** (6), 902–920.
[24] Patel, R.N. (2013) Biocatalytic synthesis of chiral alcohols and amino acids for development of pharmaceuticals. *Biomolecules*, **3** (4), 741–777.
[25] Patel, R., Hanson, R., Goswami, A. *et al.* (2003) Enzymatic synthesis of chiral intermediates for pharmaceuticals. *J. Ind. Microbiol. Biotechnol.*, **30** (5), 252–259.
[26] Zaks, A. (2001) Industrial biocatalysis. *Curr. Opin. Chem. Biol.*, **5** (2), 130–136.
[27] Nestl, B.M., Nebel, B.A., Hauer, B. (2011) Recent progress in industrial biocatalysis. *Curr. Opin. Chem. Biol.*, **15** (2), 187–193.
[28] Patel, R.N. (2008) Synthesis of chiral pharmaceutical intermediates by biocatalysis. *Coord. Chem. Rev.*, **252** (5–7), 659–701.
[29] Patel, R., Banerjee, A., Szarka, L. (1995) Synthesis of four chiral pharmaceutical intermediates by biocatalysis. *J. Am. Oil Chem. Soc.*, **72** (11), 1247–1264.
[30] Choi, J.-M., Han, S.-S., Kim, H.-S. Industrial applications of enzyme biocatalysis: Current status and future aspects. *Biotechnol. Adv.*, **33** (7), 1443–1454.
[31] Tufvesson, P., Lima-Ramos, J., Haque, N.A. *et al.* (2013) Advances in the process development of biocatalytic processes. *Org. Process Res. Dev.*, **17** (10), 1233–1238.
[32] Ciriminna, R., Pagliaro, M. (2013) Green chemistry in the fine chemicals and pharmaceutical industries. *Org. Process Res. Dev.*, **17** (12), 1479–1484.
[33] Wu, Q., Tao, J. (2012) Biocatalysis, in *Green Techniques for Organic Synthesis and Medicinal Chemistry* (eds W. Zhang and B. Cue Jr.), John Wiley & Sons, Ltd, Chichester, pp. 217–240.
[34] Moody, T., Brown, G. (2012) Making API synthesis greener. *Pharm. Technol. Eur.*, **24** (9), 44–46.
[35] Birmingham, W.R., Starbird, C.A., Panosian, T.D. *et al.* (2014) Bioretrosynthetic construction of a didanosine biosynthetic pathway. *Nat. Chem. Biol.*, **10** (5), 392–399.
[36] Turner, N.J., O'Reilly, E. (2013) Biocatalytic retrosynthesis. *Nat. Chem. Biol.*, **9** (5), 285–288.
[37] http://www.chem.qmul.ac.uk/iubmb/enzyme/ (accessed 19 April 2016).
[38] McDonald, A.G., Boyce, S., Tipton, K.F. (2015) Enzyme classification and nomenclature, in *Encyclopeida of Life Sciences*, John Wiley & Sons, Ltd, Chichester.
[39] McDonald, A.G., Tipton, K.F. (2014) Fifty-five years of enzyme classification: advances and difficulties. *FEBS J.*, **281** (2), 583–592.
[40] Tipton, K., Boyce, S. (2000) History of the enzyme nomenclature system. *Bioinformatics*, **16** (1), 34–40.
[41] http://www.enzyme-database.org/ (accessed 19 April 2016)
[42] http://www.epa.gov/greenchemistry/pubs/epa_gc.html.2010 (accessed 21 April 2016).

[43] Anastas, P., Eghbali, N. (2010) Green chemistry: principles and practice. *Chem. Soc. Rev.*, **39** (1), 301–312.
[44] Wolfenden, R., Snider, M.J. (2001) The depth of chemical time and the power of enzymes as catalysts. *Acc. Chem. Res.*, **34** (12), 938–945.
[45] Menger, F.M. (1993) Enzyme reactivity from an organic perspective. *Acc. Chem. Res.*, **26** (4), 206–212.
[46] Zechel, D.L., Withers, S.G. (1999) Glycosidase mechanisms: anatomy of a finely tuned catalyst. *Acc. Chem. Res.*, **33** (1), 11–18.
[47] Tanaka, A., Tosa, T., Kobayashi, A. (eds) (1993) *Industrial Application of Immobilized Biocatalysts (Biotechnology and Bioprocessing)*, Marcel Dekker Inc., New York.
[48] Calvin, S.J., Mangan, D., Miskelly, I. *et al.* (2012) Overcoming equilibrium issues with carbonyl reductase enzymes. *Org. Process Res. Dev.*, **16** (1), 82–86.
[49] Sutton, P.W., Adams, J.P., Archer, I. *et al.* (2012) Biocatalysis in the fine chemical and pharmaceutical industries, in *Practical Methods for Biocatalysis and Biotransformations 2* (eds J. Whithall and P. Sutton), John Wiley & Sons, Ltd, Chichester, pp. 1–59.
[50] Ogawa, J., Shimizu, S. (2002) Industrial microbial enzymes: their discovery by screening and use in large-scale production of useful chemicals in Japan. *Curr. Opin. Biotechnol.*, **13** (4), 367–375.
[51] Rieckenberg, F., Ardao, I., Rujananon, R., Zeng, A.-P. (2014) Cell-free synthesis of 1, 3-propanediol from glycerol with a high yield. *Eng. Life Sci.*, **14** (4), 380–386.
[52] Sabra, W., Dietz, D., Tjahjasari, D., Zeng, A.-P. (2010) Biosystems analysis and engineering of microbial consortia for industrial biotechnology. *Eng. Life Sci.*, **10** (5), 407–421.
[53] Abbasov, M.E., Romo, D. (2014) The ever-expanding role of asymmetric covalent organocatalysis in scalable, natural product synthesis. *Nat. Prod. Rep.*, **31** (10), 1318–1327.
[54] Dalko, P.I. (ed) (2013) *Comprehensive Enantioselective Organocatalysis: Catalysts, Reactions, and Applications*, Wiley-VCH Verlag GmbH & Co. KGaA, Weinheim.
[55] Jacobsen, E.N., Pfaltz, A., Yamamoto, H. (eds) (2011) *Comprehensive Asymmetric Catalysis*, Springer, Berlin.
[56] Christmann, M., Braese, S. (eds) (2008) *Asymmetric Synthesis – The Essentials*, 2nd edn, Wiley-VCH, Verlag GmBH&Co., Weinheim.
[57] Stecher, H., Faber, K. (1997) Biocatalytic deracemization techniques: dynamic resolutions and stereoinversions. *Synthesis*, **01**, 1–16.
[58] Strauss, U.T., Felfer, U., Faber, K. (1999) Biocatalytic transformation of racemates into chiral building blocks in 100% chemical yield and 100% enantiomeric excess. *Tetrahedron: Asymmetry*, **10** (1), 107–117.
[59] Schmid, A., Dordick, J.S., Hauer, B. *et al.* (2001) Industrial biocatalysis today and tomorrow. *Nature*, **409**, 258–268.
[60] Kumar, P., Naidu, V., Gupta, P. (2007) Application of hydrolytic kinetic resolution (HKR) in the synthesis of bioactive compounds. *Tetrahedron*, **63** (13), 2745–2785.
[61] Ghanem, A. (2007) Trends in lipase-catalyzed asymmetric access to enantiomerically pure/enriched compounds. *Tetrahedron*, **63** (8), 1721–1754.
[62] Kamal, A., Azhar, M.A., Krishnaji, T. *et al.* (2008) Approaches based on enzyme mediated kinetic to dynamic kinetic resolutions: a versatile route for chiral intermediates. *Coord. Chem. Rev.*, **252** (5–7), 569–592.
[63] Chen, C.S., Fujimoto, Y., Girdaukas, G., Sih, C.J. (1982) Quantitative analyses of biochemical kinetic resolutions of enantiomers. *J. Am. Chem. Soc.*, **104** (25), 7294–7299.
[64] Bornscheuer, U.T., Kazlauskus, R.J. (2006) *Hydrolases in Organic Synthesis – Regio- and Stereoselective Biotransformations* 2nd edn, Wiley-VCH Verlag GmbH & Co. KGaA, Weinheim.
[65] Kroutil, W., Kleewein, A., Faber, K. (1997) A computer program for analysis, simulation and optimization of asymmetric catalytic processes proceeding through two consecutive steps. Type 1: asymmetrization-kinetic resolutions. *Tetrahedron: Asymmetry*, **8** (19), 3251–3261.
[66] http://biocatalysis.uni-graz.at/enantio/cgi-bin/enantio.pl (accessed 21 April 2016).
[67] Steinreiber, J., Faber, K., Griengl, H. (2008) De-racemization of enantiomers versus de-epimerization of diastereomers – classification of dynamic kinetic asymmetric transformations (DYKAT). *Chem. Eur. J.*, **14** (27), 8060–8072.

[68] Martín-Matute, B., Bäckvall, J.-E. (2007) Dynamic kinetic resolution catalyzed by enzymes and metals. *Curr. Opin. Chem. Biol.*, **11** (2), 226–232.
[69] Verho, O., Bäckvall, J.-E. (2015) Chemoenzymatic dynamic kinetic resolution: a powerful tool for the preparation of enantiomerically pure alcohols and amines. *J. Am. Chem. Soc.*, **137** (12), 3996–4009.
[70] Turner, N.J. (2010) Deracemization methods. *Curr. Opin. Chem. Biol.*, **14** (2), 115–121.
[71] García-Urdiales, E., Alfonso, I., Gotor, V. (2011) Update 1 of: Enantioselective enzymatic desymmetrizations in organic synthesis. *Chem. Rev.*, **111** (5), 110–180.
[72] Cantone, S., Hanefeld, U., Basso, A. (2007) Biocatalysis in non-conventional media-ionic liquids, supercritical fluids and the gas phase. *Green Chem.*, **9** (9), 954–971.
[73] Bommarius, A.S., Paye, M.F. (2013) Stabilizing biocatalysts. *Chem. Soc. Rev.*, **42** (15), 6534–6565.
[74] Griebenow, K., Klibanov, A.M. (1996) On protein denaturation in aqueous–organic mixtures but not in pure organic solvents. *J. Am. Chem. Soc.*, **118** (47), 11695–11700.
[75] Klibanov, A.M. (2001) Improving enzymes by using them in organic solvents. *Nature*, **409**, 241–246.
[76] Hudson, E.P., Eppler, R.K., Clark, D.S. (2005) Biocatalysis in semi-aqueous and nearly anhydrous conditions. *Curr. Opin. Biotechnol.*, **16** (6), 637–643.
[77] Carrea, G., Riva, S. (eds) (2008) *Organic Synthesis with Enzymes in Non-aqueous Media*, Wiley-VCH Verlag GmbH & Co. KGaA, Weinheim.
[78] Doukyu, N., Ogino, H. (2010) Organic solvent-tolerant enzymes. *Biochem. Eng. J.*, **48** (3), 270–282.
[79] Iyer, P.V., Ananthanarayan, L. (2008) Enzyme stability and stabilization – aqueous and non-aqueous environment. *Process Biochem.*, **43** (10), 1019–1032.
[80] Azevedo, A.M., Prazeres, D.M.F., Cabral, J.M.S. *et al.* (2001) Stability of free and immobilised peroxidase in aqueous–organic solvents mixtures. *J. Mol. Catal. B: Enzym.*, **15** (4–6), 147–153.
[81] Sheldon, R.A. (2007) Enzyme immobilization: the quest for optimum performance. *Adv. Synth. Catal.*, **349** (8–9), 1289–1307.
[82] Khalaf, N., Govardhan, C.P., Lalonde J.J. *et al.* (1996) Cross-linked enzyme crystals as highly active catalysts in organic solvents. *J. Am. Chem. Soc.*, **118** (23), 5494–5495.
[83] Stepankova, V., Bidmanova, S., Koudelakova, T. *et al.* (2013) Strategies for stabilization of enzymes in organic solvents. *ACS Catal.*, **3** (12), 2823–2836.
[84] Eş, I., Vieira, J., Amaral, A. (2015) Principles, techniques, and applications of biocatalyst immobilization for industrial application. *Appl. Microbiol. Biotechnol.*, **99** (5), 2065–2082.
[85] Cantone, S., Ferrario, V., Corici, L. *et al.* (2013) Efficient immobilisation of industrial biocatalysts: criteria and constraints for the selection of organic polymeric carriers and immobilisation methods. *Chem. Soc. Rev.*, **42** (15), 6262–6276.
[86] Sheldon, R.A. (2006) Immobilization of enzymes as cross-linked enzyme aggregates, in *Biocatalysis in the Pharmaceutical and Biotechnology Industries* (ed R.N. Patel), CRC Press, Boca Raton.
[87] Singh, R., Tiwari, M., Singh, R., Lee, J.-K. (2013) From protein engineering to immobilization: promising strategies for the upgrade of industrial enzymes. *Int. J. Mol. Sci.*, **14** (1), 1232–1277.
[88] Mateo, C., Palomo, J.M., Fernandez-Lorente, G. *et al.* (2007) Improvement of enzyme activity, stability and selectivity via immobilization techniques. *Enzyme Microb. Technol.*, **40** (6), 1451–1463.
[89] Welton, T. (1999) Room-temperature ionic liquids. Solvents for synthesis and catalysis. *Chem. Rev.*, **99** (8), 2071–2084.
[90] Wasserscheid, P., Welton, T. (2003) *Ionic Liquids in Synthesis*, Wiley-VCH Verlag GmbH & Co. KGaA, Weinheim.
[91] Gao, W.-W., Zhang, F.-X., Zhang, G.-X., Zhou, C.-H. (2015) Key factors affecting the activity and stability of enzymes in ionic liquids and novel applications in biocatalysis. *Biochem. Eng. J.*, **99**, 67–84.
[92] Kragl, U., Eckstein, M., Kaftzik, N. (2002) Enzyme catalysis in ionic liquids. *Curr. Opin. Biotechnol.*, **13** (6), 565–571.
[93] Sureshkumar, M., Lee, C.-K. (2009) Biocatalytic reactions in hydrophobic ionic liquids. *J. Mol. Catal. B: Enzym.*, **60** (1–2), 1–12.

[94] van Rantwijk, F., Sheldon, R.A. (2007) Biocatalysis in ionic liquids. *Chem. Rev.*, **107** (6), 2757–2785.

[95] Matzke, M., Stolte, S., Boschen, A., Filser, J. (2008) Mixture effects and predictability of combination effects of imidazolium based ionic liquids as well as imidazolium based ionic liquids and cadmium on terrestrial plants (*Triticum aestivum*) and limnic green algae (*Scenedesmus vacuolatus*). *Green Chem.*, **10** (7), 784–792.

[96] Matzke, M., Stolte, S., Arning, J. *et al.* (2008) Imidazolium based ionic liquids in soils: effects of the side chain length on wheat (*Triticum aestivum*) and cress (*Lepidium sativum*) as affected by different clays and organic matter. *Green Chem.*, **10** (5), 584–591.

[97] Egorova, K.S., Ananikov, V.P. (2014) Toxicity of ionic liquids: eco(cyto)activity as complicated, but unavoidable parameter for task-specific optimization. *ChemSusChem*, **7** (2), 336–360.

[98] Deasy, R.E., Brossat, M., Moody, T.S., Maguire, A.R. (2011) Lipase catalysed kinetic resolutions of 3-aryl alkanoic acids. *Tetrahedron: Asymmetry*, **22** (1), 47–61.

[99] Deasy, R.E., Moody, T.S., Maguire, A.R. (2013) Influence of the position of the substituent on the efficiency of lipase-mediated resolutions of 3-aryl alkanoic acids. *Tetrahedron: Asymmetry*, **24** (23), 1480–1487.

[100] Reetz, M.T. (2011) Laboratory evolution of stereoselective enzymes: a prolific source of catalysts for asymmetric reactions. *Angew. Chem. Int. Ed.*, **50** (1), 138–174.

[101] Reymond, J.-L. (2008) Colorimetric and fluorescence-based screening, in *Protein Engineering Handbook* (eds S. Lutz and U.T. Bornscheuer), Wiley-VCH Verlag GmbH & Co. KGaA, Weinheim, pp. 669–711.

[102] Woodley, J.M. (2013) Protein engineering of enzymes for process applications. *Curr. Opin. Chem. Biol.*, **17** (2), 310–316.

[103] Gupta, M.N. (1992) Enzyme function in organic solvents. *Eur. J. Biochem.*, **203** (1–2), 25–32.

[104] Rasekh, B., Khajeh, K., Ranjbar, B. *et al.* (2014) Protein engineering of laccase to enhance its activity and stability in the presence of organic solvents. *Eng. Life Sci.*, **14** (4), 442–448.

[105] Savile, C.K., Janey, J.M., Mundorff, E.C. *et al.* (2010) Biocatalytic asymmetric synthesis of chiral amines from ketones applied to Sitagliptin manufacture. *Science*, **329**, 305–309.

[106] Wang, M., Si, T., Zhao, H. (2012) Biocatalyst development by directed evolution. *Bioresour. Technol.*, **115**, 117–125.

[107] Cobb, R.E., Sun, N., Zhao, H. (2013) Directed evolution as a powerful synthetic biology tool. *Methods*, **60** (1), 81–90.

[108] Reetz, M.T. (2012) Directed evolution of enzymes, in *Enzyme Catalysis in Organic Synthesis* (eds K. Drauz, H. Gröger and O. May), Wiley-VCH Verlag GmbH & Co. KGaA, Weinheim, pp. 119–190.

[109] Timms, N., Daniels, A.D., Berry, A., Nelson, A. (2012) Directed evolution and (semi-) rational design strategies for the creation of synthetically useful, stereoselective biocatalysts, in *Comprehensive Chirality* (eds E.M. Carreira and H. Yamamoto), Elsevier, Amsterdam, pp. 21–45.

[110] Behrens, G.A., Hummel, A., Padhi, S.K. *et al.* (2011) Discovery and protein engineering of biocatalysts for organic synthesis. *Adv. Synth. Catal.*, **353** (13), 2191–2215.

[111] Turner, N.J. (2009) Directed evolution drives the next generation of biocatalysts. *Nat. Chem. Biol.*, **5** (8), 567–573.

[112] Denard, C.A., Ren, H., Zhao, H. (2015) Improving and repurposing biocatalysts via directed evolution. *Curr. Opin. Chem. Biol.*, **25**, 55–64.

[113] Valetti, F., Gilardi, G. (2004) Directed evolution of enzymes for product chemistry. *Nat. Prod. Rep.*, **21** (4), 490–511.

[114] Wen, F., McLachlan, M., Zhao, H., Begley, T.P. (2007) Directed evolution: novel and improved enzymes, in *Wiley Encyclopedia of Chemical Biology (ed T.P. Begley)*, John Wiley & Sons, Inc., Weinheim.

[115] Davids, T., Schmidt, M., Böttcher, D., Bornscheuer, U.T. (2013) Strategies for the discovery and engineering of enzymes for biocatalysis. *Curr. Opin. Chem. Biol.*, **17** (2), 215–220.

[116] Labrou, N.E. (2010) Random mutagenesis methods for in vitro directed enzyme evolution. *Curr. Protein Pept. Sci.*, **11** (1), 91–100.

[117] Lutz, S., Patrick, W.M. (2004) Novel methods for directed evolution of enzymes: quality, not quantity. *Curr. Opin. Biotechnol.*, **15** (4), 291–297.

[118] Reetz, M.T., Carballeira, J.D. (2007) Iterative saturation mutagenesis (ISM) for rapid directed evolution of functional enzymes. *Nat. Protocols*, **2** (4), 891–903.

[119] Reetz, M.T., Bocola, M., Carballeira, J.D. *et al.* (2005) Expanding the range of substrate acceptance of enzymes: combinatorial active-site saturation test. *Angew. Chem. Int. Ed.*, **44** (27), 4192–4196.

[120] Bornscheuer, U. (2013) Protein engineering as a tool for the development of novel bioproduction systems, in *Fundamentals and Application of New Bioproduction Systems* (ed A.-P. Zeng), Springer Berlin Heidelberg, Heidelberg, pp. 25–40.

[121] Esvelt, K.M., Carlson, J.C., Liu, D.R. (2011) A system for the continuous directed evolution of biomolecules. *Nature*, **472** (7344), 499–503.

[122] Badran, A.H., Liu, D.R. (2015) *In vivo* continuous directed evolution. *Curr. Opin. Chem. Biol.*, **24**, 1–10.

[123] Xiao, H., Bao, Z., Zhao, H. (2015) High throughput screening and selection methods for directed enzyme evolution. *Ind. Eng. Chem. Res.*, **54** (16), 4011–4020.

[124] van Rossum, T., Kengen, S.W.M., van der Oost, J. (2013) Reporter-based screening and selection of enzymes. *FEBS J.*, **280** (13), 2979–2996.

[125] Yang, G., Withers, S.G. (2009) Ultra high-throughput FACS-based screening for directed enzyme evolution. *ChemBioChem*, **10** (17), 2704–2715.

[126] Becker, S., Schmoldt, H.-U., Adams, T.M. *et al.* (2004) Ultra-high-throughput screening based on cell-surface display and fluorescence-activated cell sorting for the identification of novel biocatalysts. *Curr. Opin. Biotechnol.*, **15** (4), 323–329.

[127] Reetz, M.T., Kahakeaw, D., Lohmer, R. (2008) Addressing the numbers problem in directed evolution. *ChemBioChem*, **9** (11), 1797–1804.

[128] Renata, H., Wang, Z.J., Arnold, F.H. (2015) Expanding the enzyme universe: accessing non-natural reactions by mechanism-guided directed evolution. *Angew. Chem. Int. Ed.*, **54** (11), 3351–3367.

[129] Currin, A., Swainston, N., Day, P.J., Kell, D.B. (2015) Synthetic biology for the directed evolution of protein biocatalysts: navigating sequence space intelligently. *Chem. Soc. Rev.*, **44** (5), 1172–1239.

[130] Hellinga, H.W. (1997) Rational protein design: combining theory and experiment. *Proc. Natl. Acad. Sci.*, **94** (19), 10015–10017.

[131] Damborsky, J., Brezovsky, J. (2009) Computational tools for designing and engineering biocatalysts. *Curr. Opin. Chem. Biol.*, **13** (1), 26–34.

[132] Pavelka, A., Chovancova, E., Damborsky, J. (2009) HotSpot Wizard: a web server for identification of hot spots in protein engineering. *Nucleic Acids Research*, **37** (suppl 2), W376–W383.

[133] Fox, R.J., Davis, S.C., Mundorff, E.C. *et al.* (2007) Improving catalytic function by ProSAR-driven enzyme evolution. *Nat. Biotech.*, **25** (3), 338–344.

[134] Frushicheva, M.P., Mills, M.J.L., Schopf, P. *et al.* (2014) Computer aided enzyme design and catalytic concepts. *Curr. Opin. Chem. Biol.*, **21**, 56–62.

[135] Wijma, H.J., Janssen, D.B. (2013) Computational design gains momentum in enzyme catalysis engineering. *FEBS J.*, **280** (13), 2948–2960.

[136] Pleiss, J. (2012) Rational design of enzymes, in *Enzyme Catalysis in Organic Synthesis* (eds K. Drauz, Harald Gröger and O. May), Wiley-VCH Verlag GmbH & Co. KGaA, Weinheim, pp. 89–117.

[137] Strohmeier, G.A., Pichler, H., May, O., Gruber-Khadjawi, M. (2011) Application of designed enzymes in organic synthesis. *Chem. Rev.*, **111** (7), 4141–4164.

[138] Toscano, M.D., Woycechowsky, K.J., Hilvert, D. (2007) Minimalist active-site redesign: teaching old enzymes new tricks. *Angew. Chem. Int. Ed.*, **46** (18), 3212–3236.

[139] Schmid, R.D., Verger, R. (1998) Lipases: interfacial enzymes with attractive applications. *Angew. Chem. Int. Ed.*, **37** (12), 1608–1633.

[140] Reetz, M.T. (2002) Lipases as practical biocatalysts. *Curr. Opin. Chem. Biol.*, **6** (2), 145–150.

[141] Theil, F. (1995) Lipase-supported synthesis of biologically active compounds. *Chem. Rev.*, **95** (6), 2203–2227.

[142] Koeller, K.M., Wong, C.-H. (2001) Enzymes for chemical synthesis. *Nature*, **409**, 232–240.

[143] Carrea, G., Riva, S. (2000) Properties and synthetic applications of enzymes in organic solvents. *Angew. Chem. Int. Ed.*, **39** (13), 2226–2254.

[144] Siirola, E., Frank, A., Grogan, G., Kroutil, W. (2013) C-C hydrolases for biocatalysis. *Adv. Synth. Catal.*, **355** (9), 1677–1691.

[145] Gotor, V. (2000) Pharmaceuticals through enzymatic transesterification and enzymatic aminolysis reactions. *Biocatal. Biotransform.*, **18** (2), 87–103.

[146] Gotor-Fernandez, V., Gotor, C., Rebolledo, F. (2006) Preparation of chiral pharmaceuticals through enzymatic acylation of alcohols and amines, in *Biocatalysis in the Pharmaceutical and Biotechnology Industries* (ed R.N. Patel), CRC Press, Boca Raton.

[147] Li, S., Yang, X., Yang, S., Zhu, M., Wang, X. (2012) Technology prospecting on enzymes: application, marketing and engineering. *Comput. Struct. Biotechnol. J.*, **2** (3), 1–11.

[148] Stergiou, P.-Y., Foukis, A., Filippou, M. *et al.* (2013) Advances in lipase-catalyzed esterification reactions. *Biotechnol. Adv.*, **31** (8), 1846–1859.

[149] Nagarajan, S. (2012) New tools for exploring 'old friends – microbial lipases'. *Appl. Biochem. Biotechnol.*, **168** (5), 1163–1196.

[150] Bornscheuer, U.T., Flickinger, M.C. (2009) Lipases, synthesis of chiral compounds, aqueous and organic solvents, in *Encyclopedia of Industrial Biotechnology* (ed M.C. Flickinger), John Wiley & Sons, Ltd., Chichester.

[151] Ghanem, A., Aboul-Enein, H.Y. (2005) Application of lipases in kinetic resolution of racemates. *Chirality*, **17** (1), 1–15.

[152] Bordusa, F. (2002) Proteases in organic synthesis. *Chem. Rev.*, **102** (12), 4817–4868.

[153] Hidalgo, A., Bornscheuer, U.T. (2006) Directed Evolution of lipases and esterases for organic synthesis, in *Biocatalysis in the Pharmaceutical and Biotechnology Industries* (ed R.N. Patel), CRC Press, Boca Raton.

[154] Guan, Z., Li, L.-Y., He, Y.-H. (2015) Hydrolase-catalyzed asymmetric carbon-carbon bond formation in organic synthesis. *RSC Adv.*, **5** (22), 16801–16814.

[155] Truppo, M.D., Hughes, G. (2011) Development of an improved immobilized CAL-B for the enzymatic resolution of a key intermediate to Odanacatib. *Org. Process Res. Dev.*, **15** (5), 1033–1035.

[156] Ammazzalorso, A., Amoroso, R., Bettoni, G. *et al.* (2008) *Candida rugosa* lipase-catalysed kinetic resolution of 2-substituted-aryloxyacetic esters with dimethylsulfoxide and isopropanol as additives. *Chirality*, **20** (2), 115–118.

[157] Wu, Q., Soni, P., Reetz, M.T. (2013) Laboratory evolution of enantiocomplementary *Candida antarctica* lipase B mutants with broad substrate scope. *J. Am. Chem. Soc.*, **135** (5), 1872–1881.

[158] Walsgrove, T.C., Powell, L., Wells, A. (2002) A practical and robust process to produce SB-214857, Lotrafiban, ((2S)-7-(4, 4′-bipiperidinylcarbonyl)-2, 3, 4, 5-tetrahydro-4-methyl-3-oxo-1H-1,4-benzodiazepine-2-acetic Acid) utilising an enzymic resolution as the final step. *Org. Process Res. Dev.*, **6** (4), 488–491.

[159] Atkins, R.J., Banks, A., Bellingham, R.K. *et al.* (2003) The development of a manufacturing route for the GPIIb/IIIa receptor antagonist SB-214857-A. Part 2: conversion of the key intermediate SB-235349 to SB-214857-A. *Org. Process Res. Dev.*, **7** (5), 663–675.

[160] Goto, M., Kawasaki, M., Kometani, T. (2000) Enzymatic resolution of 2-phenyl-1-propanol by enantioselective hydrolysis of its ester having a bulky group in an acyl moiety. *J. Mol. Catal. B: Enzym.*, **9** (4–6), 245–250.

[161] Merabet-Khelassi, M., Houiene, Z., Aribi-Zouioueche, L., Riant, O. (2012) Green methodology for enzymatic hydrolysis of acetates in non-aqueous media via carbonate salts. *Tetrahedron: Asymmetry*, **23** (11–12), 828–833.

[162] Shen, L.-L., Wang, F., Mun, H.-S. *et al.* (2008) Solvent-dependent reactivity in porcine pancreatic lipase (PPL)-catalyzed hydrolysis. *Tetrahedron: Asymmetry*, **19** (14), 1647–1653.

[163] Igarashi, Y., Otsutomo, S., Harada, M., Nakano, S. (1997) Enzymatic resolution of indene bromohydrin acetate using immobilized lipase. *Tetrahedron: Asymmetry*, **8** (16), 2833–2837.

[164] Petersen, M., Kiener, A. (1999) Biocatalysis. *Green Chem.*, **1** (2), 99–106.

[165] Eichhorn, E., Roduit, J.-P., Shaw, N. *et al.* (1997) Preparation of (*S*)-piperazine-2-carboxylic acid, (*R*)-piperazine-2-carboxylic acid, and (*S*)-piperidine-2-carboxylic acid by kinetic resolution of the corresponding racemic carboxamides with stereoselective amidases in whole bacterial cells. *Tetrahedron: Asymmetry*, **8** (15), 2533–2536.

[166] Shaw, N.M., Naughton, A., Robins, K. *et al.* (2002) Selection, purification, characterisation, and cloning of a novel heat-stable stereo-specific amidase from *Klebsiella oxytoca*, and its application in the synthesis of enantiomerically pure (*R*)- and (*S*)-3, 3, 3-trifluoro-2-hydroxy-2-methylpropionic acids and (*S*)-3,3, 3-trifluoro-2-hydroxy-2-methylpropionamide. *Org. Process Res. Dev.*, **6** (4), 497–504.

[167] Leuchtenberger, W., Karrenbauer, M., Plöcker, U.L.F. (1984) Scale-up of an enzyme membrane reactor process for the manufacture of L-enantiomeric compounds. *Ann. N.Y. Acad. Sci.*, **434** (1), 078–086.

[168] Chênevert, R., Pelchat, N., Morin, P. (2009) Lipase-mediated enantioselective acylation of alcohols with functionalized vinyl esters: acyl donor tolerance and applications. *Tetrahedron: Asymmetry*, **20** (10), 1191–1196.

[169] van Rantwijk, F., Sheldon, R.A. (2004) Enantioselective acylation of chiral amines catalysed by serine hydrolases. *Tetrahedron*, **60** (3), 501–519.

[170] Patel, R., McNamee, C., Szarka, L. (1992) Enantioselective enzymatic acetylation of racemic [4-[4α, 6β (E)]]-6-[4, 4-bis(4-fluorophenyl)-3-(1-methyl-1H-tetrazol-5-yl)-1,3-butadienyl]-tetrahydro-4-hydroxy-2H-pyran-2-one. *Appl. Microbiol. Biotechnol.*, **38** (1), 56–60.

[171] Morrone, R., D'Antona, N., Lambusta, D., Nicolosi, G. (2010) Biocatalyzed irreversible esterification in the preparation of (*S*)-naproxen. *J. Mol. Catal. B: Enzym.*, **65** (1–4), 49–51.

[172] Saksena, A.K., Girijavallabhan, V.M., Lovey, R.G. *et al.* (1995) Highly stereoselective access to novel 2,2,4-trisubstituted tetrahydrofurans by halocyclization: practical chemoenzymatic synthesis of Sch 51048, a broad-spectrum orally active antifungal agent. *Tetrahedron Lett.*, **36** (11), 1787–1790.

[173] Martinez, C.A., Hu, S., Dumond, Y. *et al.* (2008) Development of a chemoenzymatic manufacturing process for Pregabalin. *Org. Process Res. Dev.*, **12** (3), 392–398.

[174] Jung, J.-H., Yoon, D.-H., Kang, P. *et al.* (2013) CAL-B catalyzed desymmetrization of 3-alkylglutarate: 'Olefin effect' and asymmetric synthesis of Pregabalin. *Org. Biomol. Chem.*, **11** (22), 3635–3641.

[175] Cabrera, Z., Palomo, J.M. (2011) Enantioselective desymmetrization of prochiral diesters catalyzed by immobilized *Rhizopus oryzae* lipase. *Tetrahedron: Asymmetry*, **22** (24), 2080–2084.

[176] Liu, W., Hu, Y., Jiang, L. *et al.* (2012) Synthesis of methyl (*R*)-3-(4-fluorophenyl)glutarate via enzymatic desymmetrization of a prochiral diester. *Process Biochem.*, **47** (7), 1037–1041.

[177] Choi, W. (2009) Biotechnological production of enantiopure epoxides by enzymatic kinetic resolution. *Appl. Microbiol. Biotechnol.*, **84** (2), 239–247.

[178] Kotik, M., Kyslík, P. (2006) Purification and characterisation of a novel enantioselective epoxide hydrolase from *Aspergillus niger* M200. *Biochim. Biophys. Acta*, **1760** (2), 245–252.

[179] DeSantis, G., Zhu, Z., Greenberg, W.A. *et al.* (2002) An enzyme library approach to biocatalysis: development of nitrilases for enantioselective production of carboxylic acid derivatives. *J. Am. Chem. Soc.*, **124** (31), 9024–9025.

[180] Bergeron, S., Chaplin, D.A., Edwards, J.H. *et al.* (2006) Nitrilase-catalysed desymmetrisation of 3-hydroxyglutaronitrile: Preparation of a statin side-chain intermediate. *Org. Process Res. Dev.*, **10** (3), 661–665.

[181] Chakravorty, D., Parameswaran, S., Dubey, V., Patra, S. (2012) Unraveling the rationale behind organic solvent stability of lipases. *Appl. Biochem. Biotechnol.*, **167** (3), 439–461.

[182] Yahya, A.R.M., Anderson, W.A., Moo-Young, M. (1998) Ester synthesis in lipase-catalyzed reactions. *Enzyme Microb. Technol.*, **23** (7–8), 438–450.

[183] Hari Krishna, S., Karanth, N.G. (2002) Lipases and lipase-catalyzed esterification reactions in nonaqueous media. *Catal. Rev.*, **44** (4), 499–591.

[184] Villeneuve, P., Muderhwa, J.M., Graille, J., Haas, M.J. (2000) Customizing lipases for biocatalysis: a survey of chemical, physical and molecular biological approaches. *J. Mol. Catal. B: Enzym.*, **9** (4–6), 113–148.

[185] Gomes, R., Meek, M.E. (2001) Acetaldehyde: Hazard Characterisation and Exposure–Response Analysis. *J. Environ. Sci. Health., Part C*, **19** (1), 1–21.

[186] Reiss, H., Chowdhury, M.A. (1984) Single molecule gas-phase polymerization kinetics of vinyl acetate. Non-steady measurements. *The Journal of Physical Chemistry*, **88** (26), 6667–6670.

[187] Hilterhaus, L., Thum, O., Liese, A. (2008) Reactor concept for lipase-catalyzed solvent-free conversion of highly viscous reactants forming two-phase systems. *Org. Process Res. Dev.*, **12** (4), 618–625.

[188] Ditrich, K. (2008) Optically active amines by enzyme-catalyzed kinetic resolution. *Synthesis*, **14**, 2283–2287.

[189] Goswami, A., Howell, J.M., Hua, E.Y. *et al.* (2001) Chemical and enzymatic resolution of (*R,S*)-*N*-(tert-butoxycarbonyl)-3-hydroxymethylpiperidine. *Org. Process Res. Dev.*, **5** (4), 415–420.

[190] Tamarez, M., Morgan, B., Wong, G.S.K. *et al.* (2003) Pilot-scale lipase-catalyzed regioselective acylation of ribavirin in anhydrous media in the synthesis of a novel prodrug intermediate. *Org. Process Res. Dev.*, **7** (6), 951–953.

[191] Paravidino, M., Hanefeld, U. (2011) Enzymatic acylation: assessing the greenness of different acyl donors. *Green Chem.*, **13** (10), 2651–2657.

[192] Hanefeld, U. (2003) Reagents for (ir)reversible enzymatic acylations. *Org. Biomol. Chem.*, **1** (14), 2405–2415.

[193] Theil, F. (2000) Enhancement of selectivity and reactivity of lipases by additives. *Tetrahedron*, **56** (19), 2905–2919.

[194] Alford, B.L., Hugel, H.M. (2013) Total synthesis of (+)-pentamethylsalvianolic acid C. *Org. Biomol. Chem.*, **11** (16), 2724–2727.

[195] Vaidyanathan, R., Hesmondhalgh, L., Hu, S. (2007) A chemoenzymatic synthesis of an androgen receptor antagonist. *Org. Process Res. Dev.*, **11** (5), 903–906.

[196] Araujo, Y.J.K., Avvari, N.P., Paiva, D.R. *et al.* (2015) Synthesis and enzymatic resolution of racemic 2,3-epoxy propyl esters obtained from glycerol. *Tetrahedron Lett.*, **56** (13), 1696–1698.

[197] Rn, P., Banerjee, A., Ry, K. *et al.* (1994) Enzymic preparation of (3*R*-cis)-3-(acetyloxy)-4-phenyl-2-azetidinone: a taxol side-chain synthon. *Biotechnol. Appl. Biochem.*, **20** (1), 23–33.

[198] Khmelnitsky, Y.L., Michels, P.C., Cotterill, I.C. *et al.* (2011) Biocatalytic resolution of bis-tetrahydrofuran alcohol. *Org. Process Res. Dev.*, **15** (1), 279–283.

[199] Anderson, E.M., Larsson, K.M., Kirk, O. (1998) One biocatalyst–many applications: The use of *Candida Antarctica* B-Lipase in organic synthesis. *Biocatal. Biotransform.*, **16** (3), 181–204.

[200] Martinelle, M., Hult, K. (1995) Kinetics of acyl transfer reactions in organic media catalysed by *Candida antarctica* lipase B. *Biochim. Biophys. Acta*, **1251** (2), 191–197.

[201] Muñoz Solano, D., Hoyos, P., Hernáiz, M.J. *et al.* (2012) Industrial biotransformations in the synthesis of building blocks leading to enantiopure drugs. *Bioresour. Technol.*, **115**, 196–207.

[202] Qin, B., Liang, P., Jia, X. *et al.* (2013) Directed evolution of *Candida antarctica* lipase B for kinetic resolution of profen esters. *Catal. Commun.*, **38**, 1–5.

[203] Gill, I., Patel, R. (2006) Biocatalytic ammonolysis of (5*S*)-4,5-dihydro-1H-pyrrole-1,5-dicarboxylic acid, 1-(1,1-dimethylethyl)-5-ethyl ester: preparation of an intermediate to the dipeptidyl peptidase IV inhibitor Saxagliptin. *Bioorg. Med. Chem. Lett.*, **16** (3), 705–709.

[204] Hoyos, P., Pace, V., Alcántara, A.R. (2012) Dynamic kinetic resolution via hydrolase-metal combo catalysis in stereoselective synthesis of bioactive compounds. *Adv. Synth. Catal.*, **354** (14–15), 2585–2611.

[205] Forró, E., Schönstein, L., Fülöp, F. (2011) Total synthesis of crispine A enantiomers through a *Burkholderia cepacia* lipase-catalysed kinetic resolution. *Tetrahedron: Asymmetry*, **22** (11), 1255–1260.

[206] Moreno, J., Sinisterra, J. (1995) A systematic analysis of the variables that control a highly stereoselective resolution of racemic non-steroidal antiinflammatory drugs using immobilized lipase from *Candida cylindracea*. *J. Mol. Catal. A: Chem.*, **98** (3), 171–184.

[207] Cipiciani, A., Bellezza, F., Fringuelli, F., Stillitano, M. (1999) Enantioselectivity of alcohol-treated *Candida rugosa* lipase in the kinetic resolution of racemic methyl 2-aryloxypropionates in water and aqueous organic media. *Tetrahedron: Asymmetry*, **10** (23), 4599–4605.

[208] Truppo, M.D., Journet, M., Shafiee, A., Moore, J.C. (2006) Optimization and scale-up of a lipase-catalyzed enzymatic resolution of an indole ester intermediate for a Prostaglandin D2 (DP) receptor antagonist targeting allergic Rhinitis. *Org. Process Res. Dev.*, **10** (3), 592–598.

[209] Hoekstra, M.S., Sobieray, D.M., Schwindt, M.A. *et al.* (1997) Chemical development of CI-1008, an enantiomerically pure anticonvulsant. *Org. Process Res. Dev.*, **1** (1), 26–38.

[210] Chavan, A.B., Maikap, G.C., Gurjar, M.K. (2009) An efficient process of racemization of 3-(carbamoylmethyl)-5-methylhexanoic acid: a Pregabalin intermediate. *Org. Process Res. Dev.*, **13** (4), 812–814.

[211] Dunn, P.J., Hettenbach, K., Kelleher, P., Martinez, C.A. (2010) The development of a green, energy efficient, chemoenzymatic manufacturing process for Pregabalin, in *Green Chemistry in the Pharmaceutical Industry* (eds P.J. Dunn, A. Wells and M.T. Williams, Wiley-VCH Verlag GmbH & Co. KGaA, Weinheim, pp. 161–177.

[212] Reetz, M.T., Zonta, A., Schimossek, K. *et al.* (1997) Creation of enantioselective biocatalysts for organic chemistry by *in vitro* evolution. *Angew. Chem. Int. Ed.*, **36** (24), 2830–2832.

[213] Bornscheuer, U.T. (2004) High-throughput-screening systems for hydrolases. *Eng. Life Sci.*, **4** (6), 539–542.

[214] Reetz, M.T., Wilensek, S., Zha, D., Jaeger, K.-E. (2001) Directed evolution of an enantioselective enzyme through combinatorial multiple-cassette mutagenesis. *Angew. Chem. Int. Ed.*, **40** (19), 3589–3591.

[215] Prasad, S., Bocola, M., Reetz, M.T. (2011) Revisiting the lipase from *Pseudomonas aeruginosa*: directed evolution of substrate acceptance and enantioselectivity using iterative saturation mutagenesis. *ChemPhysChem*, **12** (8), 1550–1557.

[216] Engström, K., Nyhlén, J., Sandström, A.G., Bäckvall, J.-E. (2010) Directed evolution of an enantioselective lipase with broad substrate scope for hydrolysis of α-substituted esters. *J. Am. Chem. Soc.*, **132** (20), 7038–7042.

[217] Schmidt, M., Baumann, M., Henke, E. *et al.* (2004) Directed evolution of lipases and esterases, in *Methods Enzymol.* (eds E.R. Dan and P.N. Joseph), Academic Press, pp. 199–207.

[218] Widersten, M. (2014) Protein engineering for development of new hydrolytic biocatalysts. *Curr. Opin. Chem. Biol.*, **21**, 42–47.

[219] Müller, M. (2014) Enzymatic synthesis of tertiary alcohols. *ChemBioEng Rev.*, **1** (1), 14–26.

[220] Kourist, R., Domínguez de María, P., Bornscheuer, U.T. (2008) Enzymatic synthesis of optically active tertiary alcohols: expanding the biocatalysis toolbox. *ChemBioChem*, **9** (4), 491–498.

[221] Kourist, R., Bornscheuer, U. (2011) Biocatalytic synthesis of optically active tertiary alcohols. *Appl. Microbiol. Biotechnol.*, **91** (3), 505–517.

[222] Kourist, R., Bartsch, S., Bornscheuer, U.T. (2007) Highly enantioselective synthesis of arylaliphatic tertiary alcohols using mutants of an esterase from *Bacillus subtilis*. *Adv. Synth. Catal.*, **349** (8–9), 1393–1398.

[223] Bartsch, S., Kourist, R., Bornscheuer, U.T. (2008) Complete inversion of enantioselectivity towards acetylated tertiary alcohols by a double mutant of a *Bacillus subtilis* esterase. *Angew. Chem. Int. Ed.*, **47** (8), 1508–1511.

[224] Kotik, M., Archelas, A., Wohlgemuth, R. (2012) Epoxide hydrolases and their application in organic synthesis. *Curr. Org. Chem.*, **16** (4), 451–482.

[225] de Vries, E.J., Janssen, D.B. (2003) Biocatalytic conversion of epoxides. *Curr. Opin. Biotechnol.*, **14** (4), 414–420.

[226] Steinreiber, A., Faber, K. (2001) Microbial epoxide hydrolases for preparative biotransformations. *Curr. Opin. Biotechnol.*, **12** (6), 552–558.

[227] Breuer, M., Ditrich, K., Habicher, T. *et al.* (2004) Industrial methods for the production of optically active intermediates. *Angew. Chem. Int. Ed.*, **43** (7), 788–824.

[228] Balkenhohl, F., Ditrich, K., Hauer, B., Ladner, W. (1997) Optisch aktive amine durch lipase-katalysierte methoxyacetylierung. *Journal für Praktische Chemie/Chemiker-Zeitung*, **339** (1), 381–384.

[229] Drauz, K.G., H.; May, O. (eds) (2012) *Enzyme Catalysis in Organic Synthesis*, Wiley-VCH, Weinheim.

[230] de Miranda, A.S., Miranda, L.S.M., de Souza, R.O.M.A. Lipases: valuable catalysts for dynamic kinetic resolutions. *Biotechnol. Adv.*, **33** (5), 372–393.

[231] Paetzold, J., Bäckvall, J.E. (2005) Chemoenzymatic dynamic kinetic resolution of primary amines. *J. Am. Chem. Soc.*, **127** (50), 17620–17621.

[232] Thalén, L.K., Zhao, D., Sortais, J.-B. *et al.* (2009) A chemoenzymatic approach to enantiomerically pure amines using dynamic kinetic resolution: application to the synthesis of Norsertraline. *Chem. Eur. J.*, **15** (14), 3403–3410.

[233] Liljeblad, A., Kanerva, L.T. (2006) Biocatalysis as a profound tool in the preparation of highly enantiopure β-amino acids. *Tetrahedron*, **62** (25), 5831–5854.

[234] Altenbuchner, J., Siemann-Herzberg, M., Syldatk, C. (2001) Hydantoinases and related enzymes as biocatalysts for the synthesis of unnatural chiral amino acids. *Curr. Opin. Biotechnol.*, **12** (6), 559–563.

[235] Fujii, R., Nakagawa, Y., Hiratake, J. *et al.* (2005) Directed evolution of *Pseudomonas aeruginosa* lipase for improved amide-hydrolyzing activity. *Protein. Eng. Des. Sel.*, **18** (2), 93–101.

[236] Zheng, R.-C., Wang, Y.-S., Liu, Z.-Q. *et al.* (2007) Isolation and characterization of *Delftia tsuruhatensis* ZJB-05174, capable of (*R*)-enantioselective degradation of 2, 2-dimethylcyclopropanecarboxamide. *Research in Microbiology*, **158** (3), 258–264.

[237] Zheng, R.-C., Yang, Z.-Y., Li, C.-C. *et al.* (2014) Industrial production of chiral intermediate of cilastatin by nitrile hydratase and amidase catalyzed one-pot, two-step biotransformation. *J. Mol. Catal. B: Enzym.*, **102**, 161–166.

[238] Patel, R., Banerjee, A., Chu, L. *et al.* (1998) Microbial synthesis of chiral intermediates for β-3-receptor agonists. *J. Am. Oil Chem. Soc.*, **75** (11), 1473–1482.

[239] Mahmoudian, M. (2000) Biocatalytic production of chiral pharmaceutical intermediates. *Biocatal. Biotransform.*, **18** (2), 105–118.

[240] Evans, C., McCague, R., Roberts, S.M. *et al.* (1991) Whole cell catalysed kinetic resolution of 6-azabicyclo[3.2.0]hept-3-en-7-one: synthesis of (–)-cispentacin (FR 109615). *J. Chem. Soc., Perkin Trans. 1*, (9), 2276–2277.

[241] Chen, J., Zheng, R.-C., Zheng, Y.-G., Shen, Y.-C. (2009) Microbial transformation of nitriles to high-value acids or amides, in *Biotechnology in China I* (eds J.-J. Zhong, F.-W. Bai and W. Zhang), Springer Berlin Heidelberg, pp. 33–77.

[242] Ramteke, P.W., Maurice, N.G., Joseph, B., Wadher, B.J. (2013) Nitrile-converting enzymes: an eco-friendly tool for industrial biocatalysis. *Biotechnol. Appl. Biochem.*, **60** (5), 459–481.

[243] Gong, J.-S., Lu, Z.-M., Li, H. *et al.* (2012) Nitrilases in nitrile biocatalysis: recent progress and forthcoming research. *Microb. Cell Fact.*, **11** (1), 142.

[244] Yamamoto, K., Oishi, K., Fujimatsu, I., Komatsu, K. (1991) Production of (*R*)-(−)-mandelic acid from mandelonitrile by *Alcaligenes faecalis* ATCC 8750. *Appl. Environ. Microbiol.*, **57** (10), 3028–3032.

[245] Xue, Y.-P., Xu, M., Chen, H.-S. *et al.* (2013) A novel integrated bioprocess for efficient production of (*R*)-(−)-mandelic acid with immobilized *Alcaligenes faecalis* ZJUTB10. *Org. Process Res. Dev.*, **17** (2), 213–220.

[246] Maddrell, S.J., Turner, N.J., Kerridge, A. *et al.* (1996) Nitrile hydratase enzymes in organic synthesis: enantioselective synthesis of the lactone moiety of the mevinic acids. *Tetrahedron Lett.*, **37** (33), 6001–6004.

[247] Wang, M.-X., Liu, C.-S., Li, J.-S. (2002) Enzymatic desymmetrization of 3-alkyl- and 3-arylglutaronitriles, a simple and convenient approach to optically active 4-amino-3-phenylbutanoic acids. *Tetrahedron: Asymmetry*, **12** (24), 3367–3373.

[248] Yamamoto, K., Ueno, Y., Otsubo, K. *et al.* (1990) Production of (*S*)-(+)-ibuprofen from a nitrile compound by *Acinetobacter sp.* strain AK226. *Appl. Environ. Microbiol.*, **56** (10), 3125–3129.

[249] Gradley, M., Knowles, C. (1994) Asymmetric hydrolysis of chiral nitriles by *Rhodococcus rhodochrous* NCIMB 11216 nitrilase. *Biotechnol. Lett.*, **16** (1), 41–46.

[250] Gröger, H. (2001) Enzymatic routes to enantiomerically pure aromatic α-hydroxy carboxylic acids: a further example for the diversity of biocatalysis. *Adv. Synth. Catal.*, **343** (6–7), 547–558.

[251] Monti, D., Ottolina, G., Carrea, G., Riva, S. (2011) Redox reactions catalyzed by isolated enzymes. *Chem. Rev.*, **111** (7), 4111–4140.

[252] Burton, S.G. (2003) Oxidizing enzymes as biocatalysts. *Trends Biotechnol.*, **21** (12), 543–549.

[253] Mihovilovic, M.D., Bianchi, D.A. (2008) Biooxidations in chiral synthesis, in *Asymmetric Organic Synthesis with Enzymes* (eds V. Gotor, I. Alfonso and E. Garcıa-Urdiales), Wiley-VCH Verlag GmbH & Co. KGaA, Weinheim, pp. 229–274.

[254] Wu, H., Tian, C., Song, X. *et al.* (2013) Methods for the regeneration of nicotinamide coenzymes. *Green Chem.*, **15** (7), 1773–1789.

[255] Hollmann, F., Arends, I.W.C.E., Buehler, K. *et al.* (2011) Enzyme-mediated oxidations for the chemist. *Green Chem.*, **13** (2), 226–265.

[256] Weckbecker, A., Gröger, H., Hummel, W. (2010) Regeneration of nicotinamide coenzymes: principles and applications for the synthesis of chiral compounds, in *Biosystems Engineering I* (eds C. Wittmann and R. Krull), Springer, Heidelberg, pp. 195–242.

[257] Rodriguez, C., Lavandera, I., Gotor, V. (2012) Recent advances in cofactor regeneration systems applied to biocatalyzed oxidative processes. *Curr. Org. Chem.*, **16** (21), 2525–2541.

[258] Dascier, D., Kambourakis, S., Hua, L. *et al.* (2014) Influence of cofactor regeneration strategies on preparative-scale, asymmetric carbonyl reductions by engineered *Escherichia coli*. *Org. Process Res. Dev.*, **18** (6), 793–800.

[259] Hummel, W., Gröger, H. (2014) Strategies for regeneration of nicotinamide coenzymes emphasizing self-sufficient closed-loop recycling systems. *J. Biotechnol.*, **191**, 22–31.

[260] Hollmann, F., Arends, I.W.C.E., Buehler, K. (2010) Biocatalytic redox reactions for organic synthesis: nonconventional regeneration methods. *ChemCatChem*, **2** (7), 762–782.

[261] Moore, J.C., Pollard, D.J., Kosjek, B., Devine, P.N. (2007) Advances in the enzymatic reduction of Ketones. *Acc. Chem. Res.*, **40** (12), 1412–1419.

[262] Kara, S., Schrittwieser, J., Hollmann, F., Ansorge-Schumacher, M. (2014) Recent trends and novel concepts in cofactor-dependent biotransformations. *Appl. Microbiol. Biotechnol.*, **98** (4), 1517–1529.

[263] Paul, C.E., Arends, I.W.C.E., Hollmann, F. (2014) Is simpler better? Synthetic nicotinamide cofactor analogues for redox chemistry. *ACS Catal.*, **4** (3), 788–797.

[264] Paul, C.E., Gargiulo, S., Opperman, D.J. *et al.* (2013) Mimicking nature: synthetic nicotinamide cofactors for C=C bioreduction using enoate reductases. *Org. Lett.*, **15** (1), 180–183.

[265] Prelog, V. (1964) Specification of the stereospecificity of some oxido-reductases by diamond lattice sections. *Pure Appl. Chem.*, **9** (1), 119–130.

[266] O'Connor, C.M., Wolstenholme, G.E.W. (eds) (1959) *Steric course of microbiological reactions. In honour of Prof. Dr. V. Prelog*, Little, Brown, Boston.

[267] Wohlgemuth, R. (2013) Development of sustainable biocatalytic reduction processes for organic chemists, in *Synthetic Methods for Biologically Active Molecules* (ed E. Brenna), Wiley-VCH Verlag GmbH & Co. KGaA, Weinheim, pp. 1–25.

[268] Kroutil, W., Mang, H., Edegger, K., Faber, K. (2004) Biocatalytic oxidation of primary and secondary alcohols. *Adv. Synth. Catal.*, **346** (2–3), 125–142.

[269] Hall, M., Bommarius, A.S. (2011) Enantioenriched compounds via enzyme-catalyzed redox reactions. *Chem. Rev.*, **111** (7), 4088–4110.

[270] Turner, N.J. (2011) Enantioselective oxidation of C–O and C–N bonds using oxidases. *Chem. Rev.*, **111** (7), 4073–4087.

[271] Ni, Y., Xu, J.-H. (2012) Biocatalytic ketone reduction: A green and efficient access to enantiopure alcohols. *Biotechnol. Adv.*, **30** (6), 1279–1288.

[272] Huisman, G.W., Liang, J., Krebber, A. (2010) Practical chiral alcohol manufacture using ketoreductases. *Curr. Opin. Chem. Biol.*, **14** (2), 122–129.

[273] Neuberg, C.L., Lewite, A. (1918). *Biochem. Z.*, (257), 257.

[274] Stuermer, R., Hauer, B., Hall, M., Faber, K. (2007) Asymmetric bioreduction of activated C=C bonds using enoate reductases from the old yellow enzyme family. *Curr. Opin. Chem. Biol.*, **11** (2), 203–213.

[275] Deasy, R.E., Maguire, A.R. (2014) Baker's-yeast-mediated reduction of sulfur-containing compounds. *Eur. J. Org. Chem.*, **2014** (18), 3737–3756.

[276] Truppo, M. (2014) Reduction of C=O to CHOH using enzymes and microorganisms, in *Comprehensive Organic Synthesis II (Second Edition)* (ed P. Knochel), Elsevier, Amsterdam, pp. 317–327.

[277] Servi, S. (1990) Baker's yeast as a reagent in organic synthesis. *Synthesis*, **01**, 1–25.

[278] Fuganti, C. (1990) Bakers' yeast mediated synthesis of natural products. *Pure Appl. Chem.*, **62**, 1449–1452.

[279] Csuk, R., Glaenzer, B.I. (1991) Baker's yeast mediated transformations in organic chemistry. *Chem. Rev.*, **91** (1), 49–97.

[280] Crocq, V., Masson, C., Winter, J. *et al.* (1997) Synthesis of Trimegestone: the first industrial application of bakers' yeast mediated reduction of a ketone. *Org. Process Res. Dev.*, **1** (1), 2–13.

[281] Modukuru, N.K., Sukumaran, J., Collier, S.J. *et al.* (2014) Development of a practical, biocatalytic reduction for the manufacture of (*S*)-Licarbazepine using an evolved ketoreductase. *Org. Process Res. Dev.*, **18** (6), 810–815.

[282] Ma, S.K., Gruber, J., Davis, C. *et al.* (2010) A green-by-design biocatalytic process for atorvastatin intermediate. *Green Chem.*, **12** (1), 81–86.

[283] Ren, Z.-Q., Liu, Y., Pei, X.-Q. *et al.* (2015) Bioreductive production of enantiopure (*S*)-duloxetine intermediates catalyzed with ketoreductase ChKRED15. *J. Mol. Catal. B: Enzym.*, **113**, 76–81.

[284] Liang, J., Mundorff, E., Voladri, R. *et al.* (2010) Highly enantioselective reduction of a small heterocyclic ketone: biocatalytic reduction of tetrahydrothiophene-3-one to the corresponding (*R*)-alcohol. *Org. Process Res. Dev.*, **14** (1), 188–192.

[285] Zuhse, R., Leggewie, C., Hollmann, F., Kara, S. (2015) Scaling-up of 'smart cosubstrate' 1, 4-butanediol promoted asymmetric reduction of ethyl-4,4,4-trifluoroacetoacetate in organic media. *Org. Process Res. Dev.*, **19** (2), 369–372.

[286] Homann, M.J., Previte, E. Stereoselective microbial reduction of 5-fluorophenyl-5-oxo-pentanoic acid and a phenyloxazolidinone condensation product thereof. US Patent 5618707, filed Jan. 4, 1996 and issued Apr. 8, 1997.

[287] Ma, H., Yang, L., Ni, Y. *et al.* (2012) Stereospecific reduction of methyl *o*-chlorobenzoylformate at 300 g/L without additional cofactor using a carbonyl reductase mined from *Candida glabrata. Adv. Synth. Catal.*, **354** (9), 1765–1772.

[288] Hanson, R.L., Goldberg, S.L., Guo, Z. *et al.* (2014) Enzymatic reduction of adamantanones to chiral adamantanol intermediates for the synthesis of 11-β-hydroxysteroid dehydrogenase inhibitors. *Org. Process Res. Dev.*, **18** (8), 960–968.

[289] Kara, S., Spickermann, D., Schrittwieser, J.H. *et al.* (2013) More efficient redox biocatalysis by utilising 1,4-butanediol as a 'smart cosubstrate'. *Green Chem.*, **15** (2), 330–335.

[290] Nagayama, K., Spiess, A.C., Büchs, J. (2010) Gas phase enantioselective reduction catalyzed by immobilized ketoreductase: Effects of water activity and reaction temperature. *Biochem. Eng. J.*, **52** (2–3), 301–303.

[291] Milagre, H.M.S., Milagre, C.D.F., Moran, P.J.S. *et al.* (2006) Asymmetric bioreduction of ethyl 3-halo-2-oxo-4-phenylbutanoate by *Saccharomyces cerevisiae* immobilized in ca-alginate beads with double gel layer. *Org. Process Res. Dev.*, **10** (3), 611–617.

[292] Li, H., Moncecchi, J., Truppo, M.D. (2015) Development of an immobilized ketoreductase for enzymatic (*R*)-1-(3, 5-bis(trifluoromethyl)phenyl)ethanol production. *Org. Process Res. Dev.*

[293] Gooding, O.W., Voladri, R., Bautista, A. *et al.* (2010) Development of a practical biocatalytic process for (*R*)-2-methylpentanol. *Org. Process Res. Dev.*, **14** (1), 119–126.

[294] Galletti, P., Emer, E., Gucciardo, G. *et al.* (2010) Chemoenzymatic synthesis of (2*S*)-2-arylpropanols through a dynamic kinetic resolution of 2-arylpropanals with alcohol dehydrogenases. *Org. Biomol. Chem.*, **8** (18), 4117–4123.

[295] Giacomini, D., Galletti, P., Quintavalla, A. *et al.* (2007) Highly efficient asymmetric reduction of arylpropionic aldehydes by Horse liver alcohol dehydrogenase through dynamic kinetic resolution. *Chem. Commun.*, (39), 4038–4040.

[296] Friest, J.A., Maezato, Y., Broussy, S. *et al.* (2010) Use of a robust dehydrogenase from an archael hyperthermophile in asymmetric catalysis – dynamic reductive kinetic resolution entry into (*S*)-Profens. *J. Am. Chem. Soc.*, **132** (17), 5930–5931.

[297] Bommarius, A.S., Schwarm, M., Drauz, K. (1998) Biocatalysis to amino acid-based chiral pharmaceuticals – examples and perspectives. *J. Mol. Catal. B: Enzym.*, **5** (1–4), 1–11.

[298] Busto, E., Simon, R.C., Grischek, B. *et al.* (2014) Cutting short the asymmetric synthesis of the Ramatroban precursor by employing ω-Transaminases. *Adv. Synth. Catal.*, **356** (9), 1937–1942.

[299] Kohls, H., Steffen-Munsberg, F., Höhne, M. (2014) Recent achievements in developing the biocatalytic toolbox for chiral amine synthesis. *Curr. Opin. Chem. Biol.*, **19**, 180–192.

[300] Kroutil, W., Fischereder, E.-M., Fuchs, C.S. *et al.* (2013) Asymmetric preparation of *prim*-, *sec*-, and *tert*-amines employing selected biocatalysts. *Org. Process Res. Dev.*, **17** (5), 751–759.

[301] Koszelewski, D., Tauber, K., Faber, K., Kroutil, W. ω-Transaminases for the synthesis of non-racemic α-chiral primary amines. *Trends Biotechnol.*, **28** (6), 324–332.

[302] Paul, C.E., Rodríguez-Mata, M., Busto, E. *et al.* (2014) Transaminases applied to the synthesis of high added-value enantiopure amines. *Org. Process Res. Dev.*, **18** (6), 788–792.

[303] Green, A.P., Turner, N.J., O'Reilly, E. (2014) Chiral amine synthesis using ω-transaminases: an amine donor that displaces equilibria and enables high-throughput screening. *Angew. Chem. Int. Ed.*, **53** (40), 10714–10717.

[304] Cuetos, A., Lavandera, I., Gotor, V. (2013) Expanding dynamic kinetic protocols: transaminase-catalyzed synthesis of α-substituted β-amino ester derivatives. *Chem. Commun.*, **49** (91), 10688–10690.

[305] Truppo, M.D., Rozzell, J.D., Turner, N.J. (2010) Efficient production of enantiomerically pure chiral amines at concentrations of 50 g/L using transaminases. *Org. Process Res. Dev.*, **14** (1), 234–237.

[306] Mathew, S., Yun, H. (2012) ω-Transaminases for the production of optically pure amines and unnatural amino acids. *ACS Catal.*, **2** (6), 993–1001.

[307] Shin, G., Mathew, S., Yun, H. (2015) Kinetic resolution of amines by (*R*)-selective omega-transaminase from *Mycobacterium vanbaalenii*. *J. Ind. Eng. Chem.*, **23**, 128–133.

[308] Simon, R.C., Richter, N., Busto, E., Kroutil, W. (2014) Recent developments of cascade reactions involving ω-transaminases. *ACS Catal.*, **4** (1), 129–143.

[309] Shin, G., Mathew, S., Shon, M., Kim, B.-G., Yun, H. (2013) One-pot one-step deracemization of amines using ω-transaminases. *Chem. Commun.*, **49** (77), 8629–8631.

[310] Fuchs, C.S., Simon, R.C., Riethorst, W. *et al.* (2014) Synthesis of (*R*)- or (*S*)-valinol using ω-transaminases in aqueous and organic media. *Bioorg. Med. Chem. Lett.*, **22** (20), 5558–5562.

[311] Bartoli, G., Di Antonio, G., Giovannini, R. *et al.* (2008) Efficient transformation of azides to primary amines using the mild and easily accessible $CeCl_3{\cdot}7H_2O$/NaI System. *J. Org. Chem.*, **73** (5), 1919–1924.

[312] Reynolds, A.J., Kassiou, M. (2009) Recent advances in the Mitsunobu reaction: modifications and applications to biologically active molecules. *Curr. Org. Chem.*, **13** (16), 1610–1632.

[313] Sattler, J.H., Fuchs, M., Tauber, K. *et al.* (2012) Redox self-sufficient biocatalyst network for the amination of primary alcohols. *Angew. Chem. Int. Ed.*, **51** (36), 9156–9159.

[314] Fuchs, M., Tauber, K., Sattler, J. *et al.* (2012) Amination of benzylic and cinnamic alcohols via a biocatalytic, aerobic, oxidation-transamination cascade. *RSC Adv.*, **2** (15), 6262–6265.

[315] Tauber, K., Fuchs, M., Sattler, J.H. *et al.* (2013) Artificial multi-enzyme networks for the asymmetric amination of *sec*-alcohols. *Chem. Eur. J.*, **19** (12), 4030–4035.

[316] Williams, R.E., Bruce, N.C. (2002) 'New uses for an old enzyme' – the Old Yellow Enzyme family of flavoenzymes. *Microbiology*, **148** (6), 1607–1614.

[317] Zhou, X., Chow, H.L., Wu, J.C. (2014) Bioreduction of activated alkenes by a novel 'ene'-reductase from the thermophilic strain *Bacillus coagulans* WCP10-4. *Biocatal. Biotransform.*, **32** (5–6), 267–275.

[318] Steinkellner, G., Gruber, C.C., Pavkov-Keller, T. *et al.* (2014) Identification of promiscuous ene-reductase activity by mining structural databases using active site constellations. *Nat. Commun.*, **5**, 1–9.

[319] Winkler, C.K., Clay, D., Turrini, N.G. *et al.* (2014) Nitrile as activating group in the asymmetric bioreduction of β-cyanoacrylic acids catalyzed by ene-reductases. *Adv. Synth. Catal.*, **356** (8), 1878–1882.

[320] Burda, E., Reß, T., Winkler, T. *et al.* (2013) Highly enantioselective reduction of α-methylated nitroalkenes. *Angew. Chem. Int. Ed.*, **52** (35), 9323–9326.

[321] Kille, S., Reetz, M.T. (2013) Protein engineering: development of novel enzymes for the improved reduction of C=C double Bonds, in *Synthetic Methods for Biologically Active Molecules* (ed. E. Brenna), Wiley-VCH Verlag GmbH & Co. KGaA, Weinheim, pp. 113–138.

[322] Debarge, S., McDaid, P., O'Neill, P. *et al.* (2014) Evaluation of several routes to advanced Pregabalin intermediates: synthesis and enantioselective enzymatic reduction using ene-reductases. *Org. Process Res. Dev.*, **18** (1), 109–121.

[323] Brenna, E., Gatti, F.G., Manfredi, A. *et al.* (2012) Enoate reductase-mediated preparation of methyl (*S*)-2-bromobutanoate, a useful key intermediate for the synthesis of chiral active pharmaceutical ingredients. *Org. Process Res. Dev.*, **16** (2), 262–268.

[324] Brenna, E., Gatti, F.G., Malpezzi, L. *et al.* (2013) Synthesis of Robalzotan, Ebalzotan, and Rotigotine precursors via the stereoselective multienzymatic cascade reduction of α,β-unsaturated aldehydes. *J. Org. Chem.*, **78** (10), 4811–4822.

[325] Toogood, H.S., Scrutton, N.S. (2014) New developments in 'ene'-reductase catalysed biological hydrogenations. *Curr. Opin. Chem. Biol.*, **19**, 107–115.

[326] Ni, Y., Yu, H.-L., Lin, G.-Q., Xu, J.-H. (2014) An ene reductase from *Clavispora lusitaniae* for asymmetric reduction of activated alkenes. *Enzyme Microb. Technol.*, **56**, 40–45.

[327] Kroutil, W., Mang, H., Edegger, K., Faber, K. (2004) Recent advances in the biocatalytic reduction of ketones and oxidation of *sec*-alcohols. *Curr. Opin. Chem. Biol.*, **8** (2), 120–126.

[328] Fogagnolo, M., Giovannini, P.P., Guerrini, A. *et al.* (1998) Homochiral (*R*)- and (*S*)-1-heteroaryl- and 1-aryl-2-propanols via microbial redox. *Tetrahedron: Asymmetry*, **9** (13), 2317–2327.

[329] Ramírez, M.A., Pérez, H.I., Manjarrez, N. *et al.* (2008) Biocatalytic oxidative kinetic resolution of 4-(chlorophenyl)phenylmethanol by *Nocardia corallina* B-276. *Electron. J. Biotechnol.*, **11** (4), doi: 10.2225/vol11-issue4-fulltext-3

[330] Adam, W., Lazarus, M., Boss, B. *et al.* (1997) Enzymatic resolution of chiral 2-hydroxy carboxylic acids by enantioselective oxidation with molecular oxygen catalyzed by the glycolate oxidase from spinach (*Spinacia oleracea*). *J. Org. Chem.*, **62** (22), 7841–7843.

[331] Tsuji, Y., Fukui, T., Kawamoto, T., Tanaka, A. (1994) Enantioselective dehydrogenation of β-hydroxysilanes by horse liver alcohol dehydrogenase with a novel in-situ NAD+ regeneration system. *Appl. Microbiol. Biotechnol.*, **41** (2), 219–224.

[332] Tecelão, C.S.R., van Keulen, F., da Fonseca, M.M.R. (2001) Development of a reaction system for the selective conversion of (−)-*trans*-carveol to (−)-carvone with whole cells of *Rhodococcus erythropolis* DCL14. *J. Mol. Catal. B: Enzym.*, **11** (4–6), 719–724.

[333] Voss, C.V., Gruber, C.C., Kroutil, W. (2008) Deracemization of secondary alcohols through a concurrent tandem biocatalytic oxidation and reduction. *Angew. Chem. Int. Ed.*, **47** (4), 741–745.

[334] Voss, C.V., Gruber, C.C., Faber, K. *et al.* (2008) Orchestration of concurrent oxidation and reduction cycles for stereoinversion and deracemisation of *sec*-alcohols. *J. Am. Chem. Soc.*, **130** (42), 13969–13972.

[335] Mattevi, A. (2006) To be or not to be an oxidase: challenging the oxygen reactivity of flavoenzymes. *Trends Biochem. Sci*, **31** (5), 276–283.

[336] Dijkman, W., de Gonzalo, G., Mattevi, A., Fraaije, M. (2013) Flavoprotein oxidases: classification and applications. *Appl. Microbiol. Biotechnol.*, **97** (12), 5177–5188.

[337] Yasuda, M., Ueda, M., Muramatsu, H. *et al.* (2006) Enzymatic synthesis of cyclic amino acids by *N*-methyl-L-amino acid dehydrogenase from *Pseudomonas putida*. *Tetrahedron: Asymmetry*, **17** (12), 1775–1779.

[338] Pollegioni, L., Molla, G. (2011) New biotech applications from evolved D-amino acid oxidases. *Trends Biotechnol.*, **29** (6), 276–283.

[339] Schilling, B., Lerch, K. (1995) Amine oxidases from *Aspergillus niger*: identification of a novel flavin-dependent enzyme. *Biochim. Biophys. Acta*, **1243** (3), 529–537.

[340] Sablin, S.O., Yankovskaya, V., Bernard, S. *et al.* (1998) Isolation and characterization of an evolutionary precursor of human monoamine oxidases A and B. *Eur. J. Biochem.*, **253** (1), 270–279.

[341] Alexeeva, M., Enright, A., Dawson, M.J. *et al.* (2002) Deracemization of α-methylbenzylamine using an enzyme obtained by *in vitro* evolution. *Angew. Chem. Int. Ed.*, **41** (17), 3177–3180.

[342] Ghislieri, D., Green, A.P., Pontini, M. *et al.* (2013) Engineering an enantioselective amine oxidase for the synthesis of pharmaceutical building blocks and alkaloid natural products. *J. Am. Chem. Soc.*, **135** (29), 10863–10869.

[343] Carr, R., Alexeeva, M., Dawson, M.J. *et al.* (2005) Directed evolution of an amine oxidase for the preparative deracemisation of cyclic secondary amines. *ChemBioChem*, **6** (4), 637–639.

[344] Dunsmore, C.J., Carr, R., Fleming, T., Turner, N.J. (2006) A chemo-enzymatic route to enantiomerically pure cyclic tertiary amines. *J. Am. Chem. Soc.*, **128** (7), 2224–2225.

[345] Znabet, A., Polak, M.M., Janssen, E. *et al.* (2010) A highly efficient synthesis of telaprevir by strategic use of biocatalysis and multicomponent reactions. *Chem. Commun.*, **46** (42), 7918–7920.
[346] Li, T., Liang, J., Ambrogelly, A. *et al.* (2012) Efficient, chemoenzymatic process for manufacture of the Boceprevir bicyclic [3.1.0] proline intermediate based on amine oxidase-catalyzed desymmetrization. *J. Am. Chem. Soc.*, **134** (14), 6467–6472.
[347] Clark, D.S., Geresh, S., DiCosimo, R. (1995) Enantioselective oxidation of 2-methyl-1-alkanols by alcohol oxidase from methylotrophic yeasts. *Biorg. Med. Chem. Lett.*, **5** (13), 1383–1388.
[348] Das, S., Glenn IV, J.H., Subramanian, M. (2010) Enantioselective oxidation of 2-hydroxy carboxylic acids by glycolate oxidase and catalase coexpressed in methylotrophic *Pichia pastoris*. *Biotechnol. Progr.*, **26** (3), 607–615.
[349] Adam, W., Lazarus, M., Saha-Möller, C.R., Schreier, P. (1998) Quantitative transformation of racemic 2-hydroxy acids into (*R*)-2-hydroxy acids by enantioselective oxidation with glycolate oxidase and subsequent reduction of 2-keto acids with D-lactate dehydrogenase. *Tetrahedron: Asymmetry*, **9** (2), 351–355.
[350] Yuan, B., Page, A., Worrall, C.P. *et al.* (2010) Biocatalytic desymmetrization of an atropisomer with both an enantioselective oxidase and ketoreductases. *Angew. Chem. Int. Ed.*, **49** (39), 7010–7013.
[351] Escalettes, F., Turner, N.J. (2008) Directed evolution of galactose oxidase: generation of enantioselective secondary alcohol oxidases. *ChemBioChem*, **9** (6), 857–860.
[352] Holtmann, D., Fraaije, M.W., Arends, I.W.C.E. *et al.* (2014) The taming of oxygen: biocatalytic oxyfunctionalisations. *Chem. Commun.*, **50** (87), 13180–13200.
[353] Leisch, H., Morley, K., Lau, P.C.K. (2011) Baeyer–Villiger monooxygenases: more than just green chemistry. *Chem. Rev.*, **111** (7), 4165–4222.
[354] Balke, K., Kadow, M., Mallin, H. *et al.* (2012) Discovery, application and protein engineering of Baeyer-Villiger monooxygenases for organic synthesis. *Org. Biomol. Chem.*, **10** (31), 6249–6265.
[355] Alphand, V., Wohlgemuth, R. (2011) Applications of Baeyer-Villiger monooxygenases in organic synthesis *Curr. Org. Chem.*, **14** (17), 1928–1965.
[356] Torres Pazmiño, D.E., Dudek, H.M., Fraaije, M.W. (2010) Baeyer–Villiger monooxygenases: recent advances and future challenges. *Curr. Opin. Chem. Biol.*, **14** (2), 138–144.
[357] Geitner, K., Kirschner, A., Rehdorf, J. *et al.* (2007) Enantioselective kinetic resolution of 3-phenyl-2-ketones using Baeyer–Villiger monooxygenases. *Tetrahedron: Asymmetry*, **18** (7), 892–895.
[358] Rodríguez, C., de Gonzalo, G., Torres Pazmiño, D.E. *et al.* (2009) Baeyer–Villiger monooxygenase-catalyzed kinetic resolution of racemic α-alkyl benzyl ketones: enzymatic synthesis of α-alkyl benzylketones and α-alkyl benzylesters. *Tetrahedron: Asymmetry*, **20** (10), 1168–1173.
[359] Rodríguez, C., de Gonzalo, G., Fraaije, M.W., Gotor, V. (2007) Enzymatic kinetic resolution of racemic ketones catalyzed by Baeyer–Villiger monooxygenases. *Tetrahedron: Asymmetry*, **18** (11), 1338–1344.
[360] Stewart, J.D., Reed, K.W., Martinez, C.A. *et al.* (1998) Recombinant baker's yeast as a whole-cell catalyst for asymmetric Baeyer–Villiger oxidations. *J. Am. Chem. Soc.*, **120** (15), 3541–3548.
[361] Stewart, J.D., Reed, K.W., Zhu, J. *et al.* (1996) A 'designer yeast' that catalyzes the kinetic resolutions of 2-alkyl-substituted cyclohexanones by enantioselective Baeyer–Villiger oxidations. *J. Org. Chem.*, **61** (22), 7652–7653.
[362] Kyte, B.G., Rouvière, P., Cheng, Q., Stewart, J.D. (2004) Assessing the substrate selectivities and enantioselectivities of eight novel Baeyer–Villiger monooxygenases toward alkyl-substituted cyclohexanones. *J. Org. Chem.*, **69** (1), 12–17.
[363] Wang, S., Chen, G., Kayser, M.M. *et al.* (2002) Baeyer-Villiger oxidations catalyzed by engineered microorganisms: enantioselective synthesis of δ-valerolactones with functionalized chains. *Can. J. Chem.*, **80** (6), 613–621.

[364] Mihovilovic, Marko D., Müller, B., Stanetty, P. (2002) Monooxygenase-mediated Baeyer–Villiger oxidations. *Eur. J. Org. Chem.*, **2002** (22), 3711–3730.

[365] Mihovilovic, M.D., Chen, G., Wang, S. *et al.* (2001) Asymmetric Baeyer–Villiger oxidations of 4-mono- and 4,4-disubstituted cyclohexanones by whole cells of engineered *Escherichia coli*. *J. Org. Chem.*, **66** (3), 733–738.

[366] Rial, D.V., Bianchi, D.A., Kapitanova, P. *et al.* (2008) Stereoselective desymmetrizations by recombinant whole cells expressing the Baeyer–Villiger monooxygenase from *Xanthobacter sp.* ZL5: a new biocatalyst accepting structurally demanding substrates. *Eur. J. Org. Chem.*, **2008** (7), 1203–1213.

[367] Mihovilovic, M.D., Snajdrova, R., Grötzl, B. (2006) Microbial Baeyer–Villiger oxidation of 4, 4-disubstituted cyclohexan- and cyclohexenones by recombinant whole-cells expressing monooxygenases of bacterial origin. *J. Mol. Catal. B: Enzym.*, **39** (1–4), 135–140.

[368] Mihovilovic, M.D., Rudroff, F., Grötzl, B. *et al.* (2005) Family clustering of Baeyer–Villiger monooxygenases based on protein sequence and stereo-preference. *Angew. Chem. Int. Ed.*, **44** (23), 3609–3613.

[369] Hilker, I., Gutierrez, M.C., Furstoss, R. *et al.* (2008) Preparative scale Baeyer-Villiger biooxidation at high concentration using recombinant *Escherichia coli* and *in situ* substrate feeding and product removal process. *Nat. Protocols*, **3** (3), 546–554.

[370] Hilker, I., Wohlgemuth, R., Alphand, V., Furstoss, R. (2005) Microbial transformations 59: first kilogram scale asymmetric microbial Baeyer-Villiger oxidation with optimized productivity using a resin-based *in situ* SFPR strategy. *Biotechnol. Bioeng.*, **92** (6), 702–710.

[371] Kamerbeek, N.M., Janssen, D.B., van Berkel, W.J.H., Fraaije, M.W. (2003) Baeyer–Villiger monooxygenases, an emerging family of flavin-dependent biocatalysts. *Adv. Synth. Catal.*, **345** (6–7), 667–678.

[372] de Gonzalo, G., Mihovilovic, M.D., Fraaije, M.W. (2010) Recent developments in the application of Baeyer–Villiger monooxygenases as biocatalysts. *ChemBioChem*, **11** (16), 2208–2231.

[373] Zhang, Z.-G., Parra, L.P., Reetz, M.T. (2012) Protein engineering of stereoselective Baeyer–Villiger monooxygenases. *Chem. Eur. J.*, **18** (33), 10160–10172.

[374] Kayser, M.M., Clouthier, C.M. (2006) New bioorganic reagents: evolved cyclohexanone monooxygenase-why is It more selective? *J. Org. Chem.*, **71** (22), 8424–8430.

[375] Rioz-Martínez, A., de Gonzalo, G., Torres Pazmiño, D.E. *et al.* (2010) Synthesis of chiral 3-alkyl-3, 4-dihydroisocoumarins by dynamic kinetic resolutions catalyzed by a Baeyer–Villiger monooxygenase. *J. Org. Chem.*, **75** (6), 2073–2076.

[376] Rioz-Martínez, A., Cuetos, A., Rodríguez, C. *et al.* (2011) Dynamic kinetic resolution of α-substituted β-ketoesters catalyzed by Baeyer–Villiger monooxygenases: access to enantiopure α-hydroxy esters. *Angew. Chem. Int. Ed.*, **50** (36), 8387–8390.

[377] Rioz-Martínez, A., de Gonzalo, G., Pazmiño, D.E.T. *et al.* (2010) Enzymatic synthesis of novel chiral sulfoxides employing Baeyer–Villiger monooxygenases. *Eur. J. Org. Chem.*, **2010** (33), 6409–6416.

[378] Wojaczyńska, E., Wojaczyński, J. (2010) Enantioselective synthesis of sulfoxides: 2000–2009. *Chem. Rev.*, **110** (7), 4303–4356.

[379] Bong, Y.K., Clay, M.D., Collier, S.J. *et al.* (2011) Synthesis of Prazole Compounds. WO Patent 059398, filed Dec. 8, 2010 and issued Jun. 16, 2011.

[380] Ang, E.L., Clay, M.D., Behrouzian, B. *et al.* (2012) Biocatalysts and methods for the synthesis of Armodafin. WO Patent 078800, filed Jul. 12, 2011 and issued Jun. 14, 2012.

[381] Huisman, G.W., Collier, S.J. (2013) On the development of new biocatalytic processes for practical pharmaceutical synthesis. *Curr. Opin. Chem. Biol.*, **17** (2), 284–292.

[382] Isin, E.M., Guengerich, F.P. (2007) Complex reactions catalyzed by cytochrome P450 enzymes. *Biochim. Biophys. Acta*, **1770** (3), 314–329.

[383] Bühler, B., Schmid, A. (2004) Process implementation aspects for biocatalytic hydrocarbon oxyfunctionalization. *J. Biotechnol.*, **113** (1–3), 183–210.

[384] Bernhardt, R. (2006) Cytochromes P450 as versatile biocatalysts. *J. Biotechnol.*, **124** (1), 128–145.

[385] Donova, M., Egorova, O. (2012) Microbial steroid transformations: current state and prospects. *Appl. Microbiol. Biotechnol.*, **94** (6), 1423–1447.

[386] Bureik, M., Bernhardt, R. (2007) Steroid hydroxylation: microbial steroid biotransformations using cytochrome P450 enzymes, in *Modern Biooxidation: Enzymes, Reactions and Applications* (eds R.D. Schmid and V.B. Urlacher), Wiley-VCH Verlag GmbH & Co. KGaA, Weinheim, pp. 155–176.

[387] Jung, S.T., Lauchli, R., Arnold, F.H. (2011) Cytochrome P450: taming a wild type enzyme. *Curr. Opin. Biotechnol.*, **22** (6), 809–817.

[388] Julsing, M.K., Cornelissen, S., Bühler, B., Schmid, A. (2008) Heme-iron oxygenases: powerful industrial biocatalysts? *Curr. Opin. Chem. Biol.*, **12** (2), 177–186.

[389] Torres Pazmiño, D.E., Winkler, M., Glieder, A., Fraaije, M.W. (2010) Monooxygenases as biocatalysts: Classification, mechanistic aspects and biotechnological applications. *J. Biotechnol.*, **146** (1–2), 9–24.

[390] Urlacher, V.B., Eiben, S. (2006) Cytochrome P450 monooxygenases: perspectives for synthetic application. *Trends Biotechnol.*, **24** (7), 324–330.

[391] van Beilen, J.B., Duetz, W.A., Schmid, A., Witholt, B. (2003) Practical issues in the application of oxygenases. *Trends Biotechnol.*, **21** (4), 170–177.

[392] Lundemo, M., Woodley, J. (2015) Guidelines for development and implementation of biocatalytic P450 processes. *Appl. Microbiol. Biotechnol.*, **99** (6), 2465–2483.

[393] Bernhardt, R., Urlacher, V. (2014) Cytochromes P450 as promising catalysts for biotechnological application: chances and limitations. *Appl. Microbiol. Biotechnol.*, **98** (14), 6185–6203.

[394] Behrendorff, J.B., Huang, W., Gillam, E.M. (2015) Directed evolution of cytochrome P450 enzymes for biocatalysis: exploiting the catalytic versatility of enzymes with relaxed substrate specificity. *Biochem. J.*, **467** (1), 1–15.

[395] Roiban, G.-D., Reetz, M.T. (2015) Expanding the toolbox of organic chemists: directed evolution of P450 monooxygenases as catalysts in regio- and stereoselective oxidative hydroxylation. *Chem. Commun.*, **51** (12), 2208–2224.

[396] Bormann, S., Gomez Baraibar, A., Ni, Y. *et al.* (2015) Specific oxyfunctionalisations catalysed by peroxygenases: opportunities, challenges and solutions. *Catal. Sci.Tech.*, **5** (4), 2038–2052.

[397] Nolan, L., O'Connor, K. (2008) Dioxygenase- and monooxygenase-catalysed synthesis of cis-dihydrodiols, catechols, epoxides and other oxygenated products. *Biotechnol. Lett.*, **30** (11), 1879–1891.

[398] Piontek, K., Strittmatter, E., Ullrich, R. *et al.* (2013) Structural basis of substrate conversion in a new aromatic peroxygenase: cytochrome P450 functionality with benefits. *J. Biol. Chem.*, **288** (48), 34767–34776.

[399] Churakova, E., Kluge, M., Ullrich, R. *et al.* (2011) Specific photobiocatalytic oxyfunctionalization reactions. *Angew. Chem. Int. Ed.*, **50** (45), 10716–10719.

[400] Peter, S., Kinne, M., Wang, X. *et al.* (2011) Selective hydroxylation of alkanes by an extracellular fungal peroxygenase. *FEBS J.*, **278** (19), 3667–3675.

[401] Gutiérrez, A., Babot, E.D., Ullrich, R. *et al.* (2011) Regioselective oxygenation of fatty acids, fatty alcohols and other aliphatic compounds by a basidiomycete heme-thiolate peroxidase. *Archives of Biochemistry and Biophysics*, **514** (1–2), 33–43.

[402] Kluge, M., Ullrich, R., Scheibner, K., Hofrichter, M. (2012) Stereoselective benzylic hydroxylation of alkylbenzenes and epoxidation of styrene derivatives catalyzed by the peroxygenase of *Agrocybe aegerita*. *Green Chem.*, **14** (2), 440–446.

[403] van Deurzen, M.P.J., Remkes, I.J., van Rantwijk, F., Sheldon, R.A. (1997) Chloroperoxidase catalyzed oxidations in *tert*-butyl alcohol/water mixtures. *J. Mol. Catal. A: Chem.*, **117** (1–3), 329–337.

[404] Vargas, R.R., Bechara, E.J.H., Marzorati, L., Wladislaw, B. (1999) Asymmetric sulfoxidation of a β-carbonyl sulfide series by chloroperoxidase. *Tetrahedron: Asymmetry*, **10** (16), 3219–3227.

[405] Endoma, M.A., Bui, V.P., Hansen, J., Hudlicky, T. (2002) Medium-scale preparation of useful metabolites of aromatic compounds via whole-cell fermentation with recombinant organisms. *Org. Process Res. Dev.*, **6** (4), 525–532.

[406] Chopard, C., Bertho, G., Prange, T. (2012) Naphthalene-dioxygenase catalysed *cis*-dihydroxylation of bicyclic azaarenes. *RSC Adv.*, **2** (2), 605–615.

[407] Boyd, D.R., Sharma, N.D., Allen, C.C.R. (2001) Aromatic dioxygenases: molecular biocatalysis and applications. *Curr. Opin. Biotechnol.*, **12** (6), 564–573.

[408] Bestetti, G., Bianchi, D., Bosetti, A. *et al.* (1995) Bioconversion of substituted naphthalenes to the corresponding 1,2-dihydro-1,2-dihydroxy derivatives. Determination of the regio- and stereochemistry of the oxidation reactions. *Appl. Microbiol. Biotechnol.*, **44** (3–4), 306–313.

[409] Jerina, D.M., Daly, J.W., Jeffrey, A.M., Gibson, D.T. (1971) *Cis*-1,2-dihydroxy-1,2-dihydronaphthalene: A bacterial metabolite from naphthalene. *Archives of Biochemistry and Biophysics*, **142** (1), 394–396.

[410] Jenkins, G.N., Ribbons, D.W., Widdowson, D.A. *et al.* (1995) Synthetic application of biotransformations: absolute stereochemistry and Diels–Alder reactions of the (1S,2R)-1, 2-dihydroxycyclohexa-3,5-diene-1-carboxylic acid from *Pseudomonas putida*. *J. Chem. Soc., Perkin Trans.* 1, (20), 2647–2655.

[411] R. Boyd, D., N. Sheldrake, G. (1998) The dioxygenase-catalysed formation of vicinal *cis*-diols. *Nat. Prod. Rep.*, **15** (3), 309–324.

[412] Boyd, D.R., Sharma, N.D., Dorrity, M.R.J. *et al.* (1993) Structure and stereochemistry of *cis*-dihydro diol and phenol metabolites of bicyclic azaarenes from *Pseudomonas putida* UV4. *J. Chem. Soc., Perkin Trans.* 1, (9), 1065–1071.

[413] Boyd, D.R., Sharma, N.D., Boyle, R. *et al.* (1993) Biotransformation of unsaturated heterocyclic rings by *Pseudomonas putida* to yield *cis*-diols. *J. Chem. Soc., Chem. Commun.*, (1), 49–51.

[414] Fesko, K., Gruber-Khadjawi, M. (2013) Biocatalytic methods for C-C bond formation. *ChemCatChem*, **5** (6), 1248–1272.

[415] Brovetto, M., Gamenara, D., Saenz Méndez, P., Seoane, G.A. (2011) C−C bond-forming lyases in organic synthesis. *Chem. Rev.*, **111** (7), 4346–4403.

[416] Resch, V., Schrittwieser, J.H., Siirola, E., Kroutil, W. (2011) Novel carbon–carbon bond formations for biocatalysis. *Curr. Opin. Biotechnol.*, **22** (6), 793–799.

[417] Sukumaran, J., Hanefeld, U. (2005) Enantioselective C-C bond synthesis catalysed by enzymes. *Chem. Soc. Rev.*, **34** (6), 530–542.

[418] Busto, E., Gotor-Fernandez, V., Gotor, V. (2010) Hydrolases: catalytically promiscuous enzymes for non-conventional reactions in organic synthesis. *Chem. Soc. Rev.*, **39** (11), 4504–4523.

[419] Li, C., Feng, X.-W., Wang, N. *et al.* (2008) Biocatalytic promiscuity: the first lipase-catalysed asymmetric aldol reaction. *Green Chem.*, **10** (6), 616–618.

[420] Li, C., Zhou, Y.-J., Wang, N., Feng, X.-W. *et al.* (2010) Promiscuous protease-catalyzed aldol reactions: A facile biocatalytic protocol for carbon–carbon bond formation in aqueous media. *J. Biotechnol.*, **150** (4), 539–545.

[421] Xie, Z.-B., Wang, N., Zhou, L.-H. *et al.* (2013) Lipase-catalyzed stereoselective cross-aldol reaction promoted by water. *ChemCatChem*, **5** (7), 1935–1940.

[422] Rosenthaler, L. (1908). Enzyme-effected asymmetric syntheses. *Biochem. Z.*, **14**, 238–253.

[423] Dadashipour, M., Asano, Y. (2011) Hydroxynitrile lyases: insights into biochemistry, discovery, and engineering. *ACS Catal.*, **1** (9), 1121–1149.

[424] Purkarthofer, T., Skranc, W., Schuster, C., Griengl, H. (2007) Potential and capabilities of hydroxynitrile lyases as biocatalysts in the chemical industry. *Appl. Microbiol. Biotechnol.*, **76** (2), 309–320.

[425] Wiedner, R., Kothbauer, B., Pavkov-Keller, T. *et al.* (2015) Improving the properties of bacterial (*R*)-selective hydroxynitrile lyases for industrial applications. *ChemCatChem*, **7** (2), 325–332.

[426] Holt, J., Hanefeld, U. (2009) Enantioselective enzyme-catalysed synthesis of cyanohydrins *Curr. Org. Chem.*, **6** (1), 15–37.

[427] Purkarthofer, T., Pabst, T., van den Broek, C. *et al.* (2006) Large-scale synthesis of (*R*)-2-amino-1-(2-furyl)ethanol via a chemoenzymatic approach. *Org. Process Res. Dev.*, **10** (3), 618–621.

[428] van Langen, L.M., van Rantwijk, F., Sheldon, R.A. (2003) Enzymatic hydrocyanation of a sterically hindered aldehyde. Optimization of a chemoenzymatic procedure for (*R*)-2-chloromandelic acid. *Org. Process Res. Dev.*, **7** (6), 828–831.

[429] Weis, R., Gaisberger, R., Skranc, W. *et al.* (2005) Carving the active site of almond *R*-HNL for increased enantioselectivity. *Angew. Chem. Int. Ed.*, **44** (30), 4700–4704.

[430] Hanefeld, U. (2013) Immobilisation of hydroxynitrile lyases. *Chem. Soc. Rev.*, **42** (15), 6308–6321.

[431] Okrob, D., Metzner, J., Wiechert, W. *et al.* (2012) Tailoring a stabilized variant of hydroxynitrile lyase from *Arabidopsis thaliana*. *ChemBioChem*, **13** (6), 797–802.

[432] Okrob, D., Paravidino, M., Orru, R.V.A. *et al.* (2011) Hydroxynitrile lyase from *Arabidopsis thaliana*: identification of reaction parameters for enantiopure cyanohydrin synthesis by pure and immobilized catalyst. *Adv. Synth. Catal.*, **353** (13), 2399–2408.

[433] von Langermann, J., Wapenhensch, S. (2014) Hydroxynitrile lyase-catalyzed synthesis of enantiopure cyanohydrins in biocatalytic active static emulsions (BASE) with suppression of the non-enzymatic side reaction. *Adv. Synth. Catal.*, **356** (14–15), 2989–2997.

[434] Andexer, J.N., Langermann, J.V., Kragl, U., Pohl, M. (2009) How to overcome limitations in biotechnological processes – examples from hydroxynitrile lyase applications. *Trends Biotechnol.*, **27** (10), 599–607.

[435] Windle, C.L., Müller, M., Nelson, A., Berry, A. (2014) Engineering aldolases as biocatalysts. *Curr. Opin. Chem. Biol.*, **19**, 25–33.

[436] Royer, S.F., Haslett, L., Crennell, S.J. *et al.* (2010) Structurally informed site-directed mutagenesis of a stereochemically promiscuous aldolase to afford stereochemically complementary biocatalysts. *J. Am. Chem. Soc.*, **132** (33), 11753–11758.

[437] Clapés, P., Garrabou, X. (2011) Current trends in asymmetric synthesis with aldolases. *Adv. Synth. Catal.*, **353** (13), 2263–2283.

[438] Hasnaoui-Dijoux, G., Majerić Elenkov, M. *et al.* (2008) Catalytic promiscuity of halohydrin dehalogenase and its application in enantioselective epoxide ring opening. *ChemBioChem*, **9** (7), 1048–1051.

[439] Schallmey, M., Floor, R.J., Szymanski, W., Janssen, D.B. (2012) Hydrolysis and reverse hydrolysis: halohydrin dehalogenases, in *Comprehensive Chirality* (eds E.M. Carreira and H. Yamamoto), Elsevier, Amsterdam, pp. 143–155.

[440] Elenkov, M.M., Hoeffken, H.W., Tang, L. *et al.* (2007) Enzyme-catalyzed nucleophilic ring opening of epoxides for the preparation of enantiopure tertiary alcohols. *Adv. Synth. Catal.*, **349** (14–15), 2279–2285.

[441] Elenkov, M.M., Tang, L., Meetsma, A. *et al.* (2008) Formation of enantiopure 5-substituted oxazolidinones through enzyme-catalysed kinetic resolution of epoxides. *Org. Lett.*, **10** (12), 2417–2420.

[442] Molinaro, C., Guilbault, A.-A., Kosjek, B. (2010) Resolution of 2, 2-disubstituted epoxides via biocatalytic azidolysis. *Org. Lett.*, **12** (17), 3772–3775.

[443] Prasad, S., Bhalla, T.C. (2010) Nitrile hydratases (NHases): At the interface of academia and industry. *Biotechnol. Adv.*, **28** (6), 725–741.

[444] Wang, M.-X. (2015) Enantioselective biotransformations of nitriles in organic synthesis. *Acc. Chem. Res.*, **48** (3), 602–611.

[445] Wang, M.-X., Zhao, S.-M. (2002) Synthesis of enantiomerically enriched (*S*)-(+)-2-aryl-4-pentenoic acids and (*R*)-(−)-2-aryl-4-pentenamides via microbial hydrolysis of nitriles, a chemoenzymatic approach to stereoisomers of α, γ-disubstituted γ-butyrolactones. *Tetrahedron: Asymmetry*, **13** (15), 1695–1702.

[446] Wang, M.-X., Zhao, S.-M. (2002) Highly enantioselective biotransformations of 2-aryl-4-pentenenitriles, a novel chemoenzymatic approach to (*R*)-(−)-baclofen. *Tetrahedron Lett.*, **43** (37), 6617–6620.

[447] Gao, M., Wang, D.-X., Zheng, Q.-Y. *et al.* (2007) Remarkable electronic and steric effects in the nitrile biotransformations for the preparation of enantiopure functionalized carboxylic acids and amides: implication for an unsaturated carbon–carbon bond binding domain of the amidase. *J. Org. Chem.*, **72** (16), 6060–6066.

[448] Wang, M.-X., Lin, S.-J. (2002) Practical and convenient enzymatic synthesis of enantiopure α-amino acids and amides. *J. Org. Chem.*, **67** (18), 6542–6545.

[449] Wang, M.-X., Lin, S.-J., Liu, J., Zheng, Q.-Y. (2004) Efficient biocatalytic synthesis of highly enantiopure α-alkylated arylglycines and amides. *Adv. Synth. Catal.*, **346** (4), 439–445.

[450] Wang, M.-X., Liu, J., Wang, D.-X., Zheng, Q.-Y. (2005) Synthesis of optically active α-methylamino acids and amides through biocatalytic kinetic resolution of amides. *Tetrahedron: Asymmetry*, **16** (14), 2409–2416.

[451] Wang, M.-X., Liu, C.-S., Li, J.-S., Meth-Cohn, O. (2000) Microbial desymmetrization of 3-arylglutaronitriles, an unusual enhancement of enantioselectivity in the presence of additives. *Tetrahedron Lett.*, **41** (44), 8549–8552.

[452] Resch, V., Hanefeld, U. (2015) The selective addition of water. *Catal. Sci.Tech.*, **5** (3), 1385–1399.

[453] Jin, J., Hanefeld, U. (2011) The selective addition of water to C=C bonds; enzymes are the best chemists. *Chem. Commun.*, **47** (9), 2502–2510.

[454] de Villiers, M., Puthan Veetil, V., Raj, H. *et al.* (2012) Catalytic mechanisms and biocatalytic applications of aspartate and methylaspartate ammonia lyases. *ACS Chemical Biology*, **7** (10), 1618–1628.

[455] Turner, N.J. (2011) Ammonia lyases and aminomutases as biocatalysts for the synthesis of α-amino and β-amino acids. *Curr. Opin. Chem. Biol.*, **15** (2), 234–240.

[456] Schulze, B., Wubbolts, M.G. (1999) Biocatalysis for industrial production of fine chemicals. *Curr. Opin. Biotechnol.*, **10** (6), 609–615.

[457] Botting, N.P., Akhtar, M., Cohen, M.A., Gani, D. (1988) Substrate specificity of the 3-methylaspartate ammonia-lyase reaction: observation of differential relative reaction rates for substrate-product pairs. *Biochemistry*, **27** (8), 2953–2955.

[458] Raj, H., Szymański, W., de Villiers, J., Rozeboom, H.J. *et al.* (2012) Engineering methylaspartate ammonia lyase for the asymmetric synthesis of unnatural amino acids. *Nat. Chem.*, **4** (6), 478–484.

[459] Otte, K.B., Hauer, B. (2015) Enzyme engineering in the context of novel pathways and products. *Curr. Opin. Biotechnol.*, **35**, 16–22.

[460] Palomo, J.M., Filice, M. New emerging bio-catalysts design in biotransformations. *Biotechnol. Adv.*, **33** (5), 605–613.

Index

Chirality in Supramolecular Assemblies: Causes and Consequences, First Edition. Edited by F. Richard Keene.

Printed and bound by CPI Group (UK) Ltd, Croydon, CR0 4YY
16/09/2026
14963474-0001